Problems in
Vertebrate Evolution

Reports of Linnean Symposia

Speciation in tropical environments (Lowe-McConnell)
Biological Journal of the Linnean Society Vol. *1* 1969 pp. 1–246

New research in plant anatomy (Robson, Cutler & Gregory)
Supplement 1 to the *Botanical Journal of the Linnean Society* Vol. *63* 1970

Early mammals (Kermack & Kermack)
Supplement 1 to the *Zoological Journal of the Linnean Society* Vol. *50* 1971

The biology and chemistry of the Umbelliferae (Heywood)
Supplement 1 to the *Botanical Journal of the Linnean Society* Vol. *64* 1971

Behavioural aspects of parasite transmission (Canning & Wright)
Supplement 1 to the *Zoological Journal of the Linnean Society* Vol. *51* 1972

The phylogeny and classification of the ferns (Jermy, Crabbe & Thomas)
Supplement 1 to the *Botanical Journal of the Linnean Society* Vol. *67* 1973

Interrelationships of fishes (Greenwood, Miles & Patterson)
Supplement 1 to the *Zoological Journal of the Linnean Society* Vol. *53* 1973

The biology of the male gamete (Duckett & Racey)
Supplement 1 to the *Biological Journal of the Linnean Society* Vol. *7* 1975

Continued as the Linnean Society Symposium Series

No. 1 — **The evolutionary significance of the exine** (Ferguson & Muller) (1976)

No. 2 — **Tropical trees. Variation, breeding and conservation** (Burley & Styles) (1976)

No. 3 — **Morphology and biology of reptiles** (Bellairs & Cox) (1976)

No. 4 — this volume (1977)

Also published
Botanical Journal of the Linnean Society, Vol. *73*, Nos 1–3 July/Sept./Oct. 1976 — **The biology of bracken** (Perring & Gardiner)

Linnean Society Symposium Series   Number 4

Problems in
Vertebrate Evolution

Editors
S. Mahala Andrews, R. S. Miles, A. D. Walker

Royal Scottish Museum, British Museum (Natural History)
and University of Newcastle upon Tyne

Essays presented to
Professor T. S. Westoll, F.R.S., F.L.S.

Published for the Linnean Society of London by Academic Press

ACADEMIC PRESS INC. (LONDON) LIMITED
24/28 Oval Road
London NW1
(Registered Office)
(Registered number 598514)

US edition published by
ACADEMIC PRESS INC.
111 Fifth Avenue
New York
New York 10003

Printed in Great Britain by
Henry Ling Ltd., The Dorset Press, Dorchester, Dorset

Preface

This volume is associated with the symposium held on 8th January 1976 in the rooms of the Linnean Society at Burlington House, Piccadilly, London in honour of Professor T. S. Westoll. All of the papers have been stimulated by his research, and it is gratifying to note that a third of the contributors live overseas.

Professor Westoll is a Fellow of the Linnean Society, so it gives us particular pleasure to welcome this evidence of his profound influence in several branches of comparative biology.

The Society is grateful for all the work the Editors have undertaken in arranging the Symposium and editing this volume.

January 1977

DAVID MCCLINTOCK
Editorial Secretary,
Linnean Society of London

The editors wish to acknowledge the considerable help they have received from Miss Kim Dennis in the preparation of this volume for publication.

The indexes have been prepared by Drs B. G. Gardiner and A. S. Thorley.

Contributors

ANDREWS, S. M. *Royal Scottish Museum, Chambers Street, Edinburgh EH1 1JF* (p. 271)

BARTRAM, A. W. H. *Department of Biology, Queen Elizabeth College, Campden Hill Road, London W8 7AH* (p. 227)

CARROLL, R. L. *Redpath Museum, McGill University, Montreal, Quebec, Canada* (p. 359)

GARDINER, B. G. *Department of Biology, Queen Elizabeth College, Campden Hill Road, London W8 7AH* (p. 227)

JARVIK, E. *Section of Palaeozoology, Swedish Museum of Natural History, Stockholm, Sweden* (p. 199)

MILES, R. S. *Department of Public Services, British Museum (Natural History), Cromwell Road, London SW7 5BD* (p. 123)

ØRVIG, T. *Section of Palaeozoology, Swedish Museum of Natural History, Stockholm, Sweden* (p. 53)

PANCHEN, A. L. *Department of Zoology, University of Newcastle upon Tyne, Newcastle upon Tyne 1, England* (p. 289)

PARRINGTON, F. R. *24 Birch Trees Road, Great Shelford, Cambridge CB2 5AW, England* (p. 397)

PATTERSON, C. *Department of Palaeontology, British Museum (Natural History), Cromwell Road, London SW7 5BD* (p. 77)

SCHAEFFER, B. *The American Museum of Natural History, Department of Vertebrate Paleontology, Central Park West at 79th Street, New York 10024, N.Y., U.S.A.* (p. 25)

THOMSON, K. S. *Department of Biology and Peabody Museum of Natural History, Yale University, New Haven, Connecticut 06520, U.S.A.* (p. 247)

WALKER, A. D. *Department of Geology, University of Newcastle upon Tyne, Newcastle upon Tyne 1, England* (p. 319)

WHITING, H. P. *Department of Zoology, University of Bristol, Bristol, England* (p. 1)

YOUNG, G. C. *Department of Biology, Queen Elizabeth College, Campden Hill Road, London W8 7AH* (p. 123)

Professor T. S. Westoll F.R.S.

Foreword

Although Professor Stanley Westoll's interests and research extend through all groups of lower vertebrates—and beyond—I think ichthyologists can fairly claim him as one of their number. Reading his excellent summary 'Recent advances in the palaeontology of fishes' (1960) one sees that even fifteen years ago there was not a single major group of fish-like vertebrates, gnathous or agnathous, which he had not investigated at some time or another. No doubt my classification of Westoll as an ichthyologist will be challenged by practitioners in other fields of lower vertebrate evolution. We would, however, all agree on his being an out-standing palaeozoologist and a man of considerable influence.

Most appropriately for a volume honouring Professor Westoll, the editors have chosen as its title 'Problems in Vertebrate Evolution'. Professor Westoll has on many occasions turned his talents towards some particular problem in vertebrate evolution, and always with positive results. A glance through his bibliography will show the wide spectrum of subjects he has investigated: the ancestry of tetrapods (1938), vital stages in the transition of Devonian fish to man (1962), the hyomandibula of *Eusthenopteron* and the tetrapod middle ear (1943), the origin of the primitive tetrapod limb (1943), the paired fins of placoderms (1945), the structure and classification of ostracoderms (1945), the cheek bones of teleostome fishes (1937), the evolution of the Dipnoi (1950), problems concerned with the stratigraphy of Scottish vertebrate-bearing strata, and, leaving the realm of lower vertebrates, the homology of the mammalian palate (1940). One must not forget his incisive solution (1936) of the once intractable taxonomic problems caused, as he was the first to realise, by the periodic resorption and redeposition of cosmine on the head bones of dipnoan and osteolepid fishes. This early paper clearly shows Professor Westoll's biological approach to what might be considered purely palaeontological problems.

To me this approach is no better demonstrated than in his study on the evolution and taxonomy of the Haplolepidae, a family of Carboniferous bony fishes (1943)—a paper which, incidentally, gained for Professor Westoll an honourable mention in the New York Academy of Sciences' A. Cressy Morrison Prize competition for 1942. As a taxonomic ichthyologist this paper is one I would happily give to a student of fishes, living or fossil, as a model of its kind.

All Professor Westoll's major 'problem' papers show to full advantage the application of his synthetico-critical method strongly underpinned by knowledge acquired through personal researches (and also, on several occasions, an elegant wielding of that sometimes neglected weapon, Occam's razor). The same traits, it must be said, are equally evident in his many shorter papers devoted to various taxonomic and stratigraphic problems.

This volume, the Linnean Society's tribute, speaks unmistakably of the high regard felt internationally by zoologists and palaeontologists for Stanley Westoll, and reflects the influence he has had in their varied fields of research. Its publication comes just before his retirement from the J. B. Simpson Chair of Geology at the University of Newcastle-upon-Tyne and thus, we can be sure, anticipates a period of even greater research activity on the part of its dedicatee.

P. H. GREENWOOD
President,
Linnean Society of London

Contents

Cranial nerves in lampreys and cephalaspids

H. PHILIP WHITING

Department of Zoology, University of Bristol, Bristol

The nerve-canal identified by Stensiö in 1927 as that of the *profundus ophthalmic* (V^1) of the cephalaspid head is here homologized with the next posterior dorsal nerve, the *maxillo-mandibular* (V^{2-3}) of living craniates. This implies an 'along-the-line' re-identification of all dorsal cranial nerves in cephalaspids. Their cranial nerves are here compared with those of petromyzonts to test the hypothesis: re-identification is supported, particularly by the nerve-morphology of the otic region.

Cephalaspid evidence is chiefly taken from Stensiö (1927) for reasons of space, but *Cephalaspis kozlowskii* and *C. whitei* are also discussed. For the lamprey, proammocoete, pride, and adult stages are all considered and figured.

The basis of segmentation in craniates is briefly reviewed to permit subsequent discussion of visceral anatomy and metamerism in the agnathan head.

Two criteria are then considered which the present theory, linked with the name of Damas, must satisfy. A *profundus* nerve, anterior to that so termed by Stensiö, must evidently be found in cephalaspids: a solution to this problem is proposed. Secondly, homologies between visceral arches of cephalaspids and those of modern craniates must now be proposed that fit the facts. A 'pairing' has been described which seems to give a satisfactory relation of the nerve and the blood-vascular data in the cephalaspids under review: it suggests that cephalaspids probably did have a pharynx morphologically comparable in important ways with those of prides.

CONTENTS

INTRODUCTION

The preservation of internal detail in the cephalaspid head is superb, and unique. This evidence on early, primitive craniate organization has been intensely studied for about a hundred years, but there remains serious disagreement over some aspects of its interpretation. Until the impasse is resolved, the valuable results so far achieved are difficult to use.

This essay amplifies an aspect of a previous comparison, made between the cranial anatomy of larval lampreys and that of ostracoderms generally (Whiting, 1972). The comparison is also extended now into the visceral anatomy of cephalaspids, to offer a view upon that fraught subject, the identification of the anterior visceral arches. Do the fossil relics reveal that cephalaspids were animals more primitively organized than any modern craniate, having a functional pre-mandibular arch in front of the mandibular arch of gnathostomes, with perhaps one or even two pairs of additional anterior gill-openings?

Cranial nervous system is a big subject, and the visceral part difficult. This account has been limited in two ways:—by considering lampreys almost to the exclusion of hagfish, and by centring previous description of cephalaspids very much into a single work, the main one by Stensiö (1927). By shortening the scope in this way much relevant evidence has been overlooked: it is hoped the result justifies the treatment.

What follows depends very much on illustrations of cephalaspids by Stensiö (1927), and of cranial nerves of lampreys in the treatise by Marinelli & Strenger (1954).

THE BASIS OF SEGMENTATION IN THE LAMPREY HEAD

Although in one classification lampreys are grouped with cephalaspids among the Cephalaspidomorphi, in another the lampreys are grouped with hagfishes in the taxon Cyclostomata. If the latter view correctly expresses the lineage, lampreys and hagfish should be equally considered in comparisons of living animals with cephalaspids. However the absence of a larval stage in hagfishes makes assessment of myxinoid structural characters quite difficult. For example, is the presence of only a single pair of semicircular canals to be accounted a primitive or a specialized feature? So we begin from a basis that cephalaspids are chordates, are craniates, and may properly be compared with lampreys in the first instance, although not everyone accepts the comparison with lampreys as sound procedure. (The matter has been excellently debated between Smith Woodward, Goodrich, Watson and others (Woodward, 1930).)

Secondly, is there agreement on the criteria of segmentation in the craniate head? Can data about segments in the dorsal part be applied to the ventral part of the head? A failure to have agreed on what constitutes the organization of a craniate head may explain the surprising difference between interpretations of the cephalaspid visceral arch-system.

A beginning can be made by distinguishing that anterior part of the head which is pre-chordal and not segmented; so far as we can tell it has never been segmented. The pre-chordal part contains, dorsally, the forebrain and its sensory nerves: olfactory, pineal, hypophysial, and retinal components. The chordal part of the craniate head is organized like the trunk and is essentially an anterior region of the trunk which has been cephalized. The degree of cephalization varies, increasing as followed forward to the level of the front of the notochord; also increasing as animals, more advanced in phylogeny or ontogeny, are considered.

The much-travelled ground of 'criteria of segmentation' has now to be traversed, however quickly. Consider the ontogeny of the chordal part at trunk level.

Notochord: myotome

Shortly after neurulation the upper mesoderm becomes segmented into myotomes; many experiments show that this is a causal and fundamental relationship. Meanwhile, the mesoderm is growing upward on each side of the notochord. Then the part of the myotome closest to the notochord develops the first myofibrils of that level of the embryo; that part then becomes contractile, in a direction parallel to the longitudinal axis, e.g. *Scyliorhinus* (Harris & Whiting, 1954).

Myotome: motor neurons

The first axons to penetrate the ventrolateral sector of the spinal cord's limiting membrane are future somatic motor ones. Penetration does not occur at first only at midsegmental levels; instead the growth-cones become directed forward or backward toward the middle of a myotome, on its medial aspect. Soon the early motor axons are clearly directed towards a mid-myotome position, as shown by Neal (1898: fig. 42) in *Squalus*. From there the axons grow dorsally or ventrally, keeping to the medial aspect of the myotome column. This stage can be seen in the lamprey in Plate 1A: this and earlier stages of the motor neuron are illustrated by Whiting (1948, 1957). By this time contractility extends through the myotome, to its outer side and from front to back.

Muscle-plates

The fast-muscle units of the adult lamprey, the greater part of its somatic muscle, are the muscle-plates. These flat structures extend one above another, each across the length and breadth of a myotome. Their contraction is 'fore-and-aft', as in the earliest myofibrils. The somatic muscle and its innervation is illustrated by Bone (1969), for lampreys and Amphioxus, which have a plate-like system, as well as for hagfish and jawed fishes. The anatomy of the muscle-plate system, including its extent into the somatic part of the head, is shown for lampreys by Marinelli & Strenger (1954).

There is some reason to think that the muscle-plates may be the primitive system for somatic muscle (cf. Dalcq & Gérard, 1935). If so, it may be expected in ostracoderms.

Myotome: sensory neurons

In lower craniates the early peripheral fibres grow out in an irregular fashion from the spinal cord. Originating from the now well-known Rohon-Beard cells which lie within the cord, they extend outwards at any level along it, in a lateral or even quite dorsal direction. They are described in the developing lamprey (Whiting, 1948). Meanwhile the top of each myotome grows upward until it is higher than the notochord: the pioneer Rohon-Beard fibres slip aside into one of the furrows between myotomes, i.e. into a myocomma. Thus the fibres come to run out in a pattern conforming with the shape of the myotome-interface, except for a few 'pioneers' left straddling the top of a myotome. Later fibres follow the intersegmental pioneers' pattern which has been achieved in this incidental way. *Scyliorhinus* and *Salmo* are very like *Lampetra* in their early Rohon-Beard pattern.

Later still in embryonic life, dorsal ganglion neuroblasts send out peripheral fibres. These follow the paths already made. Besides somatic sensory ganglion-cells, there are viscero-sensory ones. In the lamprey, the viscero-motor spinal fibres also follow the beginning of the dorsal root paths. The result, the relation of sensory ganglia to somatic motor nerves and to myotomes, is shown in Plate 1B.

Segmental cranial nerves

The peripheral nerve pattern of the trunk is also that of chordal levels of the head, in the main. As the segmental nerves are followed forward in the lamprey, changes due to cephalization appear gradually, chiefly because the myotomes form a continuous series up to the otic capsule, even the fourth pair (the first post-otic) being developed. In the head, as in the trunk of lampreys, the dorsal and ventral nerves do not join, and the viscero-motor component continues to run out with the dorsal root; the somatic motor component remains the only one in the ventral root.

So, ventral cranial nerves such as the oculomotor or the immediately post-otic somatic motor nerves can be said to 'belong' to a particular somite and head-segment. But the dorsal nerves are simply intersegmental for we have seen that they are not inherently more related to the somite in front than to the somite behind them. Any practical *schema* of the organization of the craniate head must show them as intersegmental. Also, it is evident that the dorsal or somatic meta-merism of the craniate head derives from a basis of the notochord/myotome relationship. Other criteria, such as the 'neuromeric' swellings along the C.N.S. which were once proposed as part of the segmental system, are not relevant to it; the neuromeres are temporary and change in position during development.

The way in which cephalization develops gradually in this way can be envisaged from a parasagittal section, Plate 1C.

The head and cranial nerves of proammocoetes

The preceding account shows a little of how the 'segmental' cranial nerves develop. It may give the right significance to the diagram of the head of a proam-mocoete stage of lamprey, Fig. 1 (cf. Plate 2B).

The ammocoete (pride) and adult are similar in organization but are difficult to depict because their muscles and visceral arches are slanted obliquely to the main planes. Figures by Alcock (1899; Gaskell, 1908) and one by Johnston (1905) show the head and many of the nerves in lateral aspect in the pride, when the animal's age is a matter of years and not weeks.

The animal whose head is portrayed is growing and changing day by day. It is quite an active swimmer. It has reality apart from any anatomical comparisons with ostracoderms that it may provide.

Only the dorsal cranial nerves are shown, for simplicity. The ventral, somatic motor, nerves can be described verbally. Although the dorsal 'segmental' nerves are like each other in so many respects and form a metameric series, each has diagnostic features by which it should be identifiable in other craniates, in spite of topographical or allometric changes. These features have been specified previously

(Whiting, 1972). The position of the myotomes can be seen in illustrations in that account, and are not shown here. Many well-known features also identify the ventral nerves of the head.

The relation of the visceral metameric units to dorsal head segmentation is still a debateable subject. The data on which the aligning of the two may be done are given most clearly by Damas (1944). It is often assumed that the visceral arches, be they few or many, will have a one-to-one relation with the dorsal somite-region. In the lamprey, a one-to-one relationship can in fact be seen for a brief period of development and the position where successive arches appear, relative to the somites above them, has been shown by Damas (1943: 223), in a table of developmental stages. The pattern is soon obscured however by the backward displacement of the more posterior arches.

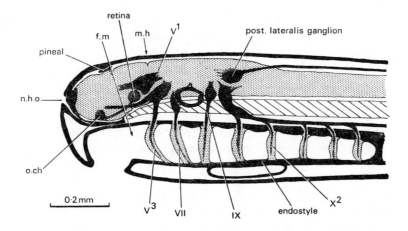

Figure 1. Medial section head of *Lampetra* (Damas stage 14) n.h.o., naso-hypophysial organ; o.ch., optic chiasma; f.m., and m.h., lines demarcating forebrain from midbrain and midbrain from hindbrain. Onto the section the dorsal cranial nerves and other structures are projected. V3 also marks the velum, VII also marks the front of pharynx.

The visceral division of a dorsal nerve, carrying the two visceral and the somatic sensory components, runs directly down a visceral arch, with no pre-trematic branch except for that of the V^{2-3} nerve. At the level of the IXth nerve, displacement is negligible, and here the nerve keeps to one transverse plane as it runs down to and then along its visceral arch: this is seen in lateral aspect in Damas' pl. II (10, 11) and in transverse aspect in Whiting's pl. 6 (3). However, Damas considers the pouch and not the arch in his table, and puts the pouch in the intersegmental position. Perhaps both pouch and arch of the visceral region were related to a given intersegment of the ancestor's head. At all events, this approximation seems to describe the facts as known today and will suffice for the present comparison with cephalaspids.

At the front of the head, the single pre-trematic branch, that equivalent to a maxillary branch, runs above and in front of the mouth and has sensory components only, just as in gnathostomes. The profundus nerve has no visceral components: it has no functional visceral arch to innervate.

The considerations which determine whether some early and more primitive craniate might have possessed an arch, anterior to the mandibular arch and with muscle innervated by a visceral division of the profundus nerve, depend on data such as Claydon (1938) and Damas (1943, 1954) provide. Such a condition seems possible though there is little evidence for it. But it is difficult to go further and hypothesize more than one visceral arch anterior to the mandibular one, without evading their evidence and much that went before, including the classical accounts, of Koltzoff (1901) and Sewertzoff (1916), of the lamprey's head-development.

IDENTIFICATION OF CRANIAL NERVES IN CEPHALASPIDS
The dorsal cranial nerves will now be considered in more detail, seriatim.

Stensiö's V^1 (profundus or ophthalmic) nerve, considered here as V^2

Stensiö states (1927: 113): 'The n. profundus certainly traversed the orbit in the normal way close to the lateral surface of the interorbital wall and close dorsally to the n. opticus, which must have entered the orbit very low.' But it can be seen from his diagram fig. 24A (p. 116), that he is showing a course for his V^1 that could not have enabled that nerve to pass dorsal to the optic tract as the latter passes between orbit and brain, however ventral the tract were. This V^1 has a path across the middle of the floor of the orbit: the optic tract must have passed medial, and so dorsal, to his 'V^1'.

Indeed the 'normal' path of the profundus or ophthalmic nerve is quite high in the orbit, near its medial wall. This can be seen in petromyzonts from figures such as those of Marinelli & Strenger (1954), Lindström (1949), Cords (1929), and here in Plate 2A.

Plates in which Stensiö shows the relation in cephalaspids between his profundus nerve and the orbit include pls 19 and 21 of specimens of *Cephalaspis hoeli*, the species on which fig. 24A is based, pl. 28 (*C. arcticus*), pl. 48 (*Thyestes verrucosus*), and pl. 49 (*Kiaeraspis auchenaspidoides*) of which a diagram is given by Whiting that perhaps helps to distinguish canals carrying the dorsal cranial metameric nerves from other canals for blood-vessels or for the electric field areas. In pl. 48 it is possible to see the antero-posterior limits within which the optic tract must have lain, well anterior to the oculomotor nerves and to the hypophysis (*III* and *fs. hyp.*). Zych (1937) provided useful data that complement some of Stensiö's plates. Zych makes the same identification as Stensiö: Zych's pl. II shows the V^1 nerve in similar fashion, coursing onto the ventral aspect of the orbit from anteriorly, in a position once again so lateral that the nerve must have underlain the optic tract.

The origin of this nerve, and its more peripheral part, will be considered later. We now evaluate the hypothesis that the V^1 of Stensiö is very probably the V^{2-3}, i.e. the maxillo-mandibular of gnathostomes, that the V^2 of Stensiö is really the facial nerve VII, and so on, to the branches of X. This means that the visceral arches of cephalaspids should for this purpose be visualized as being carried very far forward during ontogeny, further forward than proposed by Stensiö, from the embryonic position they held *vis-a-vis* the somatic system dorsal to them in the head. The nerves to the visceral arches would likewise sweep sharply forward. This is a movement of visceral arches relative to neurocranium just the opposite of that found in myxinoids; but what is possible in one direction is possible in the other, given a due cause.

Stensiö's V², considered here as the VIIth, facial nerve

The nerve labelled V², followed back, is found to begin from a fairly anterior position near the rim of the dorsal shield and lateral to V¹. This 'V²' runs toward the outer posterior edge of the orbit. It passes between the orbit and the otic capsule or labyrinth: the nerve can in some instances be seen to reach the brain. Where it does, it is close against the VIIIth nerve, and tucked close in at the medial anterior part of the otic capsule. The branches will be noticed later.

Stensiö's pl. 14 (*Boreaspis rostrata*) shows this arrangement, especially on the right; so do pls 49 and 55, of different specimens of *Kiaeraspis auchenaspidoides*. Sometimes the nerve underlies the orbit, as in pl. 28 (*C. arcticus*): sometimes, perhaps where the orbit is smaller, the nerve passes behind and clear of the orbit, pl. 14 and Zych, pl. II (*C. kozlowskii*).

This description could accord with either identification. The direction the nerve takes fits, in a topographical way, that of the V² division of the trigeminal. But the position of origin from the brain is that to be expected for the facial nerve. Two clues are unfortunately missing. The facial nerve of petromyzonts and most lower vertebrates, except elasmobranchs, carries a recurrent branch that circles the labyrinth and anastomoses with IX. Also the facial nerves should be the pair that innervate the thyroid gland. However, cephalaspid material seems to have provided no evidence in either case. These points have been mentioned in this context before (Whiting, 1972).

The detail in Stensiö's plates, for the origin of the nerve root and for the position its ganglion would have, gives weight to preferring an identification of the nerve as the facial, VII, but the evidence is not conclusive.

Stensiö's VII, considered to be the glossopharyngeal nerve, IX

This nerve is shown coming towards the brain from near the margin of the shield at about 45° to the anteroposterior axis. It usually runs close to no. 3 field-canal, which tends to be widely separate from the 1–2 and 4–6 field-canal groups, as is shown in pl. 49 (cf. Whiting, 1972: pl. V, fig. 3). This 'VIIth nerve' reaches the outer front edge of the labyrinth and begins to pass back along its *outer* side, unlike the facial nerve of any other craniate, cf. pls 24, 45 (*C. hoeli* and *Hoelaspis*) and pl. 2 of Zych. It can be followed further in pl. 25 (*C. vogti*), skirting a little way back around the outer side of the labyrinth, having penetrated what would be the limit of the otic 'capsule'. This is shown here in diagrammatic form, Fig. 2, but the original plate shows better, in a three-dimensional way, the relation of the nerve to the labyrinth. Evidently the nerve is crossing the labyrinth to a post-auditory position, consistent with its being the most anterior post-otic dorsal nerve. So there is intrinsic evidence for re-identifying this nerve as the glossopharyngeal, IX. Plate 2C, D illustrate the different position of nerves VII and IX.

Over the years, I have carefully examined Stensiö's illustrations of *C. vogti*, pls 25, 26, 27. The first two are part and counterpart of the same specimen but were not pictured at the same magnification. I have projected all three to the magnifications that gave the best 'fit' in my view, made careful drawings and superimposed them as transparencies. I used colours to distinguish the different canal systems:—arteries, veins, segmental nerves, acoustico-lateral nerves and 'field' nerves. The plates, of this species particularly, show the extreme complexity of the dorsal head-pattern in cephalaspids. Each system occurs in a metameric or pseudometameric series but with distinctive topographical detail, seen for example in comparing the dorso-lateral superficial artery sm^2 with the *spiracular* artery next in line behind it. The pattern is very constant, judged by these plates, in the minutiae of its organization; the only exception is some degree of asymmetry at the tip of the head. Altogether, Stensiö's identifications seem to have been very consistent. I pay tribute to his achievement, not simply in the serially ground sections in his *magnum opus*, but especially in the plates depicting dorsal and ventral views of cephalic shields.

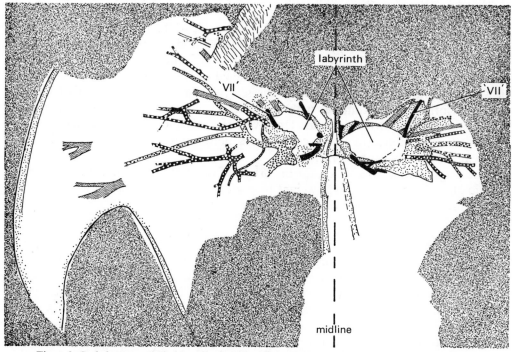

Figure 2. *Cephalaspis vogti;* diagram after Stensiö pls 25 and 26. The ventral part, with the dorsal part of his specimen 49 here shown at same magnification on the overlay. Black, nerves; wavy lines, field area canals; white spots on black, arteries; black on white, vein-canals. 'VII', the nerve-canal assigned to facial nerve by Stensiö. Magnification 'somewhat more than 5/1'.

So it seems permissible to rely on Stensiö's evidence on *C. vogti*, in the detail of the otic and post-otic region at which we now arrive, to come perhaps to different conclusions.

Estimation of the probable further course of the nerve, inward towards the brain, can now be resumed: this nerve now being regarded as the IXth nerve. There is a discontinuity in the preservation of a canal for this nerve across the middle of the labyrinth. On the medial aspect, where there was bone, there is detail again

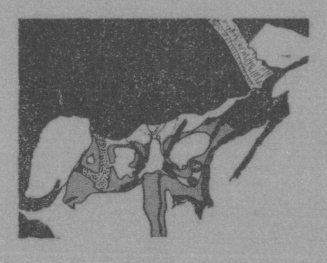

available: there are some six nerves or nerve-branches and these have to be accounted for.

When pls 25 and 26 of *C. vogti*, specimen 49, are superimposed to the same magnification from the original fossil, the pattern of the nerves in the labyrinth area is strikingly similar on the two sides. Reasonable confidence that the matching is correct is given by the detail of other canal systems lateral to the labyrinth, which also match.

Accordingly Fig. 2 (overlay) shows the relevant detail of his pl. 26. Comparing Fig. 2 and its overlay simply establishes that the data given in the plates apply to both sides, which otherwise would not be apparent. Stensiö terms the nerves on the medial aspect:—anteriorly, *1*, then *Vd*, *VII+VIIIa*, *VIIIp*, *IX+X+Xvcp;* distally, IX separates from X in the canal *IXp*. The nerve canals IX and X can be followed distally. A different identification for them will be offered. But VII+ VIIIa, and VIIIp can not be followed.

A first difficulty is that the facial nerve is normally close to but independent of the VIIIth nerve, the former having no acoustico-lateral component, merely being joined by *lateralis* nerves, in animals with a lateral-line system. The VIIIth nerve consists solely of acoustico-lateralis fibres, the most dorsal functional component in the hindbrain. *VII+VIIIa* is therefore probably VIII alone. Now lampreys and cephalaspids are alike in having only two semicircular canals, the vertical ones. In lampreys the VIIIth nerve divides into two similar branches, one directed forward to include the anterior ampulla, and one backward to include the posterior ampulla. The labyrinth and auditory nerve of the lamprey is illustrated by Marinelli & Strenger, and by Retzius (1881). The rather complex labyrinth of lampreys is similar to what can be made out of the complexities of that of cephalaspids, as Stensiö indicates.

It seems more feasible to regard the next nerve in the medial aspect, *VIIIp*, as the root of IX. Firstly, VIIIp has not the calibre nor the complementary course to that just proposed as being solely VIII, so it is unlikely to be the posterior half of VIII. Secondly, *VIIIp* is an independent nerve, next posterior to VIII. This is the condition of a primitive glossopharyngeal nerve-root, the most 'typical' dorsal cranial nerve, carrying the normal functional components—somatic and viscero-sensory and viscero-motor—and lacking any displacement forward or back by differential growth of other parts of the head (Goodrich, 1930). In the lamprey proammocoete the nerve is clearly separate from VIII in front and X behind, and very much in one transverse plane (Whiting, 1972). In the ammocoete, IX is still a very undistorted nerve, with the sensory components in a root that is slightly separate from a more ventral visceromotor root (Johnston, 1902), the separation probably itself a primitive feature. Here in the figures seems to be the nerve, in the position and with the features to make it probably the proximal part of the glossopharyngeal nerve we have been considering in its distal part. This was also Lindström's opinion.

It is just possible that *VIIIp* is only a part of the IXth nerve, and that a part took origin from the brain separately. This is Stensiö's nerve *t*, which entered the otic capsule more dorsally (fig. 17 and transverse sections A 57 and A 59). Such a dual root would resemble the condition in lampreys for this nerve; the upper root could be the somatic sensory component. Otherwise *t* remains to be explained.

On the view put forward, IX centrally, with or without nerve *t*, would apparently be a little smaller in calibre than our IX as it courses forward outside the labyrinth. This is acceptable. The nerve-fibres of lampreys (and cephalaspids?) are not myelinated. Some nerve-fibres of the lamprey increase in calibre after leaving the brain, although others retain their proximal calibre (Johnston, 1908; Lindström, 1949). A nerve-root composed of very many fibres may therefore increase in size, after leaving the brain.

There is a remote possibility that *VIIIp* was a posteriorly directed lateralis nerve, corresponding to the nerve *1* on the anterior side of the labyrinth, which Stensiö considered a forward-running lateralis branch. Whatever the identity of *1*, the posterior lateral-line nerve of lampreys takes origin from the brain, and has its large ganglion dorsal to the vagus. It then turns medially and dorsally to run caudally rather near the spinal cord, as shown by Whiting (1948, 1972), Johnston (1902) and Marinelli & Strenger (figs 51, 52) for the young and adult animals. Alcock (1899) also shows the origin of the posterior lateralis ganglion, clearly dorsal to and separate from the Xth vagus ganglion. Her elegant coloured illustration is given again by Gaskell (1908). It seems likely that the anatomy of the posterior lateralis nerve found in lampreys is the primitive form since the heterostracan *Anglaspis*, in which an impression of the relevant area can be seen, appears to show the same arrangement (Whiting, 1972). The heterostracans seem to be nearer the central craniate line, especially since their fossil record goes so much further back in time. Cephalaspids had a different shape of head from that of lampreys, so the topography of the lateralis system might have been different. But many heterostracans also had a depressed shape of head, like cephalaspids.

Before continuing with nerve-identifications, it is worth noticing that Stensiö's identifications, of his findings in the medial side of the labyrinth, did produce an awkward discrepancy. Retzius's detailed figures of the lamprey ear and auditory nerve, in the classic and immense work mentioned above, show the VIIIth nerve entering the capsule before dividing symmetrically into two branches only. It is strange that neither Stensiö, nor evidently Watson (1954) reconstructing from Stensiö's work, should have taken Retzius's description more seriously into account. Subsequently too, Lowenstein and his co-workers have investigated the structure and function of the labyrinth in lampreys. Lowenstein (1970) figures the auditory nerve entering the capsule *before* dividing into two equal parts. Further branching of the nerve is described, but this occurs further from the entry in parts of the labyrinth corresponding to regions of the labyrinth of cephalaspids where the nerve-canals have not been found preserved by Stensiö. Lowenstein also remarks that, proximal to the bifurcation, physiological analysis showed no separation of the VIIIth nerve into discrete functional bundles.

An unpublished account of the balance-system of the young ammocoete by a Bristol undergraduate (Robbins, 1971) includes figures of the VIIIth nerve in relation to the whole labyrinth, in fresh and in fixed and cleared preparations. These simply confirm Lowenstein's more sophisticated account.

The discrepancy involved in Stensiö's identification of the nerves on the medial aspect of the labyrinth appears too great to be acceptable. But there is no longer a problem here if *VIIIp* and *t* of cephalaspids are regarded as the IXth nerve of craniates and not as part of the VIIIth nerve (nor as part of the posterior lateralis system).

Stensio's IX, X¹, etc., considered as X¹, X², etc.

Most of the problems arising in re-identification have already been considered. Only the identity, not the distal distribution of the nerves, is considered at this juncture.

The peripheral parts of these nerves are not often visible from the dorsal aspect: perhaps the somatic muscle of the trunk reached forward enough to overlap them a little. Pl. 57 shows that IX and Xbr¹ ('the first branchial branch of the vagus') of Stensiö's identification did run rather deeply. But from the ventral aspect the peripheral extent of the same nerves shows well, pls 28, 44, and 49 to 52. IX shows better than Xbr¹, Xbr² hardly or not at all. Stensiö's pls 18, 24, 44 and 45 show the origin of his IX and X¹ and simply confirm the configuration of pls 25 and 26 which have been shown here in diagram form.

From these plates and identification, Stensiö gives us a picture of a glossopharyngeal nerve that sweeps somewhat forward from its origin, and has little obvious connection with the skin until the margin of the head-shield is reached. His vagus is likewise lacking in contact with dorsal skin areas, and sweeps only slightly forward.

Centrally, his view gives us a glossopharyngeal nerve and a vagus that take origin together, all the way from the brain itself apparently, and do not become separate until as far distally as the middle of the labyrinth on its posterior side. This combined root for IX and X might be sufficiently posterior as not to pass through the labyrinth cavity at all in some genera, as Zych points out when describing his species of *Cephalaspis*. Zych also mentions that these post-otic nerves did have several small dorsally-directed branches, so that a cutaneous, somatic sensory component was included evidently in the nerves we are considering.

To complete the picture, there is a branch-nerve anastomosing between IX and X, as shown in pl. 25 on the right. This quite noticeable anastomosis would be understandable on either reading of the facts, as reference to Alcock and to Gaskell (1908) or to Johnston (1905, inset in pl. 5) will show. There are choices of single components that might run back to join IX to X, or alternatively X¹ to X².

The significance of the facts given about these two nerves by Stensiö and Zych may now be considered, with the proximal part first because here there are morphological positions, on which 'bearings' may be taken.

The IXth nerve is said to leave by the same foramen as X, although a glossopharyngeal nerve is normally independent and undistorted, as remarked previously; it has a foramen anterior to that of the vagus even in the turtle. A uniting of the two seems an abnormal condition for a primitive animal unless, like *Myxine*, the whole pharyngeal pouch system were carried extraordinarily far *backward;* then it might be understandable though surprising. Furthermore, it is not a case of the vagus being forced round by a differential growth of visceral systems, so that it comes to be united with the glossopharyngeal nerve in an exit appropriate or originally belonging to the latter, but the contrary situation. The glossopharyngeal nerve of Stensiö is *less* far forward and pressed against the otic system, than in fishes; indeed as has been noted there are genera of cephalaspids where the origin of the 'glossopharyngeal-vagal' combination lies quite posterior to and separate from the whole labyrinth system.

The second hypothesis, that the two nerves are the first and second 'branches' (more correctly, dorsal segmental nerves) of the vagus, explains the observed situation of the proximal post-otic region in a very satisfactory and simple way. The first and second vagus nerves come out from the brain together in normal craniate fashion; when they extend out beyond the morphologically stable position created by the otic capsule, they both sweep forward towards their respective visceral arches which lie, on any reading of the general morphology of cephalaspids, much anterior to the somatic segments under which these arches first developed in phylogeny and no doubt in cephalaspid ontogeny. The first vagus branch would be carried or would grow noticeably further forward than the second and would therefore, once distal to the otic capsule, become separated from the second, since it extended at an angle much closer to the antero-posterior axis, towards its own more anterior visceral arch.

When the peripheral region of these nerves is considered, the identification proposed by Stensiö, and those who agreed with him, does seem the easier to envisage, because it requires a less radical change from the craniate *topographical* pattern that we know. However, there are peripherally no organs by which we can obtain a 'Fix', an Observed Position, to continue the metaphor. We can only follow out the nerves from the brain. As has been seen, there is evidence nearer the brain by which it can be seen that the second hypothesis, and not Stensiö's, is the valid one, as Plate 2A, C show.

However, the forward extension of the nerves IX, X^1 and X^2 does seem very extreme. The forward slide or slip of the visceral arches which those nerves innervated must have been caused by some very powerful 'current'; by something more fundamental than is offered by, say, the batoid body-form of the cephalaspids and the consequent in-filling of a space behind the mouth, vacant because of the stream-lining of a bottom-living animal.

The ophthalmic profundus nerve, V^1

If the profundus nerve of Stensiö's account was in fact the maxillo-mandibular nerve, what has become of the veritable profundus? In early developmental stages of lower craniates, the profundus nerve, apparently the most anterior of the dorsal cranial series, is independent of the remaining part of the fifth nerve: the coalescence to form the trigeminal complex comes of course later in ontogeny and phylogeny.

We cannot predict how prominent a nerve to expect. In living craniates only the somatic sensory component is important in the profundus; the function is chiefly exteroceptive with very little proprioceptive innervation; the nerve's branches extend over the anterior dorsal skin of the head, to tactile terminations. Lateral to the area of skin reached by the profundus, the maxillary division of the V^{2-3} nerve carries a somatic sensory component, as well as a viscero-sensory component deeper in the tissues. The latter nerve tends to replace the former which becomes reduced. This can be seen by comparing primitive sharks with a *Squalus* or a skate, and these with advanced sharks of the galeoid suborder, e.g. *Scyliorhinus*, where the dissector no longer finds a profundus nerve: a parallel change has occurred within the bony fishes. Lampreys have of course a large profundus nerve and cephalaspids may be similar. But a parallel reduction to that in gnathostomes might occur within the cephalaspids, perhaps speeded up by changes in the proportion of medial dorsal skin-area, in a line of animals continuing with a flattened shape for a long period.

But if cephalaspids were more primitive cephalaspidomorphs than are (modern) lampreys, then *a priori* the profundus should have been more prominent, with perhaps viscerosensory and visceromotor components such as Stensiö considered he had found. We cannot predict.

Also, we cannot predict that a large profundus nerve will carry with it a large profundus ganglion. Many of the cell-bodies of neurons of the craniate 'trigeminal' nerve lie in the midbrain, in the 'Mesencephalic nucleus of V'. Such cells are probably the cephalic parallel to the primitive sensory Rohon-Beard neurons of the spinal cord: Rohon-Beard neurons also have their large cell-bodies within the spinal cord, not in dorsal ganglia. So a primitive profundus nerve might have a large root at the front of the hindbrain but have no dorsal ganglion, because the cell-bodies of the nerve were inside the midbrain.

The profundus nerve of lampreys arises at the front of the hindbrain, high in the trigeminal complex but clearly separate from the maxillo-mandibular nerve. Close to its origin the root bears a large but elongate ganglion. The ganglion and the nerve-fibres continuing beyond the ganglion are directed anteriorly high along the medial face of the orbit. This condition is shown in Plate 2A, and has been described in detail many times, especially at embryonic stages. For the adult animal, more relevant here, there are illustrations by Cords (1929), several by Marinelli & Strenger (figs 51–53), and by Lindström (who also describes the condition in ammocoetes) as already mentioned.

In cephalaspids, Stensiö has been able to describe this region effectively in his reconstructions from the serial ground sections. This is fortunate, because the canal-system would by itself be inadequate in so central a part. In the position of the lamprey profundus root and ganglion, Stensiö portrays a similar outgrowth from the brain, passing into the orbit at what seems the same place, but which he terms the trochlear nerve. This is shown in dorsal aspect for *Kiaeraspis* (figs 20, 27), and for *C. hoeli* (fig. 23). The lateral aspect of the *Kiaeraspis* reconstruction (fig. 22) is perhaps the most easily comparable with the morphology of lampreys.

The trochlear nerve of lampreys, after leaving the brain quite dorsally at the junction of midbrain and cerebellum, runs towards the orbit on a joining course with the profundus nerve. The path by which the very small trochlear passes through the fibres of the very large profundus nerve can vary. Lindström gives a clear account of the details and elucidates some of the confusions among previous workers on this topic. It is probable—in fact it is difficult to avoid stating that it is certain—that Stensiö was depicting a large profundus nerve in his reconstructions here. It is not necessary to determine exactly where the trochlear nerve crossed through the profundus, for whatever its variations may have been, it is certain that it was always a very small nerve.

In fig. 32, Stensiö suggests what the brain of *Kiaeraspis* may have been like. His reasoning is perfectly logical, here and in the text, except that he is identifying a very large nerve, which he must account for, with an inherently small one. Looking at the matter at a much later time and with more information available, investigators today have an altogether easier task. It should be added that Watson concurred with Stensiö's identifications:—'These structures taken together are of such complexity that it is impossible not to believe that Stensiö's restoration is correct in essence; even the minute IVth cranial nerve has its own independent canal.' The last statement may be true, in that the *origin* of what seems to be the true trochlear appears as a small foramen on pl. 78, labelled as the trochlear nerve. The nerve-foramen seems to lead back to the dorsal aspect of the midbrain-cerebellar junction as it should.

To enter the orbit, the profundus nerve of lampreys passes through a spherical connective-tissue capsule, tough and easily found in dissection. The capsule is described by Walls in his book on the vertebrate eye (1963). Within the capsule, the profundus passes above all the extrinsic muscles of the eye. Marinelli & Strenger (fig. 53) show the orbit seen in side view, after removal of the eye-ball: V^2 nerve is seen to run *beneath* the capsule.

Returning to the evidence in cephalaspids, there seems to be a fairly large nerve, in the expected morphological position, leading towards the orbit (to the capsule?). It is shown in Plate 3A, a drawing of the *Tremataspis schmidti*, Stensiö's pl. 60. Stensiö describes here a canal 'Either for the n. trochlearis or for a branch of the post-orbital superficial artery'. This canal probably originally showed on both sides in this specimen, pl. 59. Likewise, the dorsal view of *Boreaspis rostrata*, pl. 13, shows a similar nerve in apparently the same position, on both sides. This Stensiö describes as 'Either the myodome or the canal for the majority of the trigeminus-profundus roots' and labels it 'V'. A similar nerve, again entering the orbit in a dorsal position at the ventromedial 'angle' is that indicated as 'V' in pl. 47, of *Hoelaspis angulata*. Stensiö's own naming of these examples suggests that he had a rather similar view of their identity. But it is hardly possible to reconcile the course of these nerves with that elsewhere attributed to the profundus by him. For example, the profundus nerve shown in pls 48 and 49 (*Thyestes* and *Kiaeraspis*) is running (followed back) below and on the outer side of the orbit to reach the crucial position at about the same depth from the upper surface as the oculomotor nerve (shown in both examples). These, previously identified in this account as part of the maxillo-mandibular system, are evidently different nerves from the profundus of pls 13, 47 and 60.

As to the profundus ganglion, there is little evidence: some indication of an elongate swelling in the nerve, proximal to the orbit. There is a possible swelling in Stensiö's text-figures of the 'trochlear' nerve.

On the other hand the profundus ganglion in the lamprey is in part lateral to the medial wall of the orbit, where in the fossil it might leave little trace.

If the evidence about the profundus of cephalaspids is accepted so far, it remains to look for any indication of how it left the orbit and continued forward. The situation in the lamprey is nicely shown in fig. 53 of Marinelli & Strenger and gives some clues for examination of cephalaspids. The anterodorsal aspect of the orbit capsule is open in lampreys, so that the profundus nerve simply extends forward over a lip between the front wall and the higher medial wall. The nerve continues forward, parallel and close to the midline, at a subcutaneous level; it remains large during most of this course, because most nerve-terminations are near the sensitive skin above the suctorial mouth. Of course this quite dorsal position means that the profundus overlies the skeleton of this part of the head, namely the anterior and posterior tectal cartilages; these resemble to some degree the anterior part of the head-shield of a cephalaspid, as Gaskell among others has pointed out. The only branch given off by the profundus nerve after leaving the orbit is one on the medial side, the ethmoid branch to the nasal capsule. The branch is shown by Lindström and others, but not by Marinelli & Strenger.

Since the cephalaspids did not have an anterior sucker like lampreys, their profundus nerve would probably fan out on leaving the orbit, to innervate the anterior surface from a subcutaneous level, more evenly and so less conspicuously than happens in lampreys. There does in some cases appear traces of what seems to be an ethmoid branch: such a possible canal is visible leaving the orbit and turning anteromedially toward the nasal capsule in pl. 59, on the right, and in pl. 60 (of the same specimen) on the left. Here and in a few other cases there are indications of a canal, going forward from the orbit, that is not accounted for by the artery and vein with which Stensiö provides us in this region.

It seems possible that in the specimen of *Cephalaspis whitei*, Pl. 3B, the upper surface of the head-shield has broken away at a level equivalent to the 'subcutaneous' level of the lamprey skin, and that the minute branches visible, in front of the naso-hypophysial opening and the orbits in particular, are in part capillary vessels and in part tactile branches of the profundus.

They may correspond to the layer of 'Canals of the subaponeurotic (subcutaneous) vascular plexus' shown in section as *dplx* in Stensiö's pl. 64, fig. 2 of *Cephalaspis*. This possibility is offered very tentatively. Outside the superficial layer *sl* shown in Stensiö's section in this plate, I should expect there to have been in the living animal a thick epithelial layer, with mucous protection, as in lampreys and many bony fish. The sensory terminations would enter this layer and thus pass beyond the topmost layer that is preserved in the fossil material.

Nervus terminalis

It is not surprising that this nerve, running from the forebrain to the nasal region, should not be found in cephalaspids or lampreys, in spite of its occurrence in primitive living gnathostomes. For in lampreys the dorsal position of the nasohypophysial opening is achieved during ontogeny; the differential growth, which this movement of the whole hypophysial area involves, is generally thought to be a secondary condition in craniates. So the small anterior region, which this nerve innervated, may well have been eliminated during the evolution of a nasohypophysis, in cephalaspids and in lampreys.

Somatic motor nerves

Turning now to the ventral cranial nerves, their identification has already been discussed in accounting for the dorsal nerves. But their anatomy, accepted from Stensiö's descriptions, has an important implication. Stensiö's description of the *oculomotor nerve* resembles that of lampreys and gnathostomes. This is not surprising. The oculomotor nerve retains a primitive form in living craniates: it runs to parts of the first pair of embryonic somites that have now become four of the six extrinsic eye-muscles; it has a direct course from the ventral and anterior aspect of the midbrain. Thus the somatic nerve originates now in craniates generally at the front of the somatic motor column of the brain and above the front of the notochord, just as would be expected in an ancestral craniate.

The *abducens nerve*, third in the series, has in modern craniates a similarly appropriate and evidently unaltered position, further along the somatic motor column. Its course forward to an ending on one posterior eye-muscle makes it unlikely to be described with any useful detail in cephalaspids.

The *trochlear nerve*, second in the series, has been described by Stensiö and discussed here. It appears to have taken origin from a dorsal part of the brain, somewhere near the midbrain-cerebellar junction. His description of the early part of the course of the nerve does not seem to be in doubt. In lampreys and craniates the nerve lies in the same position, taking origin high on the brain, dorsolaterally at the front of the cerebellum. This similarity between cephalaspid and lamprey IVth nerves is surprising.

In gnathostomes, the trochlear nerve, on entering the brain, runs contralaterally across the front of the cerebellum to an origin below the *sulcus limitans* (Pearson, 1936), i.e. in the motor columns of the side opposite to that of the superior oblique muscle that it innervates. The functional reason for this chiasma and the contralateral origin of the second of the somatic motor nerves can only be guessed.

In proammocoetes and adult lampreys, the eye-muscle innervated is probably the same one, though it lies on the posterior side of the orbit (the eye is reduced during the animal's years as a pride). The nerve, in the adult, crosses the cerebellum (Schwab, 1974) but soon arrives at the cells of origin (IV nucleus) high above the ventricle and so above the *sulcus limitans* (Pearson, 1936; Addens in Kappers, Huber & Crosby, 1936: figs 227, 229). This is evidently 'neurobiotaxis', that has not gone so far as it has in gnathostomes. In young lampreys, the trochlear cells of origin have a more primitive appearance still, for the cells of origin are homolateral, at first below, then later above the *sulcus limitans* (Larsell, 1947). So this bizarre neuronal migration can be followed back in ontogeny through three stages.

If cephalaspids had a primitive functional arrangement of their eye-muscles, we should expect from the foregoing that the nerve-foramen would be simpler than that of young lampreys: such as to lead from low on the brain, from a homolateral somatic origin that we would not see, to a position low within the orbit. The foramen would lie (as at present) halfway between the first and third motor nerves.

Instead, relying on Stensiö, we should conclude that cephalaspids were as advanced as lampreys in the evolution of their eye-muscles, since they seem to have gone as far as lampreys, probably as far as adult lampreys, toward the same strange neuromuscular functional pattern as in other craniates.

THE VISCERAL ARCH SYSTEM IN CEPHALASPIDS

Visceral arches may be identified by their innervation. Stensiö's plate and his verbal account of the nerves to the splanchnic region of the head agree, and they are supported by the rather independent approach of Zych. Stensiö describes the dorsal cranial nerves as running out, each to its visceral arch, and giving off a medial, ventrally directed branch first (fig. 43). He considers the branch to be *communis* in function, i.e. probably gustatory. Soon after this branching, the nerve and its branch appear through the floor of the 'endocranium' (figs 43, 49).

The main nerve continues without further branching, to near the rim of the headshield, where the dorsal surface is turning down to meet the flat ventral surface. Each nerve, as it too turns down in line with the upper surface, divides into two almost equal branches (figs 43, 37A). The posterior branch ran in fine rami into the exoskeleton (e.g. p. 191 for his Xth nerve), and was evidently somatic sensory. The anterior continued down along the visceral arch; presumably it innervated visceral muscle and was the only part continuing to have a visceromotor component.

In parenthesis:—fig. 43 resembles the lamprey's organisation in many ways; the 'profundus' occupies the lamprey/gnathostome position, not the sub-optic one in which it figures in his plates and in the reconstruction by Watson (1954).

Zych compares his own results with specific points given by Stensiö. Their picture seems to me consistent and clear. It may be compared with Fig. 3. This draws on Alcock and on Johnston; some use has also been made of Sterba (1952) and of my own preparations.

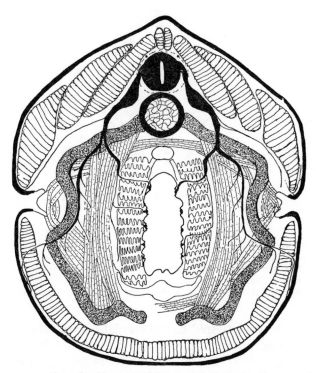

Figure 3. Transverse section of pride stage of lamprey, through the pharynx, after Alcock, and Johnston: shows a post-otic dorsal cranial nerve with branches including one to taste-buds.

The correspondence between the nerves of cephalaspids' mouth and pharynx and those of the pride stage of lampreys seems to be close. The first branch of the segmental (or intersegmental) nerve is visceral and indeed gustatory, supplying five or six large taste-buds at the narrowest part of its arch. On reaching the lateral aspect of the pharynx the nerve divides in two, one branch innervates peripheral areas and the other continues ventrally and is evidently chiefly a visceromotor branch. The similarity is satisfactory precisely because Stensiö does not seem to have taken Alcock, nor Johnston's section-figures, into account.

Anteriorly, Stensiö describes the branching of his 'profundus' nerve across the roof of the mouth in terms which fit the maxillary branch of V^{2-3}. He does not explain why the nerve he identifies as maxillary does not have the distribution of that nerve. The identification of the 'mandibular' division remains a problem, on any view. It seems possible, from his plates, that the nerve identified here as V^{2-3} gives off its mandibular half in a directly ventral direction and that both upper and lower have not been observed in any one example.

The evidence for the organization described by Stensiö and Zych is given in Zych pl. 2 and in Stensiö's plates 20, 23, 28, 29, 39, 45, 49, 51/2, 54, 55 and 57. Any part of the branching is supported by at least two of the plates, and the nerves V^{2-3} to X^2 are all described to some extent (termed by them V^1 to X^1). So the

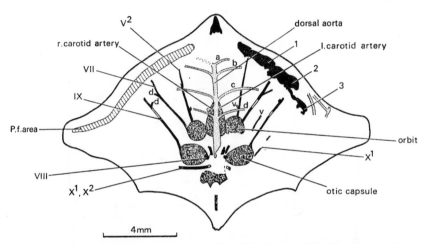

Figure 4. *Hoelaspis angulata;* diagram combining Stensiö's pls 44 and 45. Specimen 89, here seen from below, with the dorsal aorta viewed in front of the hypophysial fossa.: shows V^2 extending as far as a (mandibular artery?); b, c, d (on right side), successive efferent branchial arteries.

Nerve-identification, as Damas, not Stensiö. v, Viscero-sensory nerve-branches; d (on left side), peripheral cutaneous/visceral division: 1, 2, 3, successive branchial fossae the ventrolateral part.

visceral division of the cephalaspid dorsal cranial nerves followed in some detail the pattern along a visceral arch that is seen in the pride. It remains to establish how and why the cephalaspid visceral arches have moved so far forward relative to what must have been their primitive position.

The course of the nerves through the visceral tissues has been followed through as a metameric system, as it extends forward and downward. By using the evidence of plan-views of the visceral cranium it should now be feasible to find out to which arch a given nerve runs. Stensiö's pl. 44/45 and 46 of two specimens of *Hoelaspis* and pls 51/52 of *Kiaeraspis* are clear and useful. Some of the data given by pls 44 and 45, *H. angulata*, are shown here in Fig. 4. The visceral arches, and the lateral parts of the branchial 'fossae' between them are shown.

The V² nerve (Stensiö's V¹, but his identification is not used after this) runs to the arch in front of the first fossa, the facial in front of the second, and so on; the vagus has the 4th fossa behind it. The mouth appears to lie in front of the arch in front of the first fossa; in other words the mandibular arch lies behind the mouth. Although the pattern is consistent, caution is needed in setting down the results, from Stensiö's work of 1927 alone, in a tabular form, because the nerve runs obliquely towards its arch, which itself, as in *Hoelaspis*, is usually oblique.

Without describing the nerve-to-arch evidence further, the plan view of the blood system can be noticed now. The dorsal aorta shows in many plates. It runs forward toward the pituitary position, and then goes onward. At the pituitary and hypophysial site, a pair of carotid vessels runs dorsally to supply the brain just as in modern craniates (Goodrich 1930: 514 and figs 247, 531). This pair of canals is a *point de repère* between cephalaspids where the splanchnocranium is exposed and those where the specimen has been opened across the upper side of the braincase floor. The record of this artery-pair is visible in many of the plates and of course in other workers' accounts (compare e.g. Miles, 1971: 16). It marks the front of the notochord, the embryonic commissure between the craniate lateral head-veins, and the opening of the hypophysial sense-organ, where this is patent.

The extension of the cephalaspid dorsal aorta forward from this key-position where it gives off the carotids dorsally seems to continue in the same plane, as if the central part of the splanchnocranium slid forward on the neurocranium during growth. There are three pairs of efferent arteries entering the forwardly projected section of the aorta: these are presumably vessels which occurred in metameric order behind the carotid artery 'marker' in the embryo and have slipped forward as the cephalaspid developed its adult form of head. They are named the rostral, first and second efferent pairs by Stensiö. He gives his reasons for this nomenclature; the rostral enters the dorsal aorta at a slightly different angle and is smaller than those behind. Nevertheless, all three pairs can be seen to draw from successive arches, and the angle of entry of all three shows that the blood-flow was in the same direction in them all.

From their position relative to the arches as identified by nerve-supply, it is consistent to name them, at least as a first hypothesis, the mandibular, hyoidean and first branchial efferent-artery homologues of the gnathostome vessels of those names. This naming fits the appreciable facts we have available, so far as the three specimens mentioned.

On this view, there would be a functional hemibranch on the anterior side of the 'spiracular' gill-cavity behind the mouth (Damas (1954) goes into this matter in detail). The 'spiracular' gill-cleft in front of the hyoid arch would then have the same feeding and respiratory functions as the gill-clefts or gill-pouches behind it: less effective than the three pairs just behind it, because it was a shorter cleft, more like those following the arches innervated by vagus², vagus³ nerves and the subsequent branches of the vagus in cephalaspids.

Cephalaspids appear from the foregoing to have been more primitive than living vertebrates because these either lose their first or mandibular arches during development, or as in elasmobranches for example, the mandibular efferent vessel is retained although no respiration is effected in the elasmobranch mandibular arch, but the blood-flow is reversed and the vessel becomes an afferent vessel

(Goodrich, 1930: fig. 531C, 531D). The angle at which the mandibular arch meets the dorsal aorta alters when the blood-flow is reversed, as Goodrich's figures show. This change in angle has not happened in the cephalaspids mentioned.

The smaller of the two *Cephalaspis whitei* has a fracture on its left side, visible in Plate 4A, B, which exposes the branchial skeleton in its outer part and makes it possible to draw arch extremities. This makes possible a comparison with the visceral system shown by Stensiö's specimens and by Watson in his lateral-aspect reconstruction of *Kiaeraspis*.

It is difficult to hazard an estimate on how cephalaspids obtained their food, and whether the shape of the branchial skeleton favours the form of gill-pouch proposed by Stensiö, or by Watson, or a different form: Denison (1961) and Thomson (1971) have reviewed the various proposals. We can say that the central pharyngeal cavity was probably quite large since it is (again probably) outlined in plan view by the gustatory branches of the dorsal nerve, and these branches enter the splanchnocranium at some distance from the mid-line.

It can be said, negatively, that cephalaspids did not draw in their food by the action of a large muscular velum. Although the paired velar organ of the pro-ammocoete and pride is large, it does not occupy a very large space since it passes, on retraction, between the outer parts of the hyoidean skeleton, due to the velum having a funnel-shaped form when drawn back. But a larger animal, such as many cephalaspids were, would require a relatively larger velar system, which would require a different shape of splanchnocranium anteriorly from that found in cephalaspids. Secondly, the innervation of the velum involves a nerve conformation quite different from that supplying the visceral arches of the pharyngeal system, whereas the successive dorsal nerves of cephalaspids are alike, and like the pharyngeal pattern of prides. The oblique pattern of the arches shown in the reconstruction by Stensiö and Watson is more like that of the pride stage (cf. Sterba, 1952) than that of the adult lamprey's gill-pouches. The absence of an upper and lower division of the pharynx also argues against a gill-pouch mechanism. Probably an understanding of the respiratory and feeding structure must await an answer to the question of why the cephalaspid visceral anatomy is brought forward in so specialised a manner. This answer will probably depend on the true nature of the field-areas of the dorsal surface of cephalaspids.

CONCLUSIONS

The morphological comparison of the head of lampreys with the head in those cephalaspids described by Stensiö (1927) and Zych shows a strong similarity between the 'maxillo-mandibular' (V^{2-3}) of the lamprey and the profundus nerve (V^1) of their account. Following back along the series of dorsal nerves, the facial (VII) nerve of 'their' cephalaspid head has too lateral and posterior a position to reach the anterior aspect of the otic system and the VIIIth nerve; it appears on the contrary to be in the cranial position of the IXth nerve. The alternative identification put forward by Damas, Lindström and myself is therefore strongly supported by this study.

Further, the IXth nerve of Stensiö's account is no longer closely combined with the nerve behind it, the Xth of his account, which comparison with the lamprey for example shows to be an improbable condition. Instead, on the alternative reading, the X^1 is combined with X^2, etc.

A 'new' true profundus has to be found in cephalaspids, as a required corollary of the Damas identification. It is very probable that the profundus is described as a large root (called IVth nerve by Stensiö), near the brain. This can then sometimes be traced as a canal leading to the orbit, through a cephalaspid head-region that is often ill-preserved; Stensiö did usually identify this part as the profundus. Beyond the orbit the profundus nerve may have been large, but spread through the cutaneous layer as in lampreys. A search for it at more superficial levels of fossil material is needed. Within the orbit, the profundus nerve perhaps entered a capsular chamber, as it does in lampreys, within which it was not preserved.

The investigation of the splanchocranium, by means of the Damas identification of the dorsal nerves reaching the visceral head, has been begun. So far, on a narrow basis of cephalaspid material, the picture of their head as an entirety becomes easier to understand.

The efferent branchial blood vessels have been examined, using the position of the carotid supply to the brain as a *point de repère*. The blood vessels add coherence to the head organization as it now appears.

Clearly, the foregoing casts further doubt on the validity of Stensiö's identification. But a new identification remains hypothetical until examined with regard to all published accounts of cephalaspids.

The general background of new development has been briefly examined here and in a previous study (Whiting, 1972). From this, one conclusion would be that a profundus nerve with visceral components could well exist in a primitive vertebrate. So far such an animal has not been found. The existence of dorsal segmental nerves anterior to the profundus nerve seems most unlikely since they would have to derive from the pre-chordal part of the brain, i.e. from the forebrain.

ACKNOWLEDGEMENTS

I am grateful for the help I received from Dr P. M. Jenkin and the late Professor J. E. Harris, F.R.S., in the early stages of this work, and from the constructive comments of many participants at Symposia on Vertebrate Palaeontology and Comparative Anatomy, when ostracoderms or lampreys were described. Professor Westoll's discussion and thorough knowledge of cephalaspids, indicated for example in his review of fish palaeontology (Westoll, 1960), and the advice of Dr N. Heintz and Dr Q. Bone have been especially helpful, and encouraged me to go on and descend into splanchnocranial regions. But they are not to be held responsible for my views.

I am greatly indebted to my wife for consistent help and to Mr G. L. E. Wing, Mr N. A. Ablett, Mr K. Woods, Mr A. R. Woolman, Mr R. Godwin, Miss L. M. Clarke and Alan Todd, J.P., F.L.S. for technical help that has often needed great patience and ability, as in microscopy of the nervous system and in photography.

Some of this work was done, appropriately, as part of my studies when I was the Sir Ray Lankester Investigator in 1966, at the Marine Laboratory at Plymouth.

Finally I have to thank Mr H. A. Toombs and Dr Roger Miles for giving the history and identity of the two specimens of *C. whitei* when these were examined at the British Museum (Nat. Hist.), and Mr Lloyd Jones of Llandenny, Gwent, for generously and trustingly lending me these beautiful cephalaspids.

REFERENCES

ALCOCK, R., 1899. The peripheral distribution of the cranial nerves of Ammocoetes. *J. Anat. Physiol., Lond.,* (*N.S.* 13): *33:* 131–53.

BONE, Q., 1969. Muscular innervation and fish classification. In *Simposio internacional de Zoofilogenia.* Salamanca: Universidad de Salamanca.

CLAYDON, G. J., 1938. The premandibular region of *Petromyzon planeri.* Part I. *Proc. zool. Soc. Lond.* (*Ser. B*), *108:* 1–16.

CORDS, E., 1929. Die Kopfnerven der Petromyzonten. (Untersuchungen an *Petromyzon marinus.*) *Z. Anat. EntwGesch., 89:* 201–49.

DALCQ, A. & GERARD, P., 1935. *Traité d'embryologie des vertébrés.* Paris: Masson.

DAMAS, H., 1944. Récherches sur le développement de *Lampetra fluviatilis* L.—contribution à l'étude de la céphalogenèse des vertébrés. *Archs Biol., Paris, 55:* 1–284.

DAMAS, H., 1954. La branchie préspiraculaire des céphalaspidés. *Annls Soc. r. zool. Belg., 85:* 89–102.

DENISON, R. H., 1961. Feeding mechanisms of Agnatha and early gnathostomes. *Am. Zool., I:* 177–81.

GASKELL, W. H., 1908. *The origin of vertebrates.* London: Longmans, Green.

GOODRICH, E. S., 1930. *Studies on the structure and development of vertebrates.* London: Macmillan.

HARRIS, J. E. & WHITING, H. P., 1954. Myogenic stage of movement in the dogfish embryo. *J. exp. Biol., 31:* 501–24.

JOHNSTON, J. B., 1902. The brain of *Petromyzon. J. comp. Neurol., 12:* 1–86.

JOHNSTON, J. B., 1905. The cranial nerve components of *Petromyzon. Morph. Jb., 34:* 149–203.

JOHNSTON, J. B., 1908. Additional notes on cranial nerves of Petromyzonts. *J. comp. Neurol., 18:* 569–608.

KAPPERS, C. U. A., HUBER, C., & CROSBY, E. C., 1936. *Comparative anatomy of the nervous system of vertebrates.* New York: Macmillan.

KOLTZOFF, N. K., 1901. Entwicklungsgeschichte des Kopfes von *Petromyzon planeri. Bull. Soc. Nat. Moscou, 15:* 259–589.

LARSELL, O., 1947. Nucleus of IVth nerve in Petromyzonts. *J. comp. Neurol., 86:* 447–66.

LINDSTRÖM, T., 1949. On the cranial nerves of the cyclostomes with special reference to N. Trigeminus. *Acta zool., Stockh., 30:* 315–458.

LOWENSTEIN, O., 1970. Responses of isolated labyrinth of the lamprey. *Proc. R. Soc.* (B), *174:* 419–34.

MARINELLI, A. M. W. & STRENGER, A. 1954. *Vergleichende Anatomie und Morphologie der Wirbeltiere, Lfg. 1,* Vienna.

MILES, R. S. 1971, in Moy-Thomas, J. A. & Miles, R. S., *Palaeozoic Fishes.* London: Chapman & Hall.

NEAL, H. V., 1898. Segmentation of the nervous system in *Squalus acanthias. Bull Mus. comp. Zool. Harv., 31:* 147–294.

PEARSON, A. A., 1936. Acoustic and cerebellar connections of fishes. *J. comp. Neurol., 65:* 201–94.

RETZIUS, G., 1881. *Das Gehörorgan der Wirbelthiere. I.* Stockholm: Samson and Wallin.

ROBBINS, C. S., 1971. (MSS filed in Zoology Dept., Bristol.).

SCHWAB, M. E., 1974. New aspects about the prosencephalon of *Lampetra. Acta anat., 86:* 353–75.

SEWERTZOFF, A. N., 1916. Etudes sur l'évolution des vertébrés inférieurs. 1. Morphologie du squelette et de la musculature de la tête des Cyclostomes. *Archs russ. anat. histol. embryol.,* 1, fasc. *1:* 1–104.

STENSIÖ, E. A., 1927. The Downtonian and Devonian vertebrates of Spitzbergen 1. Family Cephalaspidae. *Skr. Svalbard Ishavet, 12:* 1–391.

STERBA, G., 1952. *Die Neunaugen.* Leipzig: Geest & Portig.

THOMSON, K. S., 1971. The adaption and evolution of early fishes. *Q. Rev. Biol., 46:* 139–66.

WALLS, G. L., 1942. *The vertebrate eye.* Michigan: Cranbrook Institute of Science.

WATSON, D. M. S., 1954. A consideration of ostracoderms. *Phil. Trans. R. Soc.* (B), *238:* 1–25.

WESTOLL, T. S., 1960. Recent advances in the palaeontology of fishes. *Lpool Manchr geol. J., 2:* 568–96.

WHITING, H. P., 1948. Nervous structure of the spinal cord of the young larval brook-lamprey. *Q. Jl microsc. Sci., 89:* 359–83.

WHITING, H. P., 1957. Mauthner neurones in young larval lampreys (*Lampetra* spp.). *Q. Jl miscrosc. Sci., 98:* 163–78.

WHITING, H. P., 1972. Cranial anatomy of agnathan fish. In Joysey & Kemp, *Studies in vertebrate evolution.* Edinburgh: Oliver and Boyd.

WOODWARD, A. S., 1930. Relationship between Lampreys and Ostracoderms. *Proc. Linn. Soc. Lond., 142:* 44–51.

ZYCH, W., 1937. *Cephalaspis kozlowskii,* n.sp., from the Downtonian of Podole, Poland. *Archwm Tow. nauk. Lwow,* (*Sect. III*), *9:* 49–96.

Plate 1

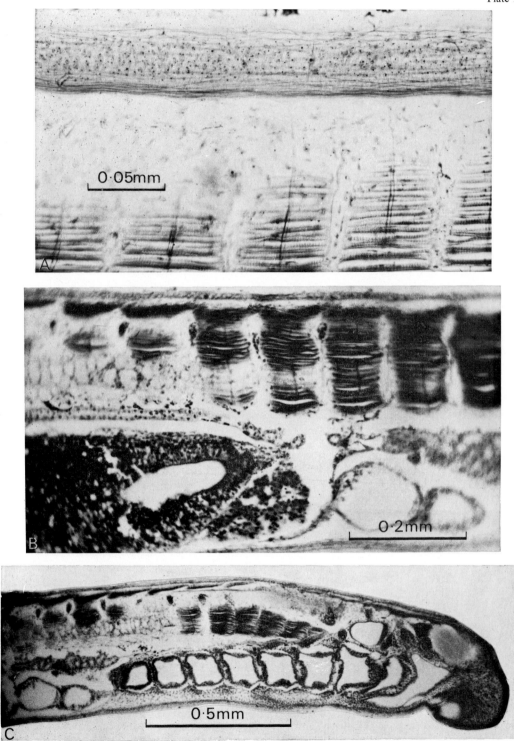

Plate 2

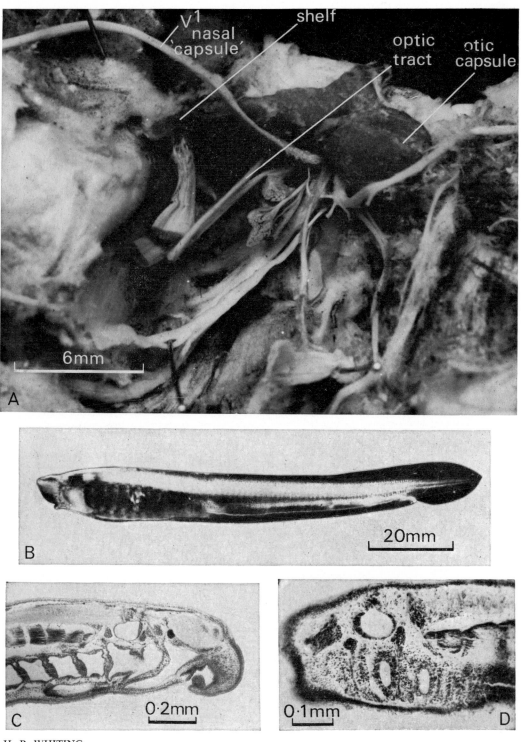

H. P. WHITING

Plate 3

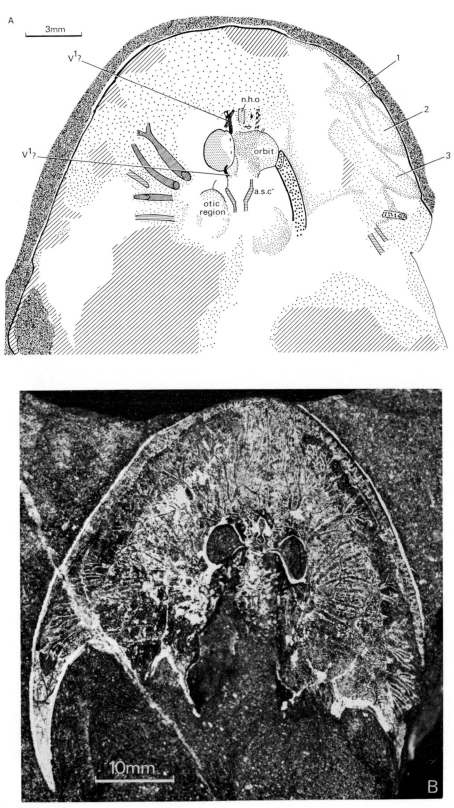

H. P. WHITING

Plate 4

25mm

15mm

A

B

H. P. WHITING

EXPLANATION OF PLATES

PLATE 1

A. Parasagittal section, silver-stained, of *Lampetra* proammocoete: spinal cord, myotomes with somatic motor nerves. Head to left.

B. Parasagittal section, silver-stained, of proammocoete: dorsal ganglia, myotomes and motor nerves. Head to right, at level of heart and liver.

C. Parasagittal section (cf. Whiting, 1972: pl. VI, 5c) of proammocoete, showing grading of trunk into head. For cranial nerves, cf. Fig. 1.

PLATE 2

A. Left orbit of *Petromyzon marinus;* the profundus nerve is lifted above its normal position, at left.

B. Living proammocoete at a stage when the nasohypophysial opening has become dorsal: eye and otic capsule visible.

C. Silver-stained parasagittal section of proammocoete (cf. Whiting, 1972: pl. VI, 5b; shows retina, V', VII, VIII, IX and X cranial nerves).

D. Parasagittal section showing relation between VII, otic capsule, IX and X nerves in a younger embryonic stage.

PLATE 3

A. *Tremataspis schmidti.* Diagram after Stensiö's pl. 60. The shading and numerals as in Figs 2 and 4.

B. *Cephalaspis whitei* from Llansoy, Monmouth. B.M. P.35679.

PLATE 4

A. A second specimen of *Cephalaspis whitei*, also from Llansoy. The fracture shows the successive outer parts of the branchial bars. (Photographed by R. Godwin.)

B. The same specimen in lateral aspect. The concave shape of the upper and outer part of the arches can be seen.

The dermal skeleton in fishes

The American Museum of Natural History and Columbia University, New York

Experimental and descriptive data on the morphogenesis and mature structure of the calcified dermal skeleton in fishes are critically reviewed. No evidence has been found to refute the hypothesis that the dermal skeleton (enameloid, enamel, dentine, membrane bone) in fishes develops from a single, integrated, modifiable morphogenetic system that is initiated by an interaction between the epithelium and the adjacent mesenchyme. Observable differences in the tissues of the dermal skeleton can be explained most economically in relation to changes in the timing or duration of steps in a morphogenetic pathway, in the elimination of one or more steps in the pathway, and in the modulation of cell activity. The systematic implications of the morphogenetic aspects remain to be evaluated.

CONTENTS

INTRODUCTION

There are few publications concerned primarily with analyses of descriptive and experimental morphogenetic data in relation to problems of vertebrate evolution and phylogeny. Exceptions are, of course, the contributions of De Beer (1937, 1951) and Schmalhausen (1949). In recent decades, however, some developmental biologists (e.g. Hall, 1975; Maderson, 1975) have given their findings and conclusions a phylogenetic orientation. But if organisms as morphologically distinct as chimpanzees and men have a 99 per cent genetic resemblance (King & Wilson, 1975), regulation of morphogenetic change is clearly of considerable evolutionary significance.

One purpose of this paper is to evaluate recent evidence on the morphogenesis and structure of the dermal skeleton in living and, by inference, extinct fishes.*

---

*The inadequate, but almost unavoidable, term 'fishes' is used here to include all major categories as defined by Moy-Thomas & Miles (1971).

By way of background information, a brief consideration of tissue interaction and of mesenchymal sources, including the neural crest, has been included. In regard to the dermal skeleton, an attempt is made to synthesize experimental evidence, which here is concerned with the elucidation of morphogenetic processes in living forms, and morphological evidence, ranging from gross form to ultrastructural detail, in both fossil and living fishes.

While we can observe the gross and microscopic structure of a fossil hard tissue, we can only conjecture through comparison with living forms about its development and about the soft tissues or cell types that may have been associated with it. This does not mean that conjecture about soft tissues or morphogenetic strategies in extinct organisms is undesirable, but only that it is not verifiable experimentally.

With these considerations in mind, the following points about the nature and utilization of morphogenetic data have been compiled:

(1) Any hypothesis concerned with the morphogenesis of living animals is testable by repetition of experiments.

(2) Any statement regarding morphogenetic mechanisms or processes in extinct animals is not directly testable (in contrast to statements regarding structure).

(3) In living organisms there is no fundamental difference between morphogenetic data and morphologic data because they merge with one another. Both provide 'characters' that can be used to test hypotheses about relationships and phylogeny, as can other attributes of living organisms (although there may be practical limits to the availability or interpretation of such data).

(4) The adult symmetry and organization of living vertebrates is a direct outcome of particular morphogenetic processes and pathways. These are known for relatively few vertebrates. Extrapolation, when stated or clearly implied, on the basis of limited experimental evidence is therefore necessary.

(5) When the morphogenetic parameters for a particular organ or structure have been established through experimentation in living forms, and when no significant deviation from these parameters has been found, we may postulate that the morphogenesis of homologous adult structure occurred in extinct forms in the same way.

A second purpose of this paper, which follows from the first, is to examine the hypotheses that the dermal skeleton develops from a single, modifiable morphogenetic system,—or from sequences of more or less independent tissue interactions that lead to the various phenotypic manifestations of the dermal skeleton in different groups of fishes.

TISSUE INTERACTION

In contrast to the early concept of embryonic induction as an essentially unidirectional, inducer-responder reaction, there is increasing experimental evidence for a reciprocal interaction between the ectodermal and the mesenchymal components of developing organ rudiments including those of the dermal skeleton. The reciprocity hypothesis, as formulated mainly by Grobstein (1956), has been repeatedly refined and corroborated (see, for example, Bernfield, Cohn & Banerjee, 1973; Hay, 1973; Hall, 1975; Maderson, 1975). Several workers have emphasized that substances concerned with an organogenetic interaction are frequently synthesized at tissue interfaces—as at the epitheliomesenchymal basement membrane.

Vertebrates generally have a multilayered epidermis and a dermis that may be more or less divided into a superficial well-vascularized layer composed of loose connective tissue and a deeper stratum that is usually less vascularized but more fibrous. The interface between the epidermis and the dermis is defined by the epithelium-derived basement membrane complex, which in fishes and amphibians is generally composed of a lucent zone just beneath the epidermis, a basement lamina (basement membrane) and a basement lamella with collagen fibrils (Nadol, Gibbins & Porter, 1969). Bernfield, Cohn & Banerjee (1973) believe that the basement lamina is continuous with embryonic epithelia, but that the fibrillar collagen of the lamella is reduced in amount in areas that are active morphogenetically.

The morphogenesis of epitheliomesenchymal structures or organs is believed to be dependent on, or initiated by, an interaction at the junction of the basement membrane and the adjacent mesenchyme. It is probable that the interaction persists throughout organogenesis—which, in the case of epitheliomesenchymal structures, could be most of the life span. Although collagen is distributed throughout the vertebrate body and is the most abundant extracellular protein, its actual role in this morphogenetic interaction is incompletely understood. Bernfield & Wessells (1970) suggest that it may be involved in stabilizing the form of developing epithelia. In this case, organ-specific morphogenesis may be dependent, in part, on the density of the collagen at specific sites below the epithelium. The studies of Berliner (1969) and Nadol, Gibbins & Porter (1969) on the basement lamella also indicate that the origin and organization of the collagen is dependent on the constant presence of the epidermis. Hay (1973) has also offered the hypothesis that differences in collagen composition and in other components of the extracellular matrix are responsible for the differentiation of the adjacent tissues. Further experimentation related to this hypothesis (Meier & Hay, 1975) has led them to conclude that the term 'embryonic induction' should be abandoned in favour of some term expressing the cell-extracellular matrix interaction. But the tissue interaction concept has been little discussed in the experimental literature on the dermal skeleton (Moss, 1972; Shellis & Miles, 1974) and I have generally followed the original investigator in using the term induction.

MESENCHYME AND THE NEURAL CREST

Vertebrate mesenchyme, which may be defined simply as embryonic connective tissue, develops from three different sources: (a) the mesoderm *sensu stricto;* (b) the prechordal plate; and (c) the neural crest. The role of 'mesodermal mesenchyme' in organogenesis is discussed in most embryology texts and will not be described here.

The prechordal plate, formed in association with endoderm in front of the notochord, can be regarded as a separate, if minor, source of mesenchyme, sometimes referred to as endomesenchyme. The so-called neuralizing factor, which may play an important role in determining the anterior (prechordal) part of the head, including the forebrain (Nieuwkoop, 1955), is most potent at the prechordal plate-notochordal junction. Prechordal plate mesenchyme may be involved in the

28

B. SCHAEFFER

formation of the posterior trabecular region of the basicranium and in the induction of other prechordal head structures, but this has not been confirmed experimentally.

Following induction of the neural plate, neural crest cells separate from the edge of the plate and collect in the area between the ectoderm and the now-formed neural tube. They are thus derived from an epithelium (ectoderm) that otherwise produces only epithelial structures such as tubes and vesicles. The mechanisms involved in the segregation of cells from the crest, their individual migration and their reassociation to form tissues or organs remain poorly understood (Hörstadius, 1950; Weston, 1970; Chibon, 1974).

As far as the author is aware, no hypothesis has been proposed for the origin of the neural crest, which is a unique vertebrate character (see also Løvtrup, 1974). The dual nature of the neural plate in vertebrates—that is as a source of both nerve tissue and neural crest cells—has not been demonstrated in the lower chordates. Because pigment cells in vertebrates arise from neural crest cells, it would be of interest to determine the origin of these cells in ascidians and cephalochordates. But even if there should be evidence that prospective pigment cells migrate from the neural plate in non-vertebrate chordates, this fact could not, by itself, explain why some crest cells in vertebrates also form dentine and visceral cartilage. We can propose the hypothesis that ectomesenchyme is mostly, or entirely, involved in the development of unique vertebrate structures. It is possible that the vertebrate ancestors evolved two additional sources of mesenchyme (prechordal plate and neural chest) before certain other unique vertebrate structures attained the vertebrate differentiation pattern? Studies on the break-up or disintegration of embryonic epithelium (such as the neural crest) to form mesenchyme, which Balinsky (1970: 324–5) regards as an important morphogenetic process, have not been made in the lower chordates.

Crest cell migration begins at the midbrain level, extends to the forebrain region and then moves posteriorly along the axis of the neural tube. The migration of individual cells is difficult to follow, mainly because they are practically indistinguishable from the stationary cells between which they are moving. Intrinsic cell markers (yolk, pigment, nuclear size, etc.) and, recently, vital dyes and protein labelling have been used with varying degrees of success.

The normal fates of neural crest cells, as recently summarized by Weston (1970: table V), may be divided into three categories: (a) pigment cells; (b) contributions to the sensory and autonomic nervous system; and (c) contributions to embryonic connective tissues and the skeletal system. In the present study, we are not concerned with the first. The second and third will be reviewed in relation to the development of the dermal skeleton.

Weston and others have emphasized that the fates of neural crest cells from successive sectors of the neural folds are different—as neatly demonstrated by Chibon (1967) for the contributions to the basicranium and the visceral skeleton in the urodele *Pleurodeles*. Nevertheless, the range of developmental capabilities for individual neural crest cells at any particular time during early development remains essentially unknown. Although there is no experimental evidence on crest cell specificity at the beginning of dispersion, there is evidence that crest cells form mesenchyme and, following interaction with other cells in their definitive environments (Noden, 1975), differentiate into a variety of neural or skeletal structures. Inductive interaction is clearly of great importance in the development of the dermal skeleton, and its analysis is a major aspect of experimental morphogenesis.

The evidence of numerous investigators regarding the specific role of ectomesenchyme in the epitheliomesenchymal interaction has been summarized by Weston (1970). Well-known examples include the formation of the sensory root ganglia of cranial nerves V, VII, IX, and X and the ganglionic neurons. The calcified tissues presumably initiated through this interaction are the main subject of the present paper. Of particular significance in this regard is the hypothesis that crest cells contribute to the dermis all over the body, as concluded by Raven (1931, 1936) from xenoplastic grafts. The experimental evidence cited by Weston suggests that this crest contribution may be diminished in higher vertebrates.

DELAMINATION

The term delamination was used by early embryologists to describe a cleavage zone between the superficial and deeper cells of a single sheet, converting it directly into two sheets. In 1959, Jarvik reviewed the observations of Goodrich (1904) on dermal fin ray development, as well as the ideas of Holmgren (1940) and proposed 'Holmgren's principle of delamination.' By this term, Jarvik (1959: 45) meant 'the capacity of the outermost parts of the undifferentiated ectomesenchyme, or, in later ontogenetic stages and in the adult, of the outermost parts of the corium, to a repeated production of laminae with potentialities for forming skeletal structures.' As examples of this phenomenon he cited Holmgren's (1940) ideas on the successive formation of neurocranial components in *Squalus*, the development of head scales in *Neoceratodus* following the sinking of certain dermal bones into the dermis (Goodrich, 1904: 50), the generations of dentine and enamel (enameloid) in the dermal bones and scales of various fishes, and, finally, several aspects of dermal fin-ray development. Jarvik emphasized that the principle applies to both endoskeletal and exoskeletal structures.

Moss (1968a, b, 1972) described delamination as 'the morphogenetic movement of the tissue products formed at the basement membrane as the result of the inductive interaction between the ectomesenchyme and the basal epidermal cell layers.' He emphasized that the process of delamination is involved in the formation of all dermal derivatives and in their topographic organization. For instance, there has been implication in past studies of dermal fin-ray development that the epidermis-derived basement membrane contributes the collagen of the (dermal) rays.

It seems inaccurate, however, to regard neural crest cell dispersion and cell movement related to mesenchymal induction as a single morphogenetic 'principle' (see Jollie, 1968). Although Moss (1968b) used the term delamination only in connection with the formation of dermal derivatives, his discussion, like that of Jarvik (1959), emphasized the movement (or delamination) aspect—which in one way or another is actually characteristic of all organogenesis. Nevertheless, the strictly sequential expression of delamination as envisioned by Holmgren is an oversimplification of various morphogenetic changes taking place at and near the epitheliomesenchymal interface. As Moss (1972) and others noted, it is now possible to reinterpret delamination in terms of the tissue interaction hypothesis. Many examples of so-called delamination represent invagination, where groups of cells push inwards from the surface epithelium, or ingression, where individual cells migrate or stream inward and lose their epithelial identity (Ballard, 1964). The term delamination should probably be eliminated in favour of a more precise description of all cell movements involved.

ENAMELOID AND ENAMEL

Turning now to the specific tissues involved in the dermal skeleton, it should be noted that the morphological and histological characteristics of enameloid (or enamel), dentine and cellular or acellular dermal bone have been described in varying degrees of detail for a great number of fossil and living vertebrates. The rather complicated and overlapping terminology applied to the structural permutations of each tissue, particularly in the early agnathans, placoderms and chondrichthyans reflects the different and changing opinions about their architecture, ontogeny and phylogeny. Rather than attempt a summary of other summaries (e.g. Ørvig, 1967; Peyer, 1968), I will comment on selected, and for the most part, recent experimental studies that offer a new or clearer understanding of the dermal calcified tissues and their interrelationships. It seems rather pointless to provide conventional definitions for these tissues when this entire paper is concerned with their characterization.

Gaunt & Miles (1967) have provided a critical review of the experimental work on tooth and dermal bone morphogenesis in *Ambystoma*, *Triturus*, and *Bombina*—the only vertebrates in which tooth morphogenesis is well documented. In these amphibians neural crest cells migrate by well-defined pathways to specific parts of the mouth ectoderm where they form the dental papillae. The papillae induce the overlying epithelium to form the dental organs. The inner cells of the dental organs differentiate into the inner dental epithelium, which in turn induces (or interacts with) the subjacent papillar cells to modulate into odontoblasts. The odontoblasts form the organic matrix of the dentine and are concerned with its mineralization. During dentinogenesis, the basal cells of the inner dental epithelium are modified into ameloblasts, which are responsible for enamel. The odontoblasts thus move 'centripetally' and the ameloblasts 'centrifugally.' This well corroborated hypothesis for ambystomid tooth morphogenesis has, in effect, become the model for gnathostomes in general. But there is still no experimental evidence that the dental papilla in fishes is formed from ectomesenchyme.

The development of the outer enamel-like layer (here called enameloid) covering the tooth crowns in sharks and actinopterygians, has been a matter of contention for nearly a century (Lison, 1954; Kerr, 1955, 1960; Peyer, 1968; Moss, 1970). Three different hypotheses have been proposed: (a) that the organic matrix of this layer is formed by the inner dental epithelium and is therefore ectodermal; (b) that it is formed by odontoblasts and is therefore mesodermal or ectomesenchymal; (c) that it is formed through the interaction of the inner dental epithelium and the mesenchyme (or ectomesenchyme) and is therefore epitheliomesenchymal. Repeated testing of the first two hypotheses have yielded inconclusive evidence, and it seems more meaningful to discuss the third and most recent hypothesis, particularly in light of autoradiographic studies by Shellis & Miles (1974) on tooth development in the eel *Anguilla* and in the wrasse *Labrus*, as well as in certain elasmobranchs (Shellis, pers. comm.). Protein labelling in these fishes corroborates the opinion of numerous workers that the odontoblasts synthesize the enameloid matrix and that this matrix reaches its definitive size and form before enameloid mineralization is initiated. Matrix growth occurs at the papillary surface and is therefore centripetal (inward). The odontoblasts retreat centripetally, leaving behind, in the enameloid matrix, dentinal tubules that contain the odontoblast processes.

Although various workers (e.g., Kvam, 1946; Kerr, 1955, 1960; Peyer, 1968; Moss, 1968b; Poole, 1971; Herold, 1974) have noted that the inner dental epithelium in fishes is involved in enameloid formation, Shellis & Miles (1974) have come up with the first hypothesis that is based on experimental data. Their protein labelling studies indicate that the inner dental epithelium secretes a special protein that interacts with the enameloid matrix (formed by the odontoblasts) so that this matrix becomes labile during enameloid mineralization. Crystal growth, which may be initially influenced by the collagen fibre arrangement, continues as the collagen is removed. In consequence, mature enameloid in the eel and wrasse is highly mineralized and has very little organic matter. Enameloid matrix removal and mineralization begin next to the predentine (Kerr, 1960; Shellis & Miles, 1974), not in the vicinity of the inner dental epithelium. In the first figure (Plate 1A) of the sequence demonstrating tooth development in *Amia* (Plates 1 and 2), enameloid matrix (em) is present in the tooth germ. In all subsequent figures, the enameloid matrix is absent and the decalcified enameloid is represented by a clear triangle.

It is difficult to reconcile the electron microscope observations of Garant (1971) and Herold (1974) on odontogenesis in *Esox* with the morphogenetic model proposed above. According to Herold, pike enameloid, which is a thin and barely visible layer under the light microscope, has a granular, non-collagenous matrix that is formed after deposition of the dentinal collagen and is mineralized after the dentine. Herold believes that the inner dental epithelium contributes to enameloid formation, but he notes that the early *Esox* enameloid matrix resembles the basal lamina. Garant has suggested that the inner dental epithelium is involved in dentine formation, but he may be referring to the enameloid. On the other hand, most of Moss's (1970) observations on the developing 'enamel' of the porgy, *Stenotomus*, fit into the Shellis-Miles model.

According to Kerr (1955) and Schmidt & Keil (1971) the enameloid in a variety of fossil and living elasmobranchs, contains collagen fibres that radiate into the dentine. The orientation of the fibres varies somewhat according to the form of the tooth. In some taxa, at least, the enameloid is more heavily mineralized than the dentine, and collagen is nearly absent (Reif, 1973; also see discussion of enameloid formation in *Carcharias*, Schmidt & Keil, 1971: 232). There is an apparent correlation between the degree of calcification and the amount of organic matrix retained in the mature enameloid. Autoradiographic experiments on enameloid development in *Raja* have yielded indifferent results in regard to the role of epithelial protein (Shellis, pers. comm.). There is, however, indirect evidence of epithelial protein participation in enameloid formation—i.e. elongation of the cells composing the inner dental epithelium (Plate 3C), and the observation of Levine *et al.* (1966), that soluble protein in mature *Squalus* enameloid and bovid enamel are similar.

The problems raised in *Esox* tooth morphogenesis focus on the criteria used to identify mature enameloid (sometimes called durodentine, mantle dentine, mesodermal dentine, modified dentine or vitrodentine) in fossil and recent fish teeth of any type. Tetrapod enamel appears to be closely related chemically to enameloid in elasmobranchs and teleosts (Levine *et al.*, 1966; Shellis & Miles, 1974). The data of Kerr (1960) and Meredith Smith & Miles (1971) further support the chemical affinity of these tissues. The usefulness of birefringence for separating enamel and enameloid (Peyer, 1968) may be limited because the 'collar enamel' of *Labrus* teeth, previously identified on the basis of birefringence, turns out to be enameloid, according to protein labelling (Shellis & Miles, 1974). Positive birefringence, however, has confirmed the identification of collagen as the major matrix component in unmineralized enameloid (Shellis & Miles, 1974: 64). Other criteria that have been used to recognize adult enameloid are dentine tubules, lamellae, hypermineralization, low organic content and lack of prisms (also absent in amphibian and reptile enamel). But these characters, except for the nonprismatic condition, seem to vary among the main groups of living fishes. A survey of the literature and of sectioned teeth from a variety of teleosts suggests that it may be difficult to distinguish mature enamel from mature enameloid with the light microscope (Moss, 1970). Shellis & Miles noted that under the light microscope the only visible difference between enameloid and dentine in the eel is a smaller number of dentinal tubules in the enameloid.

According to Grady (1970) the teeth of the coelacanth *Latimeria* are covered with enamel, but the outer tissue on the spines of the scales is identified as enameloid by Meredith Smith, Hobdell & Miller (1972) on the basis of hypermineralization and low organic content. In regard to other sarcopterygians, Schultze (1969) has identified true enamel on the oral teeth of the rhipidistian crossopterygians *Laccognathus* and *Eusthenopteron*. Shellis (pers. comm.) believes that mature enamel can be recognized in sarcopterygians and other vertebrates by the sharply defined enamel-dentine boundary, the incremental-like lines and the negative birefringence in relation to the surface normal. Apart from high mineralization, positive criteria for recognizing enameloid in dipnoan teeth are elusive (Denison, 1974).

Another important aspect of the Shellis-Miles hypothesis is concerned with the morphogenetic shift from enameloid to enamel. On the basis of histochemical and autoradiographic evidence, these authors concluded that the inner dental epithelium of teleosts and the ameloblasts of tetrapods secrete similar matrix protein. Enameloid is considered to be the product of an interaction between the epithelial matrix protein and the collagen secreted by the odontoblasts. A delay (heterochronous shift) in protein secretion by the inner dental epithelium could permit the odontoblasts to form dentine before the epithelial cells become active (see Poole, 1971, for a somewhat different explanation). According to the schedule in the ambystomid odontogenesis model, the epithelial proteins would then form enamel. As Shellis & Miles noted, the work of Meredith Smith & Miles (1971) on urodele teeth also suggests that the production of enamel, as opposed to enameloid, could be dependent on a change in the timing of inner dental epithelial secretory activity in relation to odontoblast activity.

DENTINE

The morphogenesis of dentine in the various groups of living fishes has been extensively studied through use of the light microscope. Several papers by Kerr (1955, 1960) are particularly useful in understanding the arrangement of the developing collagen fibre systems in *Squalus, Scyliorhinus* and a number of teleosts. Herold (1970a, b, 1974) and Herold & Landino (1970) have studied dentine development in *Esox*, gadids and pleuronectids by various methods. Comparative autoradiographic studies on dentine formation in fishes are few, although Shellis & Miles (1974) have provided some observations.

The basic resemblance in dentine development among all toothed vertebrates is well known, but the different groups of fishes, extinct and living, show a greater diversity in dentine architecture than do the tetrapods. Two types of dentine— orthodentine (Plate 4) and trabecular dentine—are frequently recognized, but the variants of each are numerous and the terminology is correspondingly complicated.

The termination of enameloid matrix deposition is soon followed by precipitation of the insoluble, stable dentine collagen matrix. According to Shellis & Miles (1974), the labelling patterns for the odontoblasts and for the predentine in the eel and wrasse are very similar to those in amphibians and mammals. Although the causal relationships between the developing fibre bundles, odontoblast processes, capillaries and the inwardly growing dental organ remain poorly known, it is reasonable to postulate a more or less constant interaction between them during dentine development; the disposition of these soft tissues largely determines the structural features of the mature mineralized tissue. From the morphogenetic viewpoint, the variables in dentinogenesis include: (a) collagen matrix fibre size and orientation; (b) degree and pattern of pulp vascularization; (c) incorporation of vascular canals into mature tissue; (d) sites of odontoblast activity; (e) odontoblast process ramification, orientation and frequency within the matrix; (f) incorporation of cell bodies into the dentine matrix; and (g) incremental versus non-incremental growth. Certain combinations of these variables are clearly more common than others in living fishes.

Although differences in the details of collagen fibre development and orientation are evident in sharks and teleosts (Kerr, 1955, 1960; Reif, 1973), forms with orthodentine show a progressive ensheathing of the pulp chamber by the dentine matrix beginning at the level of the enameloid cap border and extending more or less simultaneously to the apex and to the expanding tooth base (Plates 1, 2 and 3C). The orthodentine matrix in sharks is 'felted' rather than more precisely orientated radially and longitudinally, as it is in teleosts. Herold (1971a), among others, has noted that orthodentine and vasodentine matrices grow incrementally in teleosts, which is not always the case in sharks.

During dentine matrix growth, the odontoblast processes are surrounded by organic matrix to form the dentine tubules. The factors influencing the number, length and spacing of the processes (and tubules) remain unknown. Various workers have noted that the tubules in viable dentine do not always contain processes—in other words, they are 'empty' canals. With one reported exception, living fishes have acellular dentine, that is, dentine in which the odontoblasts retreat centripetally as the dentine matrix is deposited. In *Amia*, cells are frequently incorporated into the late-formed matrix and remain after mineralization has occurred. These have been identified as odontoblasts (Moss, 1964) or as osteoblasts (Peyer, 1968), although they do occur with dentine tubules. In this regard, it would be interesting to follow up the suggestion of Stephan (1900) that cellular dentine may result from rapid odontogenesis.

Vasodentine, in the sense of Tomes (1923) and Herold (1970a, b), rather than that of Peyer (1968), is a variant of teleost orthodentine in that it grows appositionally and incrementally, has a distinct predentine zone and forms around the pulp cavity. It differs from orthodentine in lacking dentine tubules and in being formed around capillary loops that project outwards from the pulp. The collagen fibres are oriented concentrically around the vascular canals.

The studies of Herold (1971a, b) and of Herold & Landino (1970) on trabecular dentine in *Esox* support the hypothesis that the capillary pattern of the papilla, which is established before matrix development, influences the orientation of the collagen matrix bundles. Trabecular matrix begins to form at the apex of the previously developed orthodentine cap and, according to Herold, progresses toward the tooth base. Coarse fibres that extend from the orthodentine join together in the pulp to form large trabecular collagen bundles parallel to the longitudinal tooth axis (Lison, 1954). When these bundles reach a certain diameter, fine collagen fibrils are deposited circumferentially around them. Both sorts of fibrils are produced consecutively by the same odontoblasts (called scleroblasts by Herold). The fine fibres fill the intertrabecular areas and broadly surround the capillaries in the pulp. The odontoblasts are finally situated in the capillary canals at the conclusion of matrix formation. In *Esox*, fibre deposition is not appositional and therefore mineralization is not incremental, either around the canals or in the intertrabecular tissue (Peyer, 1968: pls 41a, b, 42a). Calcification, which is mediated by the odontoblasts, begins after most of the matrix is formed. The fine fibre matrix is evenly and heavily mineralized, the coarser matrix less so.

In most ways the observations of Herold agree with the model for trabecular dentine development proposed by Ørvig (1951, 1967), but there are several important differences. Because the same pulpar cells produce trabecular and intertrabecular matrix in *Esox*, Herold (1971a) does not identify the intertrabecular tissue as acellular bone. Peyer (1968) also questioned the assumption that modified pulpar cells in, for example, the shark *Prionace* could first form an outer layer of orthodentine matrix, then produce trabecular bone matrix and finally circumvascular dentine. In *Prionace* the trabecules appear to go through a predentine phase before mineralization. Mature dentine is usually recognized at the light microscope level, in both Recent and fossil fishes, by the dentinal tubules and the absence of cell spaces.* The tooth base in elasmobranchs is composed of mineralized tissue without dentine tubules. Although identified as trabecular dentine by James (1953) in *Squalus* and by Schmidt & Keil (1971) in *Scyliorhinus* and *Odontaspis* (*Carcharias*), the absence of tubules suggested to Schaeffer (1963) in an extinct pristid and to Moss (1970) also in *Squalus*, that the tooth base may be acellular bone. Moss has observed cells in the developing tooth base of *Squalus*, which subsequently becomes pycnotic, and Zangerl (1966) has found cell spaces in

---

*According to Bergot (1975), the odontoblasts in the developing teeth of *Salmo fario* do not form cytoplasmic processes. The mature enameloid and dentine consequently lack tubules.

the tooth base of *Ornithoprion*. In the holocephalans the spines and the tesserae of the Carboniferous forms are apparently composed only of several types of dentine (Patterson, 1965).

The more or less regular spacing pattern of the dentine tubules in the ortho-dentine and trabecular dentine of teeth (Plate 4) differs rather consistently from the dentritic pattern that occurs in the dentine layer of dermal plates and scales (Plates 5 to 8). All the tubule ramifications of one 'dentrite' (which is frequently difficult to recognize from the tubule pattern) have surrounded the cytoplasmic processes of a single cell. As the odontoblasts recede toward the pulp cavity, their cytoplasmic processes converge to a single process that is directly attached to the cell body. The resulting arborescent tubule arrangement is also evident in the exoskeleton of osteostracans (Denison, 1951), placoderms (Ørvig, 1951), acantho-dians (Gross, 1971), palaeonisciforms (Ørvig, 1951, 1957b) and rhipidistians (Gross, 1956; Ørvig, 1957a).

The dentine of extinct and living dipnoans has been well described by Denison (1974), from which it is evident that the differences in mineralization of this tissue have systematic significance. The palate and lower jaws of some Devonian dipnoans have denticle-like teeth composed of orthodentine, but most genera, including the living ones, have teeth or tooth plates composed largely of trabecular dentine (Plate 3B). The relationships of dipnoans with other osteichthyan groups suggest that separate deciduous denticles represent the primitive condition for this group. However, the denticles of most Devonian and later genera are fused to the underlying bone and become arranged in radiating rows or ridges. In *Protopterus* all the denticles of one tooth plate form under a single epithelial dental organ (Lison, 1941, 1954). This suggests that 'fusion' of the primitively separate denticles also involved union of originally separate dental organs. In elasmobranchs, teleosts and tetrapods multicuspid teeth develop from a single, although frequently complicated, epithelial dental organ. The pros and cons of the fusion hypothesis, particularly in relation to the dental organ are, unfortunately, nearly untestable.

The trabecular dentine of nearly all dipnoan teeth consists of vertical columns of collagenous dentine that surrounded the vascular canal and a highly mineralized, practically non-collagenous intertrabecular dentine (Plate 3B) called petrodentine by Lison (1941). The development of the collagenous matrix bundles in the living genera has not been described, but from Lison's (1941, 1954) description of tooth development in *Protopterus* it seems probable that the organic matrix of the intertrabecular tissue is reduced during mineralization. Also of interest is the orientation of the dentinal tubules, which extend obliquely (nearly at right angles to the crown-base tooth axis) from their convoluted ramifications in the inter-trabecular petrodentine into the trabecular dentine proper, where they converge into larger tubules that open into the pulp canals. As Denison (1974) noted, this orientation indicates that the trabecular and intertrabecular matrices are formed by the same odontoblasts—which Herold (1971a) noted in *Esox* and Radinsky (1961) in *Asteracanthus*. The seemingly complicated picture of dipnoan tooth development may be considerably simplified when the formation and orientation of the collagenous fibre bundles has been described. Most Devonian and post-Devonian dipnoan tooth plates are specialized in having abundant, highly mineralized intertrabecular dentine; otherwise, the mature, mineralized tooth crown, particularly in cross section, resembles in its general organization, the trabecular dentine in other groups of fishes. Differences of systematic significance, however, are evident, as a comparison between the trabecular dentine of *Myliobatis* and *Neoceratodus* (Plate 3A, B) indicates.

In the osteostracans and some acanthodians, Ørvig (1966, 1967) has recognized a special dentinous tissue that he named mesodentine. It is defined in terms of its cell space components, which have two or more branching and frequently inter-connected tubules of irregular shape. The shape of the cell spaces and their multiple processes indicated to Ørvig that the matrix-forming cells are scleroblasts (embryonic, mesenchymatic cells) rather than osteoblasts or odontoblasts. However, if attention is restricted to the structures (or spaces) that are actually preserved in the dentine layer of the osteostracans, there seems to be no certain basis for distinguishing the dentine tubule pattern in osteostracans from that in many taxa belonging to various groups of gnathostome fishes.

Another dentinous tissue, restricted to the gnathal plates and dermal bone tubercles of placoderms, is designated semidentine by Ørvig (1951, 1967). The dentinal tubules are single, sometimes with small distal and proximal branches, and joined to cell spaces of tear-drop form situated at varying distances from the underlying vascular canals. Semidentine may include a large number of cell spaces, or almost none at all. It occurs around vascular canals as a pallial layer of a trabecular tissue and around large cavities in the inferognathals of *Pachyosteus* and *Malerosteus* (Kulczycki, 1957). Of particular interest is the condition in the dermal tubercles of *Tollichthys* (Bystrow, 1957, fig. 19), which has an outer layer of mesodentine and an internal layer of semidentine.

Ørvig (1967: 79) regards mesodentine and semidentine as 'dentine precursors' in the sense that primitively cellular dentine may have been transformed into an entirely acellular tissue. He sees the morphological sequence as mesodentine-semidentine-metadentine (true dentine), although he stresses the uncertainty of this series, mainly because both cellular and acellular dentine are presumably 'very ancient.' The actual occurrences of cellular dentine seem to be sporadic—in osteostracans, arthrodires, and *Amia*—which may indicate an independent derivation in each case. However, the hypothesis that acellular dentine is derived from cellular dentine is not readily testable, essentially for the same reason that no satisfactory choices can be made between the relative primitiveness of cellular or acellular bone in fossil agnathans.

In regard to the morphogenesis of these different dentine manifestations, four hypotheses are worth considering. The first is that the morphogenetic mechanism responsible for dentine in teeth, dermal bones, scales and lepidotrichia is essentially the same in forms that share a nearly identical genotype (as in *Polypterus*, Plates 4 to 8), but it should be noted that this basic similarity may be far more inclusive in a systematic sense. The second is that orthodentine and trabecular dentine are successive products of the same odontoblasts, although not always in that order. The third, which might be checked with autoradiography, is that suppression of the dentine and/or the enameloid layer, which has occurred independently in various groups of fishes, may represent a relatively simple developmentally controlled modification. The fourth, which may not be fundamentally different from the third, is that a developmental shift from acellular to cellular dentine (or primitively in the opposite direction) can occur. The last three hypotheses have one important aspect in common, that is, modification in a morphogenetic pathway either through change in timing or through elimination of a particular step in the pathway.

In regard to the second hypothesis it should be noted that trabecular dentine may be peripheral to orthodentine in the dorsal spines of holocephalans and some elasmobranchs (Markert, 1896; Patterson, 1965). Almost nothing is known about chondrichthyan spine embryology. Both the embryonic and the adult tissues of these spines should be reexamined and the course of the dentine tubules, if present, determined.

DERMAL BONE AND SCALES

Experiments by Sellman (1946), Andres (1946), Wagner (1949, 1959) corroborate the hypothesis that dermal bone in urodeles develops from ectomesenchyme. The evidence of these workers is concerned mostly with tooth-bearing bones and has led to further speculation about the developmental interaction between the tooth anlage and the bone of attachment (Gaunt & Miles, 1967). Because of the enameloid, dentine and dermal bone association in the dermal skull, scales and lepidotrichia (Plates 5 to 8), it may be postulated that ectomesenchyme is present in the dermis throughout the body and, when the proper genetic signals are forthcoming, can form dermal bone as well as dental organs. There is experimental evidence that trunk neural crest cells migrate through the ectoderm and into the mesoderm to form pigment cells in urodeles (Chibon, 1967, 1974; Weston, 1970) but as yet there is none in fishes or amphibians to indicate that migrating trunk crest cells are directly involved in dermal ossification. Attempts to establish this by isotopic labelling have thus far been unsuccessful, mainly because of labelling dilution related to cell division (Weston, 1970: 71). Furthermore, the experiments of Schowing (1968) on the fowl cranium indicate that the brain and the meninges (which are partly derived from neural crest) interact with the overlying mesenchyme to produce some dermal bones.

Regardless of the induction mode, dermal bone development begins with the modulation of mesenchymal cells into osteoblasts. Populations of osteoblasts having a more or less fixed spatial distribution form a highly ordered collagen matrix that apparently controls the size and orientation of the hydroxyapatite crystals. The matrix is laid down around blood vessels, mucous canals, sensory canals, neuromasts and nerves—all of which are differentiated prior to mineralization. Coarse collagen bundles connecting the bony layers of one scale with another may remain uncalcified and the lepisosteoid tubules in some scale types may contain cell processes (Kerr, 1952). *Polypterus* scales (Sewertzoff, 1932; Kerr, 1952) have an elaborate horizontal vascular canal system under the dentine with vertical canals containing capillaries that pass into the epidermis above and the dermis below (Plate 6). Similar canals are present in the bony scales of the armoured catfishes (Ørvig, 1957b). The dermal elements of most early fishes (ostracoderms, placoderms, some palaeonisciforms, rhipidistians and dipnoans) tend to be highly canaliculated between the dentine layer and the basal laminated layer; the latter layer has fewer, larger canals. It is presumed that most of these canals contained capillaries, as they do in *Polypterus* scales.

Arguments about the phylogenetic significance of aspidin (and its resemblance to teleost acellular bone) have reached a sort of impasse. Although the distribution of aspidin in the various groups of extinct fishes (anaspids, heterostracans, ?chondrichthyans, acanthodians) has been quite thoroughly investigated, the lack of necessary phylogenetic information for the major groups of Palaeozoic fishes has confused opinions about which condition is derived. Within the teleosts, however, interrelationships are better understood, and it is increasingly probable that acellular bone is a derived condition for the teleost groups in which it occurs. There is less likelihood that this problem can be approached phylogenetically for the heterostracans and osteostracans (see Ørvig, 1965, for discussions of cellular bone and aspidin in the osteostracans), or for the acanthodians (R. S. Miles, 1966).

Osteoblasts are surely involved in the morphogenesis of acellular bone (Kobayashi, 1971) and it seems improbable that some unknown cell type, such as Tarlo's (1963) 'aspidinoblasts' could have been involved in aspidin development. According to Moss & Posner (1960), osteoblasts are incorporated into the dermal bone matrix of the turkeyfish (*Pterois*), where they become necrotic. The cell spaces are then filled with matrix and totally disappear. If this observation is confirmed, and is true for other adult teleosts with acellular bone, autoradiographic experiments may show how matrix formation and mineralization occur. Ørvig (1957b) suggested that the osteoblasts retreat to the walls of vascular canals; this seems to be the case in the catfish *Callichthys*. Except for being acellular, the fine structure of mature acellular bone is in most ways like that of cellular bone (Moss, 1961a, b; Moss & Freilich, 1963; Tarlo, 1963; Halstead, 1969), and the morphogenetic differences between the two tissues may be less than some investigators would have us believe. Although we may never work out the histogenesis of the ostracoderm or acanthodian aspidin, knowledge of what happens during acellular bone formation in the euteleosteans should give speculation about aspidin more meaning.

Dermal bone and scale patterns in most extinct and living fishes seem generally rather fixed and regular. Pattern variation is presumed to be more frequent in dermal skulls composed of small, frequently irregular ossifications than in dermal skulls that have relatively large bone units and relatively regular patterns. Actually, few studies of pattern variation in fossil and recent fishes have been made, but some examples can be given among the osteichthyans: (a) the entire dermal roof of *Acipenser* (Jarvik, 1948: fig. 19); (b) the parietal area of *Australosomus* (Lehman, 1952: figs 94–8); (c) the 'spiracular' bones of *Polypterus* (Jarvik, 1947: fig. 1); (d) the snout elements of *Eusthenopteron* (Jarvik, 1944: fig. 6); (e) the dermal roof of *Diplurus* (Schaeffer, 1952: fig. 5); and (f) the dermal roof of *Dipterus* (Parrington, 1950). Among a sample of 50 *Amia* skulls from the same locality, one has a single median parietal instead of the usual pair. These examples focus on the still unanswered question of what controls dermal bone patterns and under what conditions this control is 'relaxed.'

Pehrson (1940) and others have pointed out that adult dermal bones, both anamestic and along sensory canals, may have more than one ossification centre. The hypothesis has been proposed that a gradient system involving a competition-dominance phenomenon develops in and around ossification centres (Spiegelman, 1945; Devillers, 1965; Schaeffer, 1967). According to this concept, one or several 'cooperating' centres exert influence (dominance) over a particular amount or area of osteogenic tissue before and/or during deposition of the organic matrix. Other non-cooperating centres compete with or repel one another.

There is some experimental evidence in the teleosts to reinforce the concept of bone rudiment interaction as a factor in the development of a regular dermal bone pattern. Tatarko's (1934) experiments involving partial and total removal of opercular bones in *Cyprinus* indicate that an absent element (or part of an element) may be 'replaced' by the abnormal enlargement of a neighbouring one. Extirpation, presumably without disturbing neighbouring elements, of the developing frontal bone in fry of *Salmo* results in a regenerated frontal approximately the size and shape of the normal element, but lacking a sensory canal (Pinganaud-Perrin, 1973). Devillers & Corsin (1968) have discussed similar experiments.

Another aspect of the dermal bone pattern problem is related to the number of elements occupying a particular area (as distinct from the spatial arrangement of elements). De Beer (1937: 503–9) discussed dermal bone fusion and elimination in connection with his hypothesis that, primitively, one dermal element is formed from one ossification centre. He noted, for instance, that the frontal of *Amia* arises from four ossification centres, while 'in other forms' it arises from one. But, as he also implied, it is nearly impossible to test the hypothesis that the *Amia* frontal arose from an ancestor with four separate elements. Even when the number of ossification centres for a particular dermal element is known, the phylogenetic meaning of the number is obscure. Other factors, such as developing tooth buds or sensory placodes, may also be involved in determining the number of centres. Moy-Thomas (1934) and Herold (1971b) concluded that the dentine in *Esox* teeth stimulates the formation of the bone of attachment, which appears as a separate collar below each tooth. Each bony anlage maybe formed from a single ossification centre, which subsequently enlarges and fuses with others to form the dermal tooth-bearing elements of the visceral skeleton in the osteichthyans (Nelson, 1969). In *Amia* there is no apparent causal relationship between the developing tooth and the bone of attachment (Plates 1 and 2), but this can only be ascertained by experiment.

Nelson (1970) has presented cogent reasons for assuming that the micromeric condition (many small dermal bone units) is primitive for vertebrates and that the macromeric condition (few large dermal bone units) arose independently, perhaps four times. The hypothesis that the micromeric dermal skull of the earliest dipnoans represents a retention of the primitive vertebrate condition is difficult to test by out-group comparison. R. S. Miles (1975) has provided corroboration for the hypothesis that the Dipnoi are the sister group of the Crossopterygii, which have a bone pattern that is simultaneously micro-, meso-, and macromeric. Knowing more about ossification centre interaction will probably not resolve specific problems in dermal bone pattern analysis, but it may considerably influence our thinking about the presumed fusion, fragmentation or elimination of adult elements. Because there is no experimental basis for discussing these problems, stated bone homologies are frequently arbitrary, and are often seemingly meaningless.

The developmental interaction between lateral line placodes and dermal bones remains enigmatic in both fishes and amphibians. Although palaeontologists and embryologists have considered various aspects of this problem for nearly a century (Parrington, 1949; Ørvig, 1972), it is clear that more experimental analysis is needed to provide critical evidence beyond the demonstration by Devillers (1947) that there is some sort of a causal relationship involving neuromast organs and the initiation of membrane bone ossification.

A central problem in these deliberations is the nature of the sensory placode-mesenchyme interaction. The lateral line system first develops as a series of ectodermal placodes situated both anterior and posterior to the inner ear placodes. The medial cells of the lateral line placodes differentiate into ganglionic masses composed of sensory neurons, the axons of which extend into the acousticolateral area of the medulla. The outer epithelial cells of the placodes migrate through the ectoderm along routes apparently determined by the neighbouring tissues. At intervals along the migration routes, clusters of cells are left behind that differentiate into the lateral line neuromasts. The medial cells of the lateral line placodes, which form the lateral line (trunk) ganglia, become associated with the root ganglia of cranial nerves VII, IX and X (but not of VIII). Neural crest cells enter into the formation of the cranial nerve root ganglia but the lateral line placodes (and hence the trunk ganglia and neuromasts) are formed from the lateral ectoderm or, in part, from the neural fold (Weston, 1970). Neuron processes from the cells of the lateral line ganglia are 'towed along' by the migrating neuro-mast cells to their final locations; these two components—nerves and neuromast organs—form the lateral line system (Weiss, 1939; Ballard, pers. comm.).

There is good evidence that the embryonic neuromasts, once in their definitive position, induce the epithelium to form enclosing pits or canals. Because the osteoblasts of lateral line-related bones tend to appear around each placode (Pehrson, 1940; Devillers, 1947), it has been assumed that the neuromasts function as inductors of these bones. Devillers (1947) suggested that in teleosts with the lateral line in a superficial position, ossification of membrane bones and of over-lying tubes around lateral line placodes and nerves may be more or less independent. The complete regeneration of extirpated frontal bones in *Salmo* in the absence of neuromast organs is regarded as corroboration of this hypothesis (Pinganaud-Perrin, 1973).

Hörstadius (1950: 94–5) has summarized the earlier experimental evidence, all indirect, suggesting that the lateral line placodes develop through interaction between the ectodermal placode cells and ectomesenchyme. Coupled with the possibility that all dermal ossification may involve ectomesenchyme, this explana-tion is a reasonable one, but further experimentation is obviously necessary. Merrilees & Crossman (1973) have carried out regeneration experiments on *Esox*, which indicate that neuromast and nonsensory pit cells are responsible for the formation of the notch in lateral line scales. However, there is no evidence that these cells are responsible for the scales *per se*.

In summary, any hypothesis about dermal bone ossification should take into account the following: (a) the embryonic tissue of origin and possible interaction of this tissue with other tissues; (b) the roles of dentine and lateral line placodes in early osteogenesis; (c) the interaction among separate centres of ossification and among adjacent growing bone rudiments in relation to dermal bone (or scale) size, shape and pattern. Pilot experiments should be focused on *Polypterus*, *Acipenser*, *Lepisosteus* and *Amia*.

The orientation of the tubules in a dentine covered dermal bone or scale (as in a tooth) indicates that the migrating odontoblasts and the processes they leave behind converge on a vascular canal that presumably was present in the mesenchyme before dentinogenesis began. The convergence of the odontoblasts and their processes toward a particular part of a pulp cavity or a particular vascular space during dentine matrix formation is probably responsible for the discrete dentine units that have inspired Ørvig's (1967) odontode concept. According to Ørvig, odontodes may form discrete tubercles or be combined into ridges and continuous dentine layers on dermal bones or scales. The conformation of the ectodermal dental organ in dermal bones and scales, as in multicuspid teeth, is the visible guide to the final shape and extent of the dentine unit. Ørvig has postulated that odontodes form in the dermis anywhere on the body surface—in other words, wherever the mesenchyme is instructed to form a dental papilla and to interact with the overlying ectoderm. This hypothesis is not refuted by experimental data, but its phylogenetic implications require further investigation, particularly in regard to the presumed primitiveness of the odontode.

Periodic reactivation of the formative dental tissues, either for growth or replacement, is indicated in the dermal plates and scales of various groups of fishes by the retention of vertical and/or partly lateral generations of successive laminae or odontode-like units. Out-group comparison suggests that this retention of dental tissues may be a derived condition, but further investigation is needed. Although there is considerable variation in the expression of this phenomenon (Ørvig, 1957b; 1967; Sewertzoff, 1932; Kerr, 1952), one relationship remains constant: a contact between the enameloid layer and the dentine unit of the same generation. There is some evidence that the dentine tubules in *Polypterus* scales meet the enameloid lamellae at the interdigitations of the lamellae with dentine. The tubules then turn toward the vascular canals as shown in Plate 6B. However, it is difficult to understand how odontoblasts could be involved in the formation of successive enameloid laminae in the dermal bones, scales, spines and lepidotrichia after the development of the first lamina.* On the basis of present evidence, the simplest explanation for this condition is the conventional one—that the laminae are formed by the inner dental epithelium, which means that the tissue is true enamel. Fortunately, *Polypterus* can now be raised in an aquarium (Arnoult, 1964) and autoradiographic studies of dermal skeleton development should be possible.

In recent years the term cosmine has been used by Gross (1956), Ørvig (1969) and Thomson (1975) to designate a combination of enameloid (or enamel), dentine and trabecular bone around the pore-canal system in some osteostracans, osteolepid rhipidistians and the early dipnoans. As these workers have noted, the histology of the tissues involved differs in no significant way from that in the dermal skeleton of other fishes. Even the elaborate pore-canal system, to the extent that it can be understood by examining a maze of empty tubules, is not basically different from the sensory canal system of cephalaspids and early actinopterygians. There is usually a single layer of enameloid (or enamel) and a single

---

*According to illustrations in Moy-Thomas & Miles (1971) and Gross (1971), the 'dentine' laminae forming the crown of the scales in the acanthodians *Acanthodes* and *Gomphonchus* contain long branching dentinal tubules that are parallel to the crown surface until they approach the vascular canals. The position and length of these tubules suggest that odontoblasts are somehow involved in the formation of the crown tissue.

zone of dentine, but remnants of one or two older generations of these tissues may be present in some porolepid rhipidistians (Jarvik, 1950; Gross, 1966) and in at least one Devonian dipnoan (Denison, 1968). The frequent absence of cosmine in limited areas of both dermal bones and scales and the disposition of whatever enamel and dentine are present have been regarded as evidence for periodic resorption and redeposition of the cosmine complex. Although the evidence for this cycle is indirect, it is convincing. More troublesome is speculation about why it occurred. Growth is the usual explanation; Thomson (1975) has suggested increased calcium utilization during reproduction as another reason.

In regard to the morphogenetic mechanism involved in the cosmine cycle, it is probable that there was a periodic cessation and renewal of the epitheliomesenchymal interaction along with an additional mechanism that brought about synchronous resorption of the mature mineralized tissues. This mechanism may have been the same one that caused partial resorption of earlier generation enameloid-dentine denticles in some palaeonisciform and holostean scales (Ørvig, 1967; Schultze, 1966).

Possibly the most dramatic morphogenetic shift in the history of the osteichthyan squamation was the change from scales with enameloid (or enamel), dentine and bone to scales with only weak ossification to epithelium without scales. On the basis of out-group comparison, the loss of any of the three original calcified tissues may be regarded as a derived condition in any group of fishes. A few extinct actinopterygians (Schultze, 1966) are known to have well-calcified rhomboidal scales anteriorly and weakly calcified cycloidal scales posteriorly. The pycnodonts and one semionotid, *Hemicalypterus* (Schaeffer, 1967), have rhomboidal scales anteriorly and no scales posteriorly. We may conclude from this that enameloid and dentine production were independently eliminated in various actinopterygian and sarcopterygian lineages. Dentine is not formed in the lepisosteoid type of actinopterygian scale, but the enameloid may be well developed and lamellated, as it is in *Lepisosteus* (Kerr, 1952). Loss of enameloid and dentine in both scales and dermal bones may be associated with ossification that occurs well below the epitheliomesenchymal junction, as in the teleost dermal skull.

FIN FOLDS AND FIN RAYS

Paired or lateral fin folds begin as thickenings of the somatic layers of the lateral plate mesoderm. In most fishes they form at about the level of the renal duct, as diagrammed by Bouvet (1974a) for *Salmo*. Mesenchymal cells from the mesodermal thickenings migrate laterally to the epithelium where they concentrate into an anteroposteriorly elongated cell mass that induces the adjacent epithelium to thicken and form the lateral fin fold. As the pectoral and pelvic fins develop, the fold or ridge between them disappears. Lopashov (1950) has shown that pieces of lateral plate mesoderm transplanted to the abdominal region in various teleosts (*Perca, Gobio, Misgurnus, Cyprinus, Carassius*) will induce fin fold development in the overlying ectoderm, but without musculature, which requires tissue from neighbouring somites (Geraudie & Francois, 1973). Lateral mesoderm must therefore interact with the ectoderm to initiate the paired fins, as well as tetrapod limbs. Competence for limb development in *Triturus* (Balinsky, 1933, 1935, 1937) is present along the entire flank between the normal fore- and hindlimb regions. However, the factors that localize fin or limb development remain elusive.

The phylogenetic significance of the lateral fin folds has been much debated (Devillers, 1954; Westoll, 1958; Jarvik, 1965; Maderson, 1967). Although the elongated, undivided paired fins of *Pharyngolepis* (Ritchie, 1964) and the lateral plates and finlets of other anaspids and some acanthodians suggest that the lateral folds in these forms are more or less morphogenetically active through their length, the supposition that continuous, uninterrupted lateral fins represent the primitive adult vertebrate condition is difficult to corroborate. On the other hand, Devillers (1954) and Nursall (1962) have suggested that the lateral folds are basically embryonic structures that may represent a larval adaptation.

In contrast to the lateral folds, the median fold begins as an epidermal thickening, or ridge, that extends from behind the skull back to the caudal area, where it turns forward to the anus. There is abundant evidence in the amphibians (Weston, 1970: 69–70) that trunk neural crest cells induce the median fin fold and also provide the mesenchyme that occupies the centre of the fold. Although this has not yet been demonstrated in the sharks or bony fishes, it may be regarded as a testable hypothesis. On the basis of an ultrastructure study, Uwa (1974) thinks that scleroblasts in the anal fin of the killifish *Oryzias* arise from the mesenchymal cells within the fold.

The fact that the morphogenetic programmes for the median and lateral fin folds are different may reflect the different sources of mesenchyme available in different parts of the embryo. Both lateral plate and neural crest cells can function as inductors and contribute mesenchyme to their respective folds, and both can induce limb folds in amphibians when transplanted under ectoderm in other areas of the body (Du Shane, 1935; Terentiev, 1941).

Our understanding of dermal fin ray development begins essentially with the well known paper by Goodrich (1904), in which he corroborated an observation of Harrison (1895) that the basement membrane of the fin is somehow involved in the formation of the ray rudiments, and he concluded that actinotrichia, ceratotrichia and lepidotrichia arise from this source. Goodrich's conclusion that successive thickenings of the basement membrane become detached and sink into the mesenchyme to form the rays has played a major role in the delamination hypothesis. Recently Nadol, Gibbins & Porter (1969) described the locus of the developing dorsal fin rays in *Fundulus* as being between the basement lamella and the sublamellar dermal cells. They have also noted that the collagen-rich basement lamella extends into the dorsal fin and that it diminishes toward the fin edge.

Ultrastructure investigation of developing fin ray rudiments in the pectoral and pelvic fins of *Salmo* (Bouvet, 1974a, b) suggests a causal relationship between the basement lamella and the rudiments of the actinotrichia. In this teleost, the actinotrichia begin to form in the apex of the fold as short rods set obliquely to the basement lamella, projecting slightly into the mesenchymal core of the fold. These structures continue to develop beyond the distal end of the endoskeletal radials as the fin fold elongates. Actinotrichia, in the sense of Goodrich (1904), are composed of unjointed elastoidin fibres and do not mineralize. Elastoidin, which is a modified form of collagen, is also the principle component of the ceratotrichia in chondrichthyans (Piez & Gross, 1959). Goodrich noted that the ceratotrichia form in about the same way as the actinotrichia of actinopterygians.

Bouvet (1974a, b) agrees with Harrison (1895) and Goodrich (1904) that *Salmo* lepidotrichia arise independently of the actinotrichia. However, Bouvet does not follow Goodrich and various later investigators in believing that the lepidotrichia form as thickenings of the basement membrane that are subsequently separated from the membrane by insinuating scleroblasts. According to Bouvet's interpretation, the collagenous matrix of the lepidotrichia is formed by mesenchymal cells resembling osteoblasts that are situated between the endoskeletal radials and the more distal actinotrichia. The basal lamella is superficial to these cells and presumably is not directly involved.

Unlike the actinotrichia, the lepidotrichia develop as paired concave segments enclosing loose vascularized connective tissue. In early stages mesenchymal cells cover the surface of each concave ray moiety (as Uwa, 1974, observed in *Oryzias*). It is thus possible that the lepidotrichia arise entirely from mesenchyme. In Bouvet's (1974b) opinion, they will not differentiate unless the distal radial cartilages are present (this may represent another instance of epitheliomesenchymal interaction). Following extirpation of presumptive distal radial tissue, the lepidotrichia do not differentiate.

The mature lepidotrichia of *Salmo* are made up of an outer zone of longitudinal collagen fibres, a middle zone of bone and an inner core of dense hydroxyapatite crystals (Bouvet, 1974b). Osteoblasts are incorporated into the lepidotrichia matrix only at the end of the yolk sac resorption stage. At this time the bone becomes cellular. Elongation of the lepidotrichia is accomplished by terminal ossification. In *Polypterus* (Plates 7 and 8) the dorsal spines and lepidotrichia have the same tissue sequence as the dermal bones and scales. The enamel (or enameloid) is laminated, and the resemblance of the spine in cross-section to the scale of the Carboniferous palaeonisciform *Elonichthys* (Aldinger, 1937: fig. 3) is remarkably close.

The proposal of Goodrich (1904) and Jarvik (1959) that lepidotrichia represent modified scales is based on histological resemblances and on a morphological transition from scales to ray segments in, for example, the osteolepid crossopterygian *Gyroptychius*. Lepidotrichia and scales in the early actinopterygians and sarcopterygians frequently look alike and they may be covered with enameloid (enamel) and dentine. But this does not necessarily indicate that the lepidotrichia literally evolved from scales—there is little evidence for that. It is perhaps more meaningful to propose that scales and lepidotrichia composed of enameloid, dentine and bone are somewhat differently shaped manifestations of the same morphogenetic system.

DISCUSSION AND CONCLUSIONS

Experimental and descriptive data from many sources have not refuted the hypothesis that the calcified dermal skeleton in living fishes develops from a single modifiable morphogenetic system that is established by the interaction of the epithelium and the adjacent mesenchyme. Indeed, it is this constant developmental interrelationship that delimits what is meant here by the 'single' system hypothesis. Out-group comparison indicates that the arrangement of enameloid (or enamel), dentine and membrane bone in layers more or less parallel to the surface epithelium is the primitive vertebrate condition. Actually the continuity of the enameloid (or enamel) and dentine layers may be interrupted in the sense that these tissues are in the form of tubercles, and tubercles arranged along a jaw margin or in the oral cavity are structurally teeth. The differences in the expression of each calcified dermal tissue can also be explained most economically as modulations of a single morphogenetic system. The morphogenetic shifts from enameloid to enamel, from acellular to cellular dentine and from cellular to acellular bone (which are conceivably reversible) may be caused by changes in the duration or timing of particular steps in the system, as noted by Shellis & Miles (1974) for enameloid and enamel. In view of this temporal shift explanation, which is theoretically testable in living fishes, arguments about whether aspidin, mesodentine, and semidentine are primitive or derived seem less compelling. This does not mean, of course, that these tissues, as preserved, have less systematic value, but rather that the morphogenetic system that has produced them is, within certain canalized limits, subject to modulation, and this does have phylogenetic implications.

Additional examples of morphogenetic system modification include the effective elimination of one or more stages or steps in the developmental pathway. This condition could lead to the suppression of enameloid (or enamel), dentine and finally bone. One final example is the alteration in cell (odontoblast) activity that may result in a fairly abrupt change from orthodentine to intertrabecular dentine to trabecular dentine. There is no evidence that identical cells produce the organic matrix of ortho- and trabecular dentine, as apparently is the case with the intertrabecular and trabecular components of trabecular dentine, but the same cell type is involved.

Any hypothesis about the origin of the neural crest in vertebrates should take into account the 'striking diversity of differentiated cells and tissues' derived from the migratory neural crest cells. Accordingly, a primary function (but not necessarily the only one) of the neural crest may be to provide additional mesenchyme. Although crest cells are known to migrate all over the body and to differentiate into pigment cells, there is no experimental evidence to show that they are involved in the development of body scales and lepidotrichia. There is, however, well-established evidence that neural crest cells in certain amphibians form the dental papillae, and indirect evidence that they are responsible for at least some membrane bones in the skull. Crest cells also induce the amphibian median fin fold and form the mesenchymal core of the fold. In view of this evidence, and the similarity in the tissue composition of teeth, dermal bones, scales and lepidotrichia in all groups of fishes in which they occur, it is reasonable to propose that ectomesenchyme is involved in the formation of all dermal calcifications. The representation of the dermal skeleton in any particular area of the body is probably dependent on the spatial distribution of epithelium and mesenchyme.

The organization of the collagen matrix, the disposition of the accompanying blood vessels and the relative abundance and orientation of the odontoblast processes (or dentine tubules)—all of which are determined prior to mineralization —play an essential role in the structure of mature tissues. Although the arrangement of collagen may be somewhat altered during mineralization, each tissue pattern is essentially formed prior to calcium phosphate deposition.

Differences in the histology and gross form of the teeth, dermal skull elements, scales and fin components obviously have systematic significance. If we are concerned with extinct animals, we can compare directly only structure in attempting to infer relationships. Nevertheless, increasing knowledge of the morphogenesis of the components of the dermal skeleton in living fishes (and other vertebrates) is bound to increase our understanding of the tissues involved and the scope of their adult expression. Morphogenetic characters may be used in postulating the relationships of living organisms in the same way that physiological, behavioural or soft tissue characters are used.

Without wishing to belabour the homology problem, we may propose that enameloid, enamel, dentine and dermal bone are homologous as specific types of calcified dermal tissue in all groups of vertebrates in which they occur. By contrast, the various occurrences of trabecular dentine (as in batoids and dipnoans) apparently developed independently. Although the basic design of this dentine type is the same in both groups (probably reflecting the canalization of dentinogenesis), there are enough differences in the proportions of trabecular and intertrabecular tissues and perhaps in the relative hardness of these tissue areas to explain their distinctiveness. In terms of phylogeny, the organization of the trabecular dentine in batoids and lungfishes is not homologous even though the tissue involved may be so regarded.

The systematic implications of the calcified dermal skeleton in particular groups of fishes have been only incidentally explored in this paper, which is primarily concerned with more general problems of morphogenesis and structure. It is hoped, however, that the data and interpretations will be useful for purposes of phylogenetic inference. Also, no attempt has been made to interpret the adaptive significance of the various manifestations and combinations of dermal calcified tissues. Reif (1973) and Preuschoft, Reif & Müller (1974) have demonstrated that differences in the ultrastructure of shark teeth can be meaningfully investigated from a functional-biomechanical aspect. This sort of analysis may be extended to all of the dermal calcifications in fishes.

ACKNOWLEDGEMENTS

For valuable comments and guidance in some less familiar realms, I am indebted to Drs W. W. Ballard, W. J. Bock, C. Devillers, R. H. Denison, M. K. Hecht, P. F. A. Maderson ,M. L. Moss, C. Patterson, S. H. Salthe, R. P. Shellis, K. S. Thomson and R. Zangerl. The *Polypterus* thin sections were prepared by Mr Walter Sorensen and the photographs were taken by Mr Chester S. Tarka.

REFERENCES

ALDINGER, H., 1937. Permische Ganoidfische aus Ostgrönland. *Meddr Grønland, 102:* 1–392.
ANDRES, G., 1946. Ueber Induktion und Entwicklung von Kopforganen aus Unkenektoderm im Molch (Epidermis, Plakoden und Derivate der Neuralleiste). *Revue suisse Zool., 53:* 502–10.
ARNOULT, J., 1964. Comportement et reproduction en captivité de *Polypterus senegalus* Cuvier. *Acta zool., Stockh., 45:* 191–9.
BALINSKY, B. I., 1933. Das Extremitätenseitenfeld: seine Ausdehnung und Beschaffenheit. *Arch. EntwMech. Org., 130:* 704–46.
BALINSKY, B. I., 1935. Experimentelle Extremitäteninduktion und die Theorie des phylogenetischen Ursprungs der paarigen Extremitäten der Wirbeltiere. *Anat. Anz., 80:* 136–42.
BALINSKY, B. I., 1937. Ueber die zeitlichen Verhältnisse bei der Extremitäteninduktion. *Arch. EntwMech. Org. 136:* 250–85.

BALINSKY, B. I., 1970. *An introduction to embryology:* 3rd ed., xviii+725 pp. Philadelphia, London & Toronto: W. B. Saunders.

BALLARD, W. W., 1964. *Comparative anatomy and embryology:* viii+618 pp. New York: Ronald Press.

DE BEER, G. R., 1937. *The development of the vertebrate skull:* xxiv+552 pp. Oxford: Clarendon Press.

DE BEER, G. R., 1951. *Embryos and ancestors;* xii+159 pp. Oxford: Clarendon Press.

BERGOT, C., 1975. Morphogenèse et structure des dents d'un téléostéen (*Salmo fario* L.). *J. Biol. Buccale,* 3: 301–24.

BERLINER, J., 1969. The effects of the epidermis on the collagenous basement lamella of anuran larval skin. *Dev Biol.,* 20: 544–62.

BERNFIELD, M. R., COHN, R. H. & BANERJEE, S. D., 1973. Glycosaminoglycans and epithelial organ formation. *Am. Zool.,* 13: 1067–83.

BERNFIELD, M. R. & WESSELLS, N. K., 1970. Intra- and extracellular control of epithelial morphogenesis. In M. N. Runner (Ed.), *Changing syntheses in development:* 195–249. New York & London: Academic Press.

BERRILL, N. J., 1955. *The origin of vertebrates:* viii+257 pp. Oxford: Clarendon Press.

BOUVET, J., 1974a. Différenciation et ultrastructure du squelette distal de la nageoire pectorale chez la truite indigène (*Salmo trutta fario* L.). I. Différenciation et ultrastructure des actinotriches. *Archs Anat. microsc. Morph. exp.,* 63: 76–96.

BOUVET, J., 1974b. Différenciation et ultrastructure du squelette distal de la nageoire pectorale chex la truite indigène (*Salmo trutta fario* L.). II. Différenciation et ultrastructure des lépidotriches. *Archs Anat. microsc. Morph. exp.,* 63: 323–35.

BYSTROW, A. P., 1957. The microstructure of dermal bones in arthrodires. *Acta zool,, Stockh.,* 38: 239–75.

CHIBON, P., 1967. Marquage nucléaire par la thymidine tritiée des dérivés de la crête neurale chez l'Amphibien *Pleurodeles waltlii* Michah. *J. Embryol. exp. Morph.,* 18: 343–58.

CHIBON, P., 1974. Un système morphogénétique remarquable: la crête neurale des vertébrés. *Annls Biol.,* 13: 459–80.

DENISON, R. H., 1951. Evolution and classification of the Osteostraci. The exoskeleton of early Osteostraci. *Fieldiana, Geol.,* 11: 157–218.

DENISON, R. H., 1968. Early Devonian lungfishes from Wyoming, Utah and Idaho. *Fieldiana, Geol.,* 17: 353–413.

DENISON, R. H., 1974. The structure and evolution of teeth in lungfishes. *Fieldiana, Geol.,* 33: 31–58.

DEVILLERS, C., 1947. Recherches sur le crâne dermique des Téléostéens. *Annls Paléont.,* 33: 1–94.

DEVILLERS, C., 1954. Origine et évolution des nageoires et des membres. In P. P. Grassé (Ed.), *Traite zoologie,* 12: 710–90. Paris: Masson et Cie.

DEVILLERS, C., 1965. The role of morphogenesis in the origin of higher levels of organization. *Syst. Zool.,* 14: 259–71.

DEVILLERS, C. & CORSIN, J., 1968. Les os dermiques crâniens des Poissons et des Amphibiens; pointe de vue embryologiques sur les 'territoires osseux' et les 'fusions.' In T. Ørvig (Ed.), *Current problems of lower vertebrate phylogeny. Nobel symposium,* 4: 413–28. New York, London & Sydney: Interscience Publishers.

GARANT, P. R., 1971. Cytodifferentiation of the ectodermal component of developing teeth in the pickerel, *Esox lucius. Int. Assoc. dent. Res. Preprinted abstracts,* 49th General Meeting. Abstract 121.

GAUNT, W. A. & MILES, A. E. W., 1967. Fundamental aspects of tooth morphogenesis. In A. E. W. Miles (Ed.), *Structural and chemical organization of teeth:* 151–97. New York: Academic Press.

GERAUDIE, J. & FRANCOIS, Y., 1973. Les premiers stades de la formation de l'ébauche de nageoire pelvienne de truite (*Salmo fario* et *Salmo gairdneri*). *J. Embryol. exp. Morph.,* 29: 221–37.

GOODRICH, E. S., 1904. On the dermal fin-rays of fishes—living and extinct. *Q. Jl microsc. Sci.,* 47: 465–522.

GRADY, J. E., 1970. Tooth development in *Latimeria chalumnae* (Smith). *J. Morph.,* 132: 377–88.

GROBSTEIN, C., 1956. Inductive tissue interaction in development. *Adv. Cancer Res.,* 4: 187–236.

GROSS, W., 1956. Ueber Crossopterygier und Dipnoer aus dem baltischen Oberdevon im Zusammenhang einer vergleichenden Untersuchung des Porenkanalsystems paläozoischer Agnathen und Fische. *K. svenska VetenskAkad. Handl.,* 5: 1–140.

GROSS, W., 1966. Kleine Schuppenkunde. *Neues Jb. Geol. Paläont. Abh.,* 125: 29–48.

GROSS, W., 1971. Downtonische und dittonische Acanthodier-Reste des Ostseegebietes. *Palaeontographica, Abt. A. Palaeozool.-Stratigr.,* 136: 1–82.

HALL, B. K., 1975. Evolutionary consequences of skeletal differentiation. *Am. Zool.,* 15: 329–50.

HALSTEAD, L. B., 1969. Calcified tissues in the earliest vertebrates. *Calc. Tiss. Res.,* 3: 107–24.

HAMBURGH, M., 1971. *Theories of differentiation;* x+171 pp. New York: Elsevier.

HARRISON, R. G., 1895. Ectodermal or mesodermal origin of the bones of teleosts? *Anat. Anz.,* 10: 138–43.

HAY, E. D., 1973. Origin and role of collagen in the embryo. *Am. Zool.,* 13: 1085–107.

HEROLD, R. C., 1970a. Vasodentine and mantle dentine in teleost fish teeth. A comparative microradiographic analysis. *Archs oral Biol.,* 15: 71–85.

HEROLD, R. C., 1970b. The fine structure of vasodentine in the teeth of the white hake, *Urophycis tenuis* (Pisces, Gadidae). *Archs oral Biol.,* 15: 311–22.

HEROLD, R. C., 1971a. Osteodentinogenesis. An ultrastructural study of tooth formation in the pike, *Esox lucius*. *Z. Zellforsch. mikrosk. Anat.*, *112:* 1–14.

HEROLD, R. C., 1971b. The development and mature structure of dentine in the pike, *Esox lucius*, analyzed by microradiography. *Archs oral Biol.*, *16:* 29–41.

HEROLD, R. C., 1974. Ultrastructure of odontogenesis in the pike (*Esox lucius*). Role of dental epithelium and formation of enameloid layer. *J. Ultrastruct. Res.*, *48:* 435–54.

HEROLD, R. C. & LANDINO, L., 1970. The development and mature structure of dentine in the pike, *Esox lucius*, analyzed by bright field, phase and polarization microscopy. *Archs oral Biol.*, *15:* 747–60.

HOMGREN, N., 1940. Studies on the head in fishes. Embryological, morphological, and phylogenetical researches. Part I. Development of the skull in sharks and rays. *Acta zool. Stockh.*, *21:* 51–267.

HÖRSTADIUS, S., 1950. *The neural crest. Its properties and derivatives in the light of experimental research;* ix+111 pp. London, New York & Toronto: Oxford University Press.

HUXLEY, J. S. & DE BEER, G. R., 1934. *The elements of experimental embryology:* xiii+514 pp. Cambridge: University Press.

JAMES, W. W., 1953. The succession of teeth in elasmobranchs. *Proc. zool. Soc. Lond.*, *123:* 419–74.

JARVIK, E., 1944. On the dermal bones, sensory canals and pit-lines of the skull in *Eusthenopteron foordi* Whiteaves, with some remarks on *E. sävesöderberghi* Jarvik. *K. svenska VetenskAkad. Handl.*, *21:* 1–48.

JARVIK, E., 1947. Notes on the pit-lines and dermal bones of the head in *Polypterus*. *Zool. Bidr .Upps.*, *25:* 60–78.

JARVIK, E., 1948. On the morphology and taxonomy of the Middle Devonian osteolepid fishes of Scotland. *K. svenska VetenskAkad. Handl.*, *25:* 1–301.

JARVIK, E., 1950. Middle Devonian vertebrates from Canning Land and Wegeners Halvö (East Greenland). Part II. Crossopterygii. *Meddr Grønland*, *96:* 1–132.

JARVIK, E., 1959. Dermal fin-rays and Holmgren's principle of delamination. *K. svenska VetenskAkad. Handl.*, *6:* 1–51.

JARVIK, E., 1965. On the origin of girdles and paired fins. *Israel J. Zool.*, *14:* 141–72.

JOLLIE, M., 1962. *Chordate morphology:* xiv+478 pp. London: Chapman & Hall.

JOLLIE, M., 1968. Some implications of the acceptance of a delamination principle. In T. Ørvig (Ed.), *Current problems of lower vertebrate phylogeny. Nobel Symposium*, *4:* 89–107. New York, London & Sydney: Interscience Publishers.

KERR, T., 1952. The scales of the primitive living actinopterygians. *Proc. zool. Soc. Lond.*, *122:* 55–78.

KERR, T., 1955. Development and structure of the teeth in the dog fish, *Squalus acanthias* L. and *Scyliorhinus caniculus* (L.). *Proc. zool. Soc. Lond.*, *125:* 95–114.

KERR, T., 1960. Development and structure of some actinopterygian and urodele teeth. *Proc. zool. Soc. Lond.*, *133:* 401–22.

KING, M.-C. & WILSON, A. C., 1975. Evolution at two levels in humans and chimpanzees. *Science, N.Y.*, *188:* 107–16.

KOBAYASHI, S., 1971. Acid mucopolysaccharides in calcified tissues. *Int. Rev. Cytol.*, *30:* 257–371.

KULCZYCKI, J., 1957. Upper Devonian fishes from the Holy Cross Mountains (Poland). *Acta palaeont. pol.*, *2:* 285–380.

KVAM, T., 1946. Comparative study of the ontogenetic and phylogenetic development of dental enamel. *Norske Tandlaegeforen. Tid., Suppl.*, *56:* 1–129.

LEHMAN, J.-P., 1952. Étude complémentaire des Poissons de l'Eotrias de Madagascar. *K. svenska. VetenskAkad. Handl.*, *2:* 1–201.

LEVINE, P. T., GLIMCHER, M. J., SEYER, J. M., HUDDLESTON, J. I. & HEIN, J. W., 1966. Non-collagenous nature of the proteins of shark enamel. *Science, N.Y.*, *154:* 1192–3.

LISON, L., 1941. Recherches sur la structure et l'histogenèse des dents des Poissons Dipneustes. *Archs Biol.*, Paris, *52:* 279–320.

LISON, L., 1954. Les dents. In P. P. Grassé (Ed.), *Traité de zoologie*, *12:* 791–853. Paris: Masson et Cie.

LOPASHOV, G. V., 1950. Experimental investigations of the sources of cellular material and condition of formation of the pectoral fins in teleost fishes. *Dokl. Akad. Nauk SSSR*, *70:* 137–40.

LØVTRUP, S., 1974. *Epigenetics:* 547pp. New York: John Wiley.

MADERSON, P. F. A., 1967. A comment on the evolutionary origin of vertebrate appendages. *Am. Nat.*, *101:* 71–8.

MADERSON, P. F. A., 1975. Embryonic tissue interactions as the basis for morphological change in evolution. *Am. Zool.*, *15:* 315–27.

MARKERT, F., 1896. Die Flossenstacheln von *Acanthias*. Ein Beitrag zur Kenntniss der Hartsubstanzgebilde der Elasmobranchier. *Zool. Jahrb. Anat.*, *9:* 664–722.

MEIER, S., & HAY, E. D., 1975. Stimulation of corneal differentiation by interaction between cell surface and extracellular matrix. I. Morphometric analysis of transfilter 'induction.' *J. Cell Biol.*, *66:* 275–91.

MEREDITH SMITH, M. & MILES, A. E. W., 1971. The ultrastructure of odontogenesis in larval and adult urodeles: differentiation of the dental epithelium. *Z. Zellforsch. mikrosk. Anat.*, *121:* 470–98.

MEREDITH SMITH, M., HOBDELL, M. H. & MILLER, W. A., 1972. The structure of the scales of *Latimeria chalumnae*. *J. Zool., Lond., 167:* 501–9.

MERRILEES, M. J. & CROSSMAN, E. J., 1973. Surface pits in the Family Esocidae. I. Structure and types. *J. Morph., 141:* 307–13.

MILES, R. S., 1966. The acanthodian fishes of the Devonian Plattenkalk of the Paffrath Trough in the Rhineland, with an appendix containing a classification of the Acanthodii and a revision of the genus *Homalacanthus. Ark. Zool., 18:* 147–94.

MILES, R. S., 1975. The relationships of the Dipnoi. *Colloques int. Cent. natn Rech. scient., 218:* 133–48.

MOSS, M. L., 1961a. Osteogenesis of acellular teleost fish bone. *Am. J. Anat., 108:* 99–109.

MOSS, M. L., 1961b. Studies of the acellular bone of teleost fish. I. Morphological and systematic variations. *Acta anat., 46:* 343–462.

MOSS, M. L., 1964. Development of cellular dentine and lepidosteal tubules in the bowfin, *Amia calva. Acta anat, 58:* 333–54.

MOSS, M. L., 1968a. The origin of vertebrate calcified tissues. In T. Ørvig (Ed.), *Current problems of lower vertebrate phylogeny. Nobel Symposium, 4:* 359–71. New York, London & Sydney: Interscience Publishers.

MOSS, M. L., 1968b. Comparative anatomy of vertebrate dermal bone and teeth. I. The epidermal co-participation hypothesis. *Acta anat., 71:* 178–208.

MOSS, M. L., 1970. Enamel and bone in shark teeth: with a note on fibrous enamel in fishes. *Acta anat., 77:* 161–87.

MOSS, M. L., 1972. The vertebrate dermis and the integumental skeleton. *Am. Zool., 12:* 27–34.

MOSS, M. L. & FREILICH, M., 1963. Studies of the acellular bone of teleost fish. IV. Inorganic content of calcified tissues. *Acta anat., 55:* 1–8.

MOSS, M. L. & POSNER, A., 1960. X-ray diffraction study of acellular teleost bone. *Nature, Lond., 188:* 1037–8.

MOY-THOMAS, J. A., 1934. On the teeth of the larval *Belone vulgaris*, and the attachment of teeth in fishes. *Q. Jl microsc. Sci., 76:* 481–98.

MOY-THOMAS, J. A. & MILES, R. S., 1971. *Palaeozoic fishes:* 2nd ed., xi+259 pp. Philadelphia & Toronto: W. B. Saunders Co.

NADOL, J. B., GIBBINS, J. R. & PORTER, K. R., 1969. A reinterpretation of the structure and development of the basement lamella: an ordered array of collagen in fish skin. *Devl Biol., 20:* 304–31.

NELSON, G. J., 1969. Gill arches and the phylogeny of fishes, with notes on the classification of vertebrates. *Bull. Am. Mus. nat. Hist., 141:* 475–552.

NELSON, G. J., 1970. Pharyngeal denticles (placoid scales) of sharks, with notes on the dermal skeleton of vertebrates. *Am. Mus. Novit., 2415:* 1–26.

NIEUWKOOP, P. D., 1955. Independent and dependent development in the formation of the central nervous system in amphibians. *Expl Cell Res., Suppl., 3:* 262–73.

NIEUWKOOP, P. D., 1973. The 'organization center' of the amphibian embryo: its origin, spatial organization, and morphogenetic action. In M. Abercombie *et al.* (Eds), *Advances in morphogenesis, 10:* 1–39. New York & London: Academic Press.

NODEN, D. M., 1975. An analysis of the migratory behaviour of avian cephalic neural crest cells. *Devl Biol., 42:* 106–30.

NURSALL, J. R., 1962. Swimming and the origin of paired appendages. *Am. Zool., 2:* 127–41.

ØRVIG, T., 1951. Histologic studies of placoderms and fossil elasmobranchs. I. The endoskeleton, with remarks on the hard tissues of lower vertebrates in general. *Ark. Zool., 2:* 321–454.

ØRVIG, T., 1957a. Remarks on the vertebrate fauna of the lower Upper Devonian of Escuminac Bay, P. Q., Canada, with special reference to the porolepiform crossopterygians. *Ark. Zool., 10:* 367–426.

ØRVIG, T., 1957b. Paleohistological notes. 1. On the structure of the bone tissue in the scales of certain Palaeonisciformes. *Ark. Zool., 10:* 481–90.

ØRVIG, T., 1965. Paleohistological notes. 2. Certain comments on the phyletic significance of acellular bone tissue in early lower vertebrates. *Ark. Zool., 16:* 551–6.

ØRVIG, T., 1966. Histologic studies of ostracoderms, placoderms and fossil elasmobranchs. 2. On the dermal skeleton of two late Palaeozoic elasmobranchs. *Ark. Zool., 19:* 1–39.

ØRVIG, T., 1967. Phylogeny of tooth tissues: evolution of some calcified tissues in early vertebrates. In A. E. W. Miles (Ed.), *Structural and chemical organization of teeth;* 45–110. New York: Academic Press.

ØRVIG, T., 1969. Cosmine and cosmine growth. *Lethaia, 2:* 241–60.

ØRVIG, T., 1972. The latero-sensory components of the dermal skeleton in lower vertebrates and its phyletic significance. *Zool. Scr., 1:* 139–55.

PARRINGTON, F. R., 1949. A theory of the relations of lateral lines to dermal bones. *Proc. zool Soc. Lond., 119:* 65–78.

PARRINGTON, F. R., 1950. The skull of *Dipterus. Ann. Mag. nat. Hist., 3:* 534–47.

PATTERSON, C., 1965. The phylogeny of the chimaeroids. *Phil. Trans. R. Soc. Lond., 249:* 101–219.

PEHRSON, T., 1940. The development of dermal bones in the skull of *Amia calva. Acta zool. Stockh., 21:* 1–50.

PEYER, B., 1968. *Comparative odontology:* xiv+347 pp. Chicago & London: University of Chicago Press.

PIEZ, K. A. & GROSS, J., 1959. Amino acid composition and morphology of some invertebrate and vertebrate collagen. *Biochem. biophys. Acta, 34:* 24–39.

PINGANAUD-PERRIN, G., 1973. Conséquences de l'ablation de l'os frontal sur la forme des os du toit cranien de la Truite (*Salmo irideus* Gib. Pisces-Teleostei). *C. r. hebd. Séanc. Acad. Sci., Paris, (Ser. D. Sci. Nat.), 276:* 2809–11.

POOLE, D. F. G., 1971. An introduction to the phylogeny of calcified tissues. In A. A. Dahlberg (Ed.), *Dental morphology and evolution:* 65–79. Chicago & London: University of Chicago Press.

PREUSCHOFT, H., REIF, W.-E. & MÜLLER, W. H., 1974. Funktionsanpassungen in Form und Struktur an Haifischzähnen. *Z. Anat. EntwGesch., 143:* 315–44.

RADINSKY, L., 1961. Tooth histology as a taxonomic criterion for cartilaginous fishes. *J. Morph., 109:* 73–81.

RAVEN, C. P., 1931. Zur Entwicklung der Ganglienleiste. I. Die Kinematik der Ganglienleistenentwicklung bei den Urodelen. *Arch. EntwMech. Org., 125:* 210–92.

RAVEN, C. P., 1936. Zur Entwicklung der Ganglienleiste. V. Ueber die Differenzierung des Rumpfganglien-leistenmaterials. *Arch. EntwMech. Org., 134:* 122–46.

REIF, W.-E., 1973. Morphologie und Ultrastruktur des Hai-'Schmelzes.' *Zool. Scr., 2:* 231–50.

REVERBERI, G., ORTOLANI, G. & FARINELLA-FERRUZZA, N., 1960. The causal formation of the brain in the ascidian larva. *Acta Embryol. Morph. exp., 3:* 296–336.

RITCHIE, A., 1964. New light on the morphology of the Norwegian Anaspida. *Skr. norske Vidensk-Akad., 14:* 1–35.

RUSSELL, M. A., 1975. Differentiation. (Review of) Ciba Foundation symposium, Cell patterning. *Science, N.Y., 190:* 141–2.

SCHAEFFER, B., 1952. The Triassic coelacanth fish *Diplurus*, with observations on the evolution of the Coelacanthini. *Bull. Am. Mus. nat. Hist., 99:* 25–78.

SCHAEFFER, B., 1963. Cretaceous fishes from Bolivia, with comments on pristid evolution. *Am. Mus. Novit., 2159:* 1–20.

SCHAEFFER, B., 1967. Late Triassic fishes from the western United States. *Bull. Am. Mus. nat. Hist., 135:* 285–342.

SCHMALHAUSEN, I. I., 1949. *Factors of evolution. The theory of stabilizing selection;* xiv+327 pp. Philadelphia & Toronto: Blakiston Co.

SCHMIDT, W. J., & KEIL, A., 1971. *Polarizing microscopy of dental tissues,* 1st Engl. ed.: xix+584 pp. Oxford: Pergamon Press.

SCHOWING, J., 1968. Mise en évidence du rôle inducteur de l'encéphale dans l'ostéogenèse du crâne embryonnaire du poulet. *J. Embroyl. exp. Morph., 19:* 88–93.

SCHULTZE, H.-P., 1966. Morphologische und histologische Untersuchungen an Schuppen mesozoischer Actinopterygier (Uebergang von Ganoid- zu Rundschuppen). *N. Jb. Geol. Paläont. Abh., 126:* 232–314.

SCHULTZE, H.-P., 1969. Die Faltenzähne der rhipidistiiden Crossopterygier, der Tetrapoden und der Actinopterygier-Gattung *Lepisosteus;* nebst einer Beschreibung der Zahnstruktur von *Onychodus* (struniiformer Crossopterygier). *Palaeontogr. ital., 65:* 63–137.

SELLMAN, S., 1946. Some experiments on the determination of the larval teeth in *Ambystoma mexicanum. Odont. Tidskr., 54:* 1–128.

SEWERTZOFF, A. N., 1932. Die Entwicklung der Knochenschuppen von *Polypterus delhesi. Jena. Z. Naturw., 67:* 387–418.

DU SHANE, G. P., 1935. An experimental study of the origin of pigment cells in amphibia. *J. exp. Zool., 72:* 1–31.

SHELLIS, R. P., & MILES, A. E. W., 1974. Autoradiographic study of the formation of enameloid and dentine matrices in teleost fishes using tritiated amino acids. *Proc. R. Soc. Lond., (B), 185:* 51–72.

SPIEGELMAN, S., 1945. Physiological competition as a regulatory mechanism in morphogenesis. *Q. Rev. Biol., 20:* 121–246.

STEPHAN, P., 1900. Recherches histologiques sur la structure du tissu osseux des Poissons. *Bull. scient. Fr. Belg., 33:* 281–429.

TARLO, L. B. H., 1963. Aspidin: the precursor of bone. *Nature, Lond., 199:* 46–8.

TATARKO, K., 1934. Restitution des Kiemendeckels des Karpfens. Ein Versuch des Studiums der Formbildung durch Analyse der Wechselbeziehungen zwischen Form und Funktion bei der Restitution. *Zool. Jb. (Abt. Allg. Zool. Physiol. Tiere), 53:* 461–500.

TERENTIEV, I. B., 1941. On the role played by the neural crest in the development of the dorsal fin in Urodela. *Dokl. Akad. Nauk SSSR, 31:* 91–4.

THOMSON, K. S., 1975. On the biology of cosmine. *Bull. Peabody Mus. nat. Hist., 40:* 1–59.

TOMES, C. S., 1923. In H. W. M. Tims & C. B. Henry (Eds), *A manual of dental anatomy, human and comparative:* 8th ed., 547 pp. London: J. A. Churchill.

TUNG, T. C., WU, S. C. & TUNG, Y. Y. F., 1962. Experimental studies on the neural induction in *Amphioxus. Scientia sin., 11:* 805–20.

UWA, H., 1974. Ultrastructural study on the scleroblast of *Oryzias latipes* during ethisterone-induced anal-fin process formation. *Develop. Growth & Different., 16:* 41–53.

WAGNER, G., 1949. Die Bedeutung der Neuralleiste für die Kopfgestaltung der Amphibienlarven. Untersuchungen an Chimaeren von *Triton* und *Bombinator. Revue suisse Zool., 56:* 519–620.

WAGNER, G., 1959. Untersuchungen an *Bombinator-Triton*-Chimaeren. Das Skelett larvaler *Triton*-Köpfe mit *Bombinator*-Mesektoderm. *Arch. EntwMech. Org., 151:* 136–58.

WALL, R., 1973. Physiological gradients in development—a possible role for messenger ribonucleoprotein. In M. Abercrombie *et al.* (Eds), *Advances in morphogenesis, 10:* 41–114. New York & London: Academic Press.

WEISS, P., 1939. *Principles of development:* xix+601 pp. New York: Henry Holt.

WESTOLL, T. S., 1958. The lateral fin-fold theory and the pectoral fins of ostracoderms and early fishes. In T. S. Westoll (Ed.), *Studies on fossil vertebrates:* 180–211. London: Athlone Press.

WESTON, J. A., 1970. The migration and differentiation of neural crest cells. In M. Abercrombie *et al.* (Eds), *Advances in morphogensis, 8:* 41–114. New York & London: Academic Press.

WOLPERT, L., LEWIS, J. & SUMMERBELL, D., 1975. Morphogenesis of the vertebrate limb. In Ciba Foundation Symposium 29, *Cell patterning;* 95–130. Amsterdam: Associated Scientific Publishers.

YAMADA, T., 1937. Der Determinationszustand des Rumpfmesoderms im Molchkeim nach der Gastrulation. *Arch. EntwMech. Org., 137:* 151–270.

ZANGERL, R., 1966. A new shark of the Family Edestidae, *Ornithoprion hertwigi* from the Pennsylvanian Mecca and Logan Quarry Shales of Indiana. *Fieldiana, Geol., 16:* 1–43.

ABBREVIATIONS USED IN PLATES

bl	bony lamellae	itd	intertrabecular dentine
bo	membrane bone of jaw	lf	longitudinal connecting collagen fibres
de	dentine	od	odontoblasts
dt	dentinal tubules	ode	outer dental epithelium
ec	collagen of unmineralized enameloid	pc	pulp cells or pulp cavity
el	enameloid or enamel (decalcified in *Amia* series)	pd	predentine
ide	inner dental epithelium	vc	vascular canals

EXPLANATION OF PLATES

PLATE 1
Tooth development in *Amia calva*.
A. Early tooth germ (at right of nearly erupted tooth) showing enameloid matrix surrounded by columnar cells of the inner dental epithelium.
B. Later tooth germ with mineralized enameloid (decalcified).
C, D. Still later stages showing dentine formation and elongation of tooth shaft. × 300; **original.**

PLATE 2
Tooth development in *Amia calva*.
A, B. Further development of dentine with fusion to bone of attachment in B. × 300; original.

PLATE 3
A. Trabecular dentine in tooth plate of *Myliobatis* sp. Transverse section; × 60.
B. Trabecular dentine in tooth plate of *Neoceratodus* sp. from Miocene of Australia. Transverse section; × 60; original.
C. Tooth germs of *Scyllium* sp. Field Museum slide 4947; × 165.

PLATE 4
Tooth structure in *Polypterus ornatipinnis*.
A. Off-centre longitudinal section of marginal tooth with enameloid cap. Vertical lines near cap base are apparent dentine tubules in different plane from those of orthodentine below; × 60.
B. Longitudinal section of marginal tooth showing partial enameloid cap, orthodentine and pulp cavity; × 60.
C. Marginal tooth in transverse section; × 75. Ground sections; original.

PLATE 5
Membrane bone in *Polypterus ornatipinnis*.
A. Dermal skull roof of *Polypterus ornatipinnis;* × 40.
B. Detail of same element to demonstrate dentine tubules; × 60. Ground sections; original.

PLATE 6
Scale structure in *Polypterus ornatipinnis*.
A. Transverse section of flank scale; × 60.
B. Polarized transverse section showing dentine tubules extending from enamel (or enameloid) to vascular canal; × 300. Ground sections; original.

PLATE 7
Dorsal spine structure in *Polypterus ornatipinnis*.
A. Transverse section of partial dorsal spine. Anterior border is at top of photograph; × 60.
B. Enlarged portion of transverse section to show details of enamel (or enameloid) and dentine layer; × 100. Ground sections; original.

PLATE 8
Lepidotrichial structure in *Polypterus ornatipinnis*.
A. Transverse section of lepidotrichium from caudal fin. Outer border is at top of photograph; × 60.
B, C. Transverse sections of lepidotrichium showing details of enamel (or enameloid) and presumed dentine layers; × 300. Ground sections; original.

Plate 1

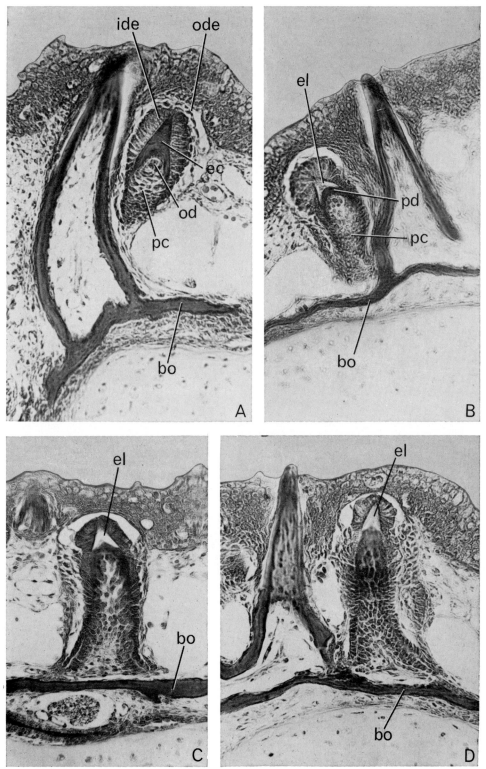

Plate 2

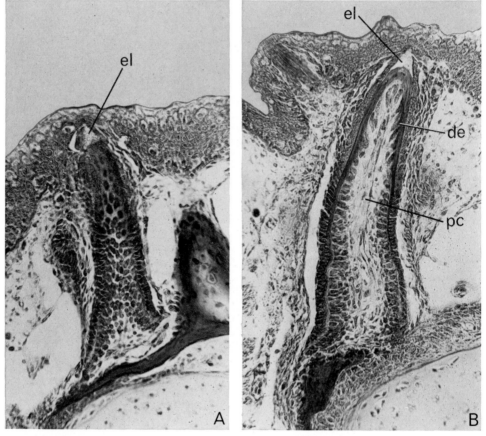

B. SCHAEFFER

Plate 3

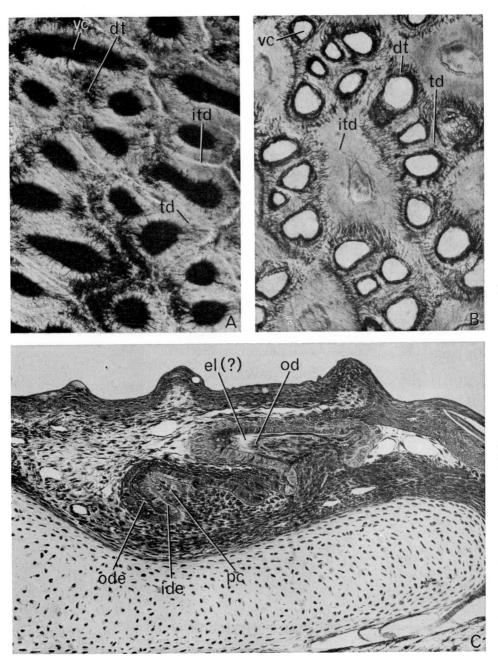

B. SCHAEFFER

Plate 4

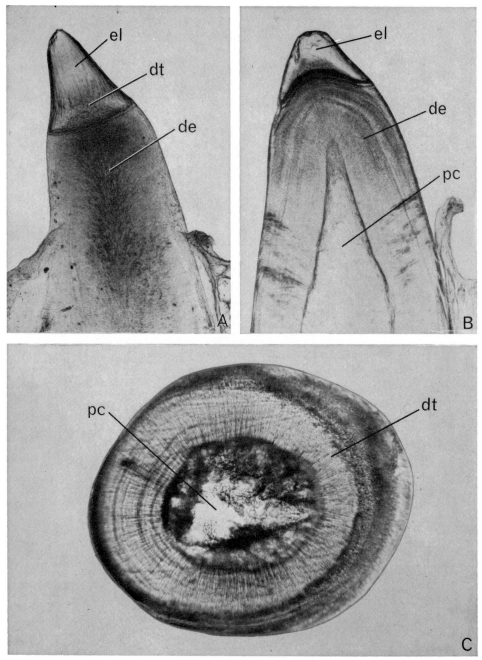

Plate 5

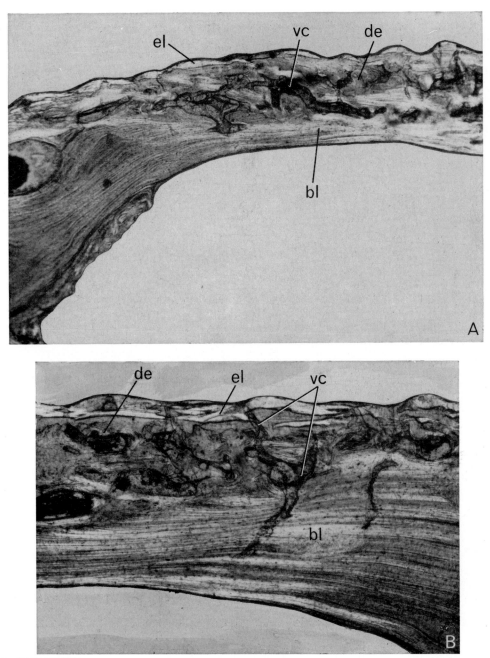

Plate 6

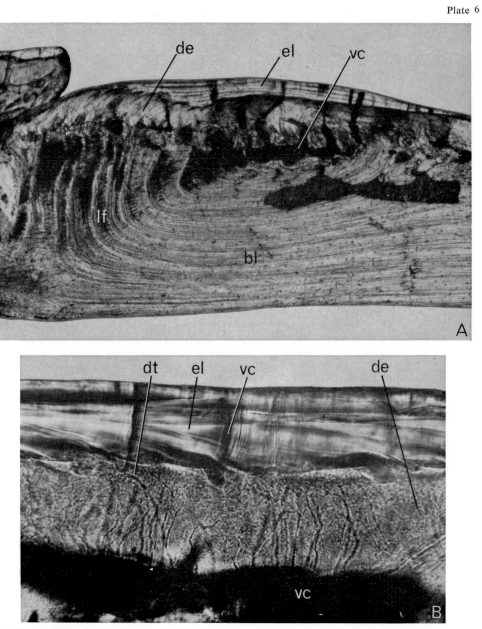

B. SCHAEFFER

Plate 7

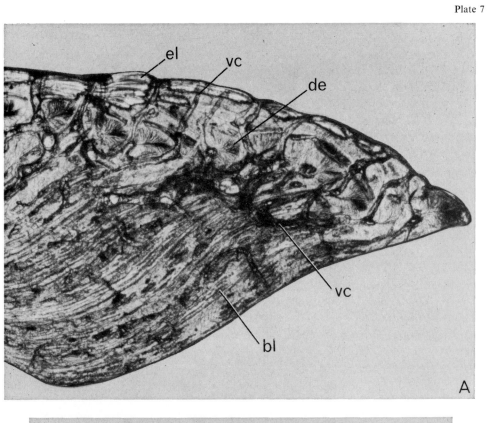

A

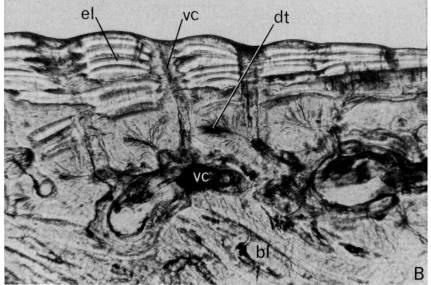

B

B. SCHAEFFER

Plate 8

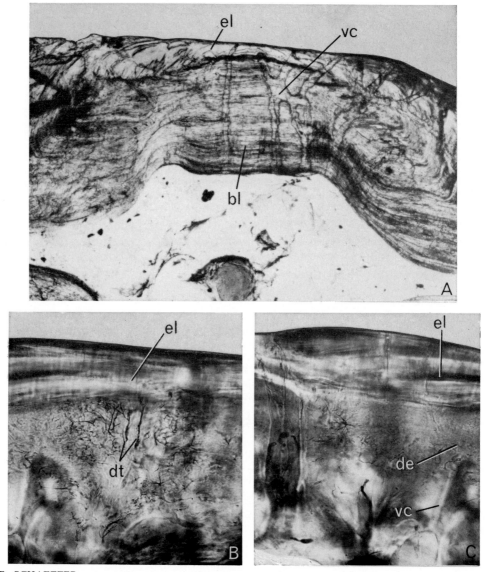

B. SCHAEFFER

A survey of odontodes ('dermal teeth') from developmental, structural, functional, and phyletic points of view

TOR ØRVIG

Section of Palaeozoology, Swedish Museum of Natural History, Stockholm, Sweden

The odontodes of the dermal skeleton in vertebrates below tetrapod level are reviewed with reference to their developmental and histological properties, relations to dermal bones and scales, functional adaptations and variations in shape, size and distribution. An attempt has been made to specify the characters by which they are distinguishable from jaw-teeth proper. Remarks are also given on the way in which odontodia, viz. small dermal elements consisting of one or more odontodes on a basal plate, could have participated in the phyletic formation of more comprehensive dermal elements like, for example, dermal bones.

CONTENTS

INTRODUCTION

The term *odontodes* (singular *odontode*) was introduced by the writer in 1967 (see also Ørvig, 1968, 1969b, 1972, 1973a, 1975) for those structures most of which were earlier referred to more or less vaguely as 'dermal teeth' or 'denticles'. Originally (Ørvig, 1967 : 47) the term was stated to mean 'hard tissue units corresponding very closely to teeth and difficult to distinguish from teeth by any rational criteria', but obviously, there is much more to say on the subject. The present paper offers a more detailed explanation. Its aim has been (a) to give a definition of the term odontodes, (b) to sketch the various shapes and structural properties odontodes may assume, and (c) to indicate in what way and in what circumstances odontodes may be distinguished from jaw-teeth proper (see also

53

Ørvig, 1973a). Remarks are also added on how, in early phyletic stages, odontodes may have entered as constituent parts of larger dermal elements as, for instance, dermal bones. As far as the latter point is concerned, the writer has profited by certain observations he was able to make on Triassic coelacanthid material during a visit in 1971 to the Palaeontological Institute of the Zürich University (abbreviated PIZ), and on certain early Devonian heterostracans during another visit to the Institute of Sedimentary and Petroleum Geology of the Geological Survey of Canada, Calgary, Alberta (GSC). Material belonging to the Mineralogical Museum of the Copenhagen University (MMHVP) and the Swedish Museum of Natural History, Stockholm (SMNH) has also been figured.

For editorial reasons, the illustration material accompanying this paper has been kept to an absolute minimum. References to figures in the literature may, it is hoped, serve as a compensation.

DEFINITIONS

In the dermal skeleton of vertebrates below tetrapod level, odontodes may be defined as special hard tissue units, or dental units, which generally speaking have those developmental and structural properties in common with the teeth of the jaws (whenever such are present) that they (a) each form ontogenetically in a *single, undivided dental papilla* of mesenchymal soft tissue, bounded at its circumference by an *epithelial dental organ* in the adjoining epidermis, (b) consist of *dentine* or, in some forms, *dentinous tissue* (see Ørvig, 1967) and (c) frequently (but not always) possess a superficial layer of *enameloid*. At the same time, however, odontodes are in general terms also characterized by (d) not belonging to the dentition sensu stricto but to other parts of the dermal skeleton (sometimes including those immediately adjoining the dentition), (e) not, as a rule, fulfilling similar *functions* as teeth, (f) not forming in a submerged position in connection with a dental lamina or single, ingrowing epidermis digitations but always in the *superficial part of the corium* (cf. Fig. 1, and Ørvig, 1967: fig. 14; 1973a: fig. 3A), because of which they are not replaced from below or sideways in the manner of teeth in the great majority of fishes, and (g) in many cases, not reaching nearly the same height (from top to base) as teeth. A critical appraisal of most of these points follows in the sequel.

We shall here refer to columns or clusters of odontodes which, during consecutive stages of growth of the dermal elements to which they belong, have developed *directly upon or beside each other* (each, of course, in a dental papilla of its own) as *odontocomplexes*. Examples of such complexes are met with in many groups of lower vertebrates including, for instance, actinopterygians (cf. Fig. 1G, and Ørvig, 1968: fig. 4). One should even regard as odontocomplexes the crowns of the periodically growing scales of acanthodians, some elasmobranchs, *Polypterus* and several ganoid fishes, consisting as they do of series (or groups) of odontodes lying close together and corresponding to zones of growth or parts of such zones (see e.g. Ørvig, 1966: figs 3A, 4; etc.). On the other hand, the odontodes of successive generations which in the dermal skeleton of e.g. some coccosteomorph arthrodires and holoptychiid crossopterygians show a column-like arrangement but are separated from each other by layers of bone tissue (see, among others, Ørvig, 1957: fig. 9B, D–F), or the sheets of cosmine in various osteostracans, crossopterygians and dipnoans (see below) are not taken as odontocomplexes in the present comprehension.

A scale which consists of an odontode situated on a small bony basal plate, or a group of odontodes on such a plate, has for practical reasons been named an *odontodium*. A further development of this, if necessary or desirable, could be to speak of scales whose crown is made up of one odontode only (like those of thelodonts and many selachians) as *monodontodia*, and of cases when the crown contains a number of separate odontodes (like those of the early selachian *Ohiolepis*) as *polyodontodia*.

It deserves to be emphasized at this point that the terms now defined, viz. odontodes, odontocomplexes and odontodia, are all intended to be purely descriptive. They do not involve interpretation and can be applied irrespective of whether or not one adopts lines of reasoning such as those of e.g. the lepidomorial theory (sensu Stensiö, 1961). Odontodes, for instance, can just as well be single lepidomorial crowns as multi-lepidomorial formations of all degrees of complexity when they are analysed on the basis of that theory (an analysis which lies outside the scope of the present paper).

ONTOGENY

The above statement that odontodes form each in a separate dental papilla may need a few explanatory remarks. When the writer first introduced the term odontodes, this point was not made fully clear. On that occasion (Ørvig, 1967: 86) he spoke of groups of odontodes fusing in such a way as to develop in a common dental papilla. Fusion of odontodes has no doubt taken place innumerable times in the history of the dermal skeleton. When a group of odontodes actually fuse, however, and lose their original individuality to the extent of forming in one common dental papilla, there is no longer a group of them any more, but *one single odontode* of larger size (odontocomplexes, on the other hand, still consist of a number of independent odontodes). Furthermore, a dental papilla in which an odontode develops refers to a well-defined portion of corium mesenchyme bounded by the epidermis, without at all implying that this portion necessarily always assumes one and the same *shape*. The typical bud- or bell-like shape one is accustomed to associate with the developing dental papillae of tooth-germs in general, cannot be displayed by the particular dental papillae with which one is here concerned other than those where tubercle-like (rounded or more or less conical) odontodes form: such odontodes as the ornamental ridges of dermal elements clearly develop in papillae with the same elongated shape they themselves possess. The same obviously also applies for example to the dental papillae originally forming for odontodes such as the ring-like zones of growth in the scale-crown of acanthodians, *Holmesella*? sp., *Polypterus* and *Cheirolepis* (see e.g. Ørvig, 1966: figs 3A, 4) and the corresponding parallel or subparallel ones in a variety of fossil actinopterygians. It may also be mentioned in this context that in some odontocomplexes, the dental papillae for the successively developing component odontodes do not as a rule show quite the same shape as those of odontodes forming well apart; in these cases each papilla may be truncated on one side by the peripheral part of the adjoining, previously developed odontode against which it lies (see further Fig. 1). Furthermore, it is characteristic of some ganoid odonto-complexes (not figured here) that each dental papilla for a new component odontode has in its superficial part a larger extent than normal, since it has here to provide space for the development of a ganoin layer wide enough to cover, in part or completely, the corresponding, already existing layer of the adjoining

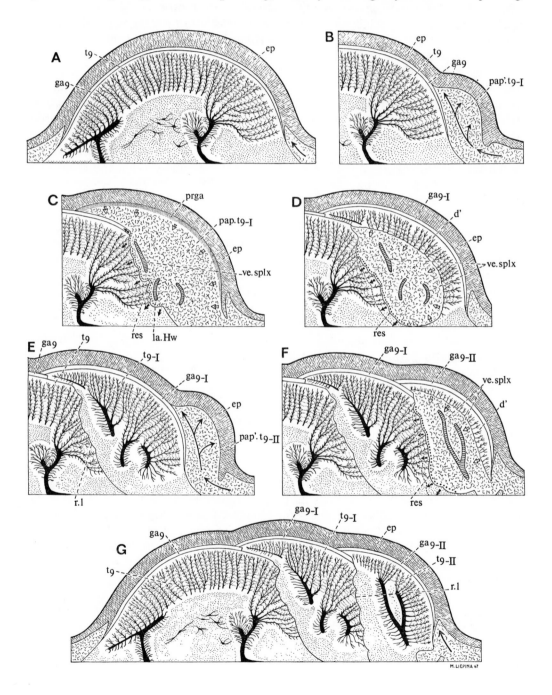

odontode of the complex. There may be occasion to return in more detail to these matters in another connection.

Turning now to the epithelial dental organ (sometimes named, inappropriately, the 'enamel organ') of developing odontodes, this has naturally to correspond exactly in shape and extent to the dental papilla with which in each individual case it is associated: the above remarks on the shape of the papillae, therefore, also apply to this organ. As far as odontodes are concerned, the epithelial dental organ does not seem to have quite the same complicated structure that it frequently displays in tooth-germs (at least not the sort of tooth-germs generally figured in histological text-books) where it is made up of an inner dental epithelium, a stellate reticulum and an outer dental epithelium (see e.g. Lehner & Plenk, 1936: fig. 20; Kvam, 1950: fig. 1). In the case of developing odontodes, it consists of no more than a single layer of turgid, columnar cells in the basal part of the epidermis, corresponding to the inner dental epithelium only (see illustrations of these cells in e.g. Studnička, 1909: pl. 7, pl. 8: 59). An epithelial dental organ of this (no doubt primitive, as frequently pointed out) kind is well known from developing placoid scales of selachians and developing spinules on the scutes of certain siluroids (see, e.g. Klaatsch, 1890: pl. 6: 1–5; Bhatti, 1938: pl. 2: 23–25; and others) and one has every reason for the assumption that it existed in the ontogeny of odontodes in fossil forms as well, from ostracoderms upwards. Finally, because of the site of formation of odontodes referred to above it also follows that the epithelial dental organ surrounding each odontode papilla always arises in a superficial position and never in a submerged one as is generally the case as far as tooth-germs are concerned.

Mention should, at this point, also be made of the intimate relations which, in the individual tooth-germs, are generally held to exist—in one way or another—between the epithelial dental organ and the dental papilla, and which to all intents and purposes also existed between the corresponding parts of odontode primordia from early stages of vertebrate history onwards. According to one school of

---

Figure 1. *Boreosomus piveteaui* Nielsen. Diagrams illustrating the ontogeny of an odontocomplex forming on the vomer by local areal growth of odontodes belonging to the ninth generation. Long arrows indicate growth of corium mesenchyme in developing odontode papillae, short white arrows growth of dentine, and short black arrows resorption of hard tissue. Areas of active resorption marked off by black dots. The epidermis covering the complex is here reconstructed as continuous, even in phases between the formation of new odontodes (a point which will deserve further discussion in another connection). A, A stage where the first (ontogenetically oldest) odontode of the complex was fully formed and still single. In B, the papilla for the second odontode had begun to form at the side of the first. C, Shows the new papilla fully developed with preganoin matrix in its external part and vessels in its interior; at this stage resorption of hard tissue had presumably commenced along the inner wall of the papilla. In D, a layer of dentine had arisen at the same time as resorption still went on opposite it. E, shows the new odontode fully formed and the initial development of a new papilla for a third odontode at its side. In F, dentine formation in the new papilla was proceeding together with resorption of hard tissue along its inner wall and floor (this corresponds to the stage actually preserved in the specimen from which the sequence of events has been reconstructed). G, Illustrates the complex with three fully developed odontodes; it could naturally include a fourth, fifth, and so on by further growth.

d', developing dentine; ep, epidermis; $ga_9, ga_{9-I}, ga_{9-II}$, ganoin layers of the first to third odontode of the complex; $la. Hw$, lacunae of Howship; $pap'.t_{9-I}, pap'.t_{9-II}$, odontode papillae under formation, and $pap.t_{9-I}$, a fully developed papilla; $prga$, preganoin matrix; res, areas of resorption; $r.l.$, resorption line; t_9, t_{9-I}, t_{9-II}, fully developed odontodes of the complex; $ve. splx.$, vessels of the subepidermal vascular plexus.

thought, dating back to the end of the last century (see e.g. Röse, 1894: 551) and adopted by several subsequent writers with reference especially to the tooth development in higher vertebrates, it is the epithelial dental organ which is the active partner in this connection. More precisely, from this organ should emanate the first induction for the adjoining mesenchyme to form a dental papilla, and for cells of this mesenchyme to differentiate into matrix-producing odontoblasts (see Lehner & Plenk, 1936: 480–1, and references cited there; Bargmann, 1956: 398–9; Ruch & Karcher-Djuricic, 1971; and others). There have occasionally been suggestions to the effect that similar conditions are, in fact, met with throughout the vertebrate series so that even in fishes (Poole, 1956: 107) the existence of an epithelial dental organ is somehow tied up with, or possibly the necessary prerequisite for, the initial emergence of a dental papilla underneath, and of active scleroblasts in that papilla. As far as odontodes are concerned, one should then assume that even an epithelial dental organ consisting of the inner dental epithelium only should be ascribed similar importance. On the other hand, it has been maintained from investigations in experimental zoology (especially such carried out on amphibian material) that the mesenchyme is the real activator in the initial stages of tooth-germ formation (see Hörstadius, 1950: 69–70, and references cited there as well as in Gaunt & Miles, 1967; see further e.g. de Beer, 1947; Colefax, 1952: xliii); this mesenchyme should, to be more explicit, possess the ability to induce the differentiation, superficially to it, of an epithelial dental organ of ectodermal (sometimes even endodermal, see Chibon, 1970, and others) origin, a point of view which should, of course, apply equally well to developing odontode primordia. To the dicussion of these alternative explanations (which still goes on, see e.g. Ruch & Karcher-Djuricic, 1971), nothing further can be added here. For our present purposes the remark suffices that it is probably as unwarranted, if one favours the leading role of the epithelial dental organ, to consider the adjoining mesenchyme forming the papilla as wholly passive, as it is, if the leading role should be that of the mesenchyme, to disregard completely the epithelial dental organ as a source of directive influences. With Moss (1969), one may appropriately characterize the sequence of events in terms of an *inductive interaction* between the epithelial dental organ and the emerging dental papilla which are surely both essential parts of the tooth-germs and both have essential functions to fulfill, separately as well as in relation to each other. There is every reason to assume that this is so even in developing odontodes, of whatever kind they are, and in whatever lower vertebrates they occur. The potentiality for such inductive interactions between the epidermis and the adjoining mesenchyme was originally without doubt equally distributed throughout the whole morphogenetic field where odontodes formed, viz. the whole of the skin outside the oralobranchial chamber (here taken as the *extraoral* part of the field) and in various forms and to a varying degree also that chamber itself (the *oralopharyngial* part of the field): in this morphogenetic field odontodes could surely, from the very outset, form anywhere.

The formation of enameloid (when present) and dentine (or dentinous tissue) in odontodes in general is surely in all respects the same process as in teeth of fishes consisting of these hard tissues, and needs no further comment here. Of interest even as far as odontodes are concerned, is the active participation in enameloid development of the cells of the inner dental epithelium, recently demonstrated by Shellis & Miles (1974) in the teeth of certain teleosts.

OCCURRENCE, DISTRIBUTION, SHAPE AND SIZE

In these respects, odontodes show a wide range of variation which cannot be more than briefly touched upon here. It is well known that odontodes were abundantly developed throughout the dermal skeleton of a great many Palaeozoic lower vertebrates, including ostracoderms (from the Ordovician onwards) and many different fishes: this wide extent of the morphogenetic field where odontodes formed has obviously to be taken as a primitive condition. It is also well known that in several groups they were subjected to a successive phyletic reduction process with the result that in various instances they became vestigially developed and eventually disappeared completely, or nearly so. An illustrative example of this is found, for instance, in the history of actinopterygians. Odontodes are normal constituents of the dermal skeleton in palaeonisciforms, various sub-holosteans and even a few holosteans whereas in the teleosts, including the host of such fishes of the present day, there are with few exceptions (denticipitids, xiphiids, certain siluroids, see below) no traces left of such dental units in the dermal elements (exclusive, that is, of those that may still persist in the oral cavity or the pharynx). Odontode reduction surely also took place in those lineages of crossopterygians which gave rise to the early tetrapods as in none of the latter odontodes are known to have survived. It is not invariable, however, that the odontodes tended to be reduced with time and disappear. In the osteostracans, for instance, odontodes are retained even in the geologically youngest forms known so far (Ørvig, 1968: fig. 2C, D) although in various other ways the dermal skeleton had then undergone a regressive development. And in the coelacanthids, odontodes are as richly represented in the entire dermal skeleton of the extant *Latimeria* (Figs 2B and 3; Plate 3A, see also Roux, 1942; Millot & Anthony, 1958; Bernhauser, 1961; Smith, Hobdell & Miller, 1972; Castanet, Meunier, Bergot & Francois, 1975) as ever they were in that of Devonian or, for that matter, Triassic forms (Plates 2 and 3B). Even the recent brachiopterygian *Polypterus* still retains the ability to form odontodes in its dermal skeleton (Ørvig, 1967: fig. 12). One may also in this connection mention the many elasmobranchs of modern times where odontodes (i.e. crowns of placoid scales) are as abundant in all parts of the organism as in the ancestral forms of long ago. Why odontodes should persist in these cases but disappear in so many others still remains to be satisfactorily explained.

In shape, odontodes may be of several different kinds. For instance, the crowns of some selachian placoid scales and certain thelodont scales are more or less tooth-like. Other examples of odontodes with this shape are the so-called 'teeth' on the sword of *Xiphias* (Carter, 1919; Rauther, 1929: 324, tended to regard these as rudimentary jaw-teeth) and *Histiophorus* (Carter, 1927: pl. 9: 13, 14), and further the spinules on some of the dermal bones in denticipitid teleosts ('extra-oral dentition', Clausen, 1959; see also Greenwood, 1960). Yet other examples are found on the dermal plates of the branchial arches in various fishes (Allis, 1922: pl. 8: 17; Rauther, 1929: fig. 270; Jarvik, 1954: figs 22, 23A, C, 24; Millot, 1954: pl. 27; Millot & Anthony, 1958: pls 41, 44; Nybelin, 1968; and many others), on the external face of the scales in *Lepisosteus*, other ganoids and some of the silur-oids (Goodrich, 1907: fig. 203, 1942; Bhatti, 1938; Lerner, 1956; etc.), and on the non-placoid scales of some of the earliest selachians (*Ohiolepis:* Wells, 1944: fig. 7a–i; Gross, 1973: pl. 30: 8, 9a, 10a, 11, 12a, 13–15, 16a, 17–21, pl. 31: 1a, 2, 3, 4a, c, 6–8). Odontodes of other shapes occur as ornamental tubercles, rounded or elongated, or ornamental ridges, sometimes rather long ones, on the dermal elements of various ostracoderms, arthrodires, acanthodians and teleostomian fishes as well as on the fin spines of several fossil elasmobranchs (concerning the latter, see e.g. Stromer, 1927; Peyer, 1946, 1957; Maisey, 1975). This does not mean, however, that the ornamental tubercles and ridges of the dermal skeleton are *always* odontodes according to the definition given above. In several cases, as a result of the phyletic process of odontode reduction referred to above, these tubercles and ridges may instead consist of bone tissue throughout their extent (in various ganoids of ganoin and bone tissue after the disappearance of the dentine originally present) and cannot then qualify as odontodes at all. In other cases, they may not individually represent single odontodes as might be expected from their external appearance, but odontocomplexes.

In individual specimens of ostracoderms and fishes, odontodes may not infre-quently show a certain variation in shape, not only when they belong to different parts of the dermal skeleton but also when they form superficially to each other in consecutive generations (see also Novitskaya, 1972). In the former case they may, for instance, be developed as tubercles in some places and as ridges in others. It is also worth mentioning in this connection that in sharks as well as in thelodonts, the scale-crown may undergo changes in shape, sometimes rather marked ones, as one proceeds from one part of the squamation to another (see e.g. Ørvig, 1969c: 397; Reif, 1973, 1974; and references in these papers), and that in certain arthrodires the odontodes ornamenting the apronic (postbranchial) lamina of the thoracic armour are clearly different from those occurring anywhere else on the dermal skeleton (see Ørvig, 1975: pls 6, 7, and references cited there). When they belong to consecutive generations, odontodes generally show a gradual increase in size, but in some instances they may at the same time also assume a shape rather different from what they had in earlier stages of growth. Examples of this are known from ostracoderms (Ørvig, 1969a: 224–5) but occur in various fishes too, including certain actinopterygians. One may be mentioned here: a maxillary of *Birgeria groenlandica* where in some places the odontodes of late generations have split off, revealing underneath them others belonging to earlier stages of growth (Plate 1). As shown by this specimen, the bone in question (and surely the dentalosplenial too) was in fairly young individuals ornamented by very small, tubercle-like odontodes, situated close to each other in regularly arranged con-centric rows (*orn″*). Later, this type of ornamentation was replaced by other odontodes developed as broader ornamental ridges of different length and distribution (*orn′*) and still later by such assuming the shape of comparatively coarse, rounded or somewhat elongated tubercles (*orn*; it may be added that, as shown by thin sections, one is here actually concerned with true odontodes consisting of ganoin and dentine, and thus not with ornamental tubercles and ridges made up of ganoin and bone tissue only). The circumstance that in *B. groenlandica* specific changes took place in the shape of the odontodes of the dermal bones with advancing age (more marked ones, in fact, than those described by Nielsen, 1949: 253–4, 286, fig. 76, pl. 17: 2, since that writer had no material showing the odontodes of early stages) is naturally of interest for the question of what phyletic changes the ornamentation of the dermal bones in the *Birgeria* species may have undergone during the Triassic (see, concerning this, Stensiö, 1932: 106, 295; Lehman, 1952: 103; Schwarz, 1970: 41–2). Examples of changes in ornamentation in individual palaeonisciform specimens, probably having to do with odontodes or, perhaps, the development of odontocomplexes, have also been recorded by Moy-Thomas (1937: 348–9, figs 1, 2) in *Styracopterus fulcratus* and *S. ottadinicus*, and something similar should apparently be expected in other forms too (see Westoll, 1937: 566).

FUNCTIONAL ADAPTATIONS

In elasmobranchs, above all, odontodes may to a varying degree undergo modifications in external morphology when they are adapted to serve specific functions in the organism. Thus in these fishes the odontodes (crowns of placoid scales) often acquire a special shape and/or size when they have definite tasks to perform, such as (a) defence against enemies (which seems to be the adequate explanation for the development of some, or several, of them into thorns and even spines in the shark *Echinorhinus brucas* as well as in skates and rays, see Bigelow & Schroeder, 1948: fig. 102A, 1953; Bendix-Almgreen, 1971; and many others; one may in this context perhaps also include the 'rasp-hooks' on the first ray of the pectoral fin in the peculiar Carboniferous iniopterygians: Zangerl & Case, 1973: 18, although the significance of this particular evidence is not yet wholly clear), (b) protection of free neuromasts in the skin (Budker, 1938; Tester & Nelson, 1967; etc.), (c) attachment during mating (then forming hooks or spurs on the genital appendages of the pelvic fins in male sharks and holocephalans and on the frontal clasper and tenacula of the latter, tenacular hooks in male iniopterygians and retractile 'alar' spines on the lateral parts of the pectoral fins in male rajids, see among others Dean, 1906; Backman, 1915; Bigelow & Schroeder, 1953: 134; Applegate, 1967; Zangerl & Case, 1973), and (d) sieving of planktonic food (then developed as enormously elongated, rod-like elements on the branchial arches of *Cetorhinus*: Hendricks, 1908; Gross-Lerner, 1957; etc. cf. also the condition of *Rhinodon*). To this list one should, perhaps, also add (e) hunting for prey by slashing or mud-grubbing in which activities the large 'teeth' on the lateral side of the rostrum in e.g. pristids serve as effective tools; these 'teeth' are generally held to be greatly modified placoid scale-crowns (see e.g. Engel, 1910; Stromer, 1917; Schaeffer, 1963; Peyer, 1968: 78), but although this is probably true enough (as indicated especially by the condition of the Cretaceous *Sclerorhynchus* where there seems to be a gradual transition between them and more 'normal' crowns of placoid scales, see Woodward, 1889: 76–7, pl. 3: 1), they have at the same time also acquired certain peculiarities of their own not shared by odontodes in general, viz. their initial stages of formation in a submerged position in the corium and presumably also (as generally maintained in extant representatives of the group) their capability of continuous growth. With regard to special usage or special shape/size of placoid scale-crowns in elasmobranchs, mention should be made also of the condition of *Cephaloscyllium* where (according to Grover, in Applegate,

1967: 44) new-born individuals are released from their enveloping egg-cases by means of large crowns of this kind with a row-like arrangement on the dorsal side of the flanks (cf. also in *Heterodontus:* Reif, 1974: 31), and of that of e.g. rajids where sexual dimorphism may be reflected to a certain extent in the proportional size of some of these crowns in males and females (see Bigelow & Schroeder, 1953: 134).

As far as teleostomian fishes are concerned, the ability to produce modified odontodes for utilization in a variety of special capacities seems generally to be not nearly as pronounced as in many of the elasmobranchs. Why this should be so, is difficult to say: a possible explanation could be that, as crowns of *separate* placoid scales, odontodes may individually possess a higher degree of morphological changeability and functional adaptability than when they belong, in groups, to more comprehensive dermal elements of the teleostomian kind. Conditions like the development in the pharynx of cyprinids and certain other teleosts of large, often peculiarly shaped 'Rachenzähne' or 'Schlundzähne' (see e.g. Rauther, 1929: 319–23), sometimes acting against each other as a 'second pair of jaws' (Gosline, 1971: 64–5), may come to mind in the present connection but do not actually figure there. In these cases one is to all intents and purposes concerned with dental units which, because of their replacement mechanism, should be interpreted as real teeth and not as odontodes (cf. also in *Pseudoscarus*, Tomes, 1923: 274). To what extent, in teleosts or other teleostomians, the odontodes *sensu stricto* which may form in oralopharyngial locations are liable to undergo modifications to meet specific functional demands is largely unknown. But so much can at any rate be said that in those teleostomians where they still persist, *extraoral* odontodes as a rule do not show modifications in shape and/or size to suit special purposes in similar ways or to similar degrees as in elasmobranchs in general. There are to the best of the writer's knowledge no perceptible signs of such modifications in, for instance, the coelacanthids which throughout their known history (from the Devonian onwards) possess dermal bones and scales rich in odontodes, and there are none in other crossopterygians either, as e.g. the holoptychiid porolepiforms. Even in dipnoans and brachiopterygians—whenever extraoral odontodes are in evidence—the situation seems to be much the same. It is, in fact, only in actinopterygians that we have a few instances of such odontodes which may have undergone modification for what seems to be special usage. One such case is in the catopterids (or if one prefers, redfieldoids) where forms like *Atocephala*, *Cionichthys*, *Redfieldius*, *Lasalichthys* and *Synorichthys* show what appears to be exceptionally large, tooth-like odontodes (or, perhaps, odontocomplexes) on some of the dermal bones of the snout and/or cheek (Lehman, 1966: fig. 101; Schaeffer, 1967: figs 5–11, pl. 12: 1, 4, pls 13, 14, pl. 15: 2; Schaeffer & Mangus, 1970: fig. 1, pl. 6). These odontodes (or odontocomplexes) obviously fulfilled some function but whether this (as suggested by Schaeffer, 1967: 331, for those of them which lie on the snout) was to support a fleshy upper lip, is difficult to decide. An interesting parallel is the development of especially large tubercles (single semidentine odontodes or, perhaps, columns of such) on the rostral and postnasal plates and the anteroventral part of the suborbital plate in the coccosteomorph arthrodire *Millerosteus minor* (Heintz, 1938: fig. 1: 1, fig. 2, pl. 1: 1, pl. 2: 1; Desmond, 1974: 284, fig. 2A), a condition which seems to be as unusual in arthrodires as that of the catopterids just referred to among more primitive actinopterygians. The 'serrated appendages' of the shoulder girdle in *Amia* (see Liem & Woods, 1973, and references cited there) form another case of odontode modification in actinopterygians. These elements, the like of which (according to information kindly supplied by Dr C. Patterson *in litt.*, 1974–06–11) also occur overlying the postbranchial lamina of the cleithrum in *Pholidophorus*, *Lepidotes*, *Caturus* and *Furo*, bear peculiar spinules, presumably representing odontodes, which clearly have some specific function to fulfill (see further discussion in Liem & Woods, 1973: 530–1), probably much the same as the tooth-like ones forming rows on the anterolateral part of the cleithrum in an undetermined actinopterygian from the Carboniferous of Kansas (Poplin, 1974: 135, fig. 45A, pl. 38: 2, pl. 40: 4). In this case also we have an interesting parallel in the often peculiarly shaped odontodes already referred to on the apronic (postbranchial) lamina of the thoracic armour of several arthrodires (see Ørvig, 1975, and references there). These latter odontodes (or, perhaps, columns of odontodes in some cases) could be assumed to have served as sieving devices to prevent e.g. parasites from entering the gill-chamber (White, 1952: 293). It is theoretically possible but, of course, far from certain, that a similar explanation also applies to the 'serrated appendages' of the actinopterygians just referred to.

RELATIONS TO SUPPORTING DERMAL ELEMENTS

Even with regard to their distribution pattern in dermal bones and scales—when they are constituent parts of such elements—odontodes may show variations. It is generally the case that the odontodes are firmly anchylosed to, and thus not easily detachable from, the supporting bony elements, but there are also cases, such as in some siluroids, where they are independent in the sense that they are attached to bony pediments underneath only by ligaments (Hertwig, 1876; Bhatti, 1938; etc.), a condition which, of course, is known also from the jaw-teeth in various teleosts. Mention should also be made of the circumstance that as new odontodes are successively added to dermal elements during their growth, so also in many cases are new bone laminae with which these odontodes are in continuity basally (see e.g. Ørvig, 1951; figs 8D, E, 14A; 1973a: fig. 2E). Among fishes in general, it is apparently only in the dermal jaw-bones of acanthodians and certain arthrodires (see further Ørvig, 1973a) that one normally finds a similar correlation between the formation of successive teeth on the one hand, and of corresponding zones of basally adjoining bone tissue on the other (in fishes with *in situ* tooth replacement naturally nothing like this can occur). Odontodes, as already pointed out, form ontogenetically in the superficial part of the corium, and may also remain in that position afterwards (as is true, for instance, of the dentine ridges on the dermal skeleton of e.g. cyathaspids and pteraspids). However, they may also (and this is a frequent occurrence) become gradually submerged underneath the external surface of the dermal elements by the subsequent formation of other hard tissue on top of them (layers of bone tissue and/or new odontodes as the case may be) with the result that they are eventually obscured completely from superficial view. When the growth of the dermal elements consists merely in the successive circumferential addition of new hard tissue, which seems to be the condition in several ostracoderms, acanthodians and arthrodires as well as

A

B

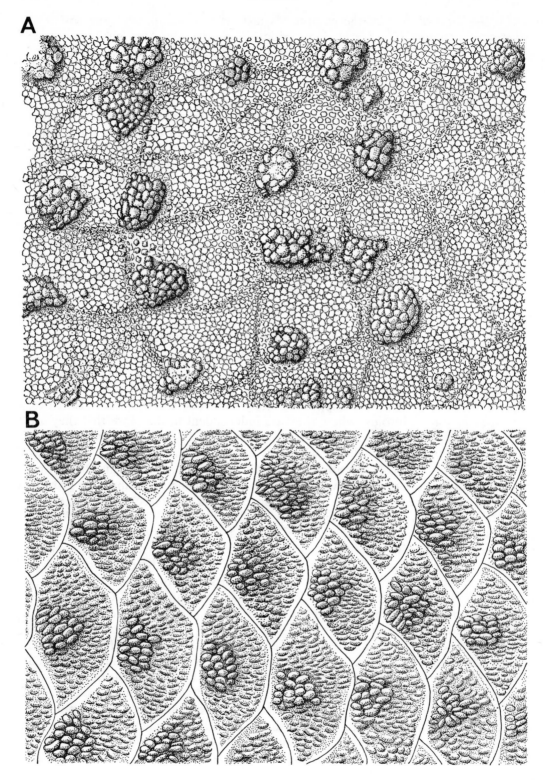

some of the earliest palaeonisciforms, such secondarily submerged odontodes may persist unaltered in the interior. On the other hand, when this growth involves not only addition, but also some degree of substitution, of hard tissue, which is true of many teleostomian fishes for example, the submerged odontodes may also be subjected to resorption in various ways and to a varying extent, and they are thus only partly preserved (see, among others, Gross, 1930: pl. 4: 3, 5); it may even happen that they disappear altogether. Processes of hard tissue resorption, incidentally, may in certain cases even take place in odontodes other than those which in the individual dermal bones and scales form beside or upon each other in consecutive areal zones of growth or consecutive generations This is true, for instance, of selachian placoid scales prior to these being shed and replaced by new ones; then the crown (as also the basal plate) is being subjected to partial resorption (Budker, 1938).

The distribution of odontodes relative to growth in area and thickness of dermal elements is a wide (and to the writer's mind fascinating) field which cannot, however, be entered upon in any detail here (see further remarks in Ørvig, 1968). It suffices to mention the circumstance that when, in the dermal bones and scales of various lower vertebrates, a new generation of odontodes is developing, it often starts at the centre of growth of these elements and spreads out, wave-like, above the odontodes of the preceeding generation, or generations (cf. in porolepiform crossopterygians, Ørvig, 1957: 402). This is for instance clearly displayed in the squamation of *Latimeria:* in a specimen of this fish at the writer's disposal, the odontodes of the youngest generation (in this case the fourth) are commencing to form as a 'blister' at the centre of growth of the individual scales (Fig. 2B: Plate 3A; incidentally, this does not take place simultaneously throughout the squamation). In this connection, comparison with the 'blisters' of new dentine tubercles (odontodes) on the scale-areas of a dermal plate in *Psammolepis paradoxa* (Fig. 2A; also figured by Stensiö, 1964: fig. 35C) is interesting. The regular distribution of these 'blisters' which contain dentine tubercles of a second generation (sometimes even of a third, Ørvig, 1976a: fig. 17), indicates that in position, they correspond to the centres of growth of the individual scale-areas (and also that they continued to form in that position after these areas had lost their original status as independent elements, and had become incorporated in the dermal plate during the marginal growth of the latter). This example and many others, show that throughout the series of lower vertebrates the odontodes of the dermal skeleton conform to a distribution pattern which may be subjected to variations of one kind and another, but always remains basically the same.

HISTOLOGICAL STRUCTURE

In histological structure the odontodes of some fishes may agree closely with the teeth of the jaws, but on the other hand, it is by no means uncommon that they

---

Figure 2. A, *Psammolepis paradoxa* Agassiz, Middle Devonian of the Baltic. Part of dermal plate with scale-areas and regularly distributed 'blisters' of new dentine tubercles (odontodes) on these areas; SMNH C 1781, approximately ×3. B, *Latimeria chalumnae* Smith, patch of scales from the dorsal side of the posterior part of the trunk, showing 'blisters' of dentine tubercles of the fourth generation located at the centres of growth of the individual scales; SMNH-specimen, approximately ×1·2.

also differ to some extent from the latter in features such as the development of their enclosed vascular canal system, the thickness and density of their dentinal tubules, etc. In odontodes, it is unusual for the system of vascular canals (when it occurs) to reach nearly the same degree of complexity as in the teeth. There are also various cases (such as those of *Scanilepis*, *Nephrotus* and *Polypterus* as well as *Glyptolepis* and other crossopterygians) where the dentine of the odontodes contains dentinal tubules which, in shape and distribution, are clearly of a type of their own as compared with those of the teeth. Furthermore, in the odontodes of various arthrodires one may find a semidentine, and in those of certain acanthodians a mesodentine, where enclosed sclerocytes are comparatively more common than in the corresponding hard tissue of the teeth of the same fishes. Structural differences like those now enumerated between odontodes and jaw-teeth are obviously of degree, not of kind. When such differences actually exist, there is every reason to expect that the odontodes (as the most primitive dental units) will, even in their histological structure, generally reflect ancestral conditions to a greater extent than their more specialized phyletic relatives, the jaw-teeth themselves. From this, however, the conclusion that odontodes are never capable of also showing specialized features is unwarranted; this they may certainly do but, one should add, only in exceptional cases. One example of this is encountered in certain fossil actinopterygians. In these fishes in general, according to the writer's investigations of comprehensive material, odontodes are generally characterized by each possessing a single superficial layer of ganoin (cf. the odontodes of the complex in Fig. 1). In some forms (as for instance *Birgeria*, *Nephrotus* and *Colobodus*) however, some of the odontodes adjoining the dentition on the external side of the lower jaw may, in addition to this superficial ganoin, also possess a portion of apical acrodin (concerning this term, see Ørvig, 1973b; 'cap enameloid', Shellis & Miles 1974: fig. 1), similar to that of the jaw-teeth. As far as one can judge from the figures by Bhatti (1938) and Lerner (1956), a cap of acrodin seems to be present also in the odontodes (spinules) of the squamation in certain siluroids. Another example of specialization is the remarkable development of enameloid in the odontodes of the dermal skeleton in the Ordovician heterostracan *Pycnaspis* (Ørvig, 1967: fig. 36A). It may be added that no odontode seems to show specializations like vasodentine or plicidentine sometimes met with in jaw-teeth of fishes.

ODONTODE REMNANTS AND ODONTODE DERIVATIVES

This survey is not complete without a brief consideration of certain histological processes where odontodes were no doubt involved in the course of phylogeny but where, as far as the end-products are concerned, the term odontode cannot always be used despite its wide applicability elsewhere. This refers to the emergence of odontode remnants, the development of cosmine in the dermal skeleton of various Osteostraci, crossopterygians and dipnoans (Ørvig, 1969b, and references cited there) and the growth of pleromin (forming composite pleromo-aspidin, see Ørvig, 1967, 1976a) in the ventral disc and branchial plates of some of the psammosteids.

Odontode remnants are products of the phyletic reduction process referred to above. For instance, in the course of this process in *Holoptychius*, odontodes may have completely lost the enameloid layer with which they were originally equipped, and consist of dentine only, more or less vestigially developed (Buistrov, 1939: fig. 10A, B; 1953: fig. 4a). Because these remnants still possess the latter hard tissue, however, it seems reasonable to regard them as odontodes, albeit greatly reduced ones. Much the same applies to the remnants of semidentine odontodes of consecutive generations which show a column-like arrangement in the ornamental tubercles of the dermal bones in some coccosteomorph arthrodires (Gross, 1930: pl. 2: 12); because they still exhibit semidentine, recognizable as such despite its vestigial development, these should still be taken as odontodes. In actinopterygians, on the other hand, where the phyletic reduction has taken another course, leading to the disappearance of the *dentine* of the odontodes and the persistence (if not indefinitely then often for a long time) of their enameloid (ganoin), the remnants will consist of the latter hard tissue directly adjoined by bone tissue, and have thus lost the distinctive character by virtue of which, above all, they should deserve the name of odontodes.

In the osteostracans, crossopterygians and dipnoans, each individual cosmine sheet is, as well known, made up of a great number of small hard tissue units (consisting of mesodentine in the osteostracans and of dentine in the other forms). These are certainly suggestive of minute odontodes in many ways ('Hautzahnparkett', Gross), and we have reason to believe that they also correspond to large groups of such, sometimes with an arrangement in areal zones of growth, which during phylogeny fused so completely with each other in a *horizontal* direction (see Ørvig, 1957; 1969b) that they lost their original ability to form separately, each in a dental papilla of its own, and all arose within *one single large dental papilla* with the same extent as the cosmine sheet as a whole (concerning this point, see Gross, 1956). In this case it seems far more appropriate to speak of component hard tissue units (cosminomoria would be one possible name) than of component odontodes. It is, in fact, each individual cosmine sheet which according to the definition above should deserve to be referred to as an odontode, but this usage is obviously awkward, and is not adopted here. Cosmine development, it may be added, has even attracted attention from other points of view (see e.g. discussion in Miles, 1975). It is even scheduled to be discussed by Thomson in a paper to appear in the same volume as the present one, but the writer has had no opportunity to see this paper beforehand, and cannot, therefore, discuss it here. As recently pointed out by the writer (Ørvig, 1976a), even the ingrowing pleromin (pleromic hard tissue) of the psammosteids may from the point of view of phylogeny be interpreted as the product of fusion of odontodes. Here, however, one may assume that the odontodes originally belonged to consecutive generations (thus not to areal zones of growth as in the case of cosmine) and fused so completely with each other in a *vertical* direction (after acquiring the ability, like some jaw-teeth, to form underneath instead of superficially to each other) that in the resulting (here also hypermineralized) hard tissue, they entirely lost all traces of the individuality they once possessed. Consequently, the term odontode no longer applies. As far as jaw-teeth are concerned, there is to the best of the writer's knowledge no direct equivalent to the fusion in a horizontal direction of numerous odontodes into continuous cosmine sheeets. The fusion of odontodes in a vertical direction, producing constantly growing pleromin, on the other hand, clearly corresponds to the fusion of consecutive tooth generations into the similar hard tissue of the tooth plates in ptyctodontids, holocephalans and dipnoans (see e.g. Ørvig, 1976b, c).

THE DISTINCTION BETWEEN ODONTODES AND TEETH

This is a comprehensive and, in some ways, a difficult question, but at the same time one which is of more than academic interest. A distinction can as a rule be made if, to express matters in a simplified way, *teeth* are regarded as dental units which are situated on (or, when not yet erupted, submerged underneath) the biting margins, or biting faces, of the jaws, and in these locations participate (or have the potentiality to participate) in the catching, crushing, etc. of food, and *odontodes* are dental units which occupy positions anywhere else in the entire dermal skeleton. A sharp line of demarcation cannot always be drawn, however, for in some of the dermal bones of the palate and lower jaw in various teleostomians, for instance, there may be zones of gradual transition between odontodes and teeth. In such instances the question of which term to apply to what dental units has to be evaluated from case to case. Furthermore, as far as the dermal bones of the oralobranchial cavity are concerned, these may not always carry dental units of the same kind in different fishes. In teleosts such as *Elops hawaiiensis* and *Albula vulpes*, material kindly demonstrated to the writer by Dr G. J. Nelson, New York, has shown that the dental units of the parasphenoid are replaced in the manner of teeth, forming first in a submerged position and rising to the surface where they take the place of others which have been resorbed in part and then shed. In this case, consequently, the units in question are appropriately referred to as teeth, a conclusion which is also clear from their relative size and structural properties. In *Boreosomus*, on the other hand, the small, more or less tooth-like protuberances of the palatal dermal bones should in the writer's opinion be classified as odontodes rather than teeth because of their mode of formation superficially to each other in consecutive generations (Ørvig, 1968: t_1-t_7, fig. 4), their shape and size and their structural features both when they occur singly and in odontocomplexes (cf. Fig. 1), and this despite the circumstance that in a way they *act* as parts of the dentition in a broad sense. As already mentioned, the pharyngeal 'Schlundzähne' of e.g. cyprinids should be taken as teeth proper because of their replacement mechanism. On the other hand, there are also quite a number of instances in teleostomians where pharyngeal dermal elements carry what seem to be true odontodes (apart from several cases where the information is far too meagre to decide with what sort of dental units one is here concerned). As to the tooth-like units present in the dentition of *Uranolophus* and *Griphognathus* among the Devonian dipnoans (Gross, 1956; Denison, 1968; Schultze, 1969; Ørvig, 1976b: figs 26–28), it is difficult to say if these should be characterized as odontodes or teeth so long as we are ignorant of how they formed relative to each other. Reference may even be made at this point to the 'teeth' on the lateral side of the rostrum in pristids mentioned above: these have frequently been regarded as enlarged crowns of placoid scales (i.e. as odontodes) but in extant forms, at any rate, they may develop in a submerged position like true jaw-teeth.

In conclusion, if one tries to summarize in what way a distinction between odontodes and teeth should be made, the outcome would be that in a great many cases such a distinction is possible by considering position and function in combination with mode and place of formation, size and structure. It is also clear, however, that there are borderline cases where not all of these criteria apply at the same time. One such case is, for instance, that of the dentigerous jaw-bones in the acanthodians *Nostolepis*, *Ischnacanthus*, *Xylacanthus*, *Atopacanthus* and *Acanthodopsis*. The teeth of these elements should to all appearances be identified as such because of their location, function and structure, but nevertheless they do not form in a submerged position like the teeth of fishes in general. Instead, like odontodes, they develop in the superficial part of the corium and show in addition the further peculiarity of not being replaced at all (see Ørvig, 1973a, and references there). That borderline cases like this and others should exist, is perhaps in itself not too surprising. Odontodes and teeth are obviously components of the dermal skeleton of fundamentally the same kind, and as odontodes gave rise to teeth when jaws first formed, it is quite possible that in some few cases (like in the acanthodians just referred to) the latter could retain some, if not all, odontode characteristics (though generally they did not).

How the small, pointed dental units situated on the 'plaque maxillaire dentée' in the foremost part of the oralobranchial chamber in certain osteostracans (Stensiö, 1964: *pl. d. spo.*, fig. 116) should be interpreted is somewhat uncertain. For the time being they are, perhaps, best regarded as odontodes.

ODONTODIA IN THE PHYLETIC HISTORY OF THE DERMAL SKELETON

The many contributions made to the question of the phylogeny of the dermal skeleton in lower vertebrates since the pioneer works of Williamson (1849) and Hertwig (1874) have been adequately summarized on more occasions than one, and need not be summarized again in the present connection. Following Hertwig, the writers dealing with these questions were for a long time preoccupied with the role which scales of the selachian placoid type could have played in the emergence of large dermal elements of the teleostomian kind, and reverberations of this still occur (see e.g. Zangerl, 1966; Peyer, 1968). When, in this century, knowledge of the dermal skeleton of early Palaeozoic lower vertebrates (including ostracoderms) increased, it became clear that a heavily ossified dermal skeleton is certainly not the *sole* property of teleostomian fishes and that selachian placoid scales are not *necessarily* the basic building blocks out of which such a dermal skeleton arose. However, the discussion went on, and an example of a new approach based on wider knowledge is the lepidomorial theory (sensu Stensiö, 1961), which has received a variety of comments (some critical) but which cannot be further considered in the limited space here available.

Much of the discussion following Hertwig's work embraced attempts to explain, theoretically, how basal plates of selachian placoid scales could, by enlargement, fusions, etc. be instrumental in producing dermal elements of larger size. Without here entering into the many variations on this theme, the phyletic process leading to the emergence of, for example, dermal bones could, in the writer's opinion, be reconstructed on less theoretical grounds with reference to the factual evidence we now possess, by disregarding placoid scales as such and instead considering the odontodia defined above (viz. small, micromeric, dermal elements consisting each of a single odontode, or a group of odontodes, situated on a thin, bony basal plate) as the basic elements involved in the process. In all likelihood, it was a dermal skeleton consisting entirely of such elements (of varying complexity from the point of view of the lepidomorial theory and not necessarily resembling typical placoid scales) that evolved in early stages of vertebrate history. Also, it was surely from a dermal skeleton of this kind that the dermal elements later to appear inherited the odontodes ornamenting their external surface. However, the question remains as to what could conceivably have happened during further stages of skeletal assimilation, when eventually a heavy dermal armour like that of ostracoderms came into existence. This has, of course, been discussed from different points of view (noteworthy, for instance, are the comments by Westoll, 1967). If the odontodes persist (as they are known to do) so also should, in some form or another, the basal plates of the odontodia to which the odontodes originally belonged. These basal plates could, of course, increase in size, fuse with each other, and in various ways become integral parts of larger elements. The bony laminae which in the scales of acanthodians a few elasmobranchs and various teleostomians, for instance, are added to the periphery of the base in continuity with new zones of growth (odontodes) in the crown could certainly be regarded as derivatives of the basal plates of original odontodia, but the basal plates in question, as the writer sees it, can hardly have been responsible *alone* for the formation of those large ossifications which in the dermal skeleton of many fossil forms assumed proportions extensive enough to reach throughout the depth of the corium (and in special cases to embrace subcutaneous portions as well). This, to all intents and purposes, was an *extra* process of ossification which phyletically succeeded, but was in itself independent of, the odontodia themselves: in general terms the latter yielded the material forming the superficial parts of thick dermal bones and scales whereas the former provided the substance of the middle and basal parts of such elements (see Ørvig, 1968: 391; cf. also in ostracoderms, Goodrich, 1907: 755). In addition, there was probably also a third component, which may have arisen quite independently of the other two just referred to, namely the latero-sensory ossifications developing in direct relation to the lateral line system (see Ørvig, 1972, and references cited there). By the interaction of these three components, the odontodia, the extra ossification in the deeper layers of the corium and the latero-sensory ossifications, which could, of course, yield a variety of different end-products, a dermal skeleton of the amphimeric (Miles, 1975) or macromeric kind eventually came into being.

The many observations of fossil and recent material on which the above considerations are based cannot be included here, but by way of illustration, it may have a certain interest to mention briefly the condition of the dermal skeleton in certain coelacanthids. In *Latimeria* Roux (1942) has figured small dermal elements which seem to correspond perfectly to odontodia as defined above (Fig. 3A in the present paper). He figured scales of this fish containing agglomerations of such odontodia, each with a crown (odontode) and a clearly defined basal plate, which had formed upon or beside each other during consecutive stages of growth, and underneath these, a basal layer of laminate bone tissue (Fig. 3B). It seems, therefore, that in the squamation of *Latimeria* it is still possible to distinguish different components from which more comprehensive dermal elements may once have formed: the superficial odontodia and the 'extra' basal corium ossification. Something similar can also be observed in certain fossil coelacanthids. In some so far undescribed Triassic forms, the dermal bones are (as in other representatives of the group) ornamented by odontodes; in the present case these are sometimes rounded and sometimes pointed at the apex, but a common feature of all of them is that at their basal circumference they are adjoined by a distinct, ring-like zone of bone tissue (Plates 2 and 3B). Together with these basal zones the odontodes form units which are certainly suggestive in many ways of odontodia (the ring-like zones corresponding to the basal plates of such elements), and there can be little doubt in the writer's opinion that this, in fact, is what they really represent. These units are of course not independent elements but integral parts of the dermal bones to which they belong. However, the circumstance that they are recognizable as odontodia, and the position they occupy, 'floating' on the

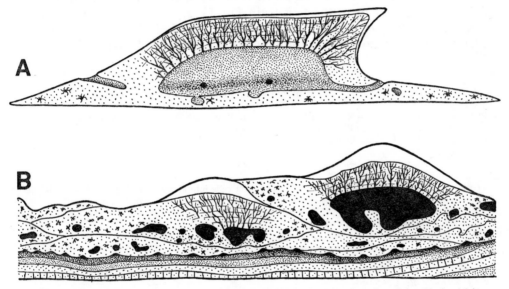

Figure 3. *Latimeria chalumnae* Smith. A, An odontodium of the squamation ('placoid denticle', Roux, 1942). B, Vertical section of a scale showing odontodia of consecutive stages of growth and underneath laminated basal bone tissue. Redrawn (with the addition in B of the enameloid of the odontodes) from Roux, 1942: figs 5 (A) and 6C (B); × 50 (B).

surface of the dermal bones in each successive stage of growth, seems to indicate that they were originally independent elements which in the course of phylogeny became *secondarily* attached to a mass of bone tissue forming underneath them in the deeper layers of the corium, and that their basal plates (still identifiable in the present case) were firmly connected to, but did not in themselves deliver material to the formation of, this mass of bone tissue. Finally, brief reference may be made here to another similar example, this time from the heterostracans. Thus, in a form as yet not more closely determined, from the early Lower Devonian of Arctic Canada (contained in the same fauna as the radotinid arthrodire recently described by the writer, Ørvig, 1975), the carapace is ornamented with elongated odontodes which are also here associated with basal plates lying in a superficial position (Plate 3C, D). These plates are remarkable in one respect: they display vascular canals belonging to the superficial part of the carapace which in each of them tend to radiate from a centre out towards the margins, a circumstance which lends further support to the contention that they represent basal parts of originally independent dermal elements of small size. Examples from other lower vertebrates (such as arthrodires) leading to the same conclusion could also be mentioned. For the time being, however, those referred to here may suffice to bring out what, in the writer's opinion is one of the more essential aspects of the phyletic history of the dermal skeleton.

ACKNOWLEDGEMENTS

For the loan of material, the writer wishes to express his indebtedness to Lic.scient.S.E.Bendix-Almgreen, Copenhagen, Professor E. Kuhn-Schnyder and Dr K. A. Hünermann, Zürich, and Dr R. Thorsteinsson, Calgary, Alberta, Canada. Information referred to in the text from Drs G. J. Nelson, New York and C. Patterson, London, is gratefully acknowledged. Technical assistance has been rendered by Mr U. Samuelson (photography) and Ms Milda Liepina (illustrations).

REFERENCES

ALLIS, E. P. Jr., 1922. The cranial anatomy of *Polypterus*, with special reference to *Polypterus bichir. J. Anat.,* 56: 189–294.

APPLEGATE, S. P., 1967. A survey of shark hard parts. In P. W. Gilbert *et al.* (Eds), *Sharks, skates and rays:* 37–67. Baltimore: J. Hopkins Press.

BACKMAN, G., 1915. Die Bauchflosse der Selachier. 2. Die Bauchflosse der Holocephali. *K. svenska Vetensk-Akad. Handl.,* 53: 1–63.

BARGMANN, W., 1956. *Histologie und mikroskopische Anatomie des Menschen:* 2nd ed. Stuttgart: G. Thieme Verl.

de BEER, G. R., 1947. The differentiation of neural crest cells into visceral cartilages and odontoblasts in *Amblystoma,* and a re-examination of the germ-layer theory. *Proc. R. Soc., (B), 134:* 377–98.

BENDIX-ALMGREEN, S. E., 1971. The anatomy of *Menaspis armata* and the phyletic affinities of the menaspid bradyodonts. *Lethaia, 4:* 21–49.

BERNHAUSER, A., 1961. Zur Knochen- und Zahnhistologie von *Latimeria chalumnae* Smith und einiger Fossilformen. *Sber. öst. Akad. Wiss., math.-naturw. Kl., Abt. 1, 170:* 119–37.

BHATTI, H. K., 1938. The integument and dermal skeleton of Siluroidea. *Trans. zool. Soc. Lond., 24:* 1–103.

BIGELOW, H. B. & SCHROEDER, W. C., 1948. Sharks. In Fishes of the Western North Atlantic. 1. *Mem. Sears Fdn mar. Res., 1:* 59–546.

BIGELOW, H. B. & SCHROEDER, W. C., 1953. Sawfishes, guitarfishes, skates and rays. In Fishes of the Western North Atlantic. 2. *Mem. Sears. Fdn mar. Res., 1:* 1–514.

BUDKER, P., 1938. Les cryptes sensorielles et les denticules cutanés des plagiostomes. *Annls Inst. Oceanogr. Monaco, (N.S.) 18:* 207–88.

BUISTROV, A. P., 1939. Zahnstruktur der Crossopterygier. *Acta zool., Stockh., 20:* 283–338.

BUISTROV, A. P., 1953. [The development of teeth in vertebrates.] *Ezheg. Vses. Paleont. Obshch., 14:* 39–60. In Russian.

CARTER, J. T., 1919. On the occurrence of denticles on the snout of *Xiphias gladius. Proc. zool. Soc. Lond., 1919:* 321–6.

CARTER, J. T., 1927. The rostrum of the fossil swordfish, *Cylindracanthus,* Leidy (*Coelorhynchus,* Agassiz), from the Eocene of Nigeria. *Occ. Pap. geol. Surv. Nigeria, 5:* 7–15.

CASTANET, I.-J., MEUNIER, F., BERGOT, C. & FRANCOIS, Y., 1975. Données préliminaires sur les structures histologiques du squelette de *Latimeria chalumnae.* 1. Dents, écailles, rayons et nageoires. *Colloques int. Cent. natn. Rech. scient., 218:* 161–7.

CHIBON, P., 1970. L'origine de l'organe adamantin des dents. Étude au moyen du marquage nucléaire de l'ectoderme stomodéal. *Annls Embryol. Morphogen., 3:* 203–13.

CLAUSEN, H. S., 1959. Scientific results from the expedition from the Zoological Museum of Copenhagen to West Africa in cooperation with University College, Ibadan. 1. Denticipitidae, a new family of primitive isospondylous teleosts from West African fresh-water. *Vidensk. Meddr dansk naturh. Foren., 121:* 141–51.

COLEFAX, A. N., 1952. Presidential address: variations on a theme. *Proc. Linn. Soc. N.S.W., 77:* x–xlvi.

A SURVEY OF ODONTODES   73

DEAN, B., 1906. Chimæroid fishes and their development. *Publs Carnegie Inst. Wash.*, *32:* 1–194.
DENISON, R. H., 1968. Early Devonian lungfishes from Wyoming, Utah and Idaho. *Fieldiana Geol.*, *17:* 351–413.
DESMOND, A. J. 1974. On the coccosteid arthrodire *Millerosteus minor*. *Zool. J. Linn. Soc. Lond.*, *54:* 277–98.
ENGEL, H., 1910. Die Zähne am Rostrum der Pristiden. *Zool. Jb. (Anat. Ontog.)*, *29:* 51–100.
GAUNT, W. A. & MILES, A. E. W., 1967. Fundamental aspects of tooth morphogenesis. In A. E. W. Miles (Ed), *Structural and chemical organization of teeth*, *1:* 151–97. New York & London: Academic Press.
GOODRICH, E. S., 1907. On the scales of fishes, living and extinct, and their importance in classification. *Proc. zool. Soc. Lond.*, *1907:* 751–74.
GOODRICH, E. S., 1942. Denticles in fossil Actinopterygii. *Q. Jl micros. Sci.*, *(N.S.)*, *83:* 459–64.
GOSLINE, W. A., 1971. *Functional morphology and classification of teleostean fishes.* Honolulu: University Press of Hawaii.
GREENWOOD, P. H., 1960. Fossil denticipitid fishes from East Africa. *Bull. Br. Mus. nat. Hist. (Geol).*, *5:* 1–11.
GROSS, W., 1930. Die Fische des mittleren Old Red Süd-Livlands. *Geol. paläont. Abh. (N.F.) 18:* 1–36.
GROSS, W., 1956. Über Crossopterygier und Dipnoer aus dem baltischen Oberdevon im Zusammenhang einer vergleichenden Untersuchung des Porenkanal-systems paläozoischer Agnathen und Fische. *K. svenska VetenskAkad. Handl.*, (4) *5:* 1–140.
GROSS, W., 1973. Kleinschuppen, Flossenstacheln und Zähne von Fischen aus europäischen und nordamerikanischen Bonebeds des Devons. *Palaeontographica (A)*, *142:* 51–155.
GROSS-LERNER, 1957. Über Bau und Entwicklung der Reusenzähne von *Cetorhinus maximus* Gunner. *Z. Zellforsch. mikrosk. Anat.*, *46:* 357–68.
GROSS-LERNER, see also LERNER, H.
GROVER, C., see APP LEGATE, S. P., 1967.
HEINTZ, A. 1938. Notes on Arthrodira. *Norsk geol. Tidsskr.*, *18:* 1–27.
HENDRICKS, K., 1908. Zur Kenntnis des gröberen und feineren Baues des Reusenapparates an den Kiemenbögen von *Selache maxima* Cuvier. *Z. wiss. Zool.*, *91:* 427–509.
HERTWIG, O., 1874. Ueber Bau und Entwickelung der Placoidschuppen und der Zähne der Selachier. *Jena. Z. Naturw.*, *8:* 331–404.
HERTWIG, O., 1876. Ueber das Hautskelet der Fische. *Morph. Jb.*, *2:* 1–68.
HÖRSTADIUS, S., 1950. *The neural crest, its properties and derivatives in the light of experimental research.* London & New York: Oxford University Press.
JARVIK, E. A. V. 1954. On the visceral skeleton in *Eusthenopteron* with a discussion of the parasphenoid and palatoquadrate in fishes. *K. svenska VetenskAkad. Handl.*, (4) *5:* 1–104.
KLAATSCH, H., 1890. Zur Morphologie der Fischschuppen und zur Geschichte der Hartsubstanzgewebe. *Morph. Jb.*, *16:* 97–196, 209–58.
KVAM, T., 1950. *The development of mesodermal enamel on piscine teeth.* Trondheim: K. norske VidenskSelsk.
LEHMAN, J.-P., 1952. Étude complémentaire des poissons de l'Eotrias de Madagascar. *K. svenska VetenskAkad. Handl.*, (4) *2:* 1–201.
LEHMAN, J.-P., 1966. Actinopterygii. In J. Piveteau (Ed.), *Traité de Paléontologie*, *4 : 3:* 1–242. Paris: Masson.
LEHNER, J. & PLENK, H., 1936. Die Zähne. In W. v. Möllendorff (Hrsg.), *Handbuch der mikroskopischen Anatomie des Menschen*, *5: 3:* 449–708. Berlin: Springer.
LERNER, H., 1956. Die histologische Natur des 'Schmelzes' der Hautzähnchen beim Panzerwels *Corydoras*. *Z. Zellforsch. mikrosk. Anat.*, *43:* 554–65.
LERNER, H., see also GROSS-LERNER, H.
LIEM, K. F. & WOODS, L. P., 1973. A probable homologue of the clavicle in the holostean fish *Amia calva*. *J. Zool., Lond.*, *170:* 521–31.
MAISEY, J. G., 1975. The interrelationships of phalacanthous selachians. *Neues Jb. Geol. Paläont. Mh.*, *1975:* 553–67.
MILES, R. S., 1975. The relationships of the Dipnoi. *Colloque int. Cent. natn. Rech. scient.*, *218:* 133–48.
MILLOT, J., 1954. Le troisième coelacanthe. Historique. Elements d'écologie. Morphologie externe. Documents divers. *Naturaliste malgache (Suppl. I):* 1–26.
MILLOT, J. & ANTHONY, J., 1958. *Anatomie de Latimeria chalumnae. 1. Squelette, muscles et formations de soutien.* Paris: Cent. natn. Rech. scient.
MOSS, M. L., 1969. Phylogeny and comparative anatomy of oral ectodermal-ectomesenchymal inductive interactions. *J. dent. Res.*, *48:* 732–7.
MOY-THOMAS, J. A., 1937. The palaeoniscids from the cement stones of Tarras Waterfoot, Eskdale, Dumfriesshire. *Ann. Mag. nat. Hist.*, (10) *20:* 345–56.
NIELSEN, E., 1949. Studies on Triassic fishes from East Greenland. 2. *Australosomus* and *Birgeria. Meddr Grønland*, *146:* 1–309.
NOVITSKAYA, L. L., 1972. Diagnostic evaluation of the ornamentation of Agnatha and Pisces. *Paleont. J. Wash.*, *5:* 494–506.

74 T. ØRVIG

NYBELIN, O., 1968. The dentition in the mouth cavity of *Elops*. In T. Ørvig, (Ed.), *Current problems of lower vertebrate phylogeny.*, *Nobel Symposium, 4:* 439–43. Stockholm: Almqvist & Wiksell.

ØRVIG, T., 1951. Histologic studies of placoderms and fossil elasmobranchs. 1. The endoskeleton, with remarks on the hard tissues of lower vertebrates in general. *Ark. Zool.*, (2) *2:* 321–454.

ØRVIG, T., 1957. Remarks on the vertebrate fauna of the Lower Upper Devonian of Escuminac Bay, P.Q. Canada, with special reference to the porolepiform crossopterygians. *Ark. Zool.*, (2) *10:* 367–426.

ØRVIG, T., 1966. Histologic studies of ostracoderms, placoderms and fossil elasmobranchs. 2. On the dermal skeleton of two late Palaeozoic elasmobranchs. *Ark. Zool.*, (2) *19:* 1–39.

ØRVIG, T., 1967. Phylogeny of tooth tissues: evolution of some calcified tissues in early vertebrates. In A. E. W. Miles (Ed.), *Structural and chemical organization of teeth, 1:* 45–110. New York & London: Academic Press.

ØRVIG, T., 1968. The dermal skeleton; general considerations. In T. Ørvig (Ed.), *Current problems of lower vertebrate phylogeny*, *Nobel Symposium, 4:* 373–97. Stockholm: Almqvist & Wiksell.

ØRVIG, T., 1969a. The vertebrate fauna of the *primaeva* beds of the Frænkelryggen Formation of Vestspitsbergen and its biostratigraphic significance. *Lethaia, 2:* 219–39.

ØRVIG, T., 1969b. Cosmine and cosmine growth. *Lethaia, 2:* 241–60.

ØRVIG, T., 1969c. Thelodont scales from the Grey Hoek Formation of Andrée Land, Spitsbergen. *Norsk geol. Tidsskr., 49:* 387–401.

ØRVIG, T., 1972. The latero-sensory component of the dermal skeleton and its phyletic significance. *Zool. Scripta, 1:* 139–55.

ØRVIG, T., 1973a. Acanthodian dentition and its bearing on the relationships of the group. *Palaeontographica, (A), 143:* 119–50.

ØRVIG, T., 1973b. Fossila fisktänder i svepelektronmikroskopet. Gamla frågeställningar i ny belysning (Fossil fish teethin the scanning electron microscope. Old questions in new light). *Fauna Flora Stockh., 68:* 166–73. In Swedish with English summary.

ØRVIG, T., 1975. Description, with special reference to the dermal skeleton, of a new radotinid arthrodire from the Gedinnian of Arctic Canada. *Colloques int. Cent. natn. Rech. scient., 218:* 41–71.

ØRVIG, T., 1976a. Palaeohistological notes. 3. The interpretation of pleromin (pleromic hard tissue) in the dermal skeleton of psammosteid heterostracans. *Zool. Scripta, 5:* 35–47.

ØRVIG, T., 1976b. Palaeohistological notes. 4. The interpretation of osteodentine, with remarks on the dentition in the Devonian dipnoan *Griphognathus*. *Zool. Scripta., 5:* 79–96.

ØRVIG, T., 1976c. Ptyctodontid tooth plates and their bearing on holocephalan ancestry (MS).

PEYER, B., 1946. Die schweizerischen Funde von *Asteracanthus* (*Strophodus*). *Schweiz. paläont. Abh., 64:* 1–101.

PEYER, B., 1957. Über die morphologische Deutung der Flossenstacheln einiger Haifische. *Mitt. naturf. Ges. Bern., (N.F.), 14:* 159–76.

PEYER, B., 1968. In R. Zangerl, (Ed.), *Comparative odontology*, Chicago & London: University of Chicago Press.

POOLE, D. F. G., 1956. The fine structure of the scales and teeth of *Raja clavata*. *Q. Jl microsc. Sci. 97:* 99–107.

POPLIN, C., 1974. Étude de quelques paléoniscidés Pennsylvaniens du Kansas. *Cah. Paléont., 1974:* 1–151.

RAUTHER, M., 1929. Echte Fische. 1. Anatomie, Physiologie und Entwicklungsgeschichte. In *Dr H. G. Bronn's Klassen und Ordnungen des Tierreichs, 6: 1: 2:* 185–328. Leipzig: Akad. Verlagsges.

REIF, W.-E., 1973. Ontogenese des Hautskelettes von *Heterodontus falcifer* (Selachii) aus dem Untertithon. Vergleichende Morphologie der Hautzähnchen der Haie. *Stuttg. Beitr. Naturk. (B), 7:* 1–16.

REIF, W.-E., 1974. Morphogenese und Musterbildung des Hautzähnchen-Skelettes von *Heterodontus*. *Lethaia, 7:* 25–42.

RÖSE, C., 1894. Das Zahnsystem der Wirbeltiere. *Anat. Hefte, 4:* 542–91.

ROUX, G. H., 1942. The microscopic anatomy of the *Latimeria* scale. *S. afr. J. med. Sci. (Biol. Suppl.), 7:* 1–18.

RUCH, J. V., & KARCHER-DJURICIC, V., 1971. Mise en évidence d'un role spécifique de l'épithelium adamantin dans la différenciation et le maintien des odontoblastes. *Annls Embryol. Morphogen., 4:* 359–66.

SCHAEFFER, B., 1963. Cretaceous fishes from Bolivia, with comments on pristid evolution. *Am. Mus. Novit., 2159:* 1–20.

SCHAEFFER, B., 1967. Late Triassic fishes from the Western United States. *Bull. am. Mus. nat. Hist., 135:* 285–342.

SCHAEFFER, B. & MANGUS, M., 1970. *Synorichthys* (Palaeonisciformes) and the Chinle-Dockum and Newark (Upper Triassic) fish faunas. *J. Paleont., 44:* 17–22.

SCHULTZE, H.-P., 1969. *Griphognathus* Gross, ein langschnauziger Dipnoer aus dem Oberdevon von Bergisch-Gladbach (Rheinisches Schiefergebirge) und von Lettland. *Geol. Palaeont., 3:* 21–79.

SCHWARZ, W., 1970. Die Triasfauna der Tessiner Kalkalpen. 20. *Birgeria stensiöi* Aldinger. *Schweiz. paläont. Abh., 89:* 1–93.

SHELLIS, R. P. & MILES, A. E. W., 1974. Autoradiographic study of the formation of enameloid and dentine matrices in teleost fishes using triturated amino acids. *Proc. R. Soc., (B) 185:* 51–72.

SMITH, M. M., HOBDELL, M. H. & MILLER, W. A., 1972. The structure of the scales in *Latimeria chalumnae*. *J. Zool. Lond., 167:* 501–9.

STENSIÖ, E. A., 1932. Triassic fishes from East Greenland collected by the Danish expeditions in 1929–1931. *Meddr. Grønland, 83:* 1–305.

Plate 1

(*Facing page 74*)

Plate 2

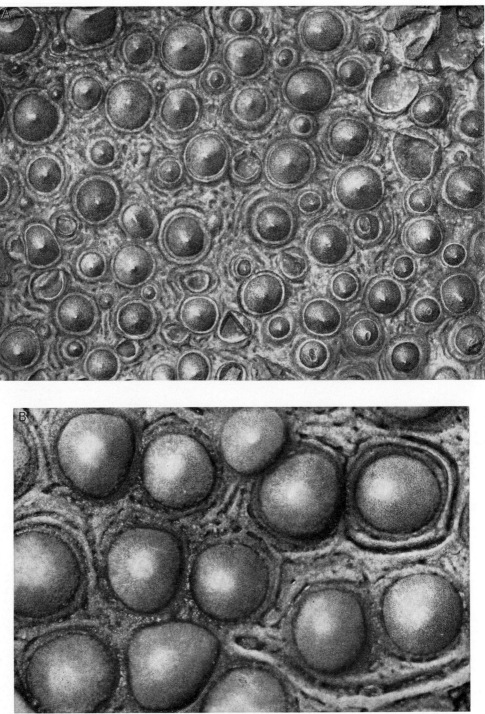

T. ØRVIG

Plate 3

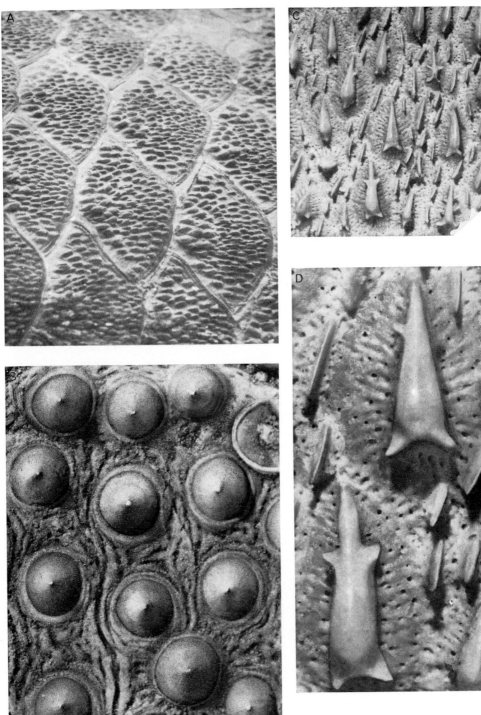

STENSIÖ, E. A., 1961. Permian vertebrates. In G. O. Raasch (Ed.), *Geology of the Arctic*, *1*: 231–47. Toronto: University of Toronto Press.

STENSIÖ, E. A., 1964. Les cyclostomes fossiles ou ostracodermes. In J. Piveteau (Ed.), *Traité de Paléontologie*, *4: 1*: 96–382. Paris: Masson.

STROMER, E., 1917. Ergebnisse der Forschungsreisen Prof. E. Stromers in den Wüsten Ägyptens. 2. Wirbeltier-Reste der Baharîje-Stufe (unteres Cenoman). 4. Die Säge des Pristiden *Onchopristis numidens* Haug sp. und über die Säge der Sägehaie. *Abh. K. bayer. Akad. Wiss. math.-phys. Kl.*, *28*: 1–28.

STROMER, E., 1927. Ergebnisse der Forschungsreisen Prof. E. Stromers in den Wüsten Ägyptens. 2. Wirbeltier-Reste der Baharîje-Stufe (unteres Cenoman). 9. Die Plagiostomen mit einem Anhang über käno- und mesozoische Rückenflossenstacheln von Elasmobranchiern. *Abh. bayer. Akad. Wiss., math.-naturw. Abt.*, *31*: 1–64.

STUDNIČKA, F. K., 1909. Vergleichende Untersuchungen über die Epidermis der Vertebraten. *Anat. Hefte.* *39*: 1–267.

TESTER, A. L. & NELSON, G. J., 1967. Free neuromasts (pit organs) in sharks. In P. W. Gilbert *et al.* (Eds), *Sharks, skates and rays*: 503–31. Baltimore: J. Hopkins Press.

TOMES, C. S., 1923. In H. W. M. Tims & C. B. Henry Eds), *A manual of dental anatomy, human and comparative*: 8th ed. (London: J. & A. Churchill.

WELLS, J. W., 1944. Fish remains from the Middle Devonian bone beds of the Cincinnati Arch Region. *Palaeontogr. Am.*, *3*: 99–161.

WESTOLL, T. S., 1937. On a remarkable fish from the Lower Permian of Autun, France. *Annls Mag. nat. Hist.*, (10) *19*: 553–78.

WESTOLL, T. S., 1967. *Radotina* and other tesserate fishes. In C. Patterson & P. H. Greenwood (Eds). *Fossil Vertebrates J. Linn. Soc. Lond.*, (*Zool.*) *47*: 83–98.

WHITE, E. I., 1952. Australian arthrodires. *Bull. Br. Mus. nat. Hist. (Geol.)* *1*: 249–304.

WILLIAMSON, W. C., 1849. On the microscopic structure of scales and dermal teeth of some ganoid and placoid fish. *Phil. Trans. R. Soc.*, *139*: 435–75.

WOODWARD, A. S., 1889. *Catalogue of the fossil fishes in the British Museum (Natural History). 1. Elasmobranchii*. London: Br. Mus. (nat. Hist.).

ZANGERL, R., 1966. A new shark of the family Edestidae, *Ornithoprion hertwigi*, from the Pennsylvanian Mecca and Logan Quarry Shales of Indiana. *Fieldiana Geol. 16*: 1–43.

ZANGERL, R. & CASE, G. R., 1973. Iniopterygia, a new order of chondrichthyan fishes from the Pennsylvanian of North America. *Fieldiana Geol. Mem.*, *6*: i–ix, 1–67.

EXPLANATION OF PLATES

PLATE 1

Birgeria groenlandica Stensiö, Eotriassic of Cape Stosch, East Greenland. Posterior part of maxillary, showing ornamentation belonging to consecutive stages of growth (*orn, orn', orn''*; for further explanation, see text). MMHVP-specimen, ×4·5.

PLATE 2

Coelacanthids not more nearly determined as to genus and species, from the Triassic 'Grenzbitumenhorizont' of Monte San Giorgio, Tessin. Ornamentation of dermal bones. A, PIZ T 1634, ×15; B, PIZ T 1633, ×27.

PLATE 3

A, *Latimeria chalumnae* Smith, patch of scales from the dorsal side of the posterior part of the trunk (cf. Fig. 2B). SMNH-specimen, slightly more than natural size. B, Enlarged detail of the ornamentation of the coelacanthid dermal bone in Plate 2A, ×27. C, D, Ornamentation of the carapace of an undetermined heterostracan from the Gedinnian of Arctic Canada, figured with the full consent of Dr Thorsteinsson who works with this material. GSC-specimen, ×10 (C) and 26 (D).

Cartilage bones, dermal bones and membrane bones, or the exoskeleton versus the endoskeleton

COLIN PATTERSON

British Museum (Natural History), Cromwell Road, London

The terms 'dermal bone' and 'membrane bone' should not be used as synonyms: 'membrane bone' is here reserved for endoskeletal bones which are no longer preformed in cartilage. The dermal skeleton and the endoskeleton of vertebrates seem always to have been distinct, for there are no known instances of interchangeability between dermal bones and cartilage bones, or of invasion of the endoskeleton by dermal bone. Cartilage in the dermal skeleton (adventitious cartilage) is restricted to birds and mammals. Supposed examples of dermal bones invading the endoskeleton are due either to fusion (ontogenetic or phylogenetic) between a dermal bone and a cartilage bone, or to loss of one or other of two separate bones, one dermal and one endoskeletal. Holmgren's principle of delamination is found to lack support as a theory about the phylogeny of the vertebrate skeleton. A new interpretation of the vertebrate urohyal/parahyoid is proposed.

CONTENTS

INTRODUCTION

The problem tackled here is one that has been most clearly set out and discussed by de Beer in section 39 of *The development of the vertebrate skull* (1937: 495–502): 'the possibility of membrane-bones becoming cartilage-bones and vice versa'. De Beer meant 'becoming' in phylogeny, not ontogeny, of course. I was led to take up this problem by my work on the braincase of actinopterygians, which convinced me that all de Beer's examples of apparent interchangeability between dermal bones and cartilage bones have other explanations. There are different ways of formulating the problem, for example, by asking whether it is ever sensible to homologize a dermal bone in one vertebrate with a cartilage bone in another, or as a test of delamination theory. But de Beer's heading asks a matter-of-fact question, which I think can be answered, and I leave delamination theory aside until the end of the paper.

De Beer (1937) decided that the evidence available to him was insufficient to answer his question, and concluded that 'the best evidence on which these questions can be solved is that provided by fossils, for which reason it is earnestly to be hoped that palaeontologists will continue to devote their attention to such problems'.

However, the recent literature suggests that de Beer's problem is solved: palaeontologists, along with comparative anatomists, histologists, embryologists and experimental zoologists are agreed that there is no valid distinction between dermal bones and cartilage bones. The following quotations, taken from the work of a palaeontologist, two comparative anatomists, a histologist and an experimental embryologist will illustrate this.

'This tendency for dermal bones to develop at deeper levels in later forms, and even to be replaced phylogenetically by cartilage-bones, is probably much more widespread than is usually accepted.' (Westoll, 1967: 96).

'There is also the question whether, in the course of phylogeny, a bone which was a cartilage bone in ancestral forms can become a membrane bone in descendants and vice versa. Modern authors are finally agreed that the answer is yes.' (Daget, 1965: 183, my translation).

'The "in vogue" view stresses the idea that there is no real difference between dermal and chondral bone and suggests that there is some sort of interchangeability of these. It is assumed that what was formerly a chondral bone may arise in ontogeny in a dermal fashion.' (Jollie, 1968: 90). 'I have come to accept a "delamination" origin of dermal bones . . . which would have as one of its corollaries the dermal origin of chondral ossifications.' (Jollie, 1975: 85).

'The classical histogenetic distinctions between the endo- and "exoskeleton" are largely artificial. Typical dermal bones can be transitorily or permanently associated with cartilages (clavicle, dentary). Incidentally, periostic ossification in the endoskeleton is identical in method to dermal ossification in the "exoskeleton", which greatly reduces the contrast between these two categories from the histological point of view.' (de Ricqles, 1975: 17, my translation).

'Cartilage cells may be transformed into bone cells, bone cells may be transformed into cartilage cells, so that [membrane] bone and cartilage form intertransformable tissues.' (Hall, 1970: 479).

Perhaps all this can be summed up in Jarvik's statement (1959: 49): 'there cannot be any fundamental differences between the various kinds of skeletal tissues . . . or between the endoskeleton and the exoskeleton.'

My thesis will be that the dermal skeleton and the endoskeleton are distinct; that there are not two categories of bones—dermal and cartilage—but three—dermal, cartilage and membrane. Dermal bones can 'become' membrane bones in conventional usage, since the two terms are treated as synonyms, but I propose to distinguish them. According to this distinction, cartilage bones can become membrane bones, through regression of cartilage. I hope to show that all supposed examples of interchangeability between dermal and cartilage bones are due either to ontogenetic or phylogenetic fusion between a dermal bone and a cartilage bone whose growth centres have become superimposed, or to suppression, in phylogeny, of one or other of two separate, superimposed ossifications, one dermal and one endoskeletal.

TERMINOLOGY

In recent textbooks and papers on anatomy, histology and embryology, usage of the terms cartilage bone, dermal bone and membrane bone is confused, and in part contradictory. The confusion arises from three main sources. First, these terms are used at three levels; for tissues (i.e., as descriptive histological and anatomical terms); for organs (e.g., a membrane bone, a cartilage bone); and for systems (e.g., the dermal skeleton). Second, dermal bone and membrane bone are treated as synonyms, and third, cartilage bone and endochondral bone are often thought to be synonymous. In this section, I will first define the terms as I propose to use them, next try to justify my interpretation by a survey of historical and recent usage, and then illustrate it by the history of the fish skeleton.

For complete bones (i.e., organs), I define the three terms as follows:

A cartilage bone is a bone which is preformed in cartilage and first ossifies, in ontogeny, as a perichondral ring or disc on the surface of the cartilage model. It may or may not later ossify endochondrally and periosteally, and may develop membrane bone outgrowths.

A dermal bone is a bone which is not preformed in cartilage, and is either tied to the ectodermal basement membrane by a surface coat of dentine and/or enameloid, or is homologous with such a bone. 'Homologous' is used here in the sense of Jardine (1970), so that the dermal sclerifications of reptiles (Moss, 1969), the theca and epitheca of turtles, the armour of armadillos, etc. are included: these dermal bones are topographic homologues of those in primitive fishes, but may or may not be their phylogenetic homologues. Plate-like, superficial dermal bones may develop internal laminae or processes of membrane bone. In birds and mammals peripheral parts of dermal bones may ossify in cartilage (adventitious cartilage, see p. 91).

A membrane bone is a bone which ossifies in membrane deep in the mesoderm, with no ontogenetic or phylogenetic connection with the ectoderm. Membrane bones are of two sorts. The first is a bone which is homologous with a cartilage bone in more primitive vertebrates: it is not preformed in cartilage in ontogeny, but is so formed in phylogeny. The second type includes neoformations: bones which are restricted to small taxonomic groups or to individuals, heterotopic (sesamoid) and pathological ossifications. Smith (1947) recommended calling this second type 'subdermal bones'.

For systems of bones, the dermal skeleton and endoskeleton can be defined simply enough. The bony dermal skeleton of a vertebrate comprises the dermal bones; the bony endoskeleton of a vertebrate comprises the cartilage bones and membrane bones.

For tissues or parts of bones, it is difficult to recommend precise usage because the terms under discussion were not originally intended as descriptive names for parts of bones (see below), and because the dermal skeleton ossifies in such a range of tissues. *Cartilage bone* occurs in two varieties, laminar perichondral bone, and cancellous or trabecular endochondral bone: these tissues may occur in the dermal skeleton in birds and mammals (p. 91). *Dermal bone* may be laminar, lamellar, cancellous, or combinations of these. It may bear superficial enamel and/ or dentine, may include varieties of dentine such as semidentine and osteodentine, and hypermineralized or compound tissues like pleromic tissue and tubular dentine (Ørvig, 1967). As tissue, dermal bone is so variable that probably the only valid use of the term is as a synonym of the dermal skeleton. *Membrane bone* may be lamellar or cancellous, and may appear as outgrowths from or ontogenetic additions to cartilage bone and dermal bone.

I cannot pretend to have traced the entire history of the terms cartilage bone, dermal bone and membrane bone. According to Huxley (1864: 296) and de Beer (1937: 1), the distinction between cartilage bones and dermal bones in the human skeleton was first noticed in 1736, by Robert Nesbitt. He wrote of bones which 'shoot either between membranes or within cartilages'. The two types of bone were next recognized by Arendt (1822) in the skull of *Esox*. The dermal bones are classed as those 'quae superne cranium tegunt ac statim subcuta occurrent' (which cover the cranium from above, and appear immediately beneath the skin). So far as I know, von Baer (1826) introduced the terms cartilage-bone (Knorpel-Knochen), cartilage skeleton (Knorpelskelet) and dermal skeleton (Hautskelet), but he named the dermal bones 'horn bones' (Horn-Knochen). Reichert (1838) referred to dermal bones as 'covering bones' (Belege-Knochen, Deckknochen), but his paper also contains the earliest mention that I have found of 'Haut-knochen', which could be translated either as 'dermal bones' or 'membrane bones'.

Within a few years, the terms 'dermal bone', 'os dermique' and 'Hautknochen' were in general use (e.g., Agassiz, 1844; Owen, 1846; Kölliker, 1849), but in a restricted sense: for the scales of fishes, some of the cranial dermal bones of sturgeons and teleosts, the dermal plates of armadillos and crocodiles, etc., and these structures were held to be different from the 'integral' or 'true' skeleton. The main reason for this failure to recognize two fundamentally distinct components of the skeleton, dermal bone and cartilage bone, seems to have been the pervasive influence of the vertebral theory of the skull. Vogt (1842: 118), an opponent of the vertebral theory, thought that the fish parasphenoid could not be a dermal bone because it was not developed beneath the skin but beneath the mucous membrane of the mouth. He also mentioned, without citing examples, bones in the fish skull which ossified in cartilage in some forms and in membrane in others. To Owen, the chief exponent of the vertebral theory at this time (1846 and many later works), the fact that the parasphenoid and vomer, or frontal and parietal, do not ossify in cartilage was irrelevant, for Oken's theory showed these bones to be parts of vertebrae: the vomer and parasphenoid were centra, and the frontal and parietal neural spines. Adherents of the vertebral theory seem to have regarded the dermal skeleton as comprising only those parts which did not find a place in the archetype vertebra of each segment.

Kölliker (1849) also argued that the 'Belegknochen' of fishes and tetrapods are not dermal bones, but integral parts of the skull, homologous with parts of vertebrae, and he even excluded the cranial sensory canal bones from the dermal skeleton. Such views prevailed until the extinction of the vertebral theory, following Huxley's 1859 analysis. In that paper, Huxley does not touch on the question of dermal bones and cartilage bones, and seems to have regarded histogenesis of bones as variable, but by 1864 he advocated an essentially modern view of the two categories, referring to the dermal bones as 'membrane bones' throughout. I have not come across an earlier usage of 'membrane bone' in English, and it is possible that Huxley adopted the alternative translation of 'Hautknochen' and harked back to Robert Nesbitt in order to avoid confusion with the idiosyncratic use of 'dermal bone' by Owen and other vertebral theorists. Since Huxley's work, dermal bone and membrane bone have been generally treated as synonyms. For example, Goodrich (1930) uses 'dermal bone' exclusively, while de Beer (1937) uses 'membrane bone' instead (but in a wider sense than Goodrich, since he included not only the dermal bones but endoskeletal structures, such as the vertebral arches of higher teleosts, which ossify in membrane without preformation in cartilage).

The original 19th century usage of the terms cartilage bone, membrane bone and dermal bone was for developmental mode, and was applied to whole bones, or to systems of bones (e.g., 'the dermal skeleton'). Use of these terms for tissues, or parts of bones, is a modern extension, due mainly to anatomists and histologists familiar with higher vertebrates, in which the dermal skeleton and endoskeleton are not readily distinguished in the adult. For example, Pritchard (1972: 16) writes 'two types of bone formation are commonly described: (1) ossification in membrane or intramembranous ossification and (2) ossification in cartilage or endochondral ossification. Bone formed by the former process is usually termed *membrane bone* and that formed by the latter *cartilage bone*. Membrane bone and cartilage bone are, of course, the names of tissues. However, in discussing the developmental history of the skeleton one refers to membrane bones and cartilage bones, where the terms refer to organs, not tissues' . . . (p. 17) 'It is therefore tempting to discard the traditional categories of intramembranous and endochondral ossification and to replace them with "ossification in fibrous tissue", etc.' This quotation, emphasizing the terms membrane bone and cartilage bone as names of tissues, synonymizing cartilage bone with endochondral bone, and suggesting discarding the terms for a series of more precise, descriptive terms, seems fairly typical of the approach of the human anatomist and histologist. Synonymizing cartilage bone and endochondral bone leads such workers to regard the lamellar periosteal bone of, for example, mammalian long bones, as membrane bone, and when, as in most textbooks on anatomy and histology, dermal bone is equated with membrane bone, the absurd position is reached where well-known textbooks imply that the greater part of the human femur, for example, is dermal bone. This usage of membrane bone for parts of cartilage bone is now often extended to include the initial lamina of perichondral bone (e.g. Hall, 1975: 337). Yet there is excellent experimental evidence that the two primary components of cartilage bone, perichondral and endochondral, are under the same genetic or epigenetic control, and that the dermal skeleton is under different control.

Evidence that perichondral and endochondral bone are successive products of a single epigenetic mechanism comes chiefly from birds (Mareel, 1967; Hinchliffe & Ede, 1968), but observations on other vertebrates indicate that this mechanism is the same in all. Hinchliffe & Ede's work on fowl embryos homozygous for *talpid*[3], a Mendelian recessive allele, is of particular interest.

In these mutants the dermal skeleton ossifies in the usual way, while the endo-skeleton is preformed in cartilage but always fails to ossify. In the endoskeleton of normal embryos the first sign of impending ossification is differentiation of the cells in the inner layer of the perichondrium into osteoblasts (recognized by cytoplasmic basophilia and alkaline phosphatase activity), which produce the initial matrix of perichondral bone. At the same time, the underlying cartilage hypertrophies. Later, the cartilage develops alkaline phosphatase activity, the chondrocytes degenerate and the cartilage is eroded and replaced by endochondral bone. In *talpid*[3] cartilage, the chondrocytes are disorientated at a time when normal chondrocytes, though not yet hypertrophic, are expanded and orientated at right angles to the long axis of the cartilage. *Talpid*[3] perichondrium fails to differentiate an inner layer of osteoblasts, and never ossifies. *Talpid*[3] cartilage may later hypertrophy and develop intense alkaline phosphatase activity, just like normal cartilage, but endochondral ossification never occurs. From this and other work it is clear that perichondral ossification is dependent on an (unknown) inductive signal from cartilage in the earliest stage of hypertrophy. In the absence of this signal, perichondral bone will not form, and in the absence of perichondral ossification, endochondral bone will not form.

In view of this evidence that the perichondral and endochondral components of cartilage bone are products of one genetic or epigenetic mechanism, and that dermal bone is the product of another, it can only obstruct progress and under-standing to conflate the products of independent developmental pathways (dermal and perichondral bone) under one name (membrane bone), and to restrict cartilage bone to endochondral bone.

As an example of the approach of the worker familiar with lower vertebrates, Jollie (1962: 55) writes 'some bones arise in or around cartilaginous precursors and are called chondral bones; other bones ossify directly in connective tissue masses and are called dermal bones. Chondral bones are of two types: those that arise at the surface of the cartilage . . . and those that arise within the cartilage. The former are perichondral, the latter endochondral . . . . A chondral bone may also have dermal extensions.' Here, the two tissues that may comprise a cartilage bone (laminar perichondral bone and trabecular endochondral bone) are correctly distinguished, but dermal bone is regarded as synonymous with membrane bone, so that any part of a cartilage bone that is not preformed in cartilage is held to be dermal. This, too, has absurd results: it would lead one to regard the greater part of the vertebral column of higher teleosts as dermal. One illustration of misunder-standings arising from this usage is provided by the uroneurals of teleosts, modified neural arches in the caudal skeleton. The fact that these bones are not preformed in cartilage was one of the reasons for the proposal (Nybelin, 1963) and acceptance (e.g., Greenwood, 1967: 595; Patterson, 1967: 96; Monod, 1968: 113–8) of the theory that they are dermal bones, scales which are secondarily associated with the endoskeleton. It was later found that the uroneurals are preformed in cartilage in primitive living teleosts (Monod, 1968: 118; Cavender, 1970: fig. 6). This observation, and the structure of the uroneurals in primitive Mesozoic teleosts, led to their reinstatement as endoskeletal structures.

Another usage of 'membrane bone' and 'dermal bone' in lower vertebrates is that of Devillers (1948, 1950, 1958), referring to dermal bones containing sensory canals. In some fishes these bones ossify in two parts, a superficial gutter- or tube-like portion carrying the sensory canal, and a deeper plate-like portion. Devillers refers to the first of these as the 'dermo' component (e.g. dermofrontal) and the second as the 'membrano' (e.g. membrano-frontal). Devillers (1948: 7) credits Allis (1935) with the invention of this nomenclature, as does de Beer (1937: 499), but this seems to be due to misreading of Allis. Allis distinguished the two components of some dermal bones, and called the deeper, plate-like portion the 'membrano' component (e.g. membranosphenotic), but he always referred to the tube- or gutter-like portion as the 'latero-sensory component', and used such terms as dermosphenotic and dermopterotic in their usual sense, for the complete bone comprising both latero-sensory and 'membrano' components. Devillers (1948: 8) noted that his own nomenclature 'présente un risque de confusion'. This is certainly so, principally because it implies that membrane bone, like dermal bone, is confined to the exoskeleton, and that where an endoskeletal element ossifies in membrane, not cartilage, it is dermal (e.g. Devillers, 1958: 614, 640, where teleostean intermuscular bones and the intercalar of *Amia* and teleosts are described as dermal bones). Where it is necessary to distinguish the two components of dermal bones, this sort of confusion can be avoided by using Daget's (1965) terms 'neurodermic' (=latero-sensory) and membranodermic.

I think that misunderstandings and errors of the kind just discussed can be avoided easily enough, without discarding the terms dermal bone and cartilage bone, and without introducing any new terms. Only one change is necessary: that dermal bone and membrane bone should not be treated as synonymous, but clearly distinguished. This distinction must be drawn at two levels, that of tissues or parts of bones, and that of whole bones. When applied to whole bones, I recommend that 'dermal bone' and 'membrane bone' should not be used in a purely descriptive sense, as when the human frontal and parietal are called membrane bones, describing their mode of formation. Rather, these terms should imply a theory of homology. Bone names like frontal and parietal can only be applied in vertebrates other than man (the type-locality of these names) by adopting a theory of homology. In the same way, the human frontal and parietal should be called dermal bones, implying their homology with enameloid and dentine-coated bones in primitive vertebrates. The term membrane bone can then be used to imply a different homology, that the bone in question arises in membrane but is homologous with a cartilage bone in more primitive vertebrates. Teleostean uroneurals and intermuscular bones, mentioned above, are examples of such bones, and others are given below. At the level of tissues or parts of bones, cartilage bone must be acknowledged to comprise two different tissues, laminar perichondral bone and cancellous endochondral bone. Where a term is required for the lamellar cortex of mature cartilage bones in higher vertebrates, membrane bone can be used, as before. Membrane bone can also be used as a descriptive term for parts of dermal bones, as illustrated below. The tissues comprising dermal bone are so varied that the term is of little use in a descriptive sense. These recommendations are summarized in the definitions on p. 79. While I cannot expect to alter the usage of anatomists and histologists working on higher vertebrates, I believe that distinguishing dermal bone and membrane bone in this way has real advantages, and will lead to greater clarity and precision. To conclude this section, I will illustrate the distinction between dermal bone, cartilage bone and membrane bone by a short history of the skeleton of fishes, especially actinopterygians.

The distinction between the bony dermal skeleton and endoskeleton of teleostomes is clearest in primitive fishes like the rhipidistians (e.g., Jarvik, 1972: fig. 71) and palaeoniscoids (e.g., Nielsen, 1942: figs 2, 58). In these fishes (Fig. 1), the head and trunk are completely covered by dermal bones, and the mouth is fully lined by them. All these dermal bones are proved to have developed in the corium immediately beneath the ectoderm by a surface coating of enameloid and dentine, in the form of teeth, tubercles, ridges or a continuous layer. The cartilage

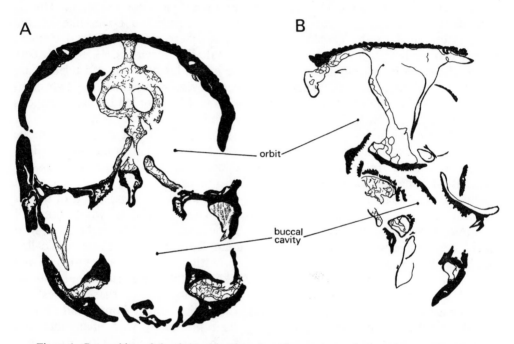

Figure 1. Composition of the skeleton in Devonian teleostomes, as shown n transverse sections through the head. A. Rhipidistian, *Eusthenopteron foordi* Whiteaves, Frasnian, Scaumenac Bay, P.Q., Canada, ground section through the front of the orbit, after Jarvik (1972: fig. 71B, q.v. for detailed interpretation), ×2 approx. B. Palaeoniscoid from the Frasnian Gogo Formation, Western Australia, ground section through the back of the orbit of a partially disarticulated skull, from a series prepared by Dr B. G. Gardiner, ×4 approx.

bones of the endoskeleton consist of a thin outer layer of laminar perichondral bone, and a sparse trabecular filling of endochondral bone, the trabeculae delimiting large, roughly spherical spaces. The dermal bones are plate-like, with plane inner faces, and the cartilage bones have smooth contours and were evidently closely modelled on their cartilage precursors. In these fishes, there are only two types of bone, dermal and cartilage, and there is nothing to which the term membrane bone can usefully be applied. In my opinion, this is the primitive condition of the teleostome or osteichthyan skeleton. Whether it is also the primitive condition of the gnathostome, or even craniate, skeleton, are questions that cannot yet be answered.*

In various primitive fishes, certain dermal bones develop internal processes or laminae. In palaeoniscoid actinopterygians examples are the descending lamina of the dermopterotic, the internal lamina of the dermometapterygoid, the horizontal lamina of the prearticular and coronoids and the ventro-medial lamina of the preopercular in *Pteronisculus* (Nielsen, 1942), and the 'lame postérieure' of the preopercular in *Kansasiella* (Poplin, 1974: 25). Similar internal laminae are developed on the roofing bones of some arthrodires (e.g., Stensiö, 1963: figs 42, 77, 112, 113), rhipidistians (e.g., Jarvik, 1954: fig. 5B) and coelacanths (Lehman, 1966: 211). These internal laminae or outgrowths lack the dentine and enameloid which coat the outer faces of the dermal bones, and if a descriptive term for them is required, they can be called membrane bone outgrowths from dermal bones. These outgrowths seem always to be developed from the ossification centre of the bone; this is particularly clear in sensory canal bones like the actinopterygian preopercular and dermopterotic, in which the sensory canal traverses the ossification centre and the descending lamina develops beneath the sensory canal.

In a few cases, the cartilage bones of palaeoniscoids also ossify partly in laminar bone, as in the hyomandibular of *Boreosomus* (Nielsen, 1942: 345) and an unnamed Carboniferous species (Poplin, 1974: 132). These laminar portions can be described as membrane bone outgrowths or components of cartilage bones.

Thus in palaeoniscoids and other primitive fishes there is evidence of the development of membrane bone outgrowths from dermal bones and (more rarely) from cartilage bones. Because of their sporadic occurrence I regard such outgrowths as derived features. In higher actinopterygians, membrane bone outgrowths from certain dermal bones are almost invariably present. Examples are the descending laminae of the frontal and dermopterotic (Patterson, 1975: 551), the internal or intercalar limb of the post-temporal (usually described as an ossified ligament, but developed beneath the sensory canal in the bone in exactly the same way as the descending lamina of the frontal and dermopterotic), and the nasal process of the premaxilla (Patterson, 1975: 512).

In teleosts and various holostean groups, as in tetrapods, there is no longer an obligatory relationship between the dermal bones and the ectoderm, because the superficial enameloid and dentine have been lost, except on tooth-bearing bones. No longer tied to the ectoderm, some of the dermal bones 'sink' beneath the skin of the fish. Examples include the opercular bones of many higher teleosts, which

---

*A well-developed dermal skeleton is certainly primitive for gnathostomes, and for a taxon comprising gnathostomes and most agnathans. I doubt that myxinoids are secondarily naked (see also Nelson, 1970). In the endoskeleton, primitive dipnoans and coelacanths have peri- and endochondral bone, like rhipidistians and actinopterygians. Acanthodians have perichondral bone, but lack endochondral bone (there may be calcified cartilage, with or without perichondral bone; see Ørvig, 1951: 412). Some placoderms (petalichthyids, rhenanids) have endochondral bone and perichondral bone, while others have perichondral bone underlain by calcified cartilage, and others calcified cartilage alone. Elasmobranchs always lack endoskeletal bone (although there is a bone-like tissue in the vertebral centra of euselachians), and have calcified cartilage of a distinctive type, made up of polygonal calcifications. Amongst Agnatha, there is no endoskeletal ossification or calcification in either extant group or in anaspids, thelodonts and most heterostracans. The Ordovician heterostracan *Eriptychius* has calcified cartilage (Denison, 1967). Many osteostracans have perichondral bone, some also have endochondral bone, and at least one has calcified cartilage beneath the perichondral bone (Ørvig, 1957).

are secondarily covered by scales; the parietal of many deep-bodied acanthopterygians, which is covered by scales and a thick layer of musculature; and many of the sensory canal bones (frontal, dermopterotic, preopercular, dentary) of the cyprinoids studied by Lekander (1949), which ossify deep in the dermis, without any connection with the epidermis or with the invaginated sensory canals. The frontal of the cobitid *Nemacheilus*, for example, consists of a sheet of bone with the relationships of a normal actinopterygian frontal, except that it contains no sensory canal, and is overlain by four or five independent tubular bones, each of which surrounds a neuromast (Lekander, 1949: figs 59, 60). In *Cobitis* the body of the frontal is similar, but there are no sensory canals and no sensory canal ossicles. It might seem legitimate to refer to a frontal of this type as a membrane bone, as has often been done in the past. But this case is no different from the human frontal (p. 83 above): the cobitid frontal is a dermal bone, homologous with the frontal of *Elops* or a palaeoniscoid. Where it is necessary to distinguish the body of the cobitid frontal from the latero-sensory component, Daget's (1965) term 'membranodermic' or 'the membranous portion' (e.g., Kapoor, 1970; Ørvig, 1972) will serve. Some might take a different view, and say that the frontal of a cobitid is only partially homologous with that of *Elops* or a palaeoniscoid, since it lacks the latero-sensory component of both of those, and the superficial dentine and enameloid of the latter. On that view, it would follow that the endopterygoid of a cobitid is only partially homologous with that of *Elops* or a palaeoniscoid, since it lacks the dentition found in more primitive actinopterygians. I think that such views are mistaken: homology does not mean identity. In teleosts, as in tetrapods, I recommend that all bones which are held to be homologous with tooth-, enameloid-, or sensory canal-bearing bones in primitive actinopterygians should be called dermal bones.

In the endoskeleton of teleosts, many cartilage bones develop extensive outgrowths of membrane bone, and other bones, topographic and phylogenetic homologues of cartilage bones in more primitive actinopterygians, develop entirely in membrane, without preformation in cartilage. Those bones, true membrane bones, are found in the skull and in the postcranial skeleton.

In the postcranial skeleton, the example cited by de Beer (1937: 496) was the vertebrae. In primitive teleosts, the centra and the bony neural and haemal arches remain independent throughout life. Teleost centra always ossify as membrane bones, without cartilage preformation (François, 1966). The neural and haemal arches are primitively ossified as cartilage bones, but even in relatively primitive living teleosts they may develop more or less extensive outgrowths of membrane bone, especially in the caudal region (e.g., Greenwood & Rosen, 1971: figs 10–16). In more advanced teleosts, including members of the Ostariophysi, Paracanthopterygii and Acanthopterygii, the vertebral arches may ossify without cartilage preformation in part of the column, or throughout the column (Emelianov, 1939). François (1966: 319) recommends caution in interpreting published data on the absence of cartilage precursors of teleostean vertebral arches, since in 5·2 mm larvae of the perciform *Crenilabrus* he observed minute cartilaginous nodules ('sous la forme de 5 or 6 chondrocytes plus ou moins lysés') in the bases of the ossified neural arches of a few of the abdominal vertebrae, but there seems little doubt that in some teleosts, such as *Gadus* (Faruqi, 1935), the entire vertebra is a membrane bone. The uroneurals in the caudal skeleton provide another example. These modified neural arches are cartilage bones in the Jurassic pholidophorids (Patterson, 1968), but amongst living teleosts they are only known to retain a cartilaginous rudiment, from the perichondral bone of which extensive membrane bone outgrowths develop, in *Elops*, *Hiodon*, *Salmo* and *Coregonus* (Monod, 1968; Cavender, 1970). In other teleosts, the uroneurals ossify as membrane bones.

Similar partial and complete replacement of cartilage bone by membrane bone occurs in the ribs and the epipleural and epineural intermuscular bones of teleosts (Emelianov, 1935).

In the skull of teleosts there are many examples of cartilage bones which develop extensive outgrowths of membrane bone, and several examples of pure membrane bones, homologous with cartilage bones in more primitive forms. Particularly striking instances of the first type are the large median crest of membrane bone which develops from the perichondral part of the supraoccipital, especially in deep-bodied teleosts, and the similar, but smaller, crest developed on the epioccipital (epiotic). In *Cyprinus*, Dornesco & Soresco (1971a, b, 1973, 1974) give a detailed account of the mode of ossification of the neurocranium, and find that most of the bones have more or less extensive membranous portions.

Examples of pure membrane bones in the skull, bones which ossify entirely in membrane but are homologous with cartilage bones in more primitive actinopterygians, include the intercalar of *Amia* and teleosts, and the basisphenoid of most teleosts. The intercalar ossifies in membrane in *Amia* and all living teleosts, but it is a cartilage bone with extensive membrane bone outgrowths in the Mesozoic pholidophorids and caturids, and a normal cartilage bone in *Pachycormus*, *Perleidus* and parasemionotids (Patterson, 1975). The basisphenoid is a massive cartilage bone in palaeoniscoids (e.g., *Cosmoptychius*, Schaeffer, 1971; Gardiner, in this volume), and ossifies as a pair of small cartilage bones in *Amia*. In primitive pholidophorids the bone still ossified in cartilage, but in advanced forms and in leptolepids the pedicel ossified in membrane (Patterson, 1975). In *Salmo* (de Beer, 1937: 129) the basisphenoid ossifies entirely in membrane, from three centres, and this is probably its mode of formation in most higher teleosts (Daget, 1965: 245). It is worth pointing out that two of the examples of 'cartilage bones' ossifying in membrane cited by de Beer (1937) have since been shown to be mistaken. These are the metapterygoid in *Polypterus* and *Syngnathus*. Subsequent work shows that in these fishes the metapterygoid portion of the palatoquadrate cartilage regresses (*Polypterus*) or fails to ossify (syngnathids), and that the supposed metapterygoid of *Polypterus* is a dermometapterygoid (Daget, 1950) and that of syngnathids is an infraorbital (Kadam, 1961).

Thus in actinopterygians there is abundant evidence in the skull and postcranial skeleton that bones which were primitively cartilage bones, ossifying peri- and endochondrally, may, in more advanced forms, develop more or less extensive membrane bone portions, either as outgrowths from a cartilage-bone primordium (e.g., supraoccipital crest of teleosts), or through regression of cartilage (e.g., pterosphenoid of many teleosts). Eventually, through regression of cartilage or changes in growth pattern, the cartilage-bone primordium may fail to appear, and the bone then ossifies only in membrane (e.g., vertebral arches of higher teleosts, intercalar of *Amia* and teleosts). Such bones, true membrane bones, are derived features, wherever they occur.

In tetrapods, the development of cartilage bones by the addition of periosteal membrane bone to the original perichondral bone is routine, while membrane bone outgrowths from cartilage bones occur quite commonly. Examples cited by de Beer (1937) include parts of the basioccipital of *Lacerta*, of the supraoccipital in man and pig, and of the alisphenoid in many mammals. Other mammalian examples are given in Roux (1947) and Kuhn (1971). But whether true membrane bones (that is, bones ossifying in membrane which are homologous with cartilage bones in more primitive forms) occur in tetrapods is doubtful. The vertebral centra of urodeles and apodans are frequently described as ossifying in membrane (Parsons & Williams, 1963: 33, describe apodan centra as dermal bone), but Wake (1970) shows that they ossify partly in cartilage, so that ossification of the centra partly in cartilage is characteristic of all three amphibian groups, and of amniotes. In the tetrapod skull, de Beer (1937: 496) cites only one example, the lamina ascendens of the ala temporalis in the marsupial *Didelphis*. In therians the lamina ascendens ossifies as a membrane bone outgrowth from the alisphenoid (e.g. Roux, 1947: 197), while in *Didelphis* it ossifies in membrane, before and independent of the body of the alisphenoid. However, in this most complex part of the mammalian skull the evidence is equivocal. Toeplich (1920: 15), who described the membrane bone in *Didelphis*, regarded it as homologous with the ossification in the membrana sphenoobturatoria of the monotreme *Tachyglossus*. But in *Tachyglossus* that membrane contains two ossifications (Kuhn, 1971: 83), a membrane-bone outgrowth from the periotic (petrosal) and an independent membrane bone, called the lamina obturans by Kuhn. Whether this independent membrane bone in *Tachyglossus* is a neoformation, as Kuhn supposes, whether it is the homologue of the lamina ascendens in *Didelphis*, or the homologue of a dermal bone, the reptilian intertemporal (as has been suggested), are questions that cannot be answered at present.* The occurrence of true membrane bones in tetrapods is therefore still a subject for investigation.

The purpose of this section has been to establish a workable terminology, and in particular to distinguish membrane bones from dermal bones. With that distinction made, de Beer's problem—'the possibility of membrane bones becoming cartilage bones and vice versa'—can be rephrased by asking whether dermal bones ever become cartilage bones or vice versa. For cartilage bones can become membrane bones—that is the definition of membrane bones adopted here— and dermal bones cannot. There are then two aspects to the problem: dermal bones becoming cartilage bones by invading the endoskeleton, and cartilage bones becoming dermal bones, through cartilage in the dermal skeleton. The second of these will be dealt with first.

CARTILAGE IN THE DERMAL SKELETON

Under this heading there are two different topics to discuss, both concerning the presence of cartilage, which may ossify, in the dermal skeleton. First, there are

---

*[Added in proof] Presley & Steel (1976) have published a review of these questions, and a new interpretation of the mammalian alisphenoid.

a few published statements implying that cartilage occurs in the dermal skeleton of fishes. And second, there are well-known and thoroughly investigated examples of such cartilage (adventitious or secondary cartilage and callus cartilage) in the dermal skeleton of birds and mammals.

Lower vertebrates

Agnathans. Halstead (1969: 21), after quoting a statement by Moss (1968a: 366—'no one has ever suggested, nor do I, that a cartilaginous dermal (exo-) skeleton ever existed in any form'), wrote 'as a consequence of Denison's (1967) discovery of calcified cartilage in the Ordovician *Eriptychius*, such a suggestion can now, in fact, be made'. Halstead does not elaborate his suggestion further, there or in later papers where it is repeated (e.g., 1973: 327; 1974: 56), and its empirical content is obscure. For Denison (1967: 142, 145) described the calcified cartilage of *Eriptychius* as 'the internal skeleton' and as having 'dermal plates applied to its surface', and wrote of *Eriptychius* as 'the only known heterostracan with a calcified endoskeleton'.

Elasmobranchs. Holmgren (1940: 246; also quoted by Jarvik, 1959: 44) wrote of superficial mesenchyme in selachian embryos 'possessing the property of forming skeletal tissue, viz. placoid scales and the envelopes of sensory canals, some of which (in the anterior part of the skull) may chondrify, especially in rays'. This remark about dermal chondrification evidently refers to Holmgren's observations, earlier in the same paper, of 'sensory canal cartilages' beneath part of the infra-orbital sensory canal in 40–90 mm *Raja* (pp. 181, 194, 198, 202), beneath the rostral commissure in 63 and 71 mm *Torpedo* (p. 225) and beneath the internasal commissure in 60 and 90 mm *Urolophus* (p. 243).

In 90 mm *Urolophus*, part of the gutter-like sensory line cartilage grows over the sensory canal and encloses it in a tube. Holmgren assumed that this cartilage is induced by the sensory canal, like the latero-sensory component of dermal bones. In adult *Raja*, Allis (1916) gives a detailed account of this sensory line cartilage (which he calls the upper labial cartilage) and its relationship to the sensory canal. Allis gives reasons, inconclusive in my view, why the cartilage cannot be regarded as developed in direct relation to the canal. I have been unable to find any detailed account of the sensory line cartilage in adult *Torpedo* or *Urolophus*. Holmgren (and Jarvik, 1959) regarded these sensory line cartilages of batoids as evidence that the sensory canals of elasmobranchs have a general capacity to induce cartilage, in the same way as those of osteichthyans induce dermal bone. This theory has not been tested experimentally, but the very limited topographic and taxonomic distribution of elasmobranch 'sensory line cartilages' seems to me to oppose it. The theory is not accepted by Ørvig, for in his review of the latero-sensory component of the dermal skeleton (1972) he makes no mention of Holmgren's observations on batoids, and maintains his earlier view (1951: 355, 378) that cartilage is restricted to the endoskeleton.

Osteichthyans. In actinopterygians, the literature contains scattered references to bones which are normally dermal consisting partly or completely of cartilage. For example, Rosen (1962: fig. 16) describes and illustrates the subopercular of the amblyopsid *Typhlichthys* as cartilage; Harrisson (1966: 451) comments on cartilaginous opercular 'bones' in elopomorphs; McAllister (1968) characterizes the branchiostegal and opercular series as dermal bones or cartilages, and describes a cartilaginous subopercular in *Acipenser* and a cartilaginous opercular in *Polyodon;* Greenwood & Rosen (1971: fig. 22) illustrate a 'large cartilaginous appendage' on the lachrymal of *Bathylagus*. Most such observations are due to use of 'cartilage' or 'cartilaginous' to mean 'membranous' or 'poorly calcified' (D. E. McAllister, D. E. Rosen, pers. commns), but in the opercular flap, there is good evidence of the presence of cartilage in primitive osteichthyans.

Latimeria has a large opercular cartilage in the distal part of the operculum, behind and beneath the opercular bone (Millot & Anthony, 1958: 24; 1965: fig. 26). Dipnoans have cartilages on the inner surfaces of the opercular and subopercular bones (Goodrich, 1930: fig. 315; Holmgren & Stensiö, 1936: fig. 289). Larval *Polypterus* has a small opercular cartilage (Daget, Bauchot, Bauchot & Arnoult, 1964: fig. 19), and in the adult the articular facet of the opercular is cartilage-lined (Allis, 1922: 240). No opercular cartilages are described in living holosteans or chondrosteans. Amongst teleosts, there is a small opercular cartilage in larval *Anguilla* (Norman, 1926) and in leptocephali of other eels (Bauchot, 1959), whilst in saccopharyngoids the cartilage may be quite large (Orton, 1963). Harrisson (1966: 451) takes these opercular cartilages of anguilliforms to be the rudiment of the opercular bone, but Norman (1926: 398, fig. 32) shows that the opercular bone ossifies ventro-lateral to the cartilage, though in contact with it dorsally.

Reviewing the distribution of opercular cartilages and bones, one could imagine a sequence in which opercular cartilages are replaced by opercular bones: from the small and numerous hyoid ray cartilages of selachians, through the large opercular cartilages of chimaeroids and *Latimeria*, the latter with a single opercular bone, to the separate opercular and subopercular cartilages and bones of dipnoans, and to the small opercular cartilage and large opercular bones of *Polypterus* and some teleosts. While that sequence implies the phylogenetic replacement of cartilage by bone in the opercular flap, there can be no question of direct replacement (i.e. of homology between opercular cartilages and opercular bones), since the opercular bones lie external to the cartilages, and the cartilage is never ossified. The only possible exception to that statement is the opercular facet of *Polypterus* and teleosts. In *Polypterus*, Daget (1950: 117, fig. 37) describes and illustrates the joint between the opercular and the hyomandibular. This joint lies between the cartilage capping the opercular process of the hyomandibular and the opercular cartilage, which is closely applied to the opercular bone and lacks a perichondral membrane where it meets the bone. In the teleost *Heterotis*, Daget & d'Aubenton (1957: 918) found a similar situation, with a layer of cartilage lining the articular facet of the opercular bone, and I have observed the same thing in sections through young *Elops*. In those sections, as in the published illustrations of the cartilage in *Polypterus* and *Heterotis*, I can see no indication that any part of the cartilage is ossified, peri- or endochondrally, and it is not hypertrophied, but it lacks a perichondrium where it contacts the bone. In *Heterotis*, Daget & d'Aubenton regarded this cartilage as a detached portion of the opercular process of the hyomandibular cartilage, but it would require sophisticated techniques to differentiate that ontogenetic sequence from the late appearance of an independent cartilage, Norman's interpretation of the situation in *Anguilla*. In *Heterotis* Daget & d'Aubenton (1957: 899) describe the same sequence—detachment of a portion of cartilage which then becomes fused to a dermal bone—in the ethmoid region of the braincase, where a cartilage lines the articular surface of the premaxilla. This attachment of cartilage to dermal bones seems to characterize mobile joints between dermal bones and cartilage bones in teleosts. In higher vertebrates, adventitious cartilage plays the same role.

Higher vertebrates: adventitious cartilage

De Beer (1937: 2, 38, 502) summarized earlier work on the 'secondary' cartilage which appears in the development of certain dermal bones in mammals and birds. He held that secondary cartilage could be distinguished histologically from 'primary' cartilage, although it resembled the hypertrophic cartilage found in ossification sites in the endoskeleton. Since de Beer's time there has been much experimental work on secondary cartilage, *in vivo* and *in vitro* (see bibliography in Hall, 1970). Following Murray (1963), such cartilage is usually referred to as adventitious cartilage. It is now agreed that there is no valid histological or histochemical difference between adventitious cartilage and primary cartilage. Adventitious cartilage always appears later in ontogeny than, and not in connection with, the cartilaginous endoskeleton. It normally arises at the periphery of a developing dermal bone, and when ossified it always forms part of a bone which began ontogeny in dermal fashion.

Adventitious cartilage of this sort seems to be confined to mammals and birds, particularly to sites where there are mobile joints between dermal bones. Murray (1963) reviewed possible adventitious cartilages in reptiles, and concluded that all are parts of the primary (palatoquadrate) endoskeleton. He failed to find any adventitious cartilage in snake embryos, despite their highly kinetic skull. According to de Beer, adventitious cartilage is absent in monotremes, but in *Tachyglossus* Kuhn (1971: 159) found that there is usually a small nodule on the parotic crest of the squamosal.

In most mammals, adventitious cartilage appears as a regular feature in the development of dermal bones like the clavicle (Andersen, 1963), dentary, squamosal, pterygoid, frontal and parietal. It also occurs in the antlers of deer (Frasier, Banks & Newbrey, 1975; Newbrey & Banks, 1975). During ontogeny, this adventitious cartilage is replaced by peri- and endochondral bone, so that in mammals some of the dermal bones have more or less extensive peripheral parts which ossify as cartilage bone. In the dermal roofing bones of the mammalian skull, adventitious cartilage appears at the sutures between the bones (Moss, 1958), and when, as sometimes happens in man, Wormian ossicles develop in the sutures between the frontals or parietals, these ossicles may ossify as cartilage bones.

In birds, adventitious cartilage develops on the articular surfaces of the squamosal, pterygoid, quadratojugal, surangular and palatine (Murray, 1963), and in the sutures between the frontals and parietals (Hall, 1970). It ossifies in the same way as in mammals.

Experimental work shows that the production of adventitious cartilage on avian and mammalian dermal bones is influenced by movement (pressure and tension) and by oxygen and blood supply. It is clear that adventitious cartilage and dermal bone are alternative products of common stem scleroblasts, and that poor blood or oxygen supply and mobility favour chondrogenesis, whereas increased blood and oxygen supply and decreased mobility favour osteogenesis. By immobilizing the embryo, the production of adventitious cartilage may be prevented (Hall, 1970, 1972).

In the repair of fractured dermal bones, adventitious (callus) cartilage may also appear in mammals and birds (Hall & Jacobson, 1975). In birds, experiments suggest that the evocation of such cartilage is favoured by mechanical stress, as in the ontogeny of avian dermal bones. So far as I know, there have been no attempts to discover whether callus cartilage is produced in the repair of fractured dermal bones in amphibians and reptiles. In fishes, the only experiments are those of Moss (1962), working with cyprinid and cichlid teleosts. Moss found (p. 53) 'that the formation of cartilage-like callus occurred only when pre-existent cartilage elements [lower jaw] were injured. No cartilage was observed in the opercular [dermal bone] callus, only membranous ossification'.

Available evidence does not contradict the hypothesis that the production of adventitious cartilage in the development of dermal bones and in the repair of damaged dermal bones are aspects of the same phenomenon, and that this is restricted to endothermous tetrapods, the birds and mammals. This hypothesis could readily be tested, by an experimentalist willing to study the repair of fractured dermal bones in amphibians and reptiles.

To sum up this section, there is no good evidence of cartilage in the dermal skeleton of fishes, amphibians or reptiles. But in birds and mammals there is abundant evidence that the peripheral parts of some dermal bones may ossify peri- and endochondrally, and that the scleroblasts of the dermal skeleton are capable of differentiating into chondroblasts. These are derived features.

DERMAL BONES IN THE ENDOSKELETON

In this section, supposed examples of dermal bones 'sinking into' or 'invading' the endoskeleton will be reviewed. Since it is my thesis that most such examples are due to phylogenetic or ontogenetic fusion between a dermal bone and a cartilage bone, it is necessary first to discuss the meaning of phylogenetic fusion and how it may be recognized.

Phylogenetic fusion and loss

There are two ways of accounting for a reduction in the number of bones in a given part of the skeleton in a phyletic line. The first is to invoke phylogenetic loss, and to regard one bone in a derived member of the lineage as the homologue of one bone in a primitive member, one or more other bones present in the primitive form having been lost, and the remaining bone having 'invaded' or 'captured' their territory. The second is to invoke phylogenetic fusion, and to regard one bone in the derived form as the homologue of two or more bones in the primitive form, the ancestral ossification centres having fused. There are also two approaches to the problem of distinguishing phylogenetic fusion from phylogenetic loss. The first, advocated by Jardine (1970), is to apply tests requiring knowledge of onto-genies and phyletic sequences (see also Jarvik 1972: 147). The second, advocated by Ørvig (1962: 59) and Nelson (1973b: 339), is to treat the supposed difference between phylogenetic loss and fusion as merely semantic. On this view, loss and fusion are alternative ways of verbalizing or visualizing one phenomenon. As Nelson (1969b: 10) wrote 'whatever the details of the phylogenetic process, which are not demonstrable, the over-all result is called fusion here, for want of a better word'.

There are three possible classes of bone fusion: fusion between dermal bones, within the dermal skeleton; fusion between cartilage bones, within the endo-skeleton; and fusion between a dermal bone and a cartilage bone. And in each case it is necessary to distinguish carefully between ontogenetic fusion and phylo-genetic fusion. Ontogenetic fusions of all three classes seem to be demonstrated beyond doubt in many vertebrates (see, for example, Warwick & Williams, 1973, on the development of the human skull, and Lekander, 1949, on the skull of cyprinoid teleosts). Clearly, two bones can only fuse during ontogeny if their growth centres become coincident (or if growth ceases, as in the human cranial vault), while two bones can only fuse during phylogeny if their ossification centres become coincident. Phylogenetic fusion between two dermal bones or two cartilage bones requires that in successive ontogenies the two ossification centres should move progressively closer to each other until they are coincident, so that only one bone arises in ontogeny: at least, I can imagine no other way in which the process might come about. This sequence of events seems unlikely to occur often, and I know of no thoroughly documented examples of it. In the dermal skeleton, it is common practice to regard certain features such as sensory canals, dentition or topographic relations as 'essential' attributes of particular bones, and when essential attributes of two bones are found together in one, and there is no evidence of ontogenetic fusion, to postulate phylogenetic fusion 'for want of a better word'. Perhaps the best candidates for phylogenetic fusion in the dermal skeleton are median bones which are interpreted as being derived, in phylogeny, from paired bones, like the median vomer and median tooth-plates on the basibranchials of teleosts (Nelson, 1969a). In the endocranium, phylogenetic fusion seems to me to be a meaningless concept (Patterson, 1975: 421). But where phylogenetic fusion between a dermal bone and a cartilage bone is suggested, it is much easier to imagine the two ossification centres becoming coincident, for the dermal centre lies superficial to the endoskeletal centre, and the two can become directly superimposed and then develop from a single ossification or growth centre.

This idea, of dermal and cartilage bone ossification centres becoming directly superimposed in phylogeny, obviously carries the implication that the dermal and endoskeletal centres are 'mobile' relative to one another. Clear evidence of this mobility comes from the back of the lower jaw in teleosts (Fig. 2; Nelson, 1973b). Primitively, the posterior end of the teleostean lower jaw contains three bones, the dermal (latero-sensory) angular and two cartilage bones, the articular and retroarticular, which lie respectively in front of and behind the facet for the quadrate articulation. Although this pattern is only found in two living teleosts, the osteoglossoids *Arapaima* and *Heterotis*, its primitive status is attested by its occurrence in various Mesozoic teleosts (Nelson, 1973a; Patterson & Rosen, 1977), and in *Amia, Lepisosteus* and *Latimeria*. Apart from *Arapaima* and *Heterotis*, all living teleosts have the angular co-ossified with one or other of the two cartilage bones, and these co-ossifications characterize major taxa. The angular/articular pattern is found in all euteleosteans and clupeomorphs, the angular/retroarticular in all elopomorphs. If these co-ossifications are interpreted as fusions, they prove that in phylogeny the dermal and endoskeletal ossification centres are mobile relative to each other: in elopomorphs either the angular ossification centre is further back than it is in euteleosts, or the retroarticular centre is further forwards.

But the back of the lower jaw does not provide fully satisfactory evidence that the different patterns are the result of fusion rather than loss. Haines (1937), who is followed by Lekander (1949), interpreted the euteleostean pattern as the result of loss of the articular, and invasion of the endoskeleton by the angular.

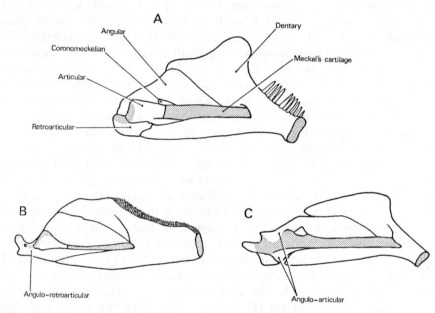

Figure 2. Left lower jaw, in medial view, of A, an osteoglossomorph teleost, *Heterotis niloticus;* B, an elopomorph, *Megalops atlantica;* and C, a euteleostean, *Coregonus* sp. After Nelson (1973b: figs 2F, 3M, 7B).

Haines's argument is inconsistent. It rests on the interpretation of living holosteans, elopiforms and euteleosts as a phylogenetic sequence, in which Haines saw a reduction in the size of the articular in elopiforms, preparatory to its loss in euteleosts. Yet Haines investigated *Elops*, and regarded the continuity between the angular and retroarticular in that fish as fusion, not invasion of the cartilage by the dermal bone, which his theory demands. In fact, the back of the lower jaw is divergently specialized in elopiforms and euteleosts, as Nelson has shown. And if the articular is reduced in elopiforms, it foreshadows loss of that bone in higher elopomorphs, the eels and their relatives, which have only one bone, the angulo-retroarticular, at the back of the lower jaw.

Nevertheless, the dermal/cartilage bone fusions at the back of the lower jaw which I postulate are not fully documented, since no living teleost is known to show distinct dermal and perichondral ossification centres in early ontogeny. In every species that has been investigated, the supposedly compound bone ossifies from a single centre, so that if fusion is involved, it is phylogenetic fusion, not ontogenetic. In my opinion, three sorts of evidence might be required before phylogenetic fusion between a dermal bone and a cartilage bone is postulated. First, evidence that the primitive condition in the group is to have a distinct dermal bone and a cartilage bone at that location. Second, evidence that those bones fuse in late ontogeny in some, presumably primitive, members of the group. Third, evidence that the bones fuse in early ontogeny in some members of the group. A fourth type of evidence might be the occurrence of two separate bones in some species or higher taxon whose relationships were such that the condition could only be interpreted parsimoniously as the secondary reappearance of the original distinct ossifications. In the back of the lower jaw of teleosts, evidence of the first sort is available; evidence of the second sort is only provided by the Mesozoic leptolepids (Patterson & Rosen, 1977: fig. 32); and evidence of the third and fourth sorts is lacking.

In general, the hypothesis of loss and invasion is very difficult to refute, for it has few, if any, testable consequences. For example, in the gill arches the primitive teleostean and actinopterygian condition is to have cartilage bones overlain by independent dermal tooth-plates. Many teleostean taxa are characterized by having a tooth-plate co-ossified with one or more of the supporting cartilage bones (e.g., Nelson, 1969a; Rosen, 1973). These co-ossifications are always interpreted as fusions, although in few cases is the ontogeny described. It would be possible to interpret all these 'fusions' as 'losses', either the cartilage bone or the dermal bone having been lost, and the remaining ossification having invaded the dermal skeleton or the gill-arch cartilage. Such a proposal is almost irrefutable, but there is one test which should discriminate between it and the fusion hypothesis. This test depends on the fact that the fusion hypothesis demands that the ossification centres of the dermal bone and cartilage bone should have become coincident or superimposed, while the loss hypothesis has no such requirement. If we imagine the dermal and endoskeletal ossification centres as mobile relative to one another (as exemplified at the back of the teleostean lower jaw), then coincidence of the centres, the prerequisite of fusion, is more probable for median bones than for paired ones. This is because median ossification centres are only free to wander in one dimension, the sagittal plane, while paired centres can wander in the two dimensions which define the envelope of the interface between the dermal skeleton and the endoskeleton. Median dermal and endoskeletal ossification centres, free to move in only one dimension, are more likely to become sufficiently close to each other for ontogenetic fusion (coincidence of growth centres) and phylogenetic fusion (coincidence of ossification centres) than are paired ossifications, free to move in two dimensions.

In the teleost skull, including the hyoid arch but excluding the gill arches, I count eight median and 21 paired cartilage bones. Five of the eight median endoskeletal bones are co-ossified with a dermal bone in more or less extensive teleostean groups, and seven of the 21 paired bones are co-ossified with a dermal bone.* This gives a 'co-ossification rate' of 62·5 % for the median bones and 33 % for paired bones. Including the gill arches increases the number of median cartilage bones by three (basibranchials), all of which exhibit co-ossification with tooth-plates in one or more teleostean groups, and increases the number of paired cartilage bones by 15, nine of which exhibit co-ossification (three hypobranchials, fifth ceratobranchial, first three epibranchials, second and third pharyngo-branchials: see Nelson 1969a for distribution of these co-ossifications). The total skull 'co-ossification rate' is then 73 % for median bones (8/11) and 44 % for paired bones (16/36). This difference, predicted by the fusion hypothesis but not by the loss hypothesis, certainly favours fusion. But since the difference is only significant at about the 0·85 level, these figures do not refute the loss hypothesis decisively.

Where primitively separate dermal and endoskeletal bones are involved, it seems that a hypothesis of loss and invasion is irrefutable, for it makes no predictions that might be contradicted. A hypothesis of phylogenetic fusion, when supported by evidence of ontogenetic fusion in some forms, is therefore preferable.

Supposed dermal bones in the endoskeleton of fishes

Neurocranium. De Beer (1937) found that the only convincing evidence of invasion of the chondrocranium by dermal bones came from the braincase of teleosts, in particular the dermal and cartilage bone on the dorsal and ventral surfaces of the ethmoid region, the lateral ethmoid, the sphenotic, pterotic and supraoccipital. But there is now good evidence that above and below the ethmoid and in the sphenotic and pterotic the primitive teleost condition is to have independent dermal and cartilage bones, and that these bones fuse ontogenetically or phylogenetically in some teleosts. The lateral ethmoid seems always to have been a cartilage bone, while the supraoccipital fuses with a latero-sensory extrascapular or with the parietals in some teleosts. Figure 3 shows these fusion sites in the teleost braincase.

In the ethmoid region, where invasion of cartilage by dermal bone was stressed by Starks (1926), and most recently by Jollie (1975), the dermal bones are the rostro-dermethmoid above and the vomer below, and the cartilage bones are the supraethmoid above and the ventral ethmoid below (see Taverne, 1974: 17 and Patterson, 1975: 514, for the distribution of the fused and unfused conditions). In both these sites there is clear evidence of originally independent bones, and there are fossil and Recent teleosts showing the unfused condition and fusion at various stages in ontogeny.

The teleostean lateral ethmoid is often taken to be a compound bone, resulting from fusion between the endoskeletal lateral ethmoid (also called parethmoid) and a dermal prefrontal. This point of view has been expressed and discussed

---

*The five median co-ossifications are (cartilage bones listed first): supraethmoid and rostro-dermethmoid, ventral ethmoid and vomer, supraoccipital and medial extrascapular and/or parietal, basihyal and its tooth-plate, urohyal and interclavicle. The seven paired co-ossifications are: sphenotic and dermosphenotic, pterotic and dermopterotic, autopalatine and dermopalatine, quadrate and quadratojugal, retroarticular and angular, articular and angular, mentomeckelian and dentary. These are all discussed in the next subsection. Two possible additional paired co-ossifications are the lateral ethmoid and antorbital in hiodontids (see p. 97) and the metapterygoid and dermometapterygoid in *Heterotis* (Daget & d'Aubenton, 1957), but in neither case is the situation well-established. Another possible median fusion is between the basisphenoid and parasphenoid of some ostariophysans (Holmgren & Stensiö, 1936: 489; Bamford, 1948: 375), but here the evidence is poor.

most recently by Taverne (1974: 21). In my opinion, there is no evidence of a dermal prefrontal in any teleost, either as an independent bone or as a component of the lateral ethmoid (Patterson, 1975: 496), while the endoskeletal lateral ethmoid appears to be a primitive actinopterygian feature. The two portions of the lateral ethmoid of *Elops* which Taverne describes as the parethmoid and prefrontal are respectively the endochondral and perichondral parts of the bone. The only fishes in which there may be fusion between the lateral ethmoid and a dermal bone are the hiodontids *Hiodon* (Nelson, 1969b: 6) and the Mesozoic *Lycoptera* (Greenwood, 1970: 27), where what appears to be the topographic homologue of the antorbital is part of the lateral ethmoid. The ontogeny is unknown, but if confirmed this fusion would be additional evidence that a prefrontal is not incorporated in the teleostean lateral ethmoid, for I find it difficult to imagine how two dermal bones could successively fuse with one cartilage bone in phylogeny.

The sphenotic (cartilage bone) and dermosphenotic are distinct and separate in the majority of teleosts, and in holosteans. The sphenotic is an ossification in the

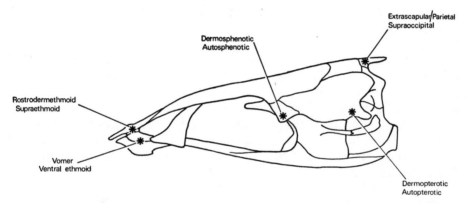

Figure 3. Diagram of a teleostean braincase in left lateral view (based on *Elops*) to show sites of fusion between dermal bones (named first) and cartilage bones in teleosts.

postorbital process of the braincase, and appears to be a primitive actinopterygian feature (Patterson, 1975: 472). It is fused with the dermosphenotic in *Polypterus* (Pehrson, 1947; the dermal bone appears first in ontogeny, but whether the cartilage bone arises in continuity with it, implying phylogenetic fusion, or arises independently, implying ontogenetic fusion, is not clear from Pehrson's account). Amongst teleosts, the dermosphenotic and sphenotic are fused in siluroids (phylogenetic fusion where the ontogeny is known—Kindred, 1919; Bamford, 1948; Kapoor, 1970), some scorpaenoids (Allis, 1909, ontogeny unknown), many beryciforms (ontogeny unknown) and *Channa* (Kapoor, 1970, phylogenetic fusion).

The pterotic (cartilage bone) and dermopterotic (or supratemporo-intertemporal) are fused in almost all teleosts, and this is one of the locations where de Beer (1937) saw indications of invasion of cartilage by dermal bone. Jollie (1975: 70) writes of the endoskeletal pterotic as 'obviously only a chondral associated outgrowth' of the dermal bone. The two bones are fully distinct in the Mesozoic pholidophorids and other primitive fossil actinopterygians (Patterson, 1975), where the bone which I interpret as the pterotic ossifies in the postero-dorsal shoulder of the otic capsule, not over the external semicircular canal as it does in leptolepids and more advanced teleosts. Amongst living teleosts, the two bones are said to remain distinct in the poorly ossified alepocephaloids (Gosline, 1969: 197; cf. Greenwood & Rosen, 1971: figs 23, 24) and in the osteoglossoid *Pantodon* (Kershaw, 1970). Fusion between the dermal and cartilage components is ontogenetic in *Salmo* (de Beer, 1937), some cyprinoids (Lekander, 1949) and some siluroids (Bamford, 1948); and is phylogenetic (fusion *ab initio*) in *Esox* (Jollie, 1975), the cyprinoid *Leuciscus* (Lekander, 1949) and the siluroid *Ictalurus* (Kindred, 1919). These facts seem to me to be decisive evidence that this case is not invasion of the endoskeleton by the dermal bone but fusion, at first ontogenetic and later phylogenetic, following from approximation of the ossification centres of the two bones, a consequence of changed ossification patterns after closure of the cranial fissure (Patterson, 1975: 422).

The supraoccipital of teleosts is often regarded as a compound bone, the result of phylogenetic fusion between a cartilage bone and a 'dermosupraoccipital' (e.g., Harrington, 1955: 274; Daget, 1965: 255), or as an originally dermal bone which has invaded the endoskeleton (Jollie, 1975: 78). The supraoccipital is a normal cartilage bone in the Mesozoic pholidophorids and leptolepids, and is certainly a primitive teleostean feature, and possibly a primitive actinopterygian feature (Patterson, 1975). There is nothing to support Jollie's (1975) suggestion that this endoskeletal bone represents 'the membrane part of the medial extrascapulars'. The extensive membrane bone supraoccipital crest of many teleosts, which Harrington interpreted as the dermosupraoccipital, develops as an outgrowth from the cartilage bone, never as an independent ossification. However, in certain teleostean groups there is evidence of fusion between the supraoccipital and a dermal bone. In living clupeomorphs the supraoccipital separates the parietals and lies beneath the central part of the supratemporal commissural sensory canal. In the few living herrings where the ontogeny of these bones has been studied, there is no indication of fusion between the supraoccipital and a dermal bone (Bamford, 1941), but here the medial part of the sensory canal is reduced and not bone-enclosed. In the Cretaceous clupeomorphs *Spratticeps* (Patterson, 1970) and *Ornategulum* (Forey, 1973) the exposed dorsal part of the supraoccipital enclosed the sensory canal, in a transverse tube which passes throughout the width of the bone. This must mean that the supraoccipital of primitive clupeomorphs fused, presumably during ontogeny, with a median extrascapular, or with the latero-sensory component of that dermal bone.

The supraoccipital also encloses the middle part of the supratemporal commissure in adult *Phractolaemus* (Daget, 1965: 272) and kneriids (Lenglet, 1974), while in *Mastacembelus* there is a plate-like median extrascapular which overlies the supraoccipital and may fuse with it in full-grown individuals (Taverne, 1973).

In siluroids, there are no parietals and the supraoccipital forms a major part of the skull roof, and carries the anterior and middle pit-lines, which are normally borne by the parietals. In the early development of *Ictalurus*, Kindred (1919) and Jollie (1975) write of fusion between the supraoccipital and paired parietals, but they do not seem to have observed separate ossification centres. In *Galeichthys*

Bamford (1948) found two pairs of dermal bone rudiments which soon fuse into a median plate that has a cartilage bone component, but again he saw no separate cartilage bone primordium. This fusion between the supraoccipital and the parietals in catfishes seems to be phylogenetic, but in occasional individuals one parietal remains as an independent bone (Alexander, 1965: fig. 13).

Although the embryological evidence is inconclusive, the supraoccipital might provide another example of relative mobility of dermal and endoskeletal ossification centres, like the back of the lower jaw (see above). If the supraoccipital fuses with the parietals in catfishes and with the medial extrascapular in herrings, kneriids and mastacembelids, the supraoccipital centre must have moved relative to the dermal bones in one group or the other. But the possibility remains (despite arguments to the contrary, Bamford, 1948: 371) that the parietal of siluroids is a parieto-extrascapular, like that of cyprinoids (Devillers, 1948; Lekander, 1949).

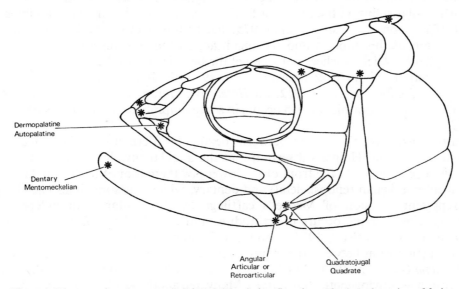

Figure 4. Diagram of a teleostean skull in left lateral view (based on *Elops*) to show sites of fusion between dermal bones (named first) and cartilage bones in the splanchnocranium (braincase fusions indicated by asterisks are labelled in Fig. 3).

In the teleost braincase, various cartilage bones are supposed to have originated by dermal bone invading the endoskeleton. But in each case there is good evidence that the cartilage bone in question is a primitive actinopterygian or teleostean feature, and there are fishes in which the cartilage bone co-exists with and is independent of the dermal bone that is supposed to have given rise to it. Where fusions between dermal bones and cartilage bones are postulated (Fig. 3) there is evidence of ontogenetic fusion in each case except the sphenotic and supraoccipital.

Splanchnocranium. In the visceral skeleton, there are few cases where a dermal bone has been held to invade the endoskeleton, but there are several examples of fusion between dermal bones and cartilage bones (Fig. 4) which serve to corroborate the theory that this is the dominant phenomenon in fishes.

In the palate, there are two examples of fusion, between the autopalatine and dermopalatine (tooth-plate), and between the quadrate and quadratojugal. Separate auto- and dermopalatines are found in *Amia*, palaeoniscoids and other primitive fossil actinopterygians. Amongst teleosts, the bones remain distinct in *Alepocephalus* (Gosline, 1969) and in elopoids and *Albula*, where the autopalatine ossifies late in ontogeny. The autopalatine may fuse with the tooth-plate in large individuals of *Elops* (Taverne, 1974: 38). In *Salmo* fusion between the two bones is ontogenetic (de Beer, 1937), whereas in *Esox* the bones are fused *ab initio* (Jollie, 1975).

The quadrate (cartilage bone) and quadratojugal (dermal) are both primitive actinopterygian (and probably osteichthyan) features. The two bones are independent in palaeoniscoids and other primitive fossil actinopterygians, but in *Lepisosteus* and the Mesozoic semionotids the quadratojugal takes the form of a splint-like prop on the posterior face of the quadrate (Patterson, 1973). Teleosts have a similar splint on the posterior face of the quadrate, but here it is an integral part of the bone. In *Esox*, Jollie (1975: fig. 8) found evidence for independent ossification of this process and the quadrate, so that the fusion is ontogenetic, but in all other teleosts where ontogeny of the bone is known the fusion seems to be phylogenetic (Holmgren & Stensiö, 1936: 497, and Jollie, 1975: 75, on *Salmo;* Daget & d'Aubenton, 1957: 912, on *Heterotis*). Ontogenetic fusion between the quadrate and quadratojugal also occurs in some amphibians and reptiles (de Beer, 1937).

In the lower jaw, the fusions between the angular (dermal) and the articular or retroarticular, and Haines's assertion that the angular invades the endoskeleton, are discussed above. At the front of the lower jaw there is another fusion, between the mentomeckelian (endoskeletal) and dentary, which has sometimes been taken to represent invasion of Meckel's cartilage by the dentary. An independent mentomeckelian bone occurs in arthrodires and acanthodians, *Latimeria*, *Polypterus* and *Amia* (Nelson, 1973b), and also in some amphibians (de Beer, 1937). Ontogenetic fusion between the mentomeckelian and dentary takes place in *Esox* and *Salmo* (Pehrson, 1944), but in other teleosts where the ontogeny is described the fusion is phylogenetic (*ab initio*). Many tetrapods exhibit the same fusion, which may be ontogenetic (especially in amphibians) or phylogenetic (see Kuhn, 1971: 66, on the situation in mammals).

The urohyal, listed on p. 96 as an instance of fusion between a dermal bone and a cartilage bone, is discussed in the next subsection.

In the gill-arch skeleton, Nelson (1969a: 519) cites two examples 'which seem to involve the transformation of dermal elements into endoskeletal ones'. Both these examples are in ostariophysan teleosts. One concerns the fourth upper pharyngeal tooth-plate, which overlies the cartilaginous fourth pharyngobranchial. In some characins this dermal bone has lost its teeth and partially surrounds the cartilage, though it lies outside the perichondrium. Some clupeoids show a similar modification. In most siluroids the single upper pharyngeal tooth-plate is supported by a perichondrally ossified cartilage which seems to be the fourth pharyngobranchial. The gonorynchiform families Kneriidae and Phractolaemidae also have an ossified fourth pharyngobranchial. Since the teleostean fourth pharyngobranchial is primitively unossified, Nelson interpreted the siluroid and gonorynchiform condition as invasion of the endoskeleton by the reduced dermal tooth-plate. Nelson's hypothesis was that primitive osteichthyans possessed an ossified fourth pharyngobranchial, and that the ossification centre had been lost early in actinopterygian history, so that its reappearance in relatively advanced teleosts would be an unacceptable reversal of evolution.

However, it now appears that the fourth pharyngobranchial is a derived structure, found only in teleosts (Patterson, 1973: 251). Ossification of that cartilage can therefore also be regarded as a derived condition, characterizing two teleostean subgroups, catfishes and some gonorynchiforms. It is not necessary to invoke invasion of the endoskeleton by dermal bone in this case.

Nelson's second example concerns ossified fourth and fifth basibranchials. The primitive teleostean condition in the basibranchials is apparently to have three basibranchial ossifications in a single basibranchial cartilage (Nelson, 1969a: 508), perhaps with a single, unossified cartilage behind this, occupying the position of fourth and fifth basibranchials. In some cyprinoids (Nelson, 1969a: 519) and in kneriid gonorynchiforms (Lenglet, 1974) there are ossified fourth and fifth basibranchials, and these basibranchials are ossified endo- as well as perichondrally. Nelson suggested that these ossifications had arisen by transformation 'from a tooth-plate or plates overlying the fourth or fifth basibranchials'. In this instance there are no known examples of teleosts with toothless or otherwise reduced tooth-plates in these positions, which might be taken as antecedent to the cyprinoid/kneriid condition. Since Nelson's general conclusion about basibranchials (1969a: 513) is that 'all aspects of segmentation are secondary, including multiplication of ossification centres', it does not seem necessary or advisable to postulate invasion of the endoskeleton by dermal bones to explain the structures found in cyprinoids and kneriids.

Supposed dermal bones in the endoskeleton of tetrapods

In tetrapods, I have only come across two supposed examples of dermal bones invading the endoskeleton (this is perhaps only a symptom of my ignorance of the literature in this field), both concerning fossil forms. These examples are the tabular of some labyrinthodont amphibians (Romer, 1947: 52) and the pelycosaur *Ophiacodon* (Romer & Price, 1940: 203—'we have here one of the few cases definitely known in tetrapods of a mixed bone'); and the cleithrum of rhizodontid rhipidistians, which is held to have given rise to the tetrapod scapular blade by Andrews & Westoll (1970b: 464).

Romer's theory about the tabular of labyrinthodonts was perhaps influenced by his earlier observations of that bone in the pelycosaur *Ophiacodon* (see below), for the theory is not documented in any detail in his monograph, and an endochondral component of the tabular, in the paroccipital process, is postulated only in *Seymouria*, and is implied in *Benthosuchus*. I am not qualified to criticize Romer's interpretation of these skulls, but Dr Alec Panchen has kindly permitted me to quote his comment (*in litt.*, March 1976) 'I very much doubt whether any labyrinthodont tabular had a cartilage bone component'. Panchen points out that Watson's original interpretation of the labyrinthodont paroccipital process was that the tabular sheathed the cartilage and cartilage bone of the process but did not ossify in it, and that this interpretation is confirmed in well-preserved capito-saurs by Howie (1970). Romer made no attempt to criticize Watson's alternative interpretation, and until evidence of an endochondral component in a labyrin-thodont tabular is presented in more detail, there seems nothing to discuss.

In the pelycosaur *Ophiacodon*, Romer & Price (1940) document their opinion about the tabular in some detail, for they observed isolated examples of the bone and sections through it, and describe it as having a stout conical structure of endochondral bone extending inwards from its centre of ossification, and contact-ing the supraoccipital, prootic and opisthotic. *Ophiacodon* is the only pelycosaur in which sutures are found in the otic and occipital regions of the braincase (Romer & Price, 1940: 64), so that the ossification pattern in other forms is unknown. It is important to note that the cartilage-bone component of the tabular in *Ophiacodon* lies dorsal to the post-temporal fossa and contacts the supra-occipital, so that it is in quite a different position from the supposed extension of the tabular in labyrinthodonts, which is in the floor of the post-temporal fossa and contacts the exoccipital or opisthotic. Again, I am hardly qualified to criticize Romer and Price's interpretation of *Ophiacodon*, that the tabular has invaded the endoskeleton, but my experience with fishes leads me to expect that this is another instance of fusion between a dermal bone and a cartilage bone.

The only living tetrapod with a tabular of known ontogeny is the insectivore *Eremitalpa* (a tabular is also reported in the insectivore *Microgale* and the seal *Phoca;* de Beer, 1937: 444). Broom (1916) described the tabular of *Eremitalpa* as a dermal bone, and remarked on its similarity to the tabular of therapsids. But Roux (1947: 313) found that the bone first ossifies perichondrally, in the antero-lateral part of the supraoccipital cartilage, and then 'expands forwards in membrane bone fashion' to cover the metotic fissure. This mode of ossification corroborates Romer's opinion about the dual nature of the pelycosaur tabular, but does not help in deciding whether this is an example of invasion of the endo-skeleton, or of phylogenetic fusion. If the latter has occurred, the position of the ossification centre in *Eremitalpa* implies that the cartilage bone concerned is an occipital bone, occupying the position of the epioccipital of primitive actinop-terygians (Patterson, 1975: fig. 118). In therapsids and fossil tabular-bearing mammals such as the multituberculates (Kielan-Jaworowska, 1971), the sutures seem never to be sufficiently clear to determine whether the tabular has a cartilage-bone component. But the similarity between the tabulars of multituberculates and *Eremitalpa* certainly supports Broom's identification of the bone in the latter, and does not favour Starck's view (1967: 484) that it is a neoformation, a 'fragment of supraoccipital'.

Assuming that the tabular of pelycosaurs and mammals has a cartilage-bone component, we can turn back to rhipidistians and primitive amphibians to try to determine whether fusion with a cartilage bone is likely. As pointed out above, in labyrinthodonts the ossification centre of the tabular lies at the tip of the paroccipital process, ventral to the post-temporal fossa and in contact with the opisthotic, ventro-lateral to its position in pelycosaurs. In the rhipidistian *Eusthen-opteron* the homologous dermal bone, the supratemporal, has two low descending laminae on its inner surface, the larger one passing beneath the ossification centre, and which are applied to the upper and lower surfaces of the parotic crest

(Jarvik, 1954: figs 5B, 21C). In rhipidistians, this crest floors the large post-temporal fossa (fossa bridgei of Jarvik; supraotic fossa of Romer, 1937) and contains the external semicircular canal: by analogy with primitive actinopterygians and tetrapods, one would expect this crest to ossify as part of the opisthotic. It is therefore clear that the ossification centre of the tabular has shifted dorso-medially in pelycosaurs, in comparison with its position in rhipidistians and most labyrinthodonts, and as discussed above (p. 93), a shift in the position of an ossification centre is usually a prerequisite for fusion between a dermal bone and a cartilage bone. As for the cartilage bone which might be involved, we are still completely ignorant of ossification patterns in the otic and occipital regions of the rhipidistian braincase, but the geometry of the ossified braincase demands that the posterodorsal shoulder of the otic capsule in *Eusthenopteron* is an ossification centre. Whether that centre is an otic bone, comparable with the pterotic of primitive actinopterygians, or an occipital bone, like the epioccipital of actinopterygians, depends on an interpretation of the cranial fissure in rhipidistians (cf. Patterson, 1975: fig. 118). It is pointless to speculate further on this bone, and on whether or not it was present in primitive tetrapods, but the shift in the ossification centre of the tabular towards the postero-dorsal shoulder of the braincase in pelycosaurs, and the probability that there was primitively an endoskeletal ossification centre at that site show that Romer and Price's interpretation of the pelycosaur tabular as an invasion of the endoskeleton is less certain than they thought.

The second supposed example of a dermal bone invading the tetrapod endoskeleton concerns the cleithrum and scapula. In cleithra of the rhipidistians *Rhizodus* and *Strepsodus*, Andrews & Westoll (1970b: 435) described a depressed lamina: an internal flange on the posterior margin of the dorsal part of the cleithrum. They interpreted this lamina as the site of origin of the latissimus dorsi and deltoid muscles, which in tetrapods originate on the blade or dorsal portion of the endoskeletal scapula. Citing the theory of delamination (Holmgren, 1940; Jarvik, 1959), Andrews & Westoll postulate that the tetrapod scapular blade arose in phylogeny from the depressed lamina of the cleithrum, perhaps after the upper parts of the cleithrum and scapulocoracoid had 'merged into one developmental field' (Andrews & Westoll, 1970a: 229). The histological structure of the dorsal part of the cleithrum of *Rhizodus* (mostly spongy bone) is adduced in support of this theory, but it is not clear why this is relevant. The idea expressed here is complex, and is not discussed further by Andrews & Westoll. Since a dermal cleithrum exists in primitive amphibians and reptiles together with a scapular blade, their thesis seems to be that mesenchyme which ossified as a part of the cleithrum in rhizodonts and other rhipidistian fishes changed its allegiance and potential in tetrapods, giving rise instead to a part of the scapular cartilage. I see no way of testing or criticizing this theory, except to say that if our mental picture of the phylogenetic process allows changes of this sort, then it is hard to see how or why the relation of homology has any meaning.

BONES OF VARIABLE HISTOGENESIS

De Beer (1937: 496) cited one example of bones 'which may vary in their method of histogenesis', the sclerotic bones. These will be discussed here, together with one other example, the urohyal/parahyoid.

The sclerotic bones ossify perichondrally, on the sclerotic cartilage, in teleosts, and as dermal bones, outside the cartilage, in sturgeons and tetrapods. If these bones are homologous, as Edinger (1929) and most later authors have supposed, then their histogenesis must vary. The answer here, as in other cases of supposed interchangeability between the endoskeleton and dermal skeleton, is that there were originally two sets of sclerotic ossifications (at least in the teleost lineage), an outer dermal ring and an inner ring or incomplete cup of cartilage bone. These separate dermal and endoskeletal sclerotic rings are present in the Mesozoic *Pholidophorus* (Patterson, 1975: 355) and in the Triassic *Australosomus* (Nielsen, 1949: 99, fig. 1). In *Pholidophorus* the four dermal and perichondral segments of the sclerotic ring agree in size and disposition, but there is also an endoskeletal basal sclerotic bone surrounding the entrance of the optic nerves and vessels into the eyeball. In *Australosomus* the perichondral ossifications are more extensive than the dermal. Separate endoskeletal and dermal sclerotics are also found in dolichothoracid, brachythoracid and rhenanid arthrodires (Stensiö, 1969: 183; R. S. Miles & E. I. White, pers. commns), suggesting that the inner and outer sets of ossifications may be a primitive gnathostome feature. Whether or not that is so, the dermal sclerotics of tetrapods and sturgeons and the endoskeletal sclerotics of teleosts are not examples of variable histogenesis, but of differential loss, the dermal sclerotic having been lost in teleosts.

The urohyal is a median bone lying between the sternohyoid muscles, beneath the basibranchials, and joined anteriorly by ligaments to the hypohyals. Amongst living fishes, there is a urohyal in teleosts (Kusaka, 1974), *Polypterus* and *Latimeria*. There is also a urohyal in Devonian coelacanths, porolepiform and osteolepiform rhipidistians (Jarvik, 1963: figs 16, 17) and lungfishes (Miles, in press). The urohyal of coelacanths, rhipidistians and lungfishes is endoskeletal, ossifying in cartilage bone. That of *Polypterus* and teleosts ossifies in membrane, and is usually (e.g., Ridewood, 1904: 76; de Beer, 1937: 416) referred to as an ossified tendon.

In amphibians there is an endoskeletal urohyal ('second basibranchial') cartilage in urodeles (Jarvik, 1963: fig. 16) and anurans (Jarvik, 1963: fig. 21), which sometimes ossifies (de Beer, 1937: 186, 188, 190, 210), and in several anurans there is a 'V'-shaped dermal bone, the parahyoid, on the underside of the hyoid plate (de Beer, 1937: 212). The distribution of the anuran parahyoid suggests that it is primitive for the group (Trueb, 1973). De Beer regarded the urohyal of *Polypterus* and teleosts as homologous with the anuran parahyoid, and recommended that it be given the latter name. Jarvik (1963) called the fish bone and urodele cartilage urohyal, and gave the same name to both the dermal parahyoid and the endoskeletal 'second basibranchial' or *Copulastiele* of anurans, which he regarded as subdivided remnants of the osteolepiform urohyal. Jarvik points out that on this interpretation the anuran parahyoid must be endoskeletal, not dermal. The possibility that traces of the urohyal persist in amniotes is discussed by Jarvik (1963: 53).

Under the name urohyal, therefore, Jarvik includes dermal (or membrane) bones, cartilages and cartilage bones, again implying variable histogenesis and some sort of interchangeability between the dermal skeleton and the endoskeleton. Light is thrown on this question by the urohyal of the Mesozoic pholidophorid fishes, and it is necessary to describe that bone briefly.

The pholidophorid urohyal is known more or less completely in *Pholidophorus germanicus*, *P. macrocephalus* and an unnamed Callovian species (cf. Patterson 1975: 284). It is similar in all, and that of *P. germanicus* is described and illustrated (Fig. 5) here. The bone consists of a median vertical plate and a horizontal, ventral

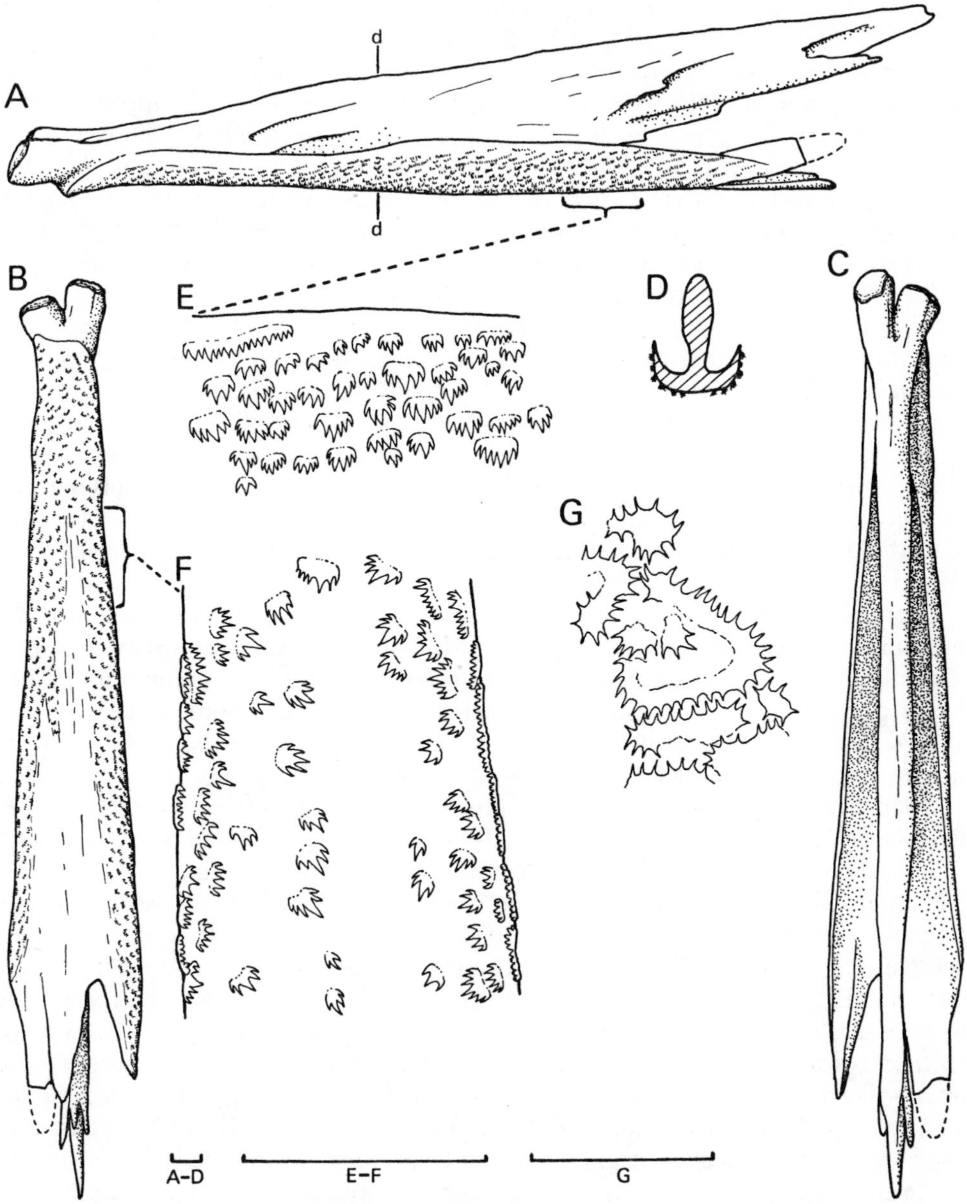

Figure 5. A–F. Urohyal of *Pholidophorus germanicus* Quenstedt, from BMNH P.3704, Upper Lias, Ilminster, Somerset: A, left lateral view; B, ventral view; C, dorsal view; D, optical section along the line d–d in A; E, F, enlarged from the areas indicated in A and B, to show the denticles. G. Denticles from the ventral surface of the urohyal in *Pholidophorus macrocephalus* Agassiz, from BMNH P.52518b, Kimmeridgian (Lithographic Stone), Bavaria. Scales at foot are 1 mm.

plate, so that it has the form of an inverted 'T' in section, like the urohyal of various primitive living teleosts (Fig. 7B; the 'ventral spread type' of Kusaka, 1974). Anteriorly the bone ends in a pair of diverging processes which are clearly the areas of insertion of the ligaments attaching the bone to the hypohyals, again as in living teleosts. The vertical plate of bone is low and broad anteriorly, but becomes taller, thinner and membranous posteriorly, where it ends in a median splint-like process with a pair of smaller splints beneath it. No sections of the pholidophorid urohyal have yet been made, because of shortage of material, but the anterior and middle portions of the vertical plate have the appearance of cartilage bone, best seen in BMNH P.1066, *P. macrocephalus*, where a break exposes the spongy interior. The ventral, horizontal part of the bone is thin and curved upwards laterally in *P. germanicus*, thicker and flatter in *P. macrocephalus*. In *P. germanicus* this plate is only joined to the vertical plate for the anterior two-thirds of its length; posteriorly the two parts are separate. The most remarkable feature of this ventral plate of bone is that its underside is ornamented with peculiar multicuspid denticles (Fig. 5E, F). In *P. germanicus* these denticles cover the entire ventral surface of the anterior part of the bone, but posteriorly they are confined to the upturned margins of the plate, so that there is a median un-ornamented area. In *P. macrocephalus* the ventral surface of the bone is covered by denticles, and their form is more irregular than in *P. germanicus* (Fig. 5G). These denticles prove that the ventral plate of the pholidophorid urohyal is dermal, and the denticles must have projected through the skin of the isthmus, so that the ventral plate was not embedded in the musculature as it is in living teleosts (Fig. 7B).

The denticles on the urohyal of *P. germanicus* are isolated and scattered near the mid-line, but are more or less coalesced in longitudinal rows towards the margin of the bone. Isolated denticles have from three to nine cusps, commonly five or seven, and the cusps are usually symmetrically disposed about a large central cusp, decreasing in size towards the ends of the denticle. On the margins of the urohyal, where the denticles are arranged in rows, the cusps are more regular in size and the denticles tend to merge into a long comb-like row of cusps. In *P. macrocephalus* the denticles are mound- or atoll-like, with flat tops and with radially-directed cusps projecting horizontally around the margin.

Multicuspid denticles of this sort are, so far as I know, otherwise found only on the shoulder girdle in non-elasmobranch fishes. In *Amia* they occur on the cleithrum and the two dermal plates forming the 'serrated appendage' in the gill aperture (Liem & Woods, 1973: pls 1, 4; Fig. 6A): Liem & Woods regard the more posterior of these plates as the clavicle. In *Lepisosteus* the same type of denticle is found on the cleithrum and the small, polygonal plates overlying the antero-ventral surface of the cleithrum which Jarvik (1944: 5) regarded as the clavicle. In *Polyodon* (McAlpin, 1947: 208) similar but poorly calcified denticles line the whole posterior margin of the gill aperture, where they lie free in the skin, not attached to the underlying dermal bones. In pholidophorids (Fig. 6B) the cleithrum bears similar denticles, and the clavicle is represented by two denticulate plates which resemble those in *Amia*, except that both are flat and did not project freely like the serrated appendage in *Amia*. I have observed denticles of the same sort on the cleithrum and a plate-like clavicle in the Mesozoic actinopterygians

Caturus, *Furo*, *Lepidotes* and *Perleidus*, indicating that they are widespread in the group. I have not found a urohyal in any of these fishes, but I cannot exclude its presence. The same type of denticle is found in many arthrodires, on the apronic lamina of the anterior lateral and intero-lateral plates, in the hind wall of the gill-chamber (Miles & Westoll, 1968: 440; Ørvig, 1975: pl. 7).

In *Pholidophorus* most of the hind margin of the gill aperture was lined by these multicuspid denticles, on the cleithrum dorsally, the clavicles postero-ventrally and the urohyal ventrally, on the isthmus (Fig. 6B). Their function is unknown. Liem & Woods (1973) suggested that the function of the serrated appendage in *Amia* might be to break the mucous seal that closes the gill aperture during aestivation, but such an interpretation cannot apply to the fixed plates which bear the same type of denticle in other fishes. The alignment of the denticles, with their cusps directed towards the gill aperture, suggests that they may serve some hydrodynamic function, or they may deter ectoparasites from entering the gill chamber.

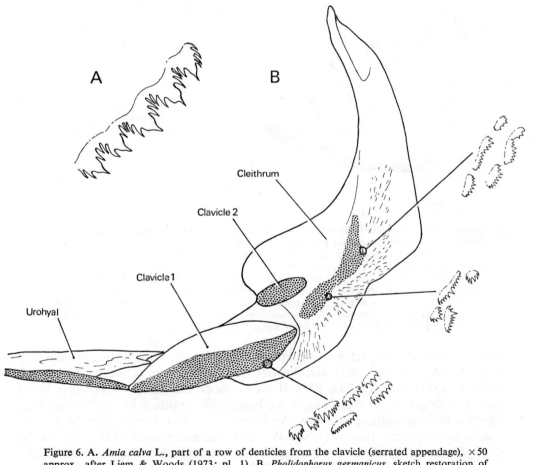

Figure 6. A. *Amia calva* L., part of a row of denticles from the clavicle (serrated appendage), ×50 approx., after Liem & Woods (1973: pl. 1). B. *Pholidophorus germanicus*, sketch restoration of posterior part of urohyal, left cleithrum and clavicles in lateral view, from BMNH P.3704, ×4 approx. The tone indicates areas ornamented with multicuspid denticles, some of which are shown enlarged (×30 approx.). The position of the small plate labelled 'Clavicle 2' is conjectural.

The structure of the pholidophorid urohyal shows that this bone in teleosts originally consisted of a dermal, denticulate plate which was exposed on the ventral surface of the isthmus and was in series with the clavicles posteriorly, and a vertical plate which may have been preformed in cartilage. Comparison with *Eusthenopteron* (Fig. 7A) suggests that this bone is another example of fusion between a cartilage bone, the homologue of the endoskeletal urohyal of rhipidistians, coelacanths and lungfishes, and a dermal bone, the interclavicle. An interclavicle is known in rhipidistians and Devonian lungfishes, but not in coelacanths. This interpretation predicts that a dermal interclavicle and an endoskeletal urohyal are primitively present in actinopterygians, as they seem to be in other osteichthyan

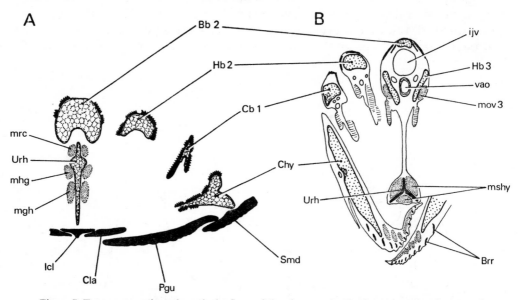

Figure 7. Transverse sections through the floor of the pharynx. A. *Eusthenopteron foordi*, restored from ground sections through the posterior part of the urohyal, after Jarvik (1963: fig. 19C), ×5 approx. B. *Elops hawaiiensis*, from a section through the same region as A in a young fish (sections lent by Dr Gareth Nelson), ×35 approx. Heavy dots in B indicate cartilage.

Bb2, Second basibranchial; Brr, branchiostegal rays; Cb1, first ceratobranchial; Chy, ceratohyal; Cla, clavicle; Hb2, Hb3, second and third hypobranchials; Icl, interclavicle; ijv, inferior jugular vein; mgh, geniohyoid muscle; mhg, hyogenioglossus muscle; mov3, obliquus ventralis muscle of third arch; mrc, rectus cervicis muscle; mshy, sternohyoid muscles; Pgu, paired gular plate; Smd, submandibular plate; Urh, urohyal; vao, ventral aorta.

groups. Amongst living actinopterygians, an interclavicle was reported in *Acipenser* sp. by MacAlpin (1947: 203), and illustrated in *Acipenser ruthenus* by Marinelli & Strenger (1973: 452, figs 232, 251). If such a bone can escape notice in a living fish for so long, it is not surprising that a bone in this position has only rarely been recorded in fossil actinopterygians. An interclavicle was described in the palaeoniscoid *Gyrolepis* by Dames (1888: 19), and was interpreted as the urohyal by Aldinger (1937: 241). In the palaeoniscoid *Acrorhabdus*, Stensiö (1921: 230) described a large urohyal (under the heading of the shoulder girdle) which is in series with the clavicles posteriorly, and is apparently a dermal bone. Watson (1925: 863) described a large urohyal in the palaeoniscoid *Mesonichthys*, again apparently a dermal bone. MacAlpin (1947: 203) described a dermal interclavicle in *Paleopsephurus*. And Lehman (1952: 105, fig. 69) described a urohyal in *Birgeria*, a dermal bone with a median crest on its anterior half. Lehman interpreted this crest as a site of muscle insertion, like the vertical plate of the teleostean urohyal, but it is not clear whether the crest is on the dorsal surface of the bone, like the teleostean vertical plate, or on the ventral surface, like the median ridge on the interclavicle of rhipidistians (Jarvik, 1944; Andrews & Westoll, 1970b). There is thus good evidence that a dermal interclavicle, homologous with the horizontal denticulate plate of the pholidophorid urohyal, is widely distributed in primitive actinopterygians.

As for an endoskeletal structure, homologous with the urohyal of rhipidistians, coelacanths and lungfishes, the only living actinopterygian with anything resembling that structure is embryo *Lepisosteus*, where there is a small, posteriorly directed cartilage between the hypohyals, which soon regresses (Hammarberg, 1937: 230, 235, fig. 12). I am not disappointed that no endoskeletal urohyal is known in primitive fossil actinopterygians, for the ventral parts of the gill-arch skeleton are known in so few forms, and are so rarely preserved in their natural position, that very little can be said. If such a bone were found, it would probably be interpreted as a basibranchial (see, for example, the bone labelled 'Co III' in sections of *Australosomus* by Nielsen, 1949: fig. 1).

My hypothesis is that the primitive teleostome condition is to have two median bones beneath the basibranchials, an endoskeletal urohyal and a dermal interclavicle. Both bones are present in osteolepiform and porolepiform rhipidistians, and in primitive dipnoans. The interclavicle is lost in coelacanths and both bones are lost in living dipnoans. Amongst actinopterygians, the interclavicle is present in various palaeoniscoids and in *Paleopsephurus* and *Acipenser*, but is lost in other chondrosteans and in holosteans. In pholidophorids the interclavicle is represented by the ventral denticulate plate of the urohyal, and the vertical plate of that bone may have been preformed in cartilage and represent the endoskeletal urohyal, the dermal and cartilage bones having fused through approximation of their centres. If so, the teleost urohyal represents these two bones in a reduced condition, the interclavicle (ventral plate) having lost its denticles and sunk beneath the skin, and the urohyal (vertical plate) having lost its cartilage precursor, and come to ossify as a membrane bone. If this interpretation is too metaphysical, the only alternative is to regard the teleostean urohyal as an interclavicle which has developed a large internal lamina of membrane bone, sunk inwards beneath the skin, and then, in many teleosts, lost its original ventral dermal component: this seems equally metaphysical. An endoskeletal urohyal may be represented by a transient rudiment in *Lepisosteus*, but is otherwise unknown in non-teleostean actinopterygians, except, perhaps, in cladistians. In *Polypterus* and *Erpetoichthys* there is a tripartite urohyal, consisting of a pair of bones which lie on the medial and ventral face of the anterior part of the sternohyoid muscle and the ligament attaching this to the hypohyal, and a median bone, behind the paired bones, which lies between the sternohyoids and is forked anteriorly (Jarvik, 1954: fig. 10C). The relative size of these bones is variable (the median bone may be larger or smaller than the paired bones), but their occurrence seems invariable. Fuchs's sections through the bones in young *Polypterus* (1929: figs 28–33) and the absence of cartilage in this region in larvae (Moy-Thomas, 1933) show that they are dermal or membrane bones. They may represent the endoskeletal urohyal, fragmented and reduced to membrane bone, as Jarvik (1963) supposed, or the posterior bone might represent the interclavicle, sunk inwards as Fuchs (1929: 447) supposed. In the absence of information on primitive conditions in cladistians, I see no way of discriminating between these hypotheses.

In tetrapods, the endoskeletal urohyal persists in urodeles and anurans, and the dermal interclavicle is present in many groups, but is held to be absent in those two amphibian taxa. According to my interpretation, the parahyoid of anurans would be the interclavicle, sunk beneath the muscles as Fuchs (1929) supposed, and not a fragment of the endoskeletal urohyal as Jarvik (1963) interpreted it.

Although my interpretation of the urohyal in teleosts and *Polypterus*, and of the parahyoid in frogs, is open to question, it is evident that in this region, as in the sclerotic, the primitive teleostome condition is to have separate dermal and endoskeletal ossifications, and that conditions in more advanced groups suggest differential loss of one bone or the other, rather than interchangeability between the dermal skeleton and the endoskeleton.

DELAMINATION THEORY

The various bones discussed in the two preceding sections provide no examples of interchangeability between the dermal skeleton and the endoskeleton. This conclusion has a bearing on 'Holmgren's principle of delamination', as formulated by Jarvik (1959), since that theory implies (or has been taken to imply) that there is some sort of interchangeability between the dermal skeleton and endoskeleton, or at least that the endoskeleton shares the same origin as the dermal skeleton.

The theory of delamination has been cited quite widely as an explanatory principle (e.g. Tarlo, 1964: 3; Daget, 1965: 311; Westoll, 1967: 96; Jollie, 1968, 1971, 1975; Moss, 1968a: 368, 1968b: 188, 1969: 529; Halstead, 1969: 121; Andrews & Westoll, 1970b: 464; Hall, 1975: 340), but the contexts of most of these citations are depressingly vague, because the facts which the theory is supposed to explain and the process which are invoked are hardly ever stated explicitly. This makes the theory difficult to criticize. The only explicit treatment is Jarvik's original (1959) application of the theory to dermal fin-rays, and the following criticism will mainly be confined to that.

The idea of ontogenetic delamination of skeletogenous tissues in vertebrates was first put forward by Holmgren (1940), on the basis of his observations of embryo selachians. Holmgren's idea was that the ectomesenchyme (neural crest) cells which form much of the selachian head skeleton reach their sites of differentiation in a series of waves or sheets, successively sinking inwards from beneath the ectodermal basement membrane. The only published evidence for this process is in sections of embryos (e.g. Jollie, 1971: fig. 10), where cells beneath the ectoderm may appear radially elongated, giving the impression that they are streaming inwards, but no distinct laminae or cell generations are seen. Marking experiments, through which migration routes of neural crest cells can be followed, have only been carried out in birds and amphibians (Weston, 1970; Le Lièvre & Le Douarin, 1975). In these experiments, especially the quail/chick grafts of Le Lièvre & Le

Douarin (1975 and references cited there: the grafted neural crest cells are recognized by nuclear features so that their development can be followed through without the dilution of the marker which occurs in experiments with dyes or tritiated thymidine), one can only say that if anything resembling delamination was observed, it was not recorded.

Jarvik (1959) elaborated Holmgren's ideas further by applying them to the dermal fin-rays of vertebrates, and in so doing he added a phylogenetic aspect to the ontogenetic theory. Like Goodrich and others, Jarvik reached the conclusion that the lepidotrichia of teleostomes are modified scale-rows. Mainly because of the condition in Palaeozoic and Recent lungfishes, he went further and interpreted

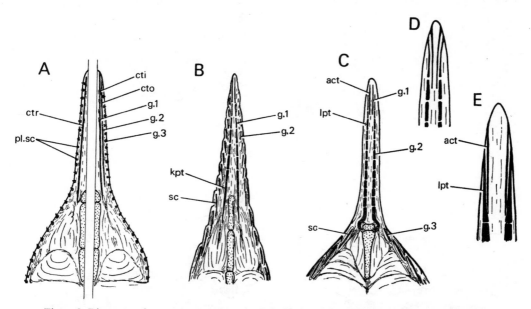

Figure 8. Diagrammatic transverse sections through the dorsal fin of fishes. A. Shark, *Scyliorhinus*, on the left after Goodrich (1904: fig. 1), on the right as modified by Jarvik (1959: fig. 20A), with two layers of ceratotrichia and three exoskeletal generations. B. Lungfish, *Neoceratodus*, after Goodrich (1904: fig. 5) and Jarvik (1959: fig. 20C). C. Teleost, after Goodrich (1904: fig. 4), with the three exoskeletal generations indicated by Jarvik (1959: fig. 20F). D. Enlarged distal part of the teleost fin diagram (C), as modified by Jarvik (1959: fig. 20F). E. Further enlargement of the distal part of the teleost fin diagram, modified to suggest the close relationship between the collagenous actinotrichia and the bony lepidotrichia.

Act, Actinotrich; cti, cto, inner and outer layers of ceratotrichia; ctr, ceratotrich; g. 1–3, oldest, next oldest and youngest generations of exoskeletal formations; kpt, camptotrich; lpt, lepidotrich; pl.sc, placoid scales; sc, cycloid scales.

the ceratotrichia of elasmobranchs and the actinotrichia of actinopterygians as modified scale-rows which have 'sunk down and have become more or less modified and degenerate' (p. 44). Jarvik published Goodrich's (1904) diagrammatic sections through the dorsal fin of a shark, a teleost and the dipnoans *Dipterus* and *Neoceratodus*, but modified the first as shown in Fig. 8A, and added a notation showing his interpretation of the generations of exoskeletal formations found in each fin (Fig. 8). According to this interpretation, the teleost fin contains three generations, the actinotrichia, lepidotrichia and scales (which cover the fins in some teleosts), the dipnoan fin contains two generations, the camptotrichia and scales, and the shark fin contains three, inner and outer ceratotrichia and placoid scales. This seems to carry the implication that dipnoan scales are not homologous with those of teleosts or sharks, but with the lepidotrichia in teleosts and the outer layer of ceratotrichia in sharks.

Jarvik's modification of Goodrich's diagrammatic section through a shark dorsal fin, the addition of an extra layer of ceratotrichia, is justified by reference to sections through the pectoral fin of adult *Squalus* and the caudal fin of embryo

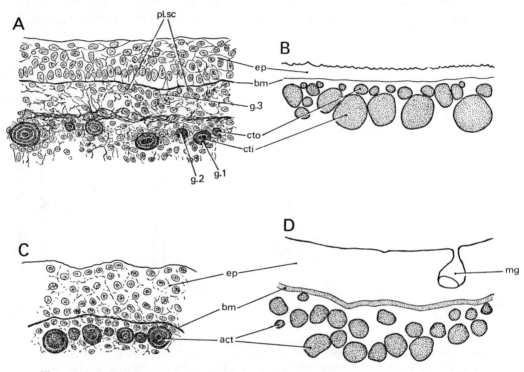

Figure 9. A, B. Evidence presented by Jarvik for the existence of two generations of ceratotrichia in sharks: A, section through the caudal fin of 60 mm *Scyliorhinus caniculus*, after Goodrich (1904: pl. 36, fig. 11) and Jarvik (1959: fig. 21C), ×300 approx.; B, section through the pectoral fin of *Squalus acanthias*, after the photograph published by Jarvik (1959: pl. 1, fig. 1), ×20 approx. C, D. Sections through teleost fins showing patterns of actinotrichia similar to the ceratotrichia in A and B: C, caudal fin of *Trichogaster sumatranus*, drawn in the same style as A after the photograph published by Haas (1962: fig. 4b), ×700 approx.; D, adipose fin of 8 cm *Salmo fario*, after Brohl (1909: pl. 29, fig. 15).

Act, Actinotrichia; bm, basement membrane; cti, cto, inner and outer layers of ceratotrichia; ep, epidermis; g. 1–3, oldest, next oldest and youngest generations of exoskeletal formations; mg, mucus gland; pl.sc, accumulation of scleroblasts which will give rise to a placoid scale.

Scyliorhinus (Fig. 9). In the latter figure, Jarvik saw 'an inner deeply embedded layer of thick ceratotrichia overlain by one or more layers of thinner rays formed later'. Interpreting these two layers as distinct generations (Fig. 8A), Jarvik wrote that the two generations 'arise almost simultaneously' (p. 40), a statement which seems only to emphasize that recognition of these generations is subjective. In teleosts the actinotrichia show a comparable variation in thickness and position

(Fig. 9; see also Kemp & Park, 1970: fig. 4, *Tilapia*; and Bouvet, 1974a: figs 7, 13, *Salmo*). The layered appearance of the ceratotrichia in Jarvik's section through the pectoral fin of adult *Squalus* seems to be matched in the adipose fin of teleosts, as shown in Fig. 9. Thus if variation in thickness and position of the ceratotrichia in sharks is evidence of distinct generations, the same generations are to be found in teleosts. I prefer to regard the congruence between shark ceratotrichia and teleost actinotrichia as evidence that these structures are homologous (cf. Kemp & Park, 1970: 338), and not that they exhibit distinct and countable generations.

Goodrich (1904) held that the actinotrichia of teleosts are quite distinct from the bony lepidotrichia, writing 'although they [actinotrichia] may occasionally become included in the bony substance forming the lepidotrichia, yet they cannot be said to give rise to the latter by coalescence or growth' (p. 476), and 'either they [actinotrichia] get re-absorbed proximally as fast as they are secreted distally, or they get bodily carried outwards in the growing edge of the fin' (p. 474). Despite modern ultrastructural work, there is still no unanimity on the relationship between teleostean actinotrichia and lepidotrichia. Haas (1962), working on *Trichogaster* and *Carassius*, concluded, like several earlier authors, that the actinotrichia are enclosed in the developing lepidotrichium, and that it is these enclosed bundles of collagen fibrils which form the bridges across the joints in mature lepidotrichia. Kemp & Park (1970), working on *Tilapia*, suggested that the collagen of the actinotrichia might be depolymerized in the zone of lepidotrich formation, the collagen being transferred to the ground substance of the developing bone. Bouvet (1974b: 326, fig. 7), working on *Salmo*, holds that the tip of the developing lepidotrich surrounds a bundle of actinotrichia, and that the actino-trichia are progressively pushed towards the margin of the fin. Pending settlement of this question, two conclusions can be suggested: that the zone of overlap between ossified lepidotrichia and free actinotrichia is very short (Haas, 1962: fig. 1d; Kemp & Park, 1970: figs 2, 3; Bouvet, 1974a: fig. 14); and that lepido-trichia do not develop except in the presence of actinotrichia. The latter conclusion is corroborated by the fact that actinotrichia develop before lepidotrichia in regenerating fins (Kemp & Park, 1970), and that lepidotrichia occur occasionally in the adipose fin of teleosts (Kosswig, 1965). It would follow that the lepidotrichia of *Latimeria* should terminate in actinotrichia, a prediction that can be tested now that young *Latimeria* are available (Smith, Rand, Schaeffer & Atz, 1975), and that the camptotrichia of living dipnoans are ceratotrichia/actinotrichia on whose outer faces ossification begins very late in ontogeny. I suggest that the relationship between actinotrichia and lepidotrichia is closer than Goodrich supposed, and modify his diagrammatic section through the teleost fin as shown in Figure 8E.

Leaving the details of Jarvik's interpretation of the fin-rays of sharks and teleosts, his central thesis is that elasmobranch ceratotrichia, actinopterygian actinotrichia and dipnoan camptotrichia are modified scale-rows, phylogenetic homologues of the outer dermal skeleton at some earlier stage in the evolution of each group. Ultrastructural work shows that teleostean actinotrichia (Nadol, Gibbins & Porter, 1969: 310; Kemp & Park, 1970; Bouvet, 1974a) and elasmo-branch ceratotrichia (Kemp & Park, 1970: 338) are giant fibres of collagen, built up of close-packed or fused fibrils aligned so that the 64 nm banding of the collagen fibrils is in phase throughout the fibre. One may ask in what sense such a structure could be homologous with a series of scales, and whether this homology is envisioned as one-to-one (one row of scales to one actinotrich/ceratotrich) or not. Until proponents of delamination clarify these and other aspects of the theory there is little basis for discussion.

In that statement, I am not insisting that delamination has no reality. It is true, as Jarvik says, that dermal bones which were primitively covered by enameloid and dentine may lose those tissues in phylogeny, and so sink beneath the epidermis. They may then be secondarily covered by bones, odontodes or other dermal structures, developing above them at the basement membrane. But in fishes, at any rate, I question that such secondary covering structures always belong to a different phyletic generation, for it seems clear that the scales which cover the head and fins in advanced actinopterygians belong to 'the same' phyletic generation as the body scales, which in turn belong to 'the same' generation as the cranial dermal bones. And when dermal bones and fin-rays have been secondarily covered by scales or odontodes in this way, I question that they may subsequently evolve into different structures, such as actinotrichia or cartilage. Some of the examples of delamination of dermal structures cited by Jarvik (1959: 45) may illustrate phylogenetic aspects of the principle when thoroughly investigated, while onto-genetic aspects may be shown by epithelial/mesenchymal interactions, or by the complexities of the basement lamella (Nadol, Gibbins & Porter, 1969) and its relationship to developing dermal bones (e.g., Waterman, 1970; Yamada, 1971). But I suggest that those who wish to invoke the principle of delamination should not treat it as an established and well-understood theory, and should be explicit about whether they apply it in an ontogenetic or phylogenetic sense, and whether the successive generations are sheets of cells, as in Holmgren's original idea, or cell products like collagen (e.g. Hall, 1975: 340).

CONCLUSIONS

The topic reviewed in this paper, the possibility of dermal bones becoming cartilage bones in phylogeny and vice versa, is rather nebulous. It also lies at the border of several subdisciplines—comparative anatomy, embryology, histology, palaeontology, experimental zoology—and spans the whole range of bony vertebrates. I believe that this partly explains the general acceptance of the idea that the dermal skeleton and endoskeleton are in some way interchangeable: specialists in one field tend to accept tenuous data from others, feeling unable to evaluate it critically. A feeling of inadequacy in fields with which I am unfamiliar has dogged me throughout this work, and I will apologize now to those whose work I have misunderstood, misinterpreted or overlooked. Nevertheless, if there are instances of interchangeability between the dermal skeleton and the endo-skeleton in vertebrates, I have not found them, except for the phenomenon of adventitious cartilage in birds and mammals. In that case the factors influencing chondrogenesis and osteogenesis are thoroughly investigated, and do not seem to be applicable elsewhere in vertebrates.

Where nomenclature is concerned, it is mistaken to synonymize cartilage bone with endochondral bone, and confusion has been caused by treating dermal bone and membrane bone as synonyms. I recommend that the term 'a membrane bone' should be reserved for endoskeletal bones which are not preformed in cartilage: membrane bones are either phylogenetic homologues of cartilage bones in more primitive forms, which no longer ossify in cartilage through regression of cartilage or changed growth patterns (e.g., ribs, intercalar, basisphenoid in teleosts), or they are neoformations (e.g., baculum and patella in mammals). Membrane bones are derived features, wherever they occur. Bones should be interpreted as membrane bones not on an *ad hoc* basis, but only when a well-established phylogeny demands that interpretation. Whether membrane bones which are phylogenetic homologues of cartilage bones occur in tetrapods is still an open question.

The term 'a dermal bone' should be used for all bones which are phylogenetic or topographic homologues of dentine- and enameloid-coated bones in fishes: mammalian dermal bones such as the frontal, parietal and clavicle should be called dermal bones, not membrane bones.

At the level of parts of bones, 'membrane bone' can be used as a descriptive term for parts of dermal and cartilage bones, as before. At the level of systems, the dermal skeleton comprises the dermal bones, and the bony endoskeleton comprises the cartilage bones and membrane bones.

Supposed examples of cartilage and cartilage bone in the dermal skeleton, and of dermal bone in the endoskeleton are reviewed. Cartilage in the dermal skeleton is restricted to mammals and birds, where it appears, and may ossify, at the margins of dermal bones in ontogeny, and in the repair of dermal bones. This adventitious cartilage is a derived feature of those groups. In every location where a dermal bone is supposed to have invaded the endoskeleton, or where the histogenesis of a bone is supposed to be variable, there is evidence that the primitive condition is to have two separate bones, one dermal and one endoskeletal. Apparent variable histogenesis (as in the sclerotic and urohyal/parahyoid) is due to differential loss of one or other of these bones. Apparent invasion of the endoskeleton by dermal bones is caused by ontogenetic or phylogenetic fusion between a dermal bone and a cartilage bone whose growth or ossification centres have become coincident. The tabular of mammals and mammal-like reptiles is the only place where primitively separate bones are not yet observed, and an endoskeletal ossification centre beneath the dermal bone is only inferred.

There are therefore no known instances of interchangeability between the dermal skeleton and the endoskeleton. The principle or theory of delamination is reviewed in the light of this conclusion. It is found that the only explicit application of the theory, to dermal fin-rays, does not stand up to criticism. Those who wish to invoke delamination theory should say clearly whether they use it in an ontogenetic or phylogenetic sense, and whether cells or cell products are delaminated.

The general conclusion of this review is that the vertebrate dermal skeleton and endoskeleton have always been distinct, so far as they are documented by known animals. It is, of course, possible to postulate some earlier episode in which the dermal skeleton gave rise to the endoskeleton or vice versa, by sinking in or delamination, but it is hard to see how such ideas would be useful.

If there is no direct phylogenetic relationship between the dermal skeleton and endoskeleton, there remains the possibility of some less direct relationship, one set of bones having induced the other. The idea that the bony endoskeleton was originally induced by the dermal skeleton is an old one. The alternative theory, that dermal bone is induced by cartilage or cartilage bone, is discussed by Hall (1975: 338). It seems that both these theories are readily disproved, for any direct induction would surely result in a one-to-one relationship between dermal bones and cartilage bones, with their ossification centres superimposed. Nothing like that is found in any known vertebrate, and when dermal and cartilage-bone centres do become directly superimposed, as in cases of fusion between the two, the approximation of the centres is always a derived feature.

ACKNOWLEDGEMENTS

I am grateful for advice to Drs P. L. Forey, M. Hills, D. E. McAllister, R. S. Miles, A. L. Panchen, Bobb Schaeffer and E. I. White, and for material to Drs B. G. Gardiner and Gareth Nelson.

REFERENCES

AGASSIZ, J. L. R., 1844. *Recherches sur les Poissons Fossiles, 1:* xlix+188 pp. Neuchâtel.

ALDINGER, H., 1937. Permische Ganoidfische aus Ostgrönland. *Meddr Grønland, 102,* 3: 1–392.

ALEXANDER, R. McN., 1965. Structure and function in the catfish. *J. Zool., Lond., 148:* 88–152.

ALLIS, E. P., 1909. The cranial anatomy of the mail-cheeked fishes. *Zoologica, Stuttg., 22,* 2: 1–219.

ALLIS, E. P., 1916. The so-called labial cartilages of *Raia clavata. Q. Jl microsc. Sci., 62:* 95–114.

ALLIS, E. P., 1922. The cranial anatomy of *Polypterus,* with special reference to *Polypterus bichir. J. Anat., 56:* 189–294.

ALLIS, E. P., 1935. On a general pattern of arrangement of the cranial roofing bones in fishes. *J. Anat., 69:* 233–91.

ANDERSEN, H., 1963. Histochemistry and development of the human shoulder and acromio-clavicular joints with particular reference to the early development of the clavicle. *Acta. anat., 55:* 124–65.

ANDREWS, S. M. & WESTOLL, T. S., 1970a. The postcranial skeleton of *Eusthenopteron foordi* Whiteaves. *Trans. R. Soc. Edinb., 68:* 207–329.

ANDREWS, S. M. & WESTOLL, T. S., 1970b. The postcranial skeleton of rhipidistian fishes excluding *Eusthenopteron. Trans. R. Soc. Edinb., 68:* 391–489.

ARENDT, E., 1822. *De capitis ossei Esocis lucii structura singulari:* 24 pp. Regiomonti.

BAER, K. E. VON, 1826. Ueber das äussere und innere Skelet. *Arch. Anat. Physiol., 9:* 327–76.

BAMFORD, T. W., 1941. The lateral line and related bones of the herring (*Clupea harengus* L.). *Ann. Mag. nat. Hist.,* (11) 8: 414–38.

BAMFORD, T. W., 1948. Cranial development of *Galeichthys felis. Proc. zool. Soc. Lond., 118:* 364–91.

BAUCHOT, M.-L., 1959. Étude des larves leptocephales du groupe *Leptocephalus lanceolatus* Strömman et identification a la famille des Serrivomeridae. *Dana Rep., 48:* 1–148.

BOUVET, J., 1974a. Différenciation et ultrastructure du squelette distal de la nageoire pectorale chez la truite indigène (*Salmo trutta fario* L.). I. Différenciation et ultrastructure des actinotriches. *Archs Anat. microsc. Morph. exp., 63:* 79–96.

BOUVET, J., 1974b. Différenciation et ultrastructure du squelette distal de la nageoire pectorale chez la truite indigène (*Salmo trutta fario* L.). II. Différenciation et ultrastructure des lepidotriches. *Archs Anat. microsc. Morph. exp. 63:* 323–35.

BROHL, E., 1909. Die sogenannten Hornfäden und die Flossenstrahlen der Fische. *Jena Z. Naturw., 45:* 345–80.

BROOM, R., 1916. On the structure of the skull in *Chrysochloris. Proc. zool. Soc. Lond., 1916:* 449–59.

CAVENDER, T. M., 1970. A comparison of coregonines and other salmonids with the earliest known teleostean fishes. In C. C. Lindsey & C. S. Woods (Eds.), *Biology of coregonid fishes:* 1–32. Winnipeg: University of Manitoba Press.

DAGET, J., 1950. Révision des affinités phylogénétiques des Polyptéridés. *Mém. Inst. fr. Afr. noire, 11:* 1–178.

DAGET, J., 1965. Le crâne des téléostéens. *Mém. Mus. natn. Hist. nat., Paris, N.S.* (A) 31: 163–341.

DAGET, J. & D'AUBENTON, F., 1957. Développement et morphologie du crâne d'*Heterotis niloticus* Ehr. *Bull. Inst. fr. Afr. noire,* (A) 19: 881–936.

DAGET, J., BAUCHOT, M.-L., BAUCHOT, R. & ARNOULT, J., 1964. Développement du chondrocrâne et des arcs aortiques chez *Polypterus senegalus* Cuvier. *Acta zool., Stockh., 45:* 201–44.

DAMES, W., 1888. Die Ganoiden des deutschen Muschelkalks. *Palaeont. Abh., 4:* 133–80.

DE BEER, G. R., 1937. *The development of the vertebrate skull:* xxiv+552 pp. London: Oxford University Press.

DENISON, R. H., 1967. Ordovician vertebrates from western United States. *Fieldiana, Geol., 16:* 131–92.

DEVILLERS, C., 1948. Recherches sur le crâne dermique des téléostéens. *Annls Paléont., 33:* 1–96.

DEVILLERS, C., 1950. Quelques aspects de l'évolution du crâne chez les poissons. *Année biol.,* (3) *26:* 145–80.

DEVILLERS, C., 1958. Le crâne des téléostéens. In P. P. Grassé (Ed.), *Traité de Zoologie, 13,* 1: 551–687. Paris: Masson.

DORNESCO, G. T. & SORESCO, C., 1971a. Développement de quelques os du neurocrâne chez *Cyprinus carpio* L. *Anat. Anz., 128:* 16–38.

DORNESCO, G. T. & SORESCO, C., 1971b. Sur le développement et la valeur morphologique de la région ethmoïdale de la carpe. *Anat. Anz., 129:* 33–52.

DORNESCO, G. T. & SORESCO, C., 1973. L'origine et le développement des os de la région otique du neurocrâne de la carpe. *Anat. Anz., 133:* 305–30.

DORNESCO, G. T. & SORESCO, C., 1974. L'origine et le développement des os de la région occipitale du neurocrâne de la carpe. *Anat. Anz., 136:* 318–33.

EMELIANOV, S. V., 1935. Die Morphologie der Fischrippen. *Zool. Jb. (Anat.), 60:* 133–262.

EMELIANOV, S. V., 1939. Sequence in the ontogenetic appearance of vertebral arches in teleosts and the omission of chondral stages in their development. *Dokl. Akad. Nauk SSSR, 23:* 978–81.

EDINGER, T., 1929. Über knöcherne Scleralringe. *Zool. Jb. (Anat.), 51:* 163–226.

FARUQI, A. J., 1935. The development of the vertebral column in the haddock (*Gadus aeglifinus*). *Proc. zool. Soc. Lond., 1935,* 2: 313–32.

FOREY, P. L., 1973. A primitive clupeomorph fish from the Middle Cenomanian of Hakel, Lebanon. *Can. J. Earth Sci., 10:* 1302–18.

FRANCILLON, H., 1974. Développement de la partie postérieure de la mandibule de *Salmo trutta fario* L. (Pisces, Teleostei, Salmonidae). *Zoologica Scr., 3:* 41–51.

FRANÇOIS, Y., 1966. Structure et développement de la vertébre de *Salmo* et des téléostéens. *Archs Zool. exp. gén., 107:* 287–328.

FRASIER, M. B., BANKS, W. J. & NEWBREY, J. W., 1975. Characterization of developing antler cartilage matrix. I. Selected histochemical and enzymatic assessment. *Calc. Tiss. Res., 17:* 273–88.

FUCHS, H., 1929. Über das Os parahyoideum der anuren Amphibien und der Crossopterygier; nebst Bemerkungen über phylogenetische Wanderungen der Haut und Deckknochen. *Morph. Jb., 63:* 408–53.

GOODRICH, E. S., 1904. On the dermal fin-rays of fishes—living and extinct. *Q. Jl microsc. Sci., 47:* 465–522.

GOODRICH, E. S., 1930. *Studies on the structure and development of vertebrates:* xxx+837 pp. London: Macmillan.

GOSLINE, W. A., 1969. The morphology and systematic position of the alepocephaloid fishes. *Bull. Br. Mus. nat. Hist. (Zool.), 18:* 183–218.

GREENWOOD, P. H., 1967. The caudal fin skeleton in osteoglossoid fishes. *Ann. Mag. nat. Hist.,* (13) *9:* 581–97.

GREENWOOD, P. H., 1970. On the genus *Lycoptera* and its relationship with the family Hiodontidae. *Bull. Br. Mus. nat. Hist. (Zool.), 19:* 259–85.

GREENWOOD, P. H. & ROSEN, D. E., 1971. Notes on the structure and relationships of the alepocephaloid fishes. *Am. Mus. Novit., 2473:* 1–41.

HAAS, H. J., 1962. Studies on the mechanisms of joint and bone formation in the skeleton rays of fish fins. *Devl. Biol., 5:* 1–34.

HAINES, R. W., 1937. The posterior end of Meckel's cartilage and related ossifications in bony fishes. *Q. Jl microsc. Sci., 80:* 1–38.

HALL, B. K., 1970. Cellular differentiation in skeletal tissues. *Biol. Rev., 45:* 455–84.

HALL, B. K., 1972. Immobilization and cartilage transformation into bone in the embryonic chick. *Anat. Rec. 173:* 391–403.

HALL, B. K., 1975. Evolutionary consequences of skeletal differentiation. *Am. Zool., 15:* 329–50.

HALL, B. K. & JACOBSON, H. N., 1975. The repair of fractured membrane bones in the newly hatched chick. *Anat. Rec., 181:* 55–70.

HALSTEAD, L. B., 1969. Calcified tissues in the earliest vertebrates. *Calc. Tiss. Res., 3:* 107–24.

HALSTEAD, L. B., 1973. The heterostracan fishes. *Biol. Rev., 48:* 279–332.

HALSTEAD, L. B., 1974. *Vertebrate hard tissues:* x+179 pp. London: Wykeham.

HAMMARBERG, F., 1937. Zur Kenntnis der ontogenetischen Entwicklung des Schädels von *Lepidosteus platystomus. Acta zool., Stockh., 18:* 209–337.

HARRINGTON, R. W., 1955. The osteocranium of the American cyprinid fish, *Notropis bifrenatus,* with an annotated synonymy of teleost skull bones. *Copeia, 1955:* 267–90.

HARRISSON, C. M. H., 1966. On the first halosaur leptocephalus: from Madeira. *Bull. Br. Mus. nat. Hist. (Zool.), 14:* 441–86.

HINCHLIFFE, J. R. & EDE, D. A., 1968. Abnormalities in bone and cartilage development in the *talpid*³ mutant in the fowl. *J. Embryol. exp. Morph., 19:* 327–39.

HOLMGREN, N., 1940. Studies on the head of fishes. Part I. Development of the skull in sharks and rays. *Acta. zool., Stockh., 21:* 51–267.

HOLMGREN, N. & STENSIÖ, E. A., 1936. Kranium und Visceralskelett der Akranier, Cyclostomen und Fische. In L. Bölk (Ed.), *Handbuch der vergleichenden Anatomie, 4:* 235–500. Berlin and Wien: Urban & Schwarzenberg.

HOWIE, A. A., 1970. A new capitosaurid labyrinthodont from East Africa. *Palaeontology, 13:* 210–53.

HUXLEY, T. H., 1859. On the theory of the vertebrate skull. *Proc. R. Soc. Lond., 9:* 381–457.

HUXLEY, T. H., 1864. *Lectures on the elements of comparative anatomy:* xi+303 pp. London: Churchill.

JARDINE, N., 1970. The observational and theoretical components of homology: a study based on the morphology of the dermal skull-roofs of rhipidistian fishes. *Biol. J. Linn. Soc., 1:* 327–61.

JARVIK, E., 1944. On the exoskeletal shoulder-girdle of teleostomian fishes, with special reference to *Eusthenopteron foordi* Whiteaves. *K. svenska VetenskAkad. Handl.,* (3) *21,* 7: 1–32.

JARVIK, E., 1954. On the visceral skeleton in *Eusthenopteron*, with a discussion of the parasphenoid and palatoquadrate in fishes. *K. svenska VetenskAkad. Handl.,* (4) *5,* 1: 1–104.

JARVIK, E., 1959. Dermal fin-rays and Holmgren's principle of delamination. *K. svenska VetenskAkad. Handl.,* (4) *6,* 1: 1–51.

JARVIK, E., 1963. The composition of the intermandibular division of the head in fish and tetrapods and the diphyletic origin of the tetrapod tongue. *K. svenska VetenskAkad. Handl.,* (4) *9,* 1: 1–74.

JARVIK, E., 1972. Middle and Upper Devonian Porolepiformes from East Greenland with special reference to *Glyptolepis groenlandica* n.sp. *Meddr Grønland, 187,* 2: 1–307.

JOLLIE, M., 1962. *Chordate morphology:* xiv+478 pp. New York: Rheinhold.

JOLLIE, M., 1968. Some implications of the acceptance of a delamination principle. *Nobel Symposium, 4:* 89–107.

JOLLIE, M., 1971. Some developmental aspects of the head skeleton of the 35–37 mm *Squalus acanthias* foetus. *J. Morph., 133:* 17–40.

JOLLIE, M., 1975. Development of the head skeleton and pectoral girdle in *Esox. J. Morph., 147:* 61–88.

KADAM, K. M., 1961. The development of the skull in *Nerophis* (Lophobranchii). *Acta zool., Stockh., 42:* 257–98.

KAPOOR, A. S., 1970. Development of dermal bones related to sensory canals of the head in the fishes *Ophicephalus punctatus* Bloch (Ophicephalidae) and *Wallago attu* Bl. & Schn. (Siluridae). *Zool. J. Linn. Soc., 49:* 69–97.

KEMP, N. E. & PARK, J. H., 1970. Regeneration of lepidotrichia and actinotrichia in the tailfin of the teleost *Tilapia mossambica. Devl. Biol., 22:* 321–42.

KERSHAW, D. R., 1970. The cranial osteology of the 'Butterfly Fish', *Pantodon buchholzi* Peters. *Zool. J. Linn. Soc., 49:* 5–19.

KIELAN-JAWOROWSKA, Z., 1971. Skull structure and affinities of the Multituberculata. *Palaeont. pol., 25:* 5–41.

KINDRED, J. E., 1919. The skull of *Amiurus. Illinois biol. Monogr., 5,* 1: 1–120.

KÖLLIKER, A. VON, 1849. Allgemeine Betrachtungen über die Entstehung des knöchernen Schädels der Wirbelthiere. *Ber. k. zootom. Anst. Würzburg, 2:* 35–52.

KOSSWIG, C., 1965. Die Fettflosse der Knochenfische (besonders der Characiden). Morphologie, Funktion, phylogenetischen Bedeutung. *Z. Zool. Syst. EvolForsch., 3:* 284–329.

KUHN, H.-J., 1971. Die Entwicklung und Morphologie des Schädels von *Tachyglossus aculeatus. Abh. senckenb. naturforsch. Ges., 528:* 1–224.

KUSAKA, T., 1974. *The urohyal of fishes:* xiv+320 pp. University of Tokyo Press.

LEHMAN, J.-P., 1952. Étude complémentaire des poissons de l'Eotrias de Madagascar. *K. svenska VetenskAkad. Handl.,* (4) *2,* 6: 1–201.

LEHMAN, J.-P., 1966. Actinopterygii. In J. Piveteau (Ed.), *Traité de Paléontologie, 4,* 3: 1–242. Paris: Masson.

LEKANDER, B., 1949. The sensory line system and the canal bones of the head of some Ostariophysi. *Acta zool., Stockh., 30:* 1–131.

LENGLET, G., 1974. Contribution à l'étude ostéologique des Kneriidae. *Annls Soc. r. zool. Belg., 104:* 51–103.

LIEM, K. F. & WOODS, L. P., 1973. A probable homologue of the clavicle in the holostean fish *Amia calva. J. Zool., Lond., 170:* 521–31.

LE LIÈVRE, C. S. & LE DOUARIN, N. M., 1975. Mesenchymal derivatives of the neural crest: analysis of chimaeric quail and chick embryos. *J. Embryol. exp. Morph., 34:* 125–54.

McALLISTER, D. E., 1968. Evolution of branchiostegals and classification of teleostome fishes. *Bull. natn. Mus. can., 221:* xiv+239 pp.

MacALPIN, A., 1947. *Paleopsephurus wilsoni*, a new polyodontid fish from the Upper Cretaceous of Montana, with a discussion of allied fish, living and fossil. *Contr. Mus. Paleont. Univ. Mich., 6:* 167–234.

MAREEL, M., 1967. Recherches sur la relation inductrice entre chondrocytes et périoste dans le tibia embryonnaire de poulet. *Archs Biol., Paris, 78:* 145–66.

MARINELLI, W. & STRENGER, A., 1973. *Vergleichende Anatomie und Morphologie der Wirbeltiere, 1, 4*: 309–460. Wien: Deuticke.

MILES, R. S., in press. Dipnoan (lungfish) skulls and the relationships of the group: a study based on new species from the Devonian of Australia.

MILES, R. S. & WESTOLL, T. S., 1968. The placoderm fish *Coccosteus cuspidatus* Miller ex Agassiz from the Middle Old Red Sandstone of Scotland. *Trans. R. Soc. Edinb., 67*: 373–476.

MILLOT, J. & ANTHONY, J., 1958. *Anatomie de Latimeria chalumnae. Tome I. Squelette, muscles et formations de soutien:* 118 pp. Paris: CNRS.

MILLOT, J. & ANTHONY, J., 1965. *Anatomie de Latimeria chalumnae. Tome II. Système nerveux et organes des sens:* 131 pp. Paris: CNRS.

MONOD, T., 1968. Le complexe urophore des poissons téléostéens. *Mem. Inst. Fond. Afr. noire, 81*: 1–705.

MOSS, M. L., 1958. Fusion of the frontal suture in the rat. *Am. J. Anat., 102*: 141–66.

MOSS, M. L., 1962. Studies on the acellular bone of teleost fish. II. Response to fracture under normal and acalcemic conditions. *Acta anat., 48*: 46–60.

MOSS, M. L., 1968a. The origin of vertebrate calcified tissues. *Nobel Symposium, 4*: 359–71.

MOSS, M. L., 1968b. Comparative anatomy of dermal bone and teeth. I. The epidermal co-participation hypothesis. *Acta anat., 71*: 178–208.

MOSS, M. L., 1969. Comparative histology of dermal sclerifications in reptiles. *Acta anat., 73*: 510–33.

MOY-THOMAS, J. A., 1933. Notes on the development of the chondrocranium of *Polypterus senegalus. Q. Jl microsc. Sci., 76*: 209–29.

MURRAY, P. D. F., 1963. Adventitious (secondary) cartilage in the chick embryo, and the development of certain bones and articulations in the chick skull. *Aust. J. Zool., 11*: 368–430.

NADOL, J. B., GIBBINS, J. R. & PORTER, K. R., 1969. A reinterpretation of the structure and development of the basement lamella: an ordered array of collagen in fish skin. *Devl. Biol., 20*: 304–31.

NELSON, G. J., 1969a. Gill arches and the phylogeny of fishes, with notes on the classification of vertebrates. *Bull. Am. Mus. nat. Hist., 141*: 475–552.

NELSON, G. J., 1969b. Infraorbital bones and their bearing on the phylogeny and geography of osteo-glossomorph fishes. *Am. Mus. Novit., 2394*: 1–37.

NELSON, G. J., 1970. Pharyngeal denticles (placoid scales) of sharks, with notes on the dermal skeleton of vertebrates. *Am. Mus. Novit., 2415*: 1–26.

NELSON, G. J., 1973a. Notes on the structure and relationships of certain Cretaceous and Eocene teleostean fishes. *Am. Mus. Novit., 2524*: 1–31.

NELSON, G. J., 1973b. Relationships of clupeomorphs, with remarks on the structure of the lower jaw in fishes. In P. H. Greenwood, R. S. Miles & C. Patterson (Eds.), *Interrelationships of fishes:* 333–49. London: Academic Press.

NEWBREY, J. W. & BANKS, W. J., 1975. Characterization of developing antler cartilage matrix. II. An ultrastructural study. *Calc. Tiss. Res., 17*: 289–302.

NIELSEN, E., 1942. Studies on Triassic fishes from East Greenland. I. *Glaucolepis* and *Boreosomus. Meddr Grønland, 138*: 1–403.

NIELSEN, E., 1949. Studies on Triassic fishes from East Greenland. II. *Australosomus* and *Birgeria. Meddr Grønland, 146*: 1–309.

NORMAN, J. R., 1926. The development of the chondrocranium of the eel (*Anguilla vulgaris*), with observations on the comparative morphology and development of the chondrocranium in bony fishes. *Phil. Trans. R. Soc. Lond., B214*: 369–464.

NYBELIN, O., 1963. Zur Morphologie und Terminologie des Schwanzskelettes der Actinopterygier. *Ark. Zool., (2) 15*: 485–516.

ØRVIG, T., 1951. Histologic studies of placoderms and fossil elasmobranchs. I: The endoskeleton, with remarks on the hard tissues of lower vertebrates in general. *Ark. Zool., (2) 2*: 321–454.

ØRVIG, T., 1957. Notes on some Paleozoic lower vertebrates from Spitsbergen and North America. *Norsk geol. Tidsskr., 37*: 285–353.

ØRVIG, T., 1962. Y a-t-il une relation directe entre les arthrodires ptyctodontides et les holocéphales? *Colloques int. Cent. natn. Res. scient., 104*: 49–61.

ØRVIG, T., 1967. Phylogeny of tooth tissues: evolution of some calcified tissues in early vertebrates. In A. E. W. Miles (Ed.), *Structural and chemical organization of teeth:* 45–110. London & New York: Academic Press.

ØRVIG, T., 1972. The latero-sensory component of the dermal skeleton in lower vertebrates and its phylogenetic significance. *Zoologica Scr., 1*: 139–55.

ØRVIG, T., 1975. Description, with special reference to the dermal skeleton, of a new radotinid arthrodire from the Gedinnian of Arctic Canada. *Colloques int. Cent. natn. Res. scient., 218*: 41–71.

ORTON, G. L., 1963. Notes on larval anatomy of fishes of the order Lyomeri. *Copeia, 1963*: 6–15.

OWEN, R., 1846. *Lectures on the comparative anatomy and physiology of the vertebrate animals. Part I. Fishes:* xi+308 pp. London: Longman.

PARSONS, T. S. & WILLIAMS, E. E., 1963. The relationships of the modern Amphibia: a re-examination. *Q. Rev. Biol., 38:* 26–53.

PATTERSON, C., 1967. Are the teleosts a polyphyletic group? *Colloques int. Cent. natn. Res. scient., 163:* 93–109.

PATTERSON, C., 1968. The caudal skeleton in Lower Liassic pholidophorid fishes. *Bull. Br. Mus. nat. Hist. (Geol.), 16:* 201–39.

PATTERSON, C., 1970. A clupeomorph fish from the Gault (Lower Cretaceous). *Zool. J. Linn. Soc., 49:* 161–82.

PATTERSON, C., 1973. Interrelationships of holosteans. In P. H. Greenwood, R. S. Miles & C. Patterson (Eds.), *Interrelationships of fishes:* 233–305. London: Academic Press.

PATTERSON, C., 1975. The braincase of pholidophorid and leptolepid fishes, with a review of the actinopterygian braincase. *Phil. Trans. R. Soc. Lond.,* B*269:* 275–579.

PATTERSON, C. & ROSEN, D. E., 1977. Review of ichthyodectiform and other Mesozoic teleost fishes and the theory and practice of classifying fossils. *Bull. Am. Mus. nat. Hist., 158:* 172–181.

PEHRSON, T., 1944. The development of the latero-sensory canal bones in the skull of *Esox lucius. Acta zool., Stockh., 25:* 135–57.

PEHRSON, T., 1947. Some new interpretations of the skull in *Polypterus. Acta zool., Stockh., 28:* 399–455.

POPLIN, C., 1974. *Étude de quelques paléoniscidés pennsylvaniens du Kansas:* 151 pp. Paris: CNRS.

PRESLEY, R. & STEEL, F. L. D., 1976. On the homology of the alisphenoid. *J. Anat., 121:* 441–59.

PRITCHARD, J. J., 1972. General histology of bone. In G. H. Bourne (Ed.), *The biochemistry and physiology of bone, 1:* 1–20. New York & London: Academic Press.

REICHERT, C., 1838. *Vergleichende Entwicklungsgeschichte des Kopfes der nackten Amphibia:* xii+275 pp. Königsberg: Börnträger.

DE RICQLES, A., 1975. Recherches paléohistologiques sur les os longs des tétrapodes. VII. Sur la classification, la signification fonctionelle et l'histoire des tissus osseux des tétrapodes. *Annls Paléont., 61:* 49–129.

RIDEWOOD, W. G., 1904. On the cranial osteology of the fishes of the families Elopidae and Albulidae, with remarks on the morphology of the skull in the lower teleostean fishes generally. *Proc. zool. Soc. Lond., 1904,* 2: 35–81.

ROMER, A. S., 1937. The braincase of the Carboniferous crossopterygian *Megalichthys nitidus. Bull. Mus. comp. Zool. Harv., 82:* 1–73.

ROMER, A. S., 1947. Review of the Labyrinthodontia. *Bull. Mus. comp. Zool. Harv., 99:* 1–368.

ROMER, A. S. & PRICE, L. I., 1940. Review of the Pelycosauria. *Spec. Pap. geol. Soc. Am., 28:* x+538 pp.

ROSEN, D. E., 1962. Comments on the relationships of the North American cave fishes of the family Amblyopsidae. *Am. Mus. Novit., 2109:* 1–35.

ROSEN, D. E., 1973. Interrelationships of higher euteleostean fishes. In P. H. Greenwood, R. S. Miles & C. Patterson (Eds), *Interrelationships of fishes:* 397–513. London: Academic Press.

ROUX, G. H., 1947. The cranial development of certain Ethiopian 'insectivores' and its bearing on the mutual affinities of the group. *Acta zool., Stockh., 28:* 165–397.

SCHAEFFER, B., 1971. The braincase of the holostean fish *Macrepistius,* with comments on neurocranial ossification in the Actinopterygii. *Am. Mus. Novit., 2459:* 1–34.

SMITH, C. L., RAND, C. S., SCHAEFFER, B. & ATZ, J. W., 1975. *Latimeria,* the living coelacanth, is ovoviviparous. *Science, N.Y., 190:* 1105–6.

SMITH, H. M., 1947. Classification of bone. *Turtox News, 25:* 234–6.

STARCK, D., 1967. Le crâne des mammifères. In P. P. Grassé (Ed.), *Traité de Zoologie, 16,* 1: 405–549. Paris: Masson.

STARKS, E. C., 1926. Bones of the ethmoid region of the fish skull. *Stanf. Univ. Publs (Biol. Sci.), 4:* 137–338.

STENSIÖ, E. A., 1921. *Triassic fishes from Spitzbergen. Part I:* xxviii+307 pp. Wien: Holzhausen.

STENSIÖ, E. A., 1963. Anatomical studies on the arthrodiran head. Part I. *K. svenska VetenskAkad. Handl.,* (4) *9,* 2: 1–419.

STENSIÖ, E. A., 1969. Arthrodires. In J. Piveteau (Ed.), *Traité de Paléontologie, 4,* 2: 71–692. Paris: Masson.

TARLO, L. B. H., 1964. The origin of bone. *Proc. First Europ. Bone and Tooth Symp.:* 3–15.

TAVERNE, L., 1973. Sur la présence d'un dermosupraoccipital chez les Mastacembelidae. *Revue Zool. Bot. afr., 87:* 825–8.

TAVERNE, L., 1974. L'ostéologie d'*Elops* Linné, C., 1766 (Pisces Elopiformes) et son intérêt phylogénétique. *Mém. Acad. r. Belg. Cl. Sci. 8°,* (2) *41,* 2: 1–96.

TOEPLICH, C., 1920. Bau und Entwicklung des Knorpelschädels vom *Didelphys marsupialis. Zoologica, Stuttg., 27,* 70: 1–83.

TRUEB, L., 1973. Bones, frogs and evolution. In J. L. Vial (Ed.), *Evolutionary biology of the Anurans:* 65–132. Columbia: University of Missouri Press.

VOGT, C., 1842. Embryologie des salmones. In J. L. R. Agassiz (Ed.), *Histoire naturelle des Poissons d'eau douce de l'Europe centrale:* vi+348 pp. Neuchâtel.

WAKE, D. B., 1970. Aspects of vertebral evolution in the modern Amphibia. *Forma Functio, 3:* 33–60.

WARWICK, R. & WILLIAMS, P. L., 1973. *Gray's anatomy,* 35th ed.: xv+1471 pp. London: Longman.

WATERMAN, R. E., 1970. Fine structure of scale development in the teleost, *Brachydanio rerio*. *Anat. Rec.*, *168:* 361–80.

WATSON, D. M. S., 1925. The structure of certain palaeoniscids and the relationships of that group with other bony fish. *Proc. zool. Soc. Lond., 1925:* 815–70.

WESTOLL, T. S., 1967. *Radotina* and other tesserate fishes. *J. Linn. Soc. (Zool), 47:* 83–98.

WESTON, J. A., 1970. The migration and differentiation of neural crest cells. *Adv. Morphogen., 8:* 41–114.

YAMADA, J., 1971. A fine structural aspect of the development of scales in the chum salmon fry. *Bull. Jap. Soc. scient. Fish., 37:* 18–29.

Placoderm interrelationships reconsidered in the light of new ptyctodontids from Gogo, Western Australia

R. S. MILES

British Museum (Natural History), London

AND

G. C. YOUNG

Queen Elizabeth College, London

Recent work on placoderm interrelationships (notably by Westoll and Denison) is critically reviewed and a new set of hypotheses is constructed. These hypotheses embrace all the major placoderm taxa as well as the basic divisions of the arthrodires. A new classification of placoderms is proposed. Two new ptyctodontids are described from the Frasnian Gogo formation: *Campbellodus decipiens* gen. et sp. nov. and *Ctenurella gardineri* sp. nov. The first has toothplates showing characters of species referred hitherto to *Ptyctodus* or *Rhynchodus*, but is nevertheless distinct from all adequately described ptyctodontids. The second is separated from the type species *C. gladbachensis* by the character of its dermal ornament.

CONTENTS

INTRODUCTION

Science is not just a collection of laws, a catalogue of unrelated facts. It is a creation of the human mind, with its freely invented ideas and concepts. [Physical] theories try to form a picture of reality and to establish its connection with the wide world of sense impression. Thus the only justification for our mental structures is whether and in what way our theories form such a link.

(Albert Einstein & Leopold Infeld)

In this paper we tackle the problem of the interrelationships of placoderms. We have been stimulated to do this by the opportunity to describe remarkable new ptyctodontids from the Upper Devonian of Western Australia and the consequent need to evaluate some of their morphological features. At the same time it has been necessary to question the soundness of some current views about placoderms. These ptyctodontids and their provenance are described in the second part of this paper.

There are several existing classifications of placoderm fishes. These classifications tend, in our view, to be almost uniformly misleading. They are normally complete in the sense that they include all known higher taxa in a prescribed system, but they give a wrong view of our knowledge of placoderm phylogeny and evolution by suggesting a far greater understanding of these subjects than we have. Many of these classifications do not appear to be based on any serious phylogenetic thinking, although equally they do not seem to be defensible as special purpose classifications of conspicuous utility.

Placoderm studies are now seriously bedevilled in our view by the existence of too many singular morphological statements and too few generalizations. Several recent works have dramatically underlined the fact that students of placoderms are in danger of amassing a pointlessly large and undiscriminating quantity of observations which, if the situation is allowed to develop further, will ultimately have a stultifying effect on new work.

We believe that our knowledge of placoderms requires organizing in a series of high level concepts which will permit generalizations to be made about structure and function. For us these concepts must take the form of an explicit phylogeny which can be expressed in a phylogenetic classification. Of course this programme may not be capable of achievement at the present time, but we suggest that observations need to be made with this end in view if the study of placoderms is to flourish. As Hempel (1966: 12) notes, 'What particular sorts of data it is reasonable to collect is not determined by the problem under study, but by a tentative answer to it that the investigator entertains in the form of a conjecture or hypothesis.' In this paper then, we re-view the problem of placoderm interrelationships with the express intention of giving direction to future research.

Part I. Placoderm interrelationships

In as much as new hypotheses originate with the criticism of old hypotheses, our poor knowledge of placoderm interrelationships may be due to the absence, over many years, of a suitable initial supposition. Many authors appear to have been more interested in buffering their conclusions against criticism than in presenting them as clear-cut, testable hypotheses. However, Westoll (1967) and Denison (1975) have now published valuable, explicit accounts of their views, which form the foundations of this paper. Our aim in the first part of this paper then, is to reopen the problem of placoderm interrelations with a critical appraisal of Westoll's and Denison's work.

Our testing procedure employs the criterion of parsimony (see e.g. Miles, 1975). An established hypothesis of the relationships of placoderms to Recent gnathostomes would undoubtedly help in testing theories of placoderm interrelations. Such a hypothesis does not exist. Instead there are two hypotheses of equal logical probability:

(A) placoderms are the collateral descendants of all other gnathostomes;
(B) placoderms are more closely related to chondrichthyans than to other gnathostomes.

We have not been able to discriminate between these hypotheses for this paper*. We shall refer to them both, as set out in Fig. 1, as hypotheses A and B.

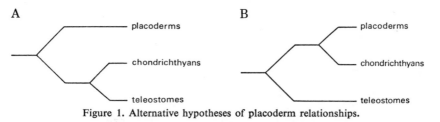

Figure 1. Alternative hypotheses of placoderm relationships.

WESTOLL'S DICHOTOMY HYPOTHESIS

Westoll (1967: 96) suggests that there are two major groups of placoderms. '(a) The group represented by Macropetalichthyida, Radotinida, Rhenanida

---

*Two initial suppositions which are fundamental to our arguments are (1) elasmobranchs and holocephalans are immediately related, i.e. the chondrichthyans are a monophyletic group (Moy-Thomas & Miles, 1971; Zangerl & Case, 1973); (2) the ptyctodontids are most closely related to other placoderms, with these last fishes also forming a monophyletic group. We reject, therefore, any suggestion that the ptyctodontids or other placoderms are directly ancestral to or are the collateral descendants of the holocephalans (Ørvig, 1960, 1962, 1971; Stensiö, 1963, 1969; Stahl, 1967). An analysis of the characters which have been put forward in support of these rejected hypotheses is beyond the scope of this paper (but see Patterson, 1965), although some new, relevant observations are included. The problem of shared ptyctodontid-holocephalan characters has been worsened in our view by a tendency to interpret some ptyctodontid characters, which are otherwise phylogenetically obscure, after a holocephalan model. That is to say, some of the evidence has been interpreted in the light of the hypothesis it is said to support.

and Stensioellida (at least in part). In this group the occipital region is long and there are both anterior and posterior paranuchals in the skull-roof.

'(b) The Euarthrodira (including Ptyctodontida), and their probable derivatives the Antiarchi. In these the occipital region is short and there is only one pair of paranuchal plates; but the earliest members (e.g. *Kujdanowiaspis*) have much longer occipital regions than the later forms.'

Initially this division is open to three possible interpretations.

(1) Both groups (a) and (b) are monophyletic taxa with uniquely derived characters. They originated from a group with unspecified primitive characters (Fig. 2A), but possibly comprising either no occipital region and no paranuchals or a very long occipital region and three or more paranuchals. This is the hypothesis favoured by Westoll (1967: 96).

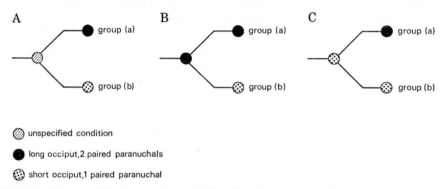

Figure 2. Three possible interpretations of Westoll's dichotomy hypothesis of placoderm interrelationships.

(2) Group (a) is a grade group characterized by primitive characters: group (b) is a monophyletic group united by uniquely derived specialized characters as set out by Westoll (Fig. 2B).

(3) Group (a) is a monophyletic group united by uniquely derived specialized characters as set out by Westoll: group (b) is a grade group characterized by primitive characters (Fig. 2C).

Hypothesis (1) requires the specification of a third, primitive condition, and evolutionary changes in two lines of descent. It is therefore unparsimonious in comparison with hypotheses (2) and (3), from which neither of these consequences arise. We reject it, therefore, without further discussion.

An attempt will be made below to discriminate between hypotheses (2) and (3), which at the outset have an equal logical probability.

DENISON'S PHYLOGENY

Denison has developed the view that an anteroposteriorly short trunk-shield is primitive and that a phyletic classification of placoderms is possible by dint of the distinctive specializations of the various groups. Denison's phylogeny is given in simplified form in Fig. 3. We have made a serious attempt to present this phylogeny in its strongest form for discussion. Denison's nomenclature for the taxa is followed in this section. The actinolepid arthrodires present a major problem in transfiguring the original, more complex phylogenetic diagram. Denison shows them as a grade group more closely related to antiarchs and phyllolepids (as direct ancestors) than to other arthrodires.

Denison lists 18 primitive placoderm characters, on the evidence of stensioellids (s.s.=stensioellids s.l. minus pseudopetalichthyids) and ptyctodontids. Only eight of these characters are immediately relevant to the purpose of this paper. They are as follows:

(1) 'The ventral shoulder girdle consists of a single pair of plates homologous either to the interolaterals or anterior ventrolaterals of Arthrodira; between them a median plate has been identified only in Ptyctodontida.

(2) 'The lateral shoulder girdle consists only of anterior laterals and anterior dorsolaterals, except in some Acanthothoraci where posterior laterals and posterior dorsolaterals are also present.

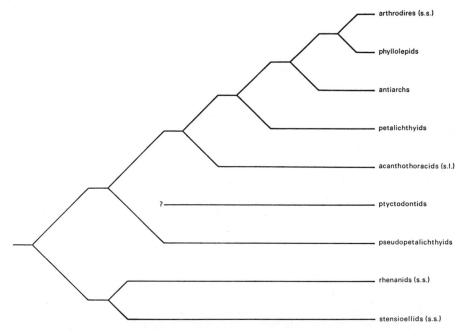

Figure 3. Denison's phylogeny of placoderms redrawn as a series of dichotomies.

(3) 'The spinal plates are absent, or small and doubtfully distinct, except in Acanthothoraci and some Ptyctodontida.

(4) 'A median dorsal plate is probably absent in Stensioellida and Pseudopetalichthyida.

(5) 'The neurocranium is long and slender with a long occipital region, except in Ptyctodontida where it must have been short.

(6) 'The dermal cranial roof bone pattern may be variable and unstable with relationships between bones and sensory canals not firmly established, except in Ptyctodontida.

(7) 'Dermal cranial roof bones may be small and part of the roof may be

covered with thin, superficial tesserae in Acanthothoraci and Rhenanida; much of the skull in Stensioellida is covered with denticles or tesserae; the central part of the cranial roof of Pseudopetalichthyida is covered with small dermal bones, but there may have been denticles or tesserae elsewhere. Denticles or tesserae are unknown in Ptyctodontida, but may have covered the snout and cheeks where dermal bones are largely absent.

(8) 'Gill covers (submarginals) may be present, though they are not known in Acanthothoraci and their dermal bones are small in Ptyctodontida'.

Accepting for the moment that these characters have been correctly evaluated, Denison's phylogeny can be criticized for its lack of parsimony. The following undesirable consequences follow from it.

(1) The anterior median ventral plate has arisen separately in ptyctodontids and the common ancestor of the petalichthyids, antiarchs, phyllolepids and arthrodires.

(2) The posterior lateral plate has arisen separately in acanthothoracids (s.l.) and arthrodires.

(3) The median dorsal plate has arisen separately in rhenanids (s.s.) and the common ancestor of ptyctodontids, acanthothoracids (s.l.), petalichthyids, antiarchs, phyllolepids and arthrodires.

(4) The trunk-armour has been shortened with the loss of plates in phyllolepids, involving a reversal of evolution.

(5) The dermal neck-joint has been lost in phyllolepids, involving a reversal of evolution.

(6) A stable cranial roof pattern with stable relations between the bones and lateral-lines has arisen separately in ptyctodontids and the common ancestor of petalichthyids, antiarchs, phyllolepids and arthrodires.

(7) Superficial roofing tesserae have been lost independently in the ptyctodontids and the common ancestor of the petalichthyids, antiarchs, phyllolepids and arthrodires.

(8) A single paired paranuchal plate has been acquired independently in ptyctodontids and the common ancestor of antiarchs, phyllolepids and arthrodires (Denison states that two pairs of paranuchals are primitive on p. 12).

(9) One further criticism of Denison's scheme is that it brings the stensioellids (s.s.) and rhenanids (s.s.) together on the evidence of primitive characters only. That is, in a grade group of unknown phylogenetic validity.

We shall attempt below to replace Denison's phylogeny with a new, more parsimonious hypothesis of placoderm interrelationships. But we wish to state now that we have not fully succeeded. It must be recognized at the present time that it is simply not possible to produce a convincing hypothesis of the relations of some placoderm taxa. Our approach appears to differ in one important respect from Denison's, in that we have been guided by a low-level evolutionary law (or rule of thumb) which predicates the regressive development of the skeleton in early vertebrates (see e.g. Jarvik, 1960: 57). Thus we have been more ready to accept the multiple loss of bones than their multiple origin.

Denison notes alternative relationships for petalichthyids, phyllolepids and ptyctodontids. His suggestions do not affect our discussions at this stage, but some of the points will be taken up below.

ANALYSIS OF SOME POSSIBLY IMPORTANT CHARACTERS

In approaching the problem of placoderm interrelations, our starting point is the statement in Moy-Thomas & Miles (1971: 198) that: 'it is not possible to say whether the trunk-shield ever extended posteriorly on the flank behind the pectoral fins in petalichthyids, ptyctodontids, phyllolepids and stensioellids, as it does in arthrodires and antiarchs, and we cannot say whether the long type of trunk-shield found in primitive arthrodires is primitive for all placoderm groups'.

Three questions are outstanding from the papers of Westoll and Denison.

(1) What do we mean by long and short trunk-shields and what is the primitive composition of the shield?

(2) What is the significance of roofing tesserae in Lower Devonian placoderms?

(3) Has the primitive skull-roof one or two pairs of paranuchal plates?

Trunk-shield

Denison suggests that the primitive, short shield comprises only single paired ventral (interolateral or anterior ventrolateral) and lateral (anterior lateral) plates. 'The justification for this assumption is the fact that a short exoskeletal shoulder girdle occurs in all other groups of fishes with bony exoskeletons, and a short scapulocoracoid is characteristic of Chondrichthyes'. This last point, concerning

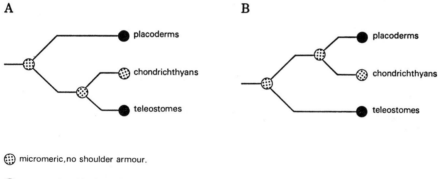

Figure 4. Evolution of the shoulder armour in alternative gnathostome phylogenies.

the chondrichthyan scapulocoracoid, appears to be irrelevant. With respect to the first point, if it is accepted that chondrichthyans are primitively micromeric and have no dermal shoulder-girdle (Nelson, 1970), then by either hypothesis A or B of placoderm relations (Fig. 4) Denison's assumption can have no *phyletic* justification. The shoulder-girdle of teleostomes and the trunk-shield of placoderms must have arisen separately. Consistent with this conclusion is the lack of one-to-one bone homologies between teleostomes and placoderms, and the absence of true, lateral exoskeletal shoulder-girdle bones in acanthodians, which are hypothesized to be primitive teleostomes (Miles, 1973b, c).

It is necessary to decide which hypothetical primitive structure will give the most parsimonious interpretation of trunk-shield evolution in the various placoderm groups. The occurrence of individual plates is given in Table 1. Certain conventions have been adopted in constructing this table, to enable us to handle unsolved problems of homology. They are: (a) the single paired plate in the anterior ventral position of ptyctodontids and rhenanids (s.s.) is termed the anterior ventrolateral, not the interolateral; (b) the anterior median dorsal of antiarchs is the median dorsal of other placoderms, the posterior median dorsal is a posterior dorsal plate secondarily incorporated from the back; (c) the semilunar plate of antiarchs is an anterior median ventral and the median ventral is a posterior median ventral plate.

Further points are that Westoll (1967) indicates the presence of a median dorsal plate in pseudopetalichthyids, and it is generally accepted that there is a spinal plate in *Pseudopetalichthys* and rhenanids (s.s.) (Gross, 1962, 1963). We believe there is no anterior median ventral plate in *Pseudopetalichthys:* the 'interclavicule possible' of Stensiö (1969: fig. 198) being the 'Rest des Endocraniums' of Gross

Table 1. Maximum number of plates present in individual placoderm groups

	MD	ADL	PDL	AL	PL	IL	SP	AVL	PVL	AV	AMV	PMV
acanthothoracids (s.l.) =palaeacanthaspids	×	×	×	×		×	×	×				
antiarchs	×	×	×		×		?	×	×		×	×
arthrodires (s.s.)	×	×	×	×	×	×	×	×	×	×	×	×
petalichthyids	×	×	×	×		×	×	×			×	
phyllolepids	×	×		×		×	×	×	×		×	
pseudopetalichthyids	×	?		×		×	×	×				
ptyctodontids	×	×		×			×	×			×	
rhenanids (s.s.)	×	×		×			×	×				
stensioellids (s.s.)				?			?	?				

(1962: fig. 8B). We do not accept (cf. p. 128 above) on present evidence that there is a posterior lateral plate in acanthothoracids (Stensiö, 1944; Gross, 1959; Ørvig, 1975). We do not accept that the two paired plates lying behind the anterior ventrolaterals in *Lunaspis* (from the Hunsrückschiefer, Gross, 1961) comprise true posterior ventrolaterals. They may be enlarged trunk scales such as occur *behind* the posterior ventrolaterals in the arthrodire *Sigaspis* (Goujet, 1973). The posterior median ventral has not been found in *Lunaspis* and is here assumed not to exist. There is no other evidence of posterior ventrolateral and posterior median ventral plates in petalichthyids. In gemuendinids, Stensiö (1959, 1969) has indicated the presence of a small posterior ventrolateral plate, but this has not been confirmed in Gross's (1963) study of *Gemuendina*. We suggest that Stensiö has merely delimited a posterior process of the anterior ventrolateral. We assume that the posterior ventrolateral plate is not developed in gemuendinids. Finally, we do not accept that anterior ventral plates have been demonstrated in acantho-thoracids (s.l.), in spite of Ørvig's (1975) account of *Romundina*.

From the observations recorded in Table 1, we propose the hypothesis that primitive placoderms possessed median dorsal, anterior dorsolateral, anterior lateral, interolateral, spinal, anterior ventrolateral and anterior median ventral plates. Posterior lateral and posterior median ventral plates appear to be unique to arthrodires and antiarchs, posterior ventrolaterals to these groups and phyllolepids, and anterior ventrals to actinolepid arthrodires. We feel strongly that observations on poorly preserved and refractory specimens of pseudopetalichthyids and stensio-ellids (s.s.) from the Hunsrückschiefer should not be given much weight in any discussion of trunk-shield or skull structure.

Our hypothesized primitive trunk-shield is closely matched by that of petalich-thyids. This type of shield is regarded as lengthened by Denison, and by that fact as specialized. To avoid misunderstanding, therefore, we propose to recognize two clear-cut types of trunk-shield, which we define as follows.

(A) A short shield with the plates listed above, primitive for placoderms, and in which there are no plates on the flank behind the pectoral fin; posterior laterals are absent.

(B) A long shield, in which the ventral and lateral plates meet on the flank to enclose the base of the pectoral fin; posterior laterals are present either as separate plates or combined with posterior dorsolaterals to form mixilaterals.

Significance of roofing tesserae

Roofing tesserae are found in rhenanids (s.s.) and acanthothoracids (s.l.). The condition of the dermal head skeleton in stensioellids (s.s.) and pseudo-petalichthyids is disputed (Gross, 1962; Westoll, 1967; Denison, 1975) but there is no unequivocal evidence of tesserae. Contrary to Denison's suggestion, there is no evidence of tesserae in ptyctodontids despite the existence of exquisitely preserved specimens of *Ctenurella* spp. (see the second part of this paper).

The tesserae around the anterior part of the skull of the petalichthyid *Lunaspis* will be implicitly included in this discussion. Strictly they belong to the cheek region (Gross, 1961), but although poorly known they are probably amenable to the same analysis as roofing tesserae.

According to either hypothesis A or B of placoderm relationships (Fig. 1), it is reasonable to assume that these fishes had primarily micromeric ancestors (Nelson, 1970). The question then, appears to be whether the tesserae of tesserate placo-derms are primitive hold-overs from primarily micromeric ancestors (Gross, 1959; Denison, 1975) or manifestations of a secondary micromerism. There is no question of each group of placoderms acquiring its own pattern of plates inde-pendently, from a micromeric ancestor, on grounds of economy of hypothesis. As Denison (1975: 11) points out, 'The pattern of dermal bones on the skull . . . shows enough similarities to suggest that . . . it was derived from a common ancestral pattern'. In most groups 'This pattern includes some or all of the follow ing: (1) median nuchal, postpineal, pineal and rostral; (2) paired centrals over the otic region; (3) paired paranuchals and marginals carrying the main lateral line forward; (4) paired pre- and postorbitals over the orbits; (5) paired postnasals beside the nostrils, and (6) paired suborbitals, postsuborbitals, postmarginals and submarginals in the cheek and opercular region'. A parallel may be noted here with the trunk–shield, which as we have already suggested, had a macromeric structure in the common ancestor of all the known groups (Fig. 4).

The tesserae of acanthothoracids (s.l.) are said by Gross (1958, 1959) to be essentially scales, homologous with those of the body. If this is true (cf. Stensiö, 1963, 1969), the hypothesis that the tesserae are secondary structures involves a reversal in evolution in the direction of the primitive state of the placoderm exoskeleton. If, however, the tesserae are primitive, we are left with a hypothesis that implies their loss in at least two different lines of evolution: in normally-plated forms like arthrodires and in acanthothoracids. This is because Ørvig (1975) has clearly demonstrated the absence of tesserae in *Romundina*, and tesserae appear to be absent in *Palaeacanthaspis* (Stensiö, 1944) and specimens of *Holopetalichthys* (Gross, 1958, 1959). Thus Occam's razor will not easily discriminate between these hypotheses because both are, to some extent, uneconomical.

Westoll (1967) and Denison (1975) have discussed the development of the tesserae, their morphological relations to deep-lying flanges of the dermal plates, and the variable relations of lateral-lines in tesserate forms. There is no need for us to review these topics here, except to note Westoll's conclusion that the specimens do not show directly whether tesserae are primitive or advanced. One problem, clearly recognized by Westoll, is that skeletogenetic (i.e. ontogenetic) processes may have been confused with phylogenetic processes, and some discussions (e.g. Gross, 1959; Denison, 1975) involve extrapolations from ontogeny to phylogeny that may not be justified.

No final decision is possible about the nature of tesserae. However for the purposes of this paper we accept that they are likely to be secondary structures (Stensiö, 1963, 1969; Westoll, 1967; Ørvig, 1975). Their occurrence is reasonably consistent with our phylogenetic conclusions although their exact phylogenetic significance remains unknown. We adopt this hypothesis because it harmonises with our conclusion that the trunk-shield was primitively built up of a well-defined series of large plates. This means that the thoracic tesserae of rhenanids (s.s.) must also be regarded as secondary structures, which is concordant with the highly-specialized habit of these fishes (Gross, 1963; Stensiö, 1969). We wish to stress Westoll's (1967: 91) conclusion that 'It is far from certain that the main dermal bones are not all present even in the most tesserate specimens', and suggest that Mark-Kurik's (1973a) account of *Kimaspis* is consistent with this view.

Paranuchal plates

The primitive number of paranuchal plates is difficult to decide on the evidence of placoderm fishes themselves, and the problem cannot be solved by reference to other groups because homologues of these plates are not known. The problem might possibly be attacked, however, if we accept that there is a primary correlation between (a) two paranuchal plates and a long occipital region, and (b) one paranuchal and a short occipital region, as Westoll (1967) implies. This correlation permits comparisons with other groups of fishes. It is clear, however, that this approach can never be fully satisfactory, for among arthrodires there are forms (albeit highly specialized) with one paranuchal and, to judge from the head-shield, a long occipital region (e.g. *Homostius, Tityosteus*). Furthermore, there are flatly contradictory accounts of the length of the occipital region in some placoderms. Thus Stensiö (1963: 39) regards the occipital region as short in rhenanids (s.s.), possibly short in radotinids and fairly long in early ('dolichothoracid') arthrodires and acanthothoracids (s.s.). Westoll (1967) regards the occipital region as long in rhenanids (s.s.), radotinids and acanthothoracids (s.s.), and fairly long in primitive arthrodires. Denison (1975) regards the occipital region as long in stensioellids (s.s.), rhenanids (s.s.), pseudopetalichthyids and acanthothoracids (s.s.).

On balance this evidence appears to support Denison's contention that a long occipital region and, therefore, two pairs of paranuchals, is primitive. However, if we turn to other gnathostomes for corroboration of this view (de Beer, 1937) there appears to be no reason for believing that a lengthened occipital region is

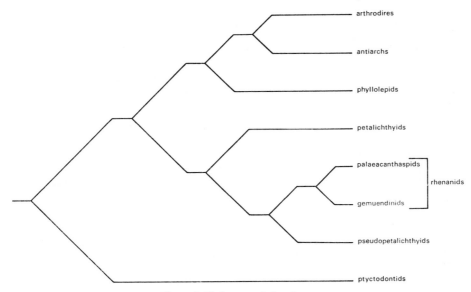

Figure 5. Phylogeny of placoderms.

primitive for placoderms, by either hypothesis A or B of placoderm relations (Fig. 1). Thus we can produce no clear answer to the question of whether one or two pairs of paranuchals is primitive. But we propose to adopt the former solution, simply on the grounds that trial and error has shown it to be the more useful hypothesis for our work. This leads to the conclusion that group (a) of Westoll (p. 125, above), with petalichthyids, radotinids, rhenanids (s.s.) and pseudo-petalichthyids, comprises fishes more closely related to each other than to other placoderms.

INTERRELATIONSHIPS OF PLACODERMS

Figure 5 is a phylogeny of the major groups of placoderms, some of whose relations can be readily determined from the foregoing discussion.

A major line of descent comprises the arthrodires and antiarchs. These fishes are specialized in having a long trunk-shield, as defined above, with posterior lateral, posterior ventrolateral and posterior median ventral plates.

They are primitive in having one paired paranuchal plate. Tesserae are absent from the skull. We are confident that both are monophyletic groups. Antiarchs are specialized, inter alia, in their pectoral appendage, two median dorsal plates, uniquely constructed dermal neck-joint (but see below) and short, broad head-shield with closely-placed, dorsal orbits. It is less easy to pick out unique specializa-tions of arthrodires, although the pectoral fenestra (modified into an incision in later forms, Miles, 1969), two pairs of upper toothplates (anterior and posterior superognathals) and large posterior postorbital processes (correlated with the width of the skull-roof across the posterolateral angles) may be examples. We suggest that additional analysis of placoderm dermal skull patterns and endo-cranial structure will provide further examples (cf. Denison, 1975: 15, 16), but this work is beyond the scope of the present paper.

A second major line of descent comprises petalichthyids, palaeacanthaspids (=acanthothoracids sensu Denison), gemuendinids (=rhenanids sensu Denison) and pseudopetalichthyids. These fishes are specialized in having two pairs of paranuchals and a long occipital region. They are primitive in having a short trunk-shield, as defined above. Tesserae are present in some members but do not have a clear-cut taxonomic distribution with phylogenetic significance.

Two caveats must be entered against the unification of these fishes as a second line of descent. One, Westoll's (1967) evidence for the presence of two paranuchals in gemuendinids (one specimen of *Gemuendina*) and pseudopetalichthyids (one specimen of *Pseudopetalichthys*) is slim, and requires strengthening by further observation. Two, problems raised by the palaeacanthaspid *Romundina* must be clarified. Here the bone pattern has been traced by radiography. Ørvig (1975) describes a mesial paranuchal bearing the endolymphatic duct and a lateral paranuchal carrying the main lateral-line. If we accept Denison's conclusion that relations between sensory lines and dermal bones were not firmly established in these fishes (p. 127, above), it is possible that the first of these bones is the anterior paranuchal and the second the posterior paranuchal plate. Alternatively, the pattern may be incorrectly shown, with true conditions closer to those of *Kosoraspis* and *Kimaspis*. But it is fruitless to speculate on this last point.

The petalichthyids are the most primitive branch of this second line of descent. Primitive features include the full complement of plates in the trunk-shield and the ventral position of the nasal openings. The dermal neck-joint is, however, an advanced feature and has, as far as can be determined at present, an unique construction (Stensiö, 1969). Tesserae are known (*Lunaspis*) but are restricted to the cheek.

The non-petalichthyid forms (palaeacanthaspids, gemuendinids and pseudo-petalichthyids) are united by the loss of the anterior median ventral plate. This plate is retained in all other groups of placoderms. Although poorly known, and not to be relied upon for trustworthy knowledge, the pseudopetalichthyids (*Pseudopetalichthys, Paraplesiobatis;* possibly the same form) may be more primitive than other non-petalichthyid members of this line in having ventral nasal openings. They are, however, highly specialized in their jaws.

The palaeacanthaspids and gemuendinids may be united as the rhenanids, which are characterized by dorsal nasal openings (Miles, 1969; Moy-Thomas & Miles, 1971). The presence of a premedian plate on the upper surface of the rostrum may be associated with this character, but it has not been clearly demonstrated in the gemuendinids. It must be regarded as a parallel development to the premedian plate of antiarchs. Roofing tesserae occur widely within the group. The palaeacanthaspids are the more primitive forms, with a high trunk-shield and lateral orbits. They are united as a monophyletic group by the possession of a skull-roof with a deeply embayed posterior margin and strongly projecting posterior paranuchal plates (Denison, 1975: 6). The gemuendinids (Stensiö, 1950, 1969; Gross, 1963) are uniquely specialized in their ray-like habit, and are also separated from palaeacanthaspids by their small, dorsal orbits and reduced trunk-shield (with loss of posterior dorsolateral and interolateral plates; Table 1). They may be unique among placoderms in having many small sclerotic plates.

Parenthetically, we note that new palaeacanthaspids described by Mark-Kurik (1973a, *Kimaspis*) and Ørvig (1975, *Romundina*) remove any doubt that acanthothoracids (s.s.), *Kolymaspis* and radotinids belong to the same monophyletic group. It is clearly nonsense to separate Acanthothoraci and Radotinida as distinct orders of placoderms. Denison notes that primitive palaeacanthaspids probably had ventral nasal openings. This view is based on the original account of poorly preserved material of *Palaeacanthaspis* (Stensiö, 1944). Professor Stensiö has kindly provided the British Museum (Natural History) with a latex cast of the original skull-roof of this form (P50036), and we are able to confirm Westoll's (1967) reinterpretation, in which the nasal openings are shown to have a dorsal position.

Three groups remain for consideration, the stensioellids (s.s.), phyllolepids and ptyctodontids. Of these, the first (comprising *Stensioella* and *Nessariostoma;* possibly the same form, Gross, 1962) is the most difficult to deal with because of the poor preservation. It is impossible to separate dermal and endoskeletal bones in the shoulder region, or to distinguish a pattern of plates or tesserae in the skull. Denison (1975) argues that stensioellids are placoderms because the gills have an anterior position under the neurocranium, there is a cranio-vertebral joint, and there are dermal bones. These reasons are not completely convincing, but in this paper we accept that stensioellids are placoderms for want of an acceptable alternative hypothesis. Denison also concludes that stensioellids (s.s.) have many primitive characters but no specializations to link them with other groups. Nevertheless, like Gross (1963), he allies them most closely with rhenanids. In our opinion, the stensioellids are too poorly known to support a feasible hypothesis of their relationships.

Gross (1937) classified the phyllolepids and ptyctodontids with 'normal' arthrodires (Euarthrodira) and petalichthyids in the Arthrodira *sensu lato*. It is doubtful that this classification has ever been defended for sound phylogenetic reasons. However it has been maintained by various authors, notably by Ritchie (1973) in recent years (p. 140, below).

Phyllolepis is the only genus of phyllolepid that we recognise. It has one paired paranuchal, no tesserae and a short trunk-shield with a posterior ventrolateral plate. The pattern of plates in the head-shield is in some respects unique, but there are no known specializations here which link this genus with any of the groups considered so far. Denison derives *Phyllolepis* from primitive arthrodires, although this would involve the secondary acquisition of a short trunk-shield (in our sense) and the secondary loss of the dermal neck-joint. He does this on the evidence of *Antarctaspis* which is said, 'in some ways to bridge the gap between *Phyllolepis* and primitive Actinolepina'. This evidence (not itemized by Denison) presumably comprises the presence of a large nuchal plate, onto which the supraorbital and central sensory lines pass from the preorbital and postorbital respectively (White, 1968). However the supraorbital canal also passes onto the nuchal plate in petalichthyids, ptyctodontids and *Wuttagoonaspis* (see below); and the central sensory line passes onto the nuchal plate in some specimens of the arthrodire *Baringaspis* and in *Wuttagoonaspis*. There are, therefore, no sound reasons for linking *Antarctaspis* and *Phyllolepis* on the evidence of the lateral-lines. The phyllolepid pattern of canals may be primitive for placoderms, or the pattern may have been subject to convergent evolution. Moy-Thomas & Miles (1971) have related *Wuttagoonaspis* to *Phyllolepis* on the evidence of the dermal ornament, large nuchal plate and sensory line pattern. We believe this suggestion to be quite wrong. The relationships of *Wuttagoonaspis* are discussed below. We have also considered that *Phyllolepis* may be most closely related to ptyctodontids by virtue of the lack of dermal bones on the snout and cheek (Obruchev, 1964), but have not been convinced by this argument.

We now tentatively suggest that phyllolepids might be most closely related to the common ancestor of arthrodires and antiarchs. These three groups are unique in having posterior ventrolateral plates. As already noted, arthrodires and antiarchs are advanced, relative to *Phyllolepis*, in having a long trunk-shield with posterior lateral and posterior median ventral plates.

The ptyctodontids are primitive in their short trunk-armour, single paranuchal plate and absence of tesserae. They are possibly specialized in the loss of the interolateral and posterior dorsolateral plates (cf. gemuendinids, pseudopetalichthyids and phyllolepids), and in the reduction of the head-shield. These characters have not led us easily to a defensible hypothesis of relationships. Ørvig (1962) related ptyctodontids to petalichthyids on the grounds that both groups have a large 'centronuchal' plate on which the supraorbital and posterior pit-line canals meet in an X. However this character is not in accord with the divergent structure of the skull in these groups. Petalichthyids are specialized in having two paired paranuchals and an elongated occipital region (above), and also differ in other respects from ptyctodontids in the dermal bone pattern (Westoll, 1962). The two groups may also be divergently specialized in the dermal neck joint (see below). Furthermore, the pattern of canals is shared with *Phyllolepis* and *Wuttagoonaspis*, as we have already noted (also *Pseudopetalichthys*), and it may be a primitive placoderm character. Denison (1975) did not reach a definite conclusion about ptyctodontid relationships. However, because 'they show many resemblances' to arthrodires (s.s.), petalichthyids and some acanthothoracids (s.l.), he concluded: 'they probably had an ancestor with a short shoulder girdle and the basic placoderm cranial bone pattern. The Radotinidae are the only known group that satisfies these conditions, but because of their elongated skull and dorsal nostrils, their relationships to Ptyctodontida will be questioned . . .'. We question Denison's proposed phylogeny principally on the grounds that ptyctodontids do not share the specialized occipital region and two paired paranuchal plates of palaeacanthaspids and their relatives.

The ptyctodontids are morphologically well known placoderms (see the second part of this paper) and we feel that it should be possible now to propose a plausible hypothesis of their relationships. But clearly this is difficult because the group exhibits, for the most part, a mixture of primitive characters and specializations not shared by any other group. A bold new conjecture is required that can be put in testable form.

We suggest that the ptyctodontids are the collateral descendants of all other placoderms (excluding stensioellids, which cannot be considered). This hypothesis

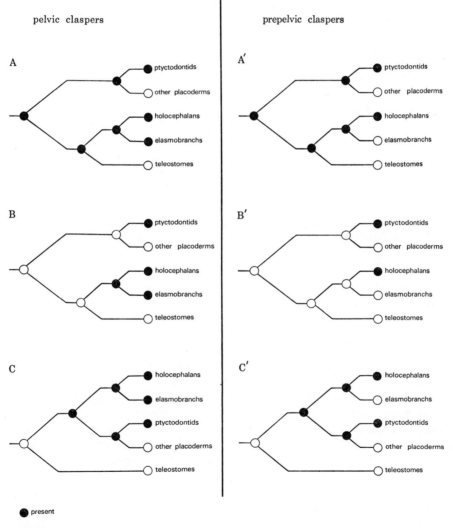

Figure 6. Evolution of pelvic and prepelvic claspers in alternative phylogenies with these structures considered as primitive (A, A¹) and derived (B, B¹; C, C¹) gnathostome characters.

is prompted by the presence of prepelvic and pelvic claspers in ptyctodontids. We suggest that non-ptyctodontid placoderms are uniquely specialized in the absence of these structures.

It is possible to examine the consequences of this hypothesis of ptyctodontid relationships against our two hypotheses of placoderm relationships. According to hypothesis A of placoderm relationships (Fig. 1A), pelvic claspers may be either primitive for gnathostomes (Fig. 6A) or independently acquired in two lines (Fig. 6B). In the first case, pelvic claspers have been lost independently in non-ptyctodontid placoderms and teleostomes. In the second case, they have evolved separately in ptyctodontids and chondrichthyans and the basis of our proposed phylogeny is destroyed. We are, however, reluctant to accept this if prepelvic claspers are also considered. The first case (Fig. 6A') requires the triple loss of these structures (in non-ptyctodontid placoderms, elasmobranchs, tele-ostomes). But the second case is still less acceptable in requiring the separate origin of prepelvic claspers in ptyctodontids and holocephalans (Fig. 6B'), or, alternatively, in ptyctodontids and primitive chondrichthyans followed by their secondary loss in elasmobranchs.

According to hypothesis B of placoderm relationships (Fig. 1B), pelvic and prepelvic claspers are most parsimoniously explained as a unique specialization of elasmobranchiomorphs (chondrichthyans plus placoderms) (Fig. 6C, C'). This has the consequence that both pelvic and prepelvic claspers are secondarily absent in non-ptyctodontid placoderms, and prepelvic claspers are secondarily absent in elasmobranchs.

We do not claim that the taxonomic distribution of claspers confirms our hypothesis of ptyctodontid relationships. Rather, claspers provide a limited means of criticizing the hypothesis on grounds of parsimony. An improved test might follow from a clear decision between alternative hypotheses of placoderm relations. Meanwhile, we suggest that our hypothesis of ptyctodontid relations offers the possibility of an economical explanation of the taxonomic distribution of claspers, and at the same time is reasonably consistent with other aspects of placoderm structure.

NECK-JOINT AND POSTERIOR DORSOLATERAL PLATE

The dermal neck-joint and posterior dorsolateral plate present problems which have not been included in the above discussion. It is important, however, that these problems should not be overlooked. The evidence from the neck-joint is the more important and has led us to a competing hypothesis of ptyctodontid relationships. Although we have come to reject this alternative hypothesis, for reasons given below, we think it has sufficient merit to be worth discussing.

Stensiö (1959: 207–8) compared the supposedly concave glenoid condyle of *Rhynchodus* with the 'reversed' dermal joint of antiarchs, but noted that the condyle 'apparently represents the glenoid process in other euarthrodires'. This interpretation of ptyctodontids is, however, not valid (see the second part of this paper). Ørvig's comparison between ptyctodontids and petalichthyids also no longer applies; the petalichthyid joint itself (Ørvig, 1957) is difficult to understand from a functional viewpoint, and we consider it poorly known. Observations on the anterior dorsolateral plate and its function in the joint are urgently required. In other placoderms the joint is well known in arthrodires and antiarchs, and there seem to be three basic types: the familiar ginglymus or hinge-joint of phlyctaenioids (Phlyctaenii plus Brachythoraci), the sliding joint of actinolepids and the reversed hinge-joint of antiarchs.

The structural differences between the antiarch and phlyctaenioid joints have generally been considered of great phyletic significance (e.g. White, 1952; Miles, 1967b), but the functional importance of the joint reversal could possibly have been over emphasized. For either joint to develop, a contact between the paranuchal and anterior dorsolateral plates would have to form in such a way as to allow a widening zone of limited movement along the transverse axis of rotation. From this condition the more stable phlyctaenioid or antiarch joint could readily develop by the differentiation of an articular fossa on the paranuchal or the anterior dorsolateral. The joint in ptyctodontids has developed to a stage where transverse elongation has occurred, but it lacks a distinct articular fossa and correspondingly convex condyle. The absence of these structures could be considered to represent either an incipient stage in the complete development of the ginglymus of phlyctaenioids, or a secondarily reduced condition.

The sliding joint of the actinolepids differs from the hinge-joint in ptyctodontids, antiarchs and phlyctaenioids in the development of flanges to effect a controlled articulation (see discussion of *Wuttagoonaspis*, below). If primitive arthrodires are assumed to lack a dermal neck-joint, as is the case in several other placoderm groups, the sliding and hinge joints developed within the arthrodires must be seen as divergent specializations (Miles, 1973a: 115).

The above observations can be interpreted in various ways, but there are only two basic interpretations which seem worth considering here.

(1) Phlyctaenioid arthrodires and antiarchs have become divergently specialized in the dermal hinge-joint from a primitive condition represented by ptyctodontids. On this evidence these three groups are more closely related to each other than to other placoderms but their own interrelationships cannot be determined.

(2) Ptyctodontids and phlyctaenioids are immediately related by dint of the possession of fundamentally the same type of specialized hinge-joint. They are jointly the collateral descendants of the antiarchs (with a specialized 'reversed' joint), all three groups being united by the possession of a transversely-elongated zone of contact between the head and trunk shields.

Reasons for rejecting both of these interpretations can be found in the preceding discussion of placoderm interrelations, and we shall not discuss them in detail. However, the following points may be noted. They lead us to conclusions which require the multiple origin of the dermal neck-joint in placoderms.

(1) Both interpretations imply that phlyctaenioids are more closely related to antiarchs and ptyctodontids than to actinolepids. That is to say, they require the sunderance of the Arthrodira, which on other grounds we regard as a monophyletic group.

(2) The second interpretation and possible phylogenetic conclusions from the first interpretation require either the secondary development of a short trunk-shield in ptyctodontids or the independent acquisition of a long shield in arthrodires and antiarchs.

(3) If the independent acquisition of a long shield in antiarchs and phlyctae-nioids is accepted it is also necessary to postulate the independent acquisition of a long shield in actinolepids.

(4) If the first interpretation is accepted in the form that phlyctaenioids and antiarchs are more closely related to each other than to ptyctodontids, it is necessary to postulate either the indepedent origin of a long shield in actinolepids and the common ancestor of antiarchs and phlyctaenioids, or—if a long shield is held to be primitive for all of these fishes—its secondary reduction in ptyctodontids.

Finally we note that this discussion of the neck-joint and placoderm inter-relations would be transfigured by reverting to the assumption (see e.g. Miles, 1969: 146–7) that the phlyctaenioid joint originated from the sliding surfaces of actinolepids.

The posterior dorsolateral plate presents a different sort of problem. It is not clear whether this plate is a primitive placoderm feature or has arisen as a subsequent specialization. Our proposed phylogeny leaves us with the following unanswered question. Has the posterior dorsolateral plate evolved independently in palaeacanthaspids, petalichthyids and the common ancestor of arthrodires and antiarchs, or has it been independently lost in phyllolepids, gemuendinids, pseudopetalichthyids and ptyctodontids? A satisfactory decision cannot be made on grounds of parsimony.

ARTHRODIRES AND THE RELATIONSHIPS OF WUTTAGOONASPIS

Denison's (1975) concept of arthrodire (s.s.) phylogeny is similar to ours in a number of important respects, although he does not represent his results clearly in a formal, phylogenetic classification. We wish only to discuss *Wuttagoonaspis fletcheri* Ritchie (1973), an arthrodire ranked purely on phenetic grounds by Denison, because this species is most directly relevant to our discussion of placoderm interrelations. The trunk-shield of this species closely resembles that of actinolepid arthrodires, and its head-shield is said to show resemblances to petalichthyids, ptyctodontids and phyllolepids (Ritchie, 1973). Ritchie concludes that *W. fletcheri* represents a new family of euarthrodires (sensu Gross, 1932), which is monotypic, and that 'the new evidence provided by the *Wuttagoonaspis* material concerning possible placoderm relationships supports the classification earlier proposed by Denison (1958)'. Thus arthrodires are classified in the following way:

Superorder Arthrodira
    Order Euarthrodira
        Suborder Arctolepida (=Dolichothoraci)
        Suborder Wuttagoonaspida
        Suborder Brachythoraci
    Order Phyllolepida
    Order Petalichthyida
    Order Ptyctodontida

If we assume that this classification is a phylogenetic arrangement, it states, inter alia, that *Wuttagoonaspis* is most closely related to dolichothoracids and brachythoracids; and that dolichothoracids and *Wuttagoonaspis* form jointly the sister group of the brachythoracids, or—just conceivably—that all three groups originated from a single ancestral species. However, Ritchie's classification is not consistent with some statements in his text, e.g. 'There is nothing to indicate a close relationship (of *Wuttagoonaspis*) with the Brachythoraci', and 'it (*Wuttagoonaspis*) may provide a link between the arctolepids and the petalichthyids'.

In the past, hypotheses about the relationships of placoderms have depended in great measure on dermal bone patterns. We agree with most of Ritchie's bone determinations in *Wuttagoonaspis*, but as a first step will re-examine his interpretation of the head-shield because we disagree with some of his phylogenetic conclusions.

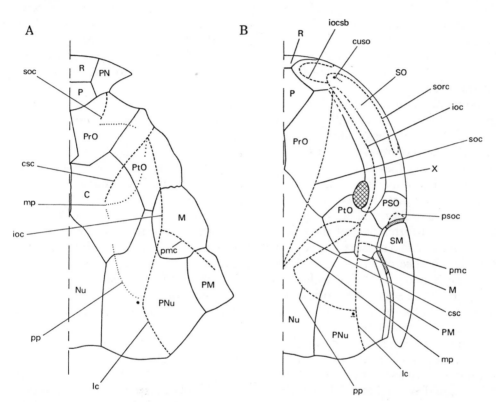

Figure 7. Dermal bone and sensory line patterns in A, *Kujdanowiaspis* and B, *Wuttagoonaspis*. After Stensiö and Ritchie.

The head of Wuttagoonaspis

The names given to dermal plates depend upon hypotheses about their phylogenetic homology. Such hypotheses can be tested by deriving propositions about the topographic relationships of the plates which can be checked against the law of constancy of morphological relations. As a standard of comparison in checking *Wuttagoonaspis*, we shall use the head-shield of actinolepid arthrodires. There are no serious disputes about the homologies of plates within this group (three recent papers are Goujet, 1973; Mark-Kurik, 1973b; Miles, 1973a), or between this group and other arthrodires (s.s.). The skulls of *Wuttagoonaspis* and an actinolepid (*Kujdanowiaspis*) are compared in Fig. 7.

The preorbital region of the skull is greatly elongated in *Wuttagoonaspis* in comparison with actinolepids. We shall assume that this has affected the proportions, but not the fundamental topographic relations of some of the plates. We agree with Ritchie's determinations for the paranuchal, preorbital, postorbital, pineal, rostral, postsuborbital, postmarginal and submarginal plates, and for all the lateral-lines and pit-lines except the 'rostral (or ethmoid) commissure'. We shall now discuss the remaining plates.

The postnasal of Ritchie's account (*SO*, Fig. 7B) shows most unusual relations for this plate in comparison with other arthrodires (Miles, 1971b). It borders the head-shield from the rostral plate to almost the hind level of the orbital fenestra, and makes contact posteriorly with the postsuborbital plate (see below). The infraorbital canal branches on its surface, and the posterior branch has been compared by Ritchie (correctly we think) with the 'supramaxillary' (supraoral) line of other arthrodires. A pit for sensory cells lies posterior to the anterior loop of this canal.

None of these characters is consistent with the interpretation of this plate as a postnasal. But they are all perfectly consistent with its interpretation as a suborbital. The conclusion that it is a suborbital enables us to interpret the 'rostral (or ethmoid) commissure' as the suborbital branch of the infraorbital sensory line; and to predict that the plate bears the same relations to the palatoquadrate and functional jaw margins (the gnathals are unknown) as the suborbital of other arthrodires (Miles, 1971b).

Ritchie's 'suborbital' (*X*, Fig. 7B) lies immediately mesial to the plate just discussed. It forms part of the margin of the orbital fenestra and carries a length of the infraorbital sensory line. These features are consistent with his interpretation, but the absence of the supraoral sensory line is not. We shall refer to this plate as bone X and are unable to homologize it with any known plate in arthrodires (s.s.). It shows some similarities to the lateral plate of antiarchs, but it may be a new formation in *Wuttagoonaspis*, correlated with the specialization of the preorbital region of the skull.

The plate boundaries are not clearly visible around the anterior region of bone X. It is this region, apparently, where the supraorbital and infraorbital sensory lines meet. We are not convinced that postnasal plates are present, but if they are, they should occur here unless their topographic relations are strongly modified. The supraorbital sensory line does not meet the suborbital branch of the infraorbital line, as might be predicted from conditions in other arthrodires, but the more posterior postorbital branch of this line. This may be a further consequence of the specialization of the snout. The relations of bone X to the suborbital and infraorbital sensory line prevent us concluding that the entire bone is the postnasal.

Immediately posterior to the orbital fenestra, Ritchie describes an area which, 'in some individuals appears to be occupied by a single plate while in others there is clear evidence for the presence of two plates'. He describes them as postorbital and marginal elements. For the purpose of our analysis we shall assume that two plates are present, and we suggest that the topographic evidence embodied in Fig. 7 fully supports Ritchie's determinations.

There remains only the large posteromedian plate of the skull-roof, the 'centronuchal'. Our disagreement with Ritchie is trivial and lies only in his assumption that elements of the centrals have fused with the nuchal plate. He makes this assumption to account for the presence of supraorbital and central lateral-lines, and the middle and posterior pit-lines. However this assumption can receive no empirical support in *Wuttagoonaspis*, as it must remain impossible in this fossil to distinguish between ontogenetic fusion and loss-and-invasion (Jardine, 1969). The name 'centronuchal' can be applied as a convention to indicate the topographic relations of the bone, but we prefer simply to name it the nuchal. For us then, the plate in question is simply a large nuchal with sensory lines converging on its radiation centre. As already noted, this disposition of the sensory lines may be primitive for placoderms. The specimens of *Wuttagoonaspis* described so far have no central plates.

Relationships

With Ritchie and Denison we accept that *Wuttagoonaspis* is an arthrodire (s.s.). The structure of the trunk-armour is consistent with this view (Denison, 1975), and the head-shield provides no contradictory evidence. However, its relationships among arthrodires can only be discussed successfully if a rigorous phylogenetic framework is adopted. This is what Ritchie and Denison have failed to do. Denison fails because his suborder Actinolepina is conceptually a grade group of primitive arthrodires, from which all other groups of arthrodires have stemmed.

In the classification of Miles (1973a), the Actinolepidoidei are construed as a monophyletic group sharing a sliding dermal neck-joint in which the anterior dorsolateral plate bears an anterior flange. Their collateral descendants, the Phlyctaenioidei, are a monophyletic group sharing a ginglymoid dermal neck-joint with a lateral articular fossa on the anterior dorsolateral plate and a lateral articular fossa on the paranuchal plate. The primitive condition is held to be the absence of a dermal neck-joint, as can be seen in some other placoderm groups.

According to this formulation, *Wuttagoonaspis* is an actinolepid by virtue of its sliding neck-joint. It is, however, specialized in the position of the orbits and dermal bone pattern in the skull. At the same time, it lacks the anterior ventral plates which characterize actinolepids sensu stricto. These plates are not found elsewhere among placoderms and appear to be an uniquely derived specialization.

Wuttagoonaspis cannot, by any economical hypothesis, be closely related to petalichthyids and phyllolepids, because it shares unique specializations with arthrodires and not with these groups (above). Ritchie's approach to the dermal skull pattern of petalichthyids appears to us to be misconceived, and therefore any results which flow from it must be suspect ('comparison of *Wuttagoonaspis* and the petalichthyids suggests how the dermal plate pattern of the latter may be interpreted in a way which brings them back firmly into the arthrodire fold'). We suggest that the disputed dermal bone pattern of petalichthyids can only be interpreted against a stable, generally-agreed model, such as the actinolepids (s.s.) we have used to interpret *Wuttagoonaspis*. It is wrong to interpret this pattern by comparison with *Wuttagoonaspis*, which, being somewhat specialized, can itself only be interpreted by careful comparison with a fixed actinolepid model.

FORMAL CLASSIFICATION

We summarize our hypotheses of placoderm relationships in the following classification.

(CLASS ELASMOBRANCHIOMORPHI?)
Subclass Placodermi
  A. Superorder Dolichothoracomorpha
      I. Order Arthrodira
          1. Suborder Actinolepidoidei
              i. Infraorder Actinolepidi
              ii. Infraorder Wuttagoonaspidi
          2. Suborder Phlyctaenioidei
             iii. Infraorder Phlyctaenii
             iv. Infraorder Brachythoraci
      II. Order Antiarcha
      III. Order Phyllolepida
  B. Superorder Petalichthyomorpha
      IV. Order Petalichthyida
      V. Order Rhenanida
          3. Suborder Palaeacanthaspidoidei
          4. Suborder Gemuendinoidei
      VI. Order Pseudopetalichthyida
  C. Superorder Ptyctodontomorpha
Placodermi *Incertae sedis: Stensioella, Nessariostoma.*

Part II. New ptyctodontids from Gogo, Western Australia

The following account of ptyctodontid structure is based on material collected in 1967 from marine limestones of Frasnian age at Gogo, Western Australia, by a joint British/Australian expedition (Brunton, Miles & Rolfe, 1969). The Gogo collection has yielded a diverse and exceptionally well-preserved fish fauna (Miles, 1971a, b; Gardiner & Miles, 1975), although ptyctodontids are relatively uncommon and only nine specimens have been identified. Of these, seven belong to a new species of *Ctenurella* Ørvig, previously known from Devonian rocks in the Bergisch Gladbach area east of Cologne (Ørvig, 1960). The new information on ptyctodontid structure presented here is largely based on these specimens. The remaining two specimens comprise fragmentary remains of a large ptyctodontid, which we describe below as **Campbellodus decipiens gen. et sp. nov.**

The Gogo fishes are preserved in calcareous nodules amenable to acetic acid preparation (Toombs & Rixon, 1959). Three ptyctodontid specimens have been completely removed from the matrix and the remainder prepared as resin transfers thus providing additional information on the mutual relationship of individual bones. All specimens are more or less disarticulated, however, although each nodule appears to have contained a complete individual. The bones are uncrushed and undistorted but their small size and fragile nature has inevitably resulted in some loss during preparation. The following descriptions are therefore incomplete, especially with regard to delicate perichondral ossifications, but it is to be expected that extensive collections awaiting preparation in Australia will greatly enlarge upon the information presented here. Our descriptions are intended to facilitate such further preparation and study.

Some points of terminology have arisen in dealing with the perichondrally ossified parts of the endocranium and visceral arch skeleton. It is important in interpreting such remains to differentiate between the space lying between peri-chondral laminae, which in life contained cartilage, and the cranial cavity and the various canals and foramina communicating with it. The former is here termed the interperichondral space. The margins of perichondrally ossified structures are said to be 'closed' if completely invested in a continuous perichondral layer, and 'open' when they exhibit two laminae separated by the interperichondral space.

The Gogo specimens are numbered with the prefix P, and are currently held by the British Museum (Natural History). The types will eventually be housed in the Western Australian Museum, Perth.

Subclass	PLACODERMI
Superorder	PTYCTODONTOMORPHA
Family	PTYCTODONTIDAE

Campbellodus gen. nov.

Etymology. After Dr K. S. W. Campbell, Canberra.

Diagnosis. A moderately large ptyctodontid with a subsemicircular pineal plate; preorbitals probably long and narrow, meeting in the midline behind the pineal and with a deeply embayed orbital margin; distinct dermal antorbital and post-orbital processes; postorbital plate strongly arched with a sharp dorsal angle and notched posteroventral margin for a postorbital fenestra in the skull-roof. Submarginal plate with flat external surface and much expanded posteriorly. Endocranium at least partially ossified perichondrally; autopalatine part of palatoquadrate perichondrally ossified and showing on its mesial face at least two areas of attachment for the ethmoidal part of the endocranium; anterior attach-ment area near the ventral margin some distance from anterior end; second attachment area of irregular, elongate shape and lying dorsal to the first; lateral face of autopalatine showing deep muscle insertion area near anterior border but lacking a depression for the dorsal process of the upper toothplate. Upper tooth-plate of variable shape with or without a beak and sometimes with a blunt dorsal process; lower toothplate longer than upper with an anterior beak and flat broad biting surface with a straight lateral edge and a convex mesial edge bearing several rounded cusps. Trunk-shield probably including reduced spinal plate and small dorsal spine behind median dorsal. Pelvic fin possibly covered with dermal scales in females. External ornament on the preorbital comprising irregular, crowded tubercles sometimes forming radiating rows near the periphery and set in a deeply-pitted cancellous surface, modified on the submarginal to longitudinally aligned struts of bone, lacking tubercles and separating elongate perforations of variable length; postbranchial ornament of multicuspid denticles extending over two-thirds of the surface of the mesial lamina of the anterior lateral plate; spinal plate and dorsal spine ornamented with scattered tubercles and occasional connecting ridges.

Type species. **Campbellodus decipiens sp. nov.**

Geological occurrence. Lower Frasnian Gogo formation, Canning Basin, Western Australia.

Remarks. On toothplate morphology alone the material dealt with here could be referred either to *Ptyctodus* Pander (1858) or *Rhynchodus* Newberry (1873), both erected for isolated teeth. Numerous species, also based on teeth, have been ascribed to both genera and it has generally been accepted (Woodward, 1891: 37; Eastman, 1908: 123; Ørvig, 1960: 316) that laterally compressed teeth with a thin sectorial biting edge should be referred to *Rhynchodus*, and those with a broad triturating surface to *Ptyctodus*. Nevertheless, Newberry's original definition of *Rhynchodus* (1873: 309) seems to have included teeth with a 'triturating or cutting edge . . . broader and fitted for crushing mollusks and other food'. The precise basis on which the two genera were distinguished by Newberry is therefore unclear.

Ptyctodontid teeth undergo marked changes in shape as a result of wear (Hussakoff & Bryant, 1918: pl. 40, figs 1–3), and our material suggests that isolated upper and lower toothplates from the same species could be assigned to different genera in the absence of other preserved remains. That many early determinations are unreliable is indicated by the fact that upper and lower plates were often not recognized as such (e.g. Eastman, 1908: pl. 3, fig. 8). The validity of *Rhynchodus* and *Ptyctodus*, even as form genera, is therefore open to question.

The first dermal plates associated with ptyctodontid teeth were described from Wildungen by Jaekel (1903, 1906, 1919, 1929) under the generic names *Rhamphodus*, *Rhamphodontus*, *Rhynchodontus* and *Rhynchognathus*. Gross (1933) reviewed the Wildungen ptyctodontids and referred all Jaekel's species to *Rhynchodus tetrodon* (Jaekel) whilst noting the presence of three types of upper toothplates in the material. The more recently described genera *Rhamphodopsis* Watson (1934) and *Ctenurella* Ørvig (1960) show numerous morphological differences and are clearly valid genera, yet they also have teeth of *Rhynchodus* type, and it is clear that teeth alone are not generally diagnostic. Although both Watson and Ørvig have implicitly regarded the Wildungen material as belonging to Newberry's genus, we feel that the doubtful validity of early determinations could necessitate resurrection of one of Jaekel's generic names. A revision of this material is however beyond the scope of the present work, and following a previous procedure (Miles, 1967a: 101) we retain Gross's nomenclature in the subsequent descriptions.

We conclude that the genera *Ptyctodus* and *Rhynchodus* can best be restricted to their type species, or used merely as (albeit unsatisfactory) form genera for isolated teeth. We therefore follow Watson (1938: 408) in erecting a new genus for the material described below. Differences in toothplate morphology in the two specimens are discussed below.

Campbellodus gen. nov. may be distinguished from all other genera by its dermal ornament, from *Chelyophorus* by the relation between the preorbitals and pineal, from *Rhynchodus* (as represented by the Wildungen species), *Rhamphodopsis* and *Ctenurella* by the shape of the preorbital, the extent of the postbranchial ornament, and probably by the morphology of the toothplates. From *Rhynchodus* and *Ctenurella* it differs in the arrangement of the attachment areas on the mesial autopalatine surface, and from *Ctenurella* in the shape of the postorbital plate, the mode of formation of the postorbital fenestra, the presence of a deep muscle insertion area on the lateral autopalatine surface, the possible enclosure within the autopalatine of the dorsal process of the upper toothplate, the presence of a dorsal spine and spinal plates in the trunk-shield, and possibly the morphology of the pelvic fins.

Campbellodus decipiens sp. nov.
(Plates 1, 2A, B, F and 4A)

1969: tooth-plates resembling those of *Rhynchodus;* Brunton, Miles & Rolfe.

1971: tooth-plates which recall those of *Ptyctodus;* Miles (1971a).

1975: *Ptyctodus* sp.; Gardiner & Miles.

Etymology. L. *decipiens*, deceiving; an allusion to the misleading role of tooth-plates in ptyctodontid classification.

Diagnosis. As for genus (only species).

Holotype. P50905, a specimen comprising disarticulated right upper (incomplete) and lower toothplates, a right preorbital (incomplete) and postorbital, the post-branchial lamina of a right anterior lateral, and ascending lamina of a right anterior ventrolateral, a right submarginal and a few indeterminable fragments.

Other material. P50907, a specimen comprising a left upper toothplate, a left spinal plate, an incomplete left interolateral plate, an incomplete dorsal spine, a pelvic fin scale, several perichondrally ossified endocranial pieces and various indeterminable exoskeletal and endoskeletal fragments.

Localities. 73 and 89 near Gogo Station, northwestern Australia (see map in Miles, 1971b: fig. 1).

Remarks. The second specimen (P50907) has a distinctive upper toothplate and is therefore only tentatively referred to this species. Tooth morphology is an unreliable generic character in ptyctodontids, and its variability within genera is unknown, although some intraspecific variation is noted below in the description of a new species of *Ctenurella*. The only skeletal part preserved in both specimens is the ascending lamina of the anterior ventrolateral, which in both cases shows a similar and distinctive postbranchial ornament. The same ornament is also seen in *Ctenurella*, however, so this resemblance is not significant. Study of further material is necessary to corroborate our hypothesis that both specimens belong to the same species.

Skull-roof and endocranium

The preorbital plate is characterized by a deeply embayed orbital margin (*om*) between distinct antorbital and postorbital angles (*aoa, ptoa*, Fig. 8B). It is incomplete mesially, but appears to be longer and narrower than the preorbital of *Rhynchodus tetrodon*: Gross (Miles 1967a: pl. 5, figs 2, 3). Both anteromesial and anterolateral corners are notched, the former for the pineal plate (*nP*); the latter shows a thick irregular margin which may be broken but is similar to this region in *Ctenurella* (see below). An anterolateral notch is not developed in *Rhynchodus* or *Rhamphodopsis* (Miles, 1967a). Closely spaced pores on the external surface show the course of the supraorbital sensory canal (*soc*). The pores open through raised tubercles which tend to coalesce into an irregular ridge anterolaterally. A second row of sensory pores runs along the posterior part of the orbital margin (*pfc*). The posterolateral margin of the plate is irregular and formed an interlocking suture with the postorbital. The preorbital is convex over the orbit, with a corresponding depression on the visceral surface (*dep*), delineated by irregular preorbital and supraorbital ridges (*rpro, rsuo*). A third low ridge (*soc,* Fig. 8A) carries the sensory canal forward to the radiation centre of the plate, where an area of cancellous bone attached to the visceral surface (*sov*) probably represents the base of the supraorbital vault.

The postorbital plate is strongly arched (Fig. 9C) and in this respect resembles that of *Rhynchodus* (Miles, 1967a: pl. 5, fig. 2), rather than the flat postorbital of

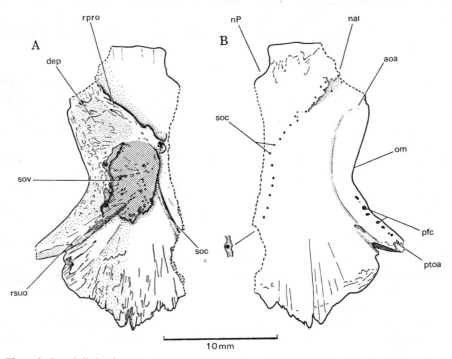

Figure 8. *Campbellodus decipiens* gen. et sp. nov. Right preorbital plate in A, visceral and B, external view. P50905.

Ctenurella (see below). A slight anterior angle (*aa*) separates the embayed orbital margin (*om*) from a short dorsal section which abutted against the postorbital angle of the preorbital plate. A most interesting feature is the deep notch (*pln*) in the posterolateral margin. The edge of the bone is thin in this region and it seems probable that the notch entered into the formation of a postorbital fenestra as described below for *Ctenurella* (*fepto,* Fig. 17A). The concave visceral surface of the plate shows a ridge (*pfc*), running down from the dorsal angle (*da*), which carried the profundus sensory canal as in *Ctenurella*. Sensory pores cannot be distinguished from the ornament on the outer surface, but the canal probably opens in an enlarged foramen just inside the orbital margin (*fpfc*). Ventrally there is a broad, low postorbital ridge (*rpto*).

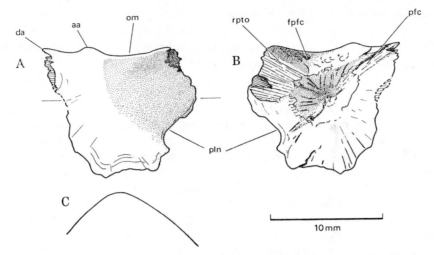

Figure 9. *Campbellodus decipiens* gen. et sp. nov. Right postorbital plate in A, external and B, visceral view. C, a profile in the plane indicated in A. P50905.

The submarginal (Fig. 10) is difficult to orientate but is thought to be a right plate. It has a slightly concave dorsal and slightly convex ventral margin. The plate broadens considerably towards the posterior end. Both outer and inner surfaces are fairly flat, with a slight inner convexity developed anteriorly where the bone reaches its maximum thickness (*th*). A thin flange (*fl*) is inflected inwards around

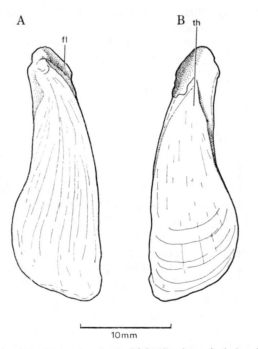

Figure 10. *Campbellodus decipiens* gen. et sp. nov. Right (?) submarginal plane in A, external and B, internal view. P50905.

the anterior margin. This plate is much broader than previously described ptyctodontid submarginals (Gross, 1933: fig. 16H; Ørvig, 1962; Miles, 1967a). It clearly
did not carry a sensory canal, as is also concluded below for the submarginal of
Ctenurella, contrary to earlier opinion (Ørvig, 1971).

Jaws and visceral arch skeleton

Three toothplates and a portion of perichondrally ossified autopalatine are
described here. As noted above, the upper toothplates in the two specimens differ
in many respects. That in P50905 (Plate 2A), probably from the right side, is
incomplete dorsally and the form of the lateral and mesial lamellae is not known.
The inner surface has been largely eroded. The tooth is laterally compressed and
produced downward anteriorly to form a beak behind which a shearing surface
slopes to a sharp cutting edge on the labial side, as seems to be the case in *Rhynchodus tetrodon* (Gross, 1933, see also Stensiö, 1969: fig. 153). In *Rhamphodopsis*
(Watson, 1938) and *Ctenurella* (Ørvig, 1960) this sharp edge lies on the lingual side
when developed (see below). Behind this surface is a rounded cusp. The preserved
form of the biting edge is similar to that of a number of previously described teeth,
for example *Ptyctodus ferox* (Eastman, 1908: fig. 21) and *Rhynchodus tetrodon*
(Gross, 1933).

The upper toothplate in the second specimen is distinctive in several respects
(Plate 1A, B). It is generally broader with a thick, rounded biting edge, and it
lacks a beak. The unusual shape of the dorsal process with its uniform breadth
and truncated dorsal margin is quite unusual. The mesial surface is grooved just
behind the anterior margin, as is generally the case in ptyctodontids (Gross, 1967:
fig. 21). To our knowledge this specimen does not closely resemble any previously
described ptyctodontid toothplate.

The lower toothplate in P50905 is considerably longer than the upper. By
comparison with *Ctenurella* (see below) its slight curvature in dorsal view (Plate
1D) and the configuration of the ventral lamella on the straight side (more extensive
posteriorly) suggest that it is a right plate. It has a broad, flat biting surface with a
straight outer margin and two low cusps on the convex inner margin. Anteriorly
it is produced upwards to form a short beak. Amongst previously described teeth
it is somewhat similar to *Ptyctodus calceolus* (Hussakof & Bryant, 1918: fig. 34).

The fragmentary autopalatine ossification (Fig. 11) may be oriented by comparison with *Ctenurella* (Fig. 27). The mesial surface shows two attachment areas
($aorb_1$, $aorb_2$). Since the only closed margin must be anterior (see below), these
may be compared with *Ctenurella* by assuming the second attachment area has (a)
a more elongate and irregular shape, and (b) a more anterodorsal position with
respect to the first attachment area. In both forms an elevated neck of bone (n,
Figs. 11, 27D) connects the two areas, and the first attachment area has an oblique
position near the ventral margin. In addition, the outer surface thus oriented shows
a deep depression (*dmad*), which although not observed in *Ctenurella* is similar
in shape and position to that figured for *Rhynchodus tetrodon* (Gross, 1933: pl. 7,
fig. 4). The adductor mandibulae muscles probably originated here.

The relationship of the dorsal process of the upper toothplate to the autopalatine
in both *Rhynchodus* and *Campbellodus* is poorly understood. The lateral autopalatine surface lacks a deep depression for its reception although such a depression is

developed in *Ctenurella* (*ddpr*, Fig. 27A). Stensiö (1963: fig. 124) assumed that the process lies within the autopalatine in *Rhynchodus*, as is perhaps suggested by Gross's figured specimen. In this specimen the disposition of attachment areas on the mesial autopalatine surface does not correspond closely to *Campbellodus;* the ventral area in *Rhynchodus* has a more anterior position and the dorsal is smaller and almost circular in shape, much as is described below for *Ctenurella*. However *Campbellodus* resembles another Wildungen specimen figured by Gross (1933: fig. 16), and reconstructed by Stensiö (1969: fig. 153), which is characterized by an elongate rather than circular dorsal attachment area. *Campbellodus* is evidently more extensively ossified than this last form.

These differences indicate some experimentation in jaw muscle arrangement and the mode of attachment between the palatoquadrate and the endocranium. Variation in ossification to strengthen different regions is probably a related factor, and these characters when better known could prove to be phylogenetically significant.

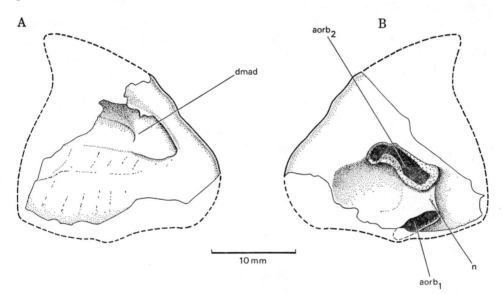

Figure 11. *Campbellodus decipiens* gen. et sp. nov. Right autopalatine in A, lateral and B, visceral view. P50905.

Trunk armour

Five examples of trunk armour plates occur in the material but only the anterior ventrolateral* is represented in both specimens.

The mesial lamina of the right anterior lateral is well preserved in P50905 (Plate 2F) and is closely comparable in size and shape to that of *Rhynchodus tetrodon* (Gross, 1933: pl. 9, fig. 5; also Stensiö, 1959: fig. 74), although the peculiar postbranchial ornament is more extensively distributed. Examination of a plaster cast (P15325) of the *Rhynchodus* specimen (the holotype of *Rhynchodontus*

---

*We acknowledge that this plate has many of the characteristics of the arthrodiran interolateral. The term anterior ventrolateral is used here without prejudice to be consistent with its use in the first part of this paper.

eximius Jaekel) suggests that the postbranchial ornament may be similar in detail to that of *Campbellodus*. A fragment of the ascending lamina of the right anterior ventrolateral also from P50905 has precisely the same overlap arrangement as in *Rhynchodus tetrodon*. In P50907 the left plate is represented by mesial parts of the ascending and ventral laminae.

The two remaining trunk armour bones come from P50907. A spinal plate (Fig. 12) shows a long, narrow, dorsal overlap area for the supraspinal lamina of the anterior lateral (*oaAL*). Ventrally this bone seems to have abutted against the infraspinal lamina of the anterior ventrolateral (*cfAVL*) although it was overlapped by that plate mesially (*oaAVL*). In *Rhynchodus tetrodon* the spinal forms a much longer contact with the supraspinal than the infraspinal lamina, and on this basis we assume this plate comes from the left side. Posteriorly it is produced into a short, hollow spine (*sp*) which presumably contained a lateral prepectoral process of the scapulocoracoid. Using the toothplate as a rough standard, the spinal is relatively smaller than in *Rhynchodus tetrodon*, and much smaller than in *Rhamphodopsis*.

Also preserved is part of a hollow dorsal spine (Fig. 13) with a deeply grooved posterior face for the first dorsal fin as in *Rhamphodopsis* (Miles, 1967a: 107). This spine is also relatively much smaller than the corresponding spine of *Rhamphodopsis*.

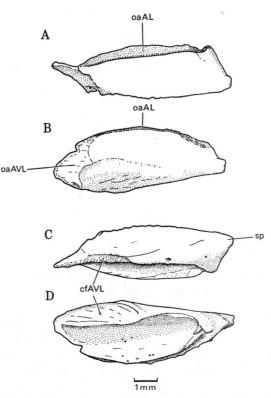

Figure 12. *Campbellodus decipiens* gen. et sp. nov. Left spinal plate in A, dorsal, B, lateral, C, ventral and D, mesial view. P50907.

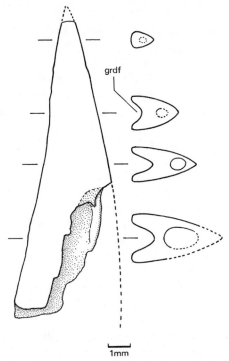

Figure 13. *Campbellodus decipiens* gen. et sp. nov. Dorsal spine in lateral view with transverse sections in the planes indicated. P50907.

Axial skeleton, paired and median fins

Perichondral fragments preserved in both specimens probably include parts of the vertebral column and endoskeleton of the paired and median fins, all well ossified in *Ctenurella* (Ørvig, 1960). They are too incomplete to be determined, but a small dermal element (Fig. 14) corresponds in size and shape to the overlapping scales seen in the pelvic fins of female *Rhynchodus* and *Rhamphodopsis* (Gross, 1933: pl. 10, fig. 7; Miles, 1967a: figs 13, 14). As in *Rhamphodopsis* this element shows a smooth overlap area in its anterior half for the scale in front. *Campbellodus* thus differs in pelvic fin morphology from *Ctenurella* (Ørvig, 1960; see below), which lacks such scales.

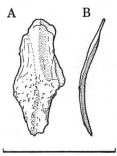

Figure 14. *Campbellodus decipiens* gen. et sp. nov. Pelvic scale in A, ventral and B, lateral view. P50907.

Ornament

The ornament on the preorbital and postorbital plates comprises small, rounded tubercles, densely but irregularly distributed over a deeply-pitted, cancellous bone surface. The tubercles tend to be aligned in radiating rows along some parts of the plate margins, and larger tubercles, sometimes coalesced into an ill-defined ridge, mark the course of the supraorbital sensory canal through the preorbital. The coarse anastomosing ridges seen in *Rhamphodopsis* (Watson, 1938), *Ctenurella* (Ørvig, 1960) and *Rhynchodus* (Ørvig, 1971: fig. 7K) are not developed, although fine struts of bone may occasionally stretch between the bases of adjacent tubercles. On the submarginal, tubercles are completely absent and the cancellous texture shows a longitudinal alignment with short, transverse struts subdividing deep, elongate grooves of variable length. As in *Rhamphodopsis* (Watson, 1938: 401), the ornament of the dorsal and lateral spines is somewhat different from that of the other plates. In *Campbellodus* it comprises small, scattered tubercles on a smooth surface with occasional faint connecting ridges.

The unusual postbranchial ornament (Plates 2F and 4A) comprises more or less horizontal rows of denticles crossing the inner laminae of the anterior lateral and anterior ventrolateral plates. Each denticle has a rounded dorsomesial surface with a fringe of denticles pointing ventrolaterally, and forms a shell over a base of cancellous bone visible beneath and inside the denticulate fringe. The middle denticle is normally large and triangular with slightly convex serrated edges; it is flanked on both sides by two or three more-elongate denticles sometimes fused at their bases. Occasionally the middle denticle shows two or three points suggesting fusion of smaller denticles.

Such an ornament has not previously been described, but it may be widespread in ptyctodontids since it occurs in *Ctenurella* from Gogo (see below) and, as stated above, is apparently also developed in specimens of *Rhynchodus tetrodon*: Gross.

Ctenurella Ørvig, 1960

Amended diagnosis. A ptyctodontid with a single nuchal element in the skull roof; pineal plate triangular to subsemicircular; preorbitals broad and short, probably meeting in the midline behind the pineal. Postorbital with a blunt dorsal angle. Small postorbital fenestra through skull-roof between postorbital, marginal and paranuchal, formed by notches in last two plates. Centrals reaching posterior border of skull-roof. Small paranuchal plate with well-developed articular process carrying flat to slightly concave articular fossa. Unornamented flange along posterior margins of paranuchal and central plates extended into a dermal posterior process. Slender submarginal and small postmarginal loosely attached to exo-cranium. Endocranium perichondrally ossified as at least three bones in addition to two or three dorsal rostral processes. Palatoquadrate perichondrally ossified as autopalatine, metapterygoid and quadrate bones. Mesial autopalatine surface showing three attachment areas for ethmoidal part of endocranium. Elongate anterior attachment area running posterodorsally from anteroventral corner, connected by neck of bone to second oval-shaped area. Posterior area probably elongate and near posteroventral corner. No distinct depression for muscle insertion on lateral autopalatine face; dorsal process of upper toothplate not enclosed in autopalatine. Meckel's cartilage ossified as a single articular bone. Upper toothplate with anteroventral beak, lateral lamella bearing a broad, pointed dorsal process, and dorsal margin of mesial lamella concave anteriorly, convex posteriorly and lacking a distinct process. Lower toothplate lacking anterior beak. Trunk armour lacking spinal plate; median dorsal with a ventral keel and low dorsal crest but lacking a dorsal spine; anterior dorsolateral plate with well-developed articular condyle and lacking a sensory groove. First four neural arches fused into crested synarcual articulating with keel of median dorsal and one or more submedian dorsal elements. Scapulocoracoid perichondrally ossified with postero-lateral, scapular and coracoid processes and coracoid fenestra but lacking a lateral prepectoral process. Two or possibly three basals in pectoral fin. Pelvic clasper in males comprising two denticulate elements made up of three dermal bones; dermal prepelvic tenacula also present. Pelvic fin of females lacking scale covering. Dermal ornament comprising tubercles and anastomosing ridges. Postbranchial ornament of multicuspid denticles, restricted to lower half of anterior lateral plate.

Type species. Ctenurella gladbachensis Ørvig, 1960.

Geological occurrence. Lower Frasnian Upper Plattenkalk, Rhineland, Western Germany.

Remarks. This diagnosis has been amended from Ørvig (1960) and is based on the Gogo material described below; some characters cannot be properly assessed at present because they are poorly shown in the type species. *Ctenurella* differs from *Chelyophorus* in the relation between the preorbital and pineal plates, from *Rhamphodopsis* in the presence of a single nuchal element in the skull-roof and the absence of a dorsal spine and spinal plate, from *Rhynchodus* (as represented by the Wildungen species) and *Campbellodus* in the shape of various skull-roof bones, the absence of a spinal plate, and the arrangement of attachment areas, and the morphology of the lateral face, of the autopalatine.

Ctenurella gardineri sp. nov.

(Plates 2C–E, 3, 4A and 5)

Etymology. After Dr B. G. Gardiner, London, who commenced a study of this species which he later kindly relinquished in our favour.

Diagnosis. Ctenurella with an ornament of reticulating ridges, modified in a clearly demarcated posteroventral area on the marginal plate to form a series of closely-spaced, subparallel ridges.

Holotype. P57637 (Plate 4B), a well-ossified specimen showing dermal bones of the head and trunk, toothplates and supporting visceral arch bones, parts of the endocranium, and synarcual.

Other material. Six specimens (P50906, 8–10; 57664, 5).

Localities. 36, 47, 73, 76 and 89, near Gogo Station, northwestern Australia (see map in Miles, 1971b: fig. 1).

Geological horizon. Lower Frasnian Gogo formation, Canning Basin, Western Australia.

Skull-roof

Bone pattern

Ørvig's (1971) reconstruction of the skull-roof of *Ctenurella gladbachensis* is reproduced in Fig. 15, and may be used as a standard of comparison for the

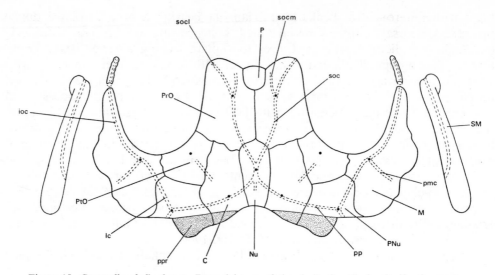

Figure 15. *Ctenurella gladbachensis*. Dermal bones of the skull after Ørvig. Ossification centres indicated by black dots.

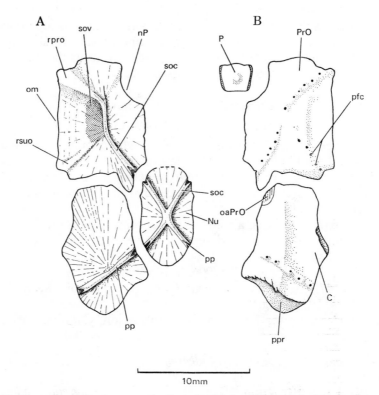

10mm

Figure 16. *Ctenurella gardineri* sp. nov. A, Central, nuchal and preorbital plates in visceral view; B, central, preorbital and pineal plates in external view. P57637.

new species from Gogo (Figs 16, 17 and 18). The skull of *Ctenurella* is distinguished from *Rhamphodopsis* (Miles, 1967a: fig. 1) by the broad preorbitals, shape of the pineal, small paranuchals and absence of a second nuchal element. The new material shows all these features.

Several problems have been encountered in attempting to reconstruct the skull-roof from the accurately-known shape of individual plates. The pineal (*P*, Fig. 16) has been recognized in only one specimen (P57637). Its small size and shape suggest that it did not fit closely into the pineal notch of the preorbital plate, although the notch is clearly defined on each preorbital studied. In addition, the nuchal plate, which is well seen in P50909 and P57637, has a broad anterior margin, and it is difficult in trial reconstructions to make the preorbitals meet in the midline in front of the nuchal without subjecting the whole roof to severe longitudinal and transverse flexure. Watson (1938) had the same difficulty with his original reconstruction of *Rhamphodopsis* (cf. Miles, 1967a), whereas Ørvig's reconstructions (1960, 1962, 1971), based on flattened Gladbach material, all show the preorbitals in contact. More complete specimens from Gogo will show whether these difficulties are real or apparent.

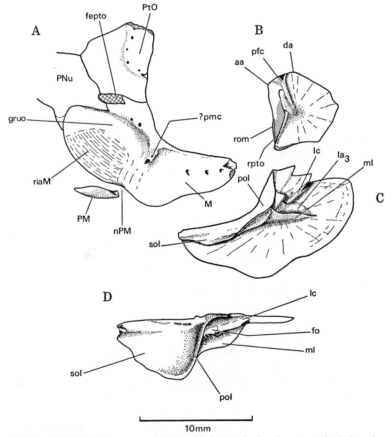

Figure 17. *Ctenurella gardineri* sp. nov. A, Postorbital, marginal and postmarginal plates in external view; B, postorbital plate in visceral view and C, marginal plate in visceral and D, dorsal view. After P57665 and P57637.

The relationship between the marginal, postorbital and paranuchal plates (Fig. 17A) is clearly shown in several specimens. The dorsal edge of the marginal is slightly overlapped by the postorbital plate anteriorly and lies against the lateral edge of the paranuchal posteriorly. Between these two sutures the marginal is distinctly notched to form the anterior and ventral borders of a small opening in the skull-roof, here termed the postorbital fenestra (*fepto*). The lower edge of the postorbital forms the dorsal, and a notch in the paranuchal the posterior border of this opening. It is possible that a small additional plate filled this gap, as occurs for example in *Notopetalichthys* (Woodward, 1941; White, 1952), but no such plate has been recognized. The marginal in *Ctenurella gladbachensis* may

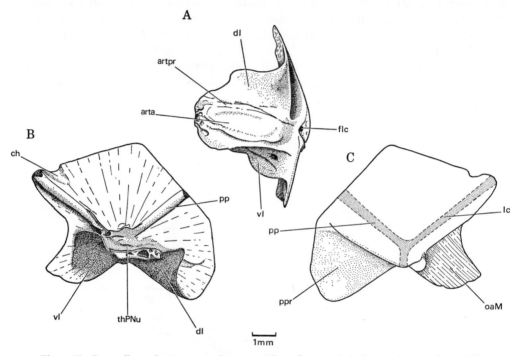

Figure 18. *Ctenurella gardineri* sp. nov. Reconstruction of paranuchal plate in A, posterior and B, internal view. After P50909, P57665 and P57637.

show a notch in a similar position (Ørvig, 1962: pl. 1, fig. 1; 1971: fig. 5B), although an opening has not previously been recognized. The occurrence of a similar fenestra in *Campbellodus decipiens* (see above) suggests, however, that this structure may be widespread in ptyctodontids.

The two distinct embayments in the ventral border of the marginal plate shown in Ørvig's (1962) reconstruction are invariably well developed. In the posterior embayment (*nPM*, Fig. 17A) the edge of the bone is slightly bevelled for the small plate reported by Ørvig (1971, caption to fig. 6). This plate, here termed the postmarginal, has been recognized in two Gogo specimens (P50906, P57637) and is also preserved *in situ* in three Gladbach specimens (P48231, P53343, P53344) in the BMNH collections. It has an elongate elliptical shape, is sometimes unornamented, and bears a short anterior process (Fig. 17A). One of the Gladbach specimens (P53344) appears to show another small, ornamented plate in the anterior embayment of the marginal.

The submarginal plate (Fig. 19) is similar to that of *Ctenurella gladbachensis*. It is elongate, slightly curved longitudinally and strongly arched transversely. Its external surface is deeply concave, and in some specimens partly unornamented. The dorsal and ventral margins may be thickened into ridges; the anterior edge is thin and bordered by a narrow smooth zone.

There are few other points of interest in the external surface of the skull. As originally illustrated by Ørvig (1962), an unornamented strip crosses the posterior margin of the paranuchal and central plates, and projects posteriorly as a distinct paired process (*ppr*, Figs 15 and 16). The marginal plate always shows a clearly defined region on the posteroventral part of its external surface with an ornament of ridges developed subparallel to the posteroventral margin of the plate (*riaM*, Fig. 17A). Dorsally this area may be separated from the pores of the infraorbital sensory line by a broad, shallow groove (*gruo*), sometimes unornamented, which runs back to the posterior corner. Anteriorly it is delineated by a dorsoventral alignment of the ornament towards the postmarginal notch.

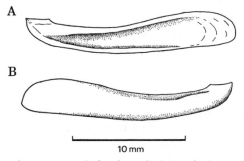

Figure 19. *Ctenurella gardineri* sp. nov. Left submarginal plate in A, lateral and B, mesial view. Based mainly on P57665.

The skull-roof bones are thin and normally show simple sutures without overlap (i.e. harmonic sutures). Exceptions are the slight overlap between the postorbital and marginal plates mentioned above, and the posteriorly directed process of the preorbital which overlaps the central plate (*oaPrO*, Fig. 16). The disarticulated paranuchal (Fig. 18) shows a distinct unornamented posterolateral process (*oaM*) which must have been overlapped by the marginal plate to some extent. However, the relationship between these plates is not clear, as the marginal characteristically shows a rounded posterior corner, and it is possible that the posterior edge of the paranuchal projected out from beneath the overlapping plate as shown in Fig. 17A.

There is some variation in bone shape and proportion, possibly correlated with the size of the individual. A small specimen (P50908) has a preorbital of somewhat triangular shape, reminiscent of Ørvig's first reconstruction of *Ctenurella gladbachensis* (1960, fig. 4A$_1$). The posterior process, slight antorbital process and anterolateral notch on the preorbital shown in Fig. 16 are absent in this specimen. The pineal notch is only slightly developed, suggesting a triangular pineal plate.

The marginal plate is much more elongate in P50909, which is only slightly smaller than the figured example (P57637). In each specimen the suborbital process of the marginal shows an unornamented flange at its anterior end, as occurs in *Rhynchodus tetrodon:* Gross (Miles, 1967a: fig. 4) and *Chelyophorus* (Ørvig, 1971: fig. 7G).

The skull has a strong transverse flexure, its dorsal and lateral parts being folded about an axis running anterolaterally across the central and postorbital plates. The paranuchal is also arched (Fig. 18A) and the orientation of its articular process indicates a near vertical position for the lateral parts of the skull. The posterolateral corner of the preorbital is inflected downward and this plate also shows a gentle rostrocaudal curvature. The marginal plate has a flat external surface. As in other ptyctodontids the pineal is not pierced by a foramen, but it has a central elevation on its external surface.

Internal surface

The Gogo material allows the visceral surface of the ptyctodontid skull-roof to be described in detail for the first time. Various tubes and ridges for the sensory canals have previously been reported (Watson, 1938; Ørvig, 1960; Miles, 1967a), but in addition the paranuchal carries a prominent articular process for the exoskeletal neck-joint on the inner surface, and the preorbital, postorbital and marginal plates exhibit a set of laminae and ridges which enclosed the orbital cavity dorsally, posteriorly and ventrally.

The conspicuous articular process for the dermal neck-joint (*artpr*, Fig. 18) arises from a thickening (*thPNu*) inside the posterior margin of the paranuchal plate, and projects mesially at right angles to the surface of the plate. On its posterior surface is a thickened pad of bone with a fairly flat to slightly concave or uneven surface (*arta*), delineated dorsally and ventrally by irregular ridges. This area corresponds in shape to the articulating surface on the condyle of the anterior dorsolateral plate. The articular process is continuous dorsally and ventrally with thin laminae which, if they can be shown to be perichondral, presumably lined the posterior face of the endocranium. The dorsal lamina (*dl*, Fig. 18) folds first inward, then outward to form a slight posterior convexity as it turns into the plane of the external surface as the more lateral part of the un-ornamented posterior process (*ppr*). The ventral lamina (*vl*) is concave posteriorly, and merges with the inner surface of the dermal bone at a low angle, just inside the thin posterior margin.

The paranuchal carries two sensory tubes across its visceral surface (*pp*, *lc*) and also shows a slight channel (*ch*) leading to the notch in the anterolateral margin. The radiation centre lies posteriorly beneath the base of the articular process.

The visceral surface of the preorbital plate is crossed diagonally by a tube for the supraorbital sensory canal (*soc*, Fig. 16). Anteriorly this tube runs within the preorbital ridge (*rpro*) as in *Campbellodus*, and the less prominent supraorbital ridge (*rsuo*) shows external pores in one specimen (P57665, Plate 2C) and therefore also carried a sensory line onto the postorbital plate (*pfc*, Fig. 16B). The supraorbital canal passes through the radiation centre, just lateral to which is an area of cancellous bone (*sov*). This probably supported a supraorbital vault like that of *Ctenurella gladbachensis* and *Rhamphodopsis* (Ørvig, 1960; Miles, 1967a). The smooth external surface layer of the vault is not normally preserved in the Gogo material.

The visceral surface of the postorbital plate (Fig. 17B) shows traces of a thin ridge (*rom*) positioned just inside its anterior (orbital) margin. Behind this a more prominent ridge (*rpto*) runs from the anteroventral corner towards the ossification centre, then turns anterodorsally. Its broad base slopes back to the central, depressed region over the postorbital radiation centre. Between these ridges is an area of cancellous bone similar to that on the visceral surface of the preorbital plate, and presumably serving the same function of supporting endocranial structures surrounding the orbit. The single sensory tube fades away as it reaches the centre of the plate, where the sensory line turns anteriorly within the bone to open at the orbital margin.

The marginal plate has a broad mesial lamina projecting inwards from its orbital margin, which enclosed the orbital cavity both ventrally and posteriorly and thus comprises subocular and postocular divisions (*sol*, *pol*, Fig. 17C, D). The radiation centre of the plate lies inside the posteroventral corner of the orbital margin, at which point the sensory tube for the lateral-line canal becomes enclosed in the thickened base of the subocular part of the lamina to run forward to the anterior end of the plate. In the same region (i.e. over the radiation centre) a second lamina, here termed the marginal lamina (*ml*, Fig. 17), joins the convex posteroventral surface of the mesial lamina. It slopes inwards and slightly upwards from the visceral surface of the plate, decreasing in width as it runs posteriorly to terminate about halfway towards the posterior margin. From its slightly thickened base a third lamina (*la₃*) extends dorsally, lying close to the visceral surface. It forms a fold (*fo*) where it crosses the sensory tube, and merges with the visceral dermal bone surface just behind the postocular lamina. A space is thus formed immediately behind the orbit, enclosed anteriorly, laterally and ventrally by these laminae. In several specimens small tubules for lateralis nerve fibres enter the adjacent section of the sensory tube (Fig. 17C, D).

The central and nuchal plates are crossed by sensory tubes but otherwise show little relief on their visceral surfaces. The central is slightly concave in visceral view with a posterior radiation centre crossed by a tube for the posterior pitline (*pp*, Fig. 16). The radiation centre of the nuchal (*Nu*) lies in the middle of the plate beneath the anastomosis of the posterior pitline and supraorbital sensory tubes.

Lateral-line system

The distribution of sensory canals is an important consideration in interpreting the skull-roof pattern in fishes. In ptyctodontids, unlike most placoderms, these canals normally run within or beneath the dermal bones, only opening to the surface through rows of pores. Difficulty in recognizing these pores has resulted in some uncertainty about the ptyctodontid sensory canal pattern. The courses of canals may also be traced as ridges on the inner surface of the dermal bones, and recently Ørvig (1971) has given a comprehensive account of the sensory system as known or inferred in ptyctodontids (Fig. 15).

Study of the Gogo material has shown that, even with excellent preservation, the courses of the canals cannot always be determined. On the visceral surface they may form distinct tubes, as on the nuchal, central and preorbital plates (Fig. 16). They may also run through the thickened basal portion of perichondral laminae attached to the visceral surface, although this is not always the case, and every internal ridge *cannot* be assumed to have carried a canal. On the outer surface the sensory pores vary in size; if large they are readily recognizable, but smaller pores are indistinguishable from the deep, circular perforations between anastomosing ridges which characterize the dermal ornament. In view of this uncertainty it is necessary to examine previous interpretations critically and to restrict discussion to those canals for which there is clear evidence.

The sensory canals observed in *Ctenurella gardineri* are as follows: the main lateral-line passes forward from the trunk-shield onto the paranuchal plate where it immediately sinks to the visceral surface through a distinct perforation (*flc*, Fig. 18A). It then divides into a mesial branch (*pp*, Figs 15 and 16), which crosses the central plate to meet its antimere in the midline on the nuchal, and a lateral branch (*lc*) which passes onto the marginal plate as does the lateral-line canal in all placoderms where these plates are recognized. A short median section on the nuchal connects the two posterior pitlines with the two supraorbital canals, thus forming the characteristic X-shaped anastomosis. The supraorbital canals pass anterolaterally onto, and obliquely across the preorbital plates, to the anterolateral corners of the skull-roof. All these sensory canals are clearly visible as tubes of circular section on the visceral surface of the skull, which open to the exterior through somewhat irregularly placed pores (Fig. 16). In addition, the main lateral-line canal continues forward beneath the orbit enclosed in the base of the subocular lamina. Its forward course is indicated by several external pores and the canal reappears anteriorly through a large foramen opening on the unornamented anterior flange of the suborbital process (Fig. 17).

The canals mentioned so far are consistently developed in each specimen studied. There are two additional canals which are either difficult to discern or variably developed. In P57665 a distinct row of pores (Plate 2C) indicates a canal passing posterolaterally from the preorbital to the postorbital plate, where it turns anteriorly to open into the orbit through a pore in the orbital margin. This canal has not been recognized in other specimens, but is not readily seen as it lies within the supraorbital ridge (*rsuo*, Fig. 16). It corresponds to the canal running around the orbital margin in *Campbellodus decipiens* (*pfc*, Fig. 8). A similar canal was reported in *Rhamphodopsis* (Watson, 1938: 399), and a superficial groove passing directly into the orbit, though not crossing the postorbital plate, occurs in *Rhynchodus tetrodon*: Gross and *Chelyophorus* (Ørvig, 1971: fig. 7E, K). This sensory line, therefore, seems generally to be present in ptyctodontids although variable in its course. It has been interpreted by Ørvig as the homologue of the profundus line in early arthrodires ('dolichothoracids').

Another sensory canal, a ventral branch of the main lateral-line passing off the marginal plate, was figured by Ørvig (1962). In most Gogo specimens a short ridge on the marginal, formed by the coalescence of several raised tubercles, occupies a similar position (? *pmc*, Fig. 17A). Each tubercle is pierced by a pore, those in P57637 being slightly larger than the surrounding pits in the ornament. Ventrally the ornament in several specimens is aligned to form one or more ill-defined grooves, passing off the edge of the marginal. We consider these structures to indicate the existence of a superficially positioned postmarginal canal in *C. gardineri*. There is no corresponding internal ridge, but in *C. gladbachensis* the marginal lamina seems to occupy a more anterior position beneath this sensory canal (Ørvig, 1962: pl. 1, fig. 2).

The remaining sensory canals described by Ørvig (1971) for ptyctodontids are not preserved in any Gogo specimen, and evidence for their existence is tenuous. The lateral and mesial branches of the supraorbital canal (*socl, socm*, Fig. 15) have only been identified in one Gladbach specimen (thought by Ørvig, 1971: fig. 6B, possibly to represent a new genus) on the basis of a double row of pores. In view of the difficulties mentioned above in distinguishing sensory pores from pits in the ornament, and the fact that a double row of pores may follow a single canal (Ørvig, 1971: 21) the evidence for these branches is insufficient. In a pre-orbital from Frohnrath, interpreted (on equivocal grounds) as a ptyctodontid, the evidence for two supraorbital canals (Ørvig, 1971: fig. 7B) is very weak; and contrary to Ørvig's (1971: 21) opinion, we consider there is no short mesial branch of the supraorbital canal shown on the preorbital of *Rhynchodus tetrodon:* Gross. The radiating arrangement of ridges in the ornament of *C. gladbachensis* (Ørvig, 1960: 316) places doubt on the existence of 'central' and 'median occipital' sensory lines, said by Ørvig (1971: 29, 30) to lie in ill-defined grooves on the surface. Furthermore, the hypothetical bifurcation of the postmarginal canal into pre-opercular and submarginal branches is not supported by the structure of the postmarginal plate, which shows no sign of a sensory line. This could, however, be attributed to the apparently superficial position of the ventral part of the postmarginal canal (see above), whereas the postmarginal plate (lacking dermal ornament in some cases) was probably deeply situated in the dermis. Finally, the submarginal plate was thought to carry an open sensory groove in *Rhamphodopsis* and *Ctenurella* (Miles, 1967a; Ørvig, 1962). In *C. gardineri* the submarginal has a deeply-grooved external surface, but this 'groove' involves the total width of the plate, results from its transverse curvature, and is clearly not a sensory groove. The surface of the submarginal may show a longitudinal ornament of subparallel or radiating ridges, as is seen in *Rhamphodopsis* (Miles, 1967a: pl. 1, fig. 2), *C. gladbachensis* (Ørvig, 1971: fig. 5) and possibly *Rhynchodus tetrodon* (Gross, 1933: fig. 16H). This ornament may have contributed to previous mis-interpretations of the plate. There is no sign of a canal on the submarginal of *Campbellodus* (see above), and we suggest that no such sensory groove was developed in ptyctodontids. The main objection to homologizing this bone with the submarginal of other placoderms (Miles, 1971b: 203, footnote) is thus re-moved. Two other sensory lines may be developed in the ptyctodontid skull, but do not occur in *C. gardineri*. A transverse groove crosses the suture between the preorbitals in *Chelyophorus*, and also the holotype of *C. gladbachensis* (Ørvig, 1960: fig. 4A$_1$, F, *mpl?*-later retracted, 1971: 27), and a short curved groove is developed on the central plate of *Rhynchodus tetrodon:* Gross (Miles, 1967a: fig. 4).

Bone homology and terminology

It is now possible to discuss the terminology for skull-roof bones used in this account. Various and often complicated terminologies have been introduced in recent years (Ørvig, 1962, 1971; Westoll, 1962; Miles, 1967a, 1971b; Stensiö, 1969). Usually the differences in terminology result from lack of agreement on the importance of fusion or loss-and-invasion in the origin of modified skull patterns. Thus Stensiö (1969: 360) believes the postorbital of this account to represent a part ('parietal paracentral anterior') of the central plate ('parietal prenuchal'), and the marginal of this account to be a compound bone (the 'postorbitomarginal'). Ørvig (1962, 1971) utilises a compound terminology for other bones of the skull; the marginal is said to include a suborbital component, the nuchal a central component, and the central a paranuchal component. The latter is termed a 'lateral paranuchal'. There is however general agreement that the pineal, preorbital and (except for Stensiö, 1969) postorbital plates correspond to those of other placoderms.

Several recent studies have re-examined the differences between the fusion and loss hypotheses of skull-roof patterns. Jardine (1969: 345) has defined the terms 'phylogenetic fusion' and 'phylogenetic loss' to differentiate clearly between phylogenetic and ontogenetic changes, and it is now apparent that it is impossible to differentiate between loss and fusion in fossils, where complete ontogenetic and phylogenetic series can never be obtained. It is also clear that there is at present no unequivocal evidence showing whether osteoblasts or neuromasts dictate the position of dermal bones in early ontogeny. There are however in placoderms, a number of cases where the courses of sensory canals vary in relation to a fixed dermal bone pattern (see first part of this paper), notably in acantho-thoracids (Gross, 1962; Westoll, 1967) but also with respect to the supraorbital (Heintz, 1962; Miles, 1971b) and central canals (Miles, 1973a) in arthrodires. The strict application of the principle of lateral-line constancy in these cases would lead to a compound bone terminology clearly controverted by more general topographic considerations.

If the differences between fusion and loss-and-invasion are largely semantic (Moy-Thomas & Miles, 1971: 85) then the choice in naming bones is between a simple or compound terminology, determined on heuristic grounds and employed as a matter of convention. Ørvig (1962: 59) justifies the use of a compound name where a bone has 'skeletogenous material normally belonging to (other) plates added to it' but we do not find this principle to be heuristic. It necessitates a new name for every change in the relative size of a bone, if interpreted uncritically, without requiring even a change in the total number of bones.

Obviously the differences in overall proportion between the ptyctodontid skull and that of other placoderms must involve major reallocations of skeleto-genic material to accommodate the much shortened skull-roof, enlarged orbits and unossified snout. Nevertheless the dermal bone pattern may be dealt with as follows: two 'nuchal' elements are present in *Rhamphodopsis* as in some palae-acanthaspids (Gross, 1958, 1959; Westoll, 1967), but since there is no way of telling whether one or the other or parts of both elements are represented in *Ctenurella*, this bone is simply termed the nuchal. The paranuchal has the same relationship to sensory canals and the dermal neck-joint as in other placoderms. The marginal receives the lateral-line canal from the paranuchal and thus also shows the normal relationship. As pointed out by Westoll (1962), its position in the orbital boundary is paralleled by various Wildungen pachyosteids with greatly enlarged orbits. The postmarginal plate of this account has unexceptional relationships to the marginal plate and postmarginal sensory canal, whilst the submarginal occupies its normal position in the cheek. The central plate reaches the posterior border of the skull, a condition otherwise shown only in palae-acanthaspids. The small U-shaped sensory groove seen in *Rhynchodus tetrodon*: Gross may be interpreted as a vestige of the paired middle pitline or central canal, both of which cross the central plate in other placoderms.

The ptyctodontid skull-roof thus exhibits a highly specialized bone pattern in which primitive features are not readily apparent. One possibly primitive character is the pattern of sensory canals on the nuchal. A similar pattern is shared by a number of other placoderm groups (see the first part of this paper), but apparently it does not indicate relationships within the placoderms (cf. Ørvig, 1962). The posterior extension of the central plates, if primitive for ptyctodontids (*Rhampho-dopsis* needs reinvestigating), may be a hitherto unevaluated specialization shared with palaeacanthaspids. The phylogeny proposed in this paper suggests that it has been independently acquired, or alternatively is a primitive placoderm character.

Endocranium

The endocranium of *Ctenurella gardineri* comprised at least three paired perichondral ossifications, here termed the occipital, orbital and ethmoidal ossifications. Ossified rostral processes are also present, as described by Ørvig (1960). Traces of the orbital and occipital ossifications have previously been recognized in the orbit of both *C. gladbachensis* (Ørvig, 1962: 61) and *Rhampho-dopsis* (Miles, 1967a: 105).

The general regions of the endocranium occupied by these ossifications are apparent from their preserved positions in several specimens, but details of their relationships to each other and adjacent dermal bones are not known. They were probably connected by cartilaginous areas of variable extent, since the available specimens show considerable differences in degree of ossification, not directly related to size. Other perichondral fragments, too incomplete for description, suggest that further ossified endocranial structures have yet to be identified. The following account is confined to descriptions of the three bones noted above, and general comparisons with other placoderms in which the endocranium is much better known (Stensiö, 1963, 1969). The sketch reconstruction (Fig. 22) illustrates possible morphological relationships discussed in the text, and is not presented as a fully accurate depiction of the skull-roof and braincase.

Occipital ossification

This is a paired bone which in life was connected to its fellow ventrally by a median strip of cartilage. The ossifications of both sides are preserved together only in P50910, but displaced examples occur in several other specimens. The reconstruction given in Fig. 20 is based on P50906, P50910 and P57664.

Each ossification bears a posterior glenoid process (*glpr*) which articulated against the concave articular facet on the synarcual (Fig. 32). The slightly concave ventral surface is crossed by a wide groove (*grra*, Fig. 20A) which passes off its anterior edge, and probably contained the radix aorta. A mesial view of the bone

(Fig. 20C) shows the cranial cavity of the occipital region and a parasagittal section of the thick endocranial floor preserved as dorsal and ventral perichondral laminae (*flcv*, *vsend*). These represent the inner ventral surface of the cranial cavity and the outer ventral surface of the endocranium respectively.

The lateral wall of the occipital region is variably developed. In a small specimen (P50908) it comprises two distally expanded processes projecting upwards and outwards from the endocranial floor, separated by a single, large opening for the passage of nerves and vessels. In larger specimens these processes tend to fuse distally to close in the lateral wall as shown in Fig. 20. A narrow slit (*sl*) is the

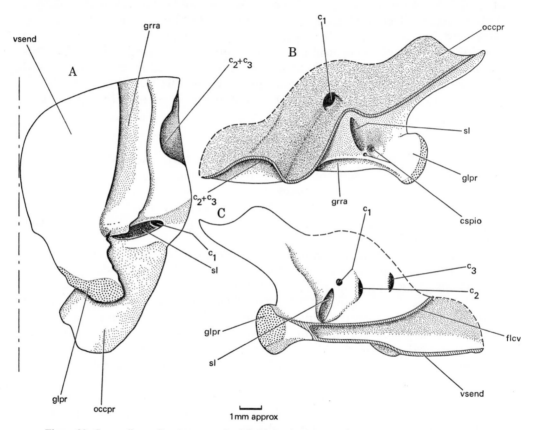

Figure 20. *Ctenurella gardineri* sp. nov. Occipital bone in A, ventral, B, lateral and C, mesial view. Based mainly on P57664 and P50910.

only vestige of the original opening between the processes, above which the completed lateral wall is pierced by several canals and foramina communicating with the central cranial cavity.

The posterior (occipital) process can still be recognized in these well ossified examples (*occpr*). It projects upwards from the thickened region just in front of the glenoid process, and curves posteromesially to form the posterior face of the endocranium. Its rounded posteromesial margin is closed, and presumably formed the border of the foramen magnum (*fm*, Fig. 22B). The dorsal and lateral sides of the process are open, as is the whole of the lateral side of the lateral wall (Fig. 20B), which is preserved as a single perichondral layer, except for canal linings, and in life must have been connected by cartilage to the inner dermal bone surface.

The anterodorsal extent of the occipital ossification is not known. As preserved the lateral wall does not curve inwards to roof over the cranial cavity, and a considerable portion of this region must have remained unossified (Fig. 22A).

The following foramina and canals pass through the lateral wall of the occipital region. Two, sometimes three, small canals originate in the posteroventral region of the cranial cavity, pierce the base of the occipital process, and emerge postero-ventral to the slit (*cspio*, Fig. 20). From their size and position they probably transmitted spino-occipital nerves. At least three large canals pass through the

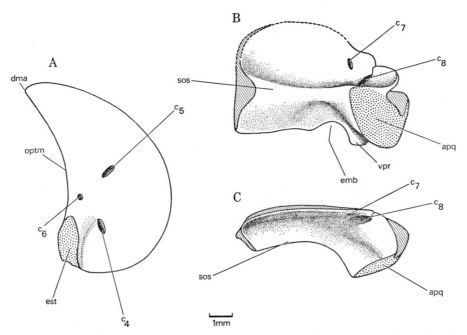

Figure 21. *Ctenurella gardineri* sp. nov. A, Left orbital bone in anterolateral view. After 50908 and P57664. B, Right ethmoidal bone in lateral and C, in dorsal view. After P57637 and P57665.

lateral wall in front of the occipital process. Two of these are preserved in P50906 and P50910, and all three are shown in P57664. The most posterior foramen (c_1, Fig. 20B, C) leaves the cranial cavity dorsal to the slit. It opens into a short, broken canal extending ventrolaterally in P50906 and P50910, but dorsolaterally and anteriorly in P57664. The second and largest canal (c_2) leaves the cranial cavity in front of the first and passes anteroventrally, lateral to the groove for the radix aorta. The third canal (c_3), only preserved in P57664, originates in front of the second, which it joins just outside the cranial cavity. Their nerves or vessels must have emerged through the single, large ventrolateral foramen (c_2+c_3).

The interpretation of these foramina is discussed below.

Orbital ossification

This bone, previously recognized in both *Ctenurella* and *Rhamphodopsis*, is well shown in two specimens from Gogo (P50908, P57664). It is of fairly simple construction (Fig. 21A), with a closed, slightly concave, anteromesial margin (*optm*), and an open, convex, posterior margin. There is a distinct dorsomesial angle (*dma*) and anteroventrally the bone is raised around an opening (*est*) continuous with the interperichondral space. Three canals (c_4, c_5, c_6) open on the anterior face of the bone, in the positions shown. All are presumed to have communicated with the cranial cavity. The posterior face of the orbital ossification is not well shown in the specimens but seems to comprise another perichondral lamina similar in form and extent to the anterior face.

Ethmoid ossification and rostral processes

An ethmoidal ossification has not previously been described in ptyctodontids. Only two good examples occur in the Gogo material (P57637, P57665), and in neither case can the bone's position in the endocranium be accurately determined. As oriented in Fig. 21B, C, the bone margins are open posteriorly and dorsally, and closed ventrally. The form of the dorsal margin is based on an impression in the matrix of P57637. The narrow ventral margin is embayed anteriorly (*emb*) behind a ventrally directed process (*vpr*). A well-developed lateral shelf (*sos*) projects from the lower half of the concave lateral face. Anteriorly the shelf is expanded to support a large articular surface (*apq*) which opens into the interperichondral space in P57637, but is filled with endochondral bone in P57665. Two foramina open on the lateral surface above the lateral shelf. The smaller posterodorsal foramen (c_7) leads to a canal passing directly through both perichondral layers. The larger foramen (c_8) leads to a short, wide canal passing anteroventrally through the bone to open on the mesial surface in front of and above the ventral process.

The rostral processes may be dealt with here since they were assumed by Ørvig to have attached to the ethmoid region of the endocranium. They are typically preserved against the anterior edge of the autopalatine in *C. gladbachensis* (Ørvig, 1960: pl. 26, fig. 1, pl. 29, fig. 1; cf. Ørvig's restorations) but have not been recognized in *Rhamphodopsis* (Miles, 1967a). In the Gogo material various perichondrally-lined, rod-like elements are preserved, but their position in the skeleton has not been established. Those that correspond most closely to Ørvig's description of the rostral processes (1960: 317) are slightly curved and expanded towards one end in the region of a blunt protruberance on the convex side (*rp*, Fig. 23; Plate 4B). Three such elements occur in P50906 and two in P57637. Their absence in other specimens may be due to lack of ossification, but could be a sexually dimorphic feature, although only one of the specimens with processes also has pelvic claspers. If correctly identified, the third element in P50906 may represent the median rostral cartilage of Recent elasmobranchs and holocephalans, which was hypothesized by Ørvig (1960: 320) to be present in *Ctenurella*.

Relation of endocranial ossifications to dermal skull-roof

In view of the difficulties encountered above in reconstructing accurately the skull-roof, a detailed restoration of dermal and perichondral ossifications in the skull has not been attempted. Figure 22, however, shows scaled outlines of the occipital and orbital ossifications against a sketch restoration of the skull-roof. The suggested position of the endocranial ossifications are based on the following considerations:

(1) The glenoid process of the occipital bone must have lain in the same axis of rotation as the articular process of the paranuchal. This places it inside the median embayment in the posterior skull-roof margin, possibly in contact with the mesial end of the paranuchal articular process. Since the occipital process (*occpr*, Fig. 20) projects behind the level of the glenoid process it must have extended upwards and slightly laterally towards the under-surface of the dermal

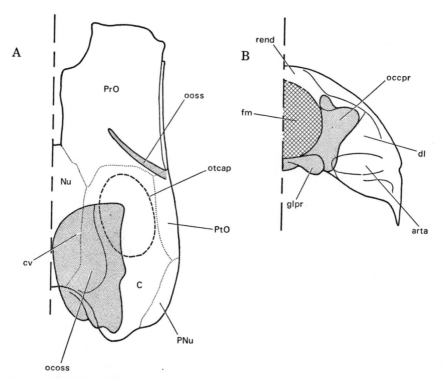

Figure 22. *Ctenurella gardineri* sp. nov. Restoration of occipital and orbital bones in relation to the skull-roof. In A, dorsal and B, posterior view.

posterior process formed by the central and paranuchal plates. Its incomplete lateral perichondral surface was probably continuous with the dorsal lamina of the paranuchal (*dl*, Figs 18 and 22), which seems to have curved in under the dermal posterior process to form an anteroventrally inclined posterior face to the endo-cranium. The occipital bone in the holotype is about two-thirds as long as the central plate. Taking into consideration the width between glenoid processes, the central cranial cavity (*cv*, Fig. 22A) must have lain beneath the nuchal and mesial parts of the central plates. The cranial cavity may have been enclosed dorsally by a cartilaginous roof (*rend*, Fig. 22B).

(2) The orbital ossification is preserved on both sides of P50908. This specimen

clearly shows that the anterior perichondral face of the bone was confluent with the ridge on the orbital margin of the postorbital plate (*rom*, Fig. 17B). The posterior perichondral face was probably joined to the postorbital ridge (*rpto*) with the intervening cancellous bone developed to support the whole attachment. The preserved position of the ossification suggests that it projected inwards at a high angle. Its general proportions are consistent with an anteromesial orientation as in Fig. 22, with its dorsal margin following the supraorbital ridge (*rsuo*, Fig. 16) and its dorsomesial angle (*dma*, Fig. 21) lying adjacent to the supposed position of the supraorbital vault (*sov*, Fig. 16).

Ventrally the orbital ossification must have been confluent with the postocular part of the mesial lamina of the marginal plate, to form the lateral part of the posterior wall of the orbit. However, its relationship to other laminae on the inner surface of the marginal is problematical. The space described above as enclosed by three laminae (*pol, ml, la$_3$*, Fig. 17C) was presumably filled with cartilage, and could represent a vestigial postorbital process of the endocranium, reduced by posterior migration of the orbit. As such its connections with both the orbital ossification and ethmoidal ossification (see below) are not readily apparent.

(3) Our interpretation of the ethmoidal ossification is based solely on the assumption that it carried an attachment surface for the palatoquadrate (*apq*, Fig. 21). That this attachment was strongly developed is shown by the morphology of the autopalatine (Fig. 27), and it would seem unlikely that the corresponding articulation in an otherwise well-ossified endocranium should remain devoid of bony support. The mutual position of the two bones in P57637 (*ethoss, Aup(r)*, Fig. 23) is consistent with this interpretation, and although further supporting evidence is lacking, a reasonable alternative position for the bone in question cannot be proposed. Nevertheless, its relationship to the skull-roof is very uncertain. The position of the jaws given in previous restorations (Watson, 1938; Ørvig, 1962; Miles, 1967a) suggests that it projected in front of the preorbitals. Its lateral shelf may have been level with, and formed an anterior continuation of, the mesial (subocular) lamina of the marginal plate, giving the jaws a more dorsal position than in Ørvig's reconstruction (1962: fig. 2A, B).

There are two problems with this interpretation. First, it is impossible to say whether the subocular lamina of the marginal (which must have formed the floor of the orbit) is of dermal or perichondral origin. If dermal, it could have rested against the lateral shelf of the ethmoidal bone, like the linguiform process rests against the autopalatine in arthrodires. However such an arrangement is not supported by the shape or extent of the lateral shelf. It could also represent the dorsal perichondral surface of an endocranial suborbital shelf, but a shelf in this position would have an anomalous relationship to the marginal lamina (*ml*, Fig. 17), especially if the later formed the ventral surface of an endocranial postorbital process as suggested above. The second problem relates to the floor of the cranial cavity in this region. With the orientation required for the attachment of the autopalatine (Fig. 21), the ethmoidal ossification of each side must have enclosed a high, narrow median space representing the anterior continuation of the cranial cavity. However, the endocranial floor could not have been continuous with these ossifications, since their ventral margins are closed.

These difficulties cannot at present be resolved.

Interpretation of endocranial cavities

The preserved endocranium is too incomplete for a comprehensive treatment of cranial nerves and vessels, but some more general problems related to the identification of its foramina will be examined.

An initial difficulty is to fix the position of the labyrinth cavities in relation to preserved structures. There are two possibilities:

(1) The cavities were confluent with the central cranial cavity as in holocephalans and teleostomes (including acanthodians).

(2) They were partitioned off from the cranial cavity by a cartilaginous wall as in elasmobranchs and other placoderms.

Although the cranial cavity is somewhat expanded posterolaterally, it is not readily interpreted as including the saccular cavities, since there are too few foramina to accommodate the semicircular canals. As stated above, the cranial cavity was positioned mainly beneath the nuchal plate, and the single (inner) preserved perichondral lamina of the lateral wall leaves adequate space for the otic capsules below the central plates, their normal position in arthrodires and petalichthyids (Stensiö, 1969). We therefore consider this to be their likely position in *Ctenurella*. A position beneath the anterior half of the central plate is suggested (*otcap*, Fig. 22) as the depth of the endocranium is much reduced posteriorly.

The two large canals passing anteroventrally through the occipital bone (c_2, c_3, Fig. 20) may now be tentatively interpreted as having transmitted the vagus and glossopharyngeal nerves respectively. The smaller posterior canal (c_1) and elongate slit (*sl*) are of unknown significance. The slit is present in all observed specimens but may open laterally, ventrally or posteriorly. Its size is variable and in P57637 it is reduced to a very narrow slit through which, apparently, a vessel or nerve of appreciable size could not pass. The posterior path of the jugular vein is not preserved, but no doubt this vein passed through a foramen lateral to the preserved part of the occipital region.

There is evidence that the cranial cavity was considerably expanded in the orbital region; the orbital ossification comprises closely spaced outer and inner perichondral laminae, and the latter must be assumed to represent the lining of the inner wall of the cavity. The foramina piercing this bone must therefore have transmitted some of the nerves and vessels normally passing from the cranial to the orbital cavity in fishes. The large anteroventral opening, continuous with the interperichondral space (*est*, Fig. 21A) is interpreted as the eyestalk attachment area, and the surrounding foramina (c_4, c_5, c_6) may have transmitted oculomotor, trochlear and profundus nerves. An eyestalk in the same relative position has been identified in several Lower Devonian arthrodires from Spitsbergen (D. Goujet, pers. comm.), and is surrounded by the same nerve foramina in *Buchanosteus* (Young, unpubl.).

The posteroventral part of the orbital wall (i.e. mesial to the postocular lamina of the marginal plate) is not known. The pituitary and jugular veins and maxillary and mandibular branches of the trigeminal nerve enter the orbit posteroventrally in *Kujdanowiaspis* and *Buchanosteus*. The complete anterior margin of the orbital ossification (*optm*, Fig. 21A) suggests a membraneous mesial orbital wall as in holocephalans. The optic nerve must have emerged in this region.

Somewhat in front of these regions lay the ethmoidal ossification, but in view of the uncertainty regarding its relationship to surrounding structures there is little point in speculating on the anterior extent of the cranial cavity. A proper understanding of the ethmoid region must await the study of better material.

Comparisons

The external form of the endocranium as described above can be compared in a general way with that of other placoderms. It is clearly specialized in the much shortened occipital region, in the apparent reduction and position of the postorbital processes (poorly known), and in the morphology of the ethmoid region and palatoquadrate attachment. Obviously the gross changes in endocranial proportions are correlated with the highly modified skull-roof. The inferred location of a vestigial postorbital process can be attributed to the modification of the marginal plate by the enlarged orbital cavity. A similar condition is developed in macropetalichthyids, where migration of the orbits back to the central plates has resulted in a large postorbital process (thought by Stensiö to represent anterior and posterior processes of other forms) bounding the orbital cavity posteriorly.

Other similarities with petalichthyids in the occipital region of the endocranium have been suggested by Ørvig (1962: 60). These require closer consideration because of their relevance to the discussion of ptyctodontid relationships in the first part of this paper. The main issue would appear to be the homology of the occipital process described above for *Ctenurella*. This process may correspond to the supravagal process of primitive arthrodires or the craniospinal process of petalichthyids, although the evidence is equivocal in both cases. We regard the petalichthyid craniospinal process as a specialization related to the great occipital elongation in this group, which is clearly not the case in ptyctodontids. In petalichthyids the craniospinal process is closely connected with a dermal lamella from the posterior paranuchal plate (Stensiö, 1969: fig. 83), and presumably it acted to support the dermal neck-joint, although the precise relationship between this process and the joint is not known. In brachythoracid arthrodires Stensiö (1963, 1969) has offered an interpretation of the occipital region based on supposed differences in the position of exit of the vagus nerve in 'dolichothoracid', 'coccosteomorph' and 'pachyosteomorph' arthrodires. Yet in *Buchanosteus* (White & Toombs, 1972) the mutual relationships of the vagus exit, posterior postorbital process and supravagal process compare closely with those in *Kujdanowiaspis* (Stensiö, 1969: fig. 10), and the supravagal process can be readily interpreted as having migrated posteriorly, preserving its relationship with the external opening of the endolymphatic duct and the ossification centre of the paranuchal plate. Thus, on the evidence of *Buchanosteus*, it is apparently the supravagal process which supports the dermal neck-joint in brachythoracids, with the craniospinal process possibly represented by the dorsal process which must have occupied the subnuchal pits of the skull-roof (Miles & Westoll, 1968: 400). Since the craniospinal process supports the neck-joint in petalichthyids, it is clear that the occipital process in *Ctenurella* has quite a different relationship to the paranuchal articular process (Fig. 22B), and its homology to either the supravagal or the craniospinal process is not immediately obvious on topographic grounds.

A non-committal terminology has therefore been used, and we stress our failure to corroborate Ørvig's (1962: 60) statement 'that the occipital region of the endocranium in *Ctenurella*, . . ., is in many ways very like that of e.g. *Macropetalichthys*, . . .'.

Jaws and visceral arch structure

Our interpretation of the jaws is based mainly on the holotype (P57637) in which upper and lower toothplates, and quadrate, articular, metapterygoid and autopalatine ossifications are preserved in association. The quadrate, articular and upper toothplate of the right side are only slightly disturbed (Fig. 23) and the remaining components are displaced upwards and forwards from their articulated positions.

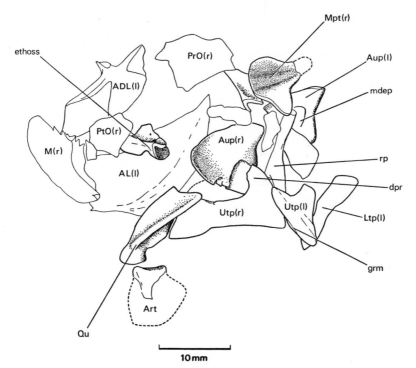

Figure 23. *Ctenurella gardineri* sp. nov. Holotype (P57637) showing the association between palato-quadrate, articular and upper toothplate; (r) and (l) indicate whether bones are from the left or right side. (Key diagram for Plate 4B.)

Quadrate

This well-ossified bone has a strongly developed, posteroventral articular surface for the mandibular joint (*artmd*, Figs 24A and 25). The main shaft of the bone (*sh*) is nearly square in cross section and is flanked by lateral and mesial ridges (*lr, mr*) which in ventral view (Fig. 25B) are seen to enclose a V-shaped cavity. From the association of the right quadrate and upper toothplate in the holotype (Fig. 23) it is clear that these ridges abutted against, or partially enclosed, the lateral and mesial lamellae of the toothplate. The dorsal surface of the quadrate is gently convex and much expanded anteriorly by the development of a prominent lateral flange (*lfl*, Fig. 25).

Articular

This bone is short and broad. It is shown in Fig. 24 in its preserved position in the holotype relative to the quadrate, but it has probably been twisted out of position so that its ventral surface faces laterally. Its mandibular joint surface is much larger than the corresponding surface on the quadrate. Like the quadrate its outer (in this case ventral) surface is gently convex, and its strongly convex dorsal surface is flanked by a large lateral and a small mesial ridge (*lr*, *mr*, Fig. 24C, D) which enclose deep grooves for the lower toothplate. The mesial ridge is thickened posteriorly (*th*) to support a circular, socket-like articular depression on the posteromesial flank of the ventral surface (*arthy*, Fig. 24E). This articulation surface also occurs in *Rhynchodontus eximius* (Jaekel, 1919) and has been interpreted by Stensiö (1969: figs 152, 154A) as a 'symplectic' articulation. Presumably the articular received an element of the hyoid arch at this point, but until the element in question has been discovered and its relationship to other visceral arch structures worked out there is little hope of determining its homology.

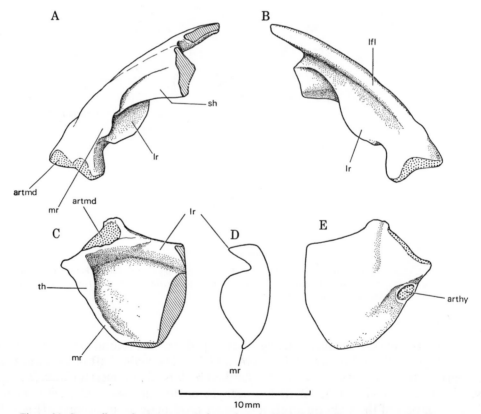

Figure 24. *Ctenurella gardineri* sp. nov. Left quadrate in A, mesial and B, lateral view. C, Articular in dorsal and E, ventral view. D, Profile of the anterior end of C. Relative positions as preserved in P57637.

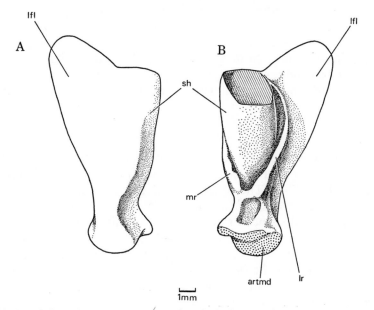

Figure 25. *Ctenurella gardineri* sp. nov. Left quadrate in A, dorsal and B, ventral view. After P57637.

Metapterygoid

The bone here interpreted as the metapterygoid has been identified only in the holotype and one other specimen (P50909). Its position in the holotype, relative to other jaw elements, is consistent with it having in life a similar position to that reconstructed by Ørvig (1962). The right metapterygoid is reconstructed in Fig. 26. It is folded around an axis passing downwards and forwards through the postero-dorsal process (*prmpt*), so as to form two almost perpendicular laminae. The mesial lamina (*mla*) has an embayed dorsal margin and a gently concave lateral surface. The lateral lamina (*lla*) is distinctly arched about a transverse axis to form a concave posterior surface. Its lateral margin is convex (Fig. 26B). Both

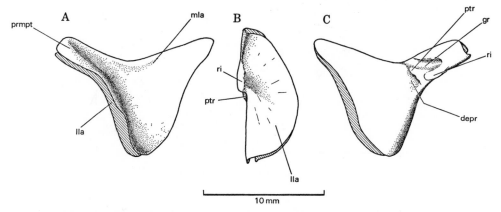

Figure 26. *Ctenurella gardineri* sp. nov. Right metapterygoid in A, lateral, B, posterior and C, mesial view. After P57637.

laminae are thickened posterodorsally to give the posterodorsal process a triangular cross-section. The individual perichondral bone laminae are also thickened in this region. The closed edge of the embayed dorsal margin is thick and rounded; all other margins are open. In mesial view (Fig. 26C) the metapterygoid shows a broad low ridge (*ri*) and a shallow groove (*gr*) which presumably played a role in its attachment to the endocranium. There is an associated low protruberance (*ptr*) and shallow depression (*depr*).

Autopalatine

This bone is only well preserved in the holotype (*Aup*(*r*), Fig. 23). It is gently concave laterally about an axis passing anteroventrally across the lower half of the lateral surface (Fig. 27A). The anteroventral corner is deeply depressed (*ddpr*) to receive the dorsal process of the lateral lamella of the upper toothplate (*dpr*, Fig. 23). The process does not fit accurately into this depression, and must have been set in connective tissue. In its ventral half the floor of the depression is further stepped down to form a shelf (*she*, Fig. 27A) running parallel to the ventral edge. There is no evidence that the dorsal process was completely enclosed within the autopalatine ossification as in Stensiö's reconstructions of '*Rhynchognathus*' and '*Ramphodontus*' (1969: figs 151, 153).

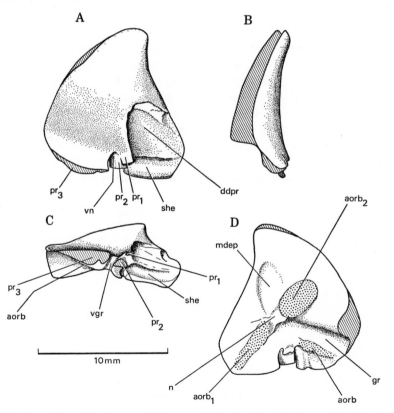

Figure 27. *Ctenurella gardineri* sp. nov. Right autopalatine in A, lateral, B, posterior, C, ventral and D, mesial view. After P57637.

The ventral margin of the autopalatine has a distinct notch (*vn*), which in ventral view (Fig. 27C) is seen to open into a complicated system of ridges and grooves separated by three open-ended, ventral projections (pr_1, pr_2, pr_3). These presumably were the main attachment areas between the autopalatine and the upper toothplate. A conspicuous groove (*vgr*) passes posteromesially across the ventral surface, giving off two smaller grooves which flank the second projection to pass on to the lateral surface. This notch in *Rhynchodus* has been compared with similar structures in *Pholidosteus* (Stensiö, 1963: figs 75, 124) and interpreted by Stensiö as transmitting a branch of the orbital artery to supply the upper jaw.

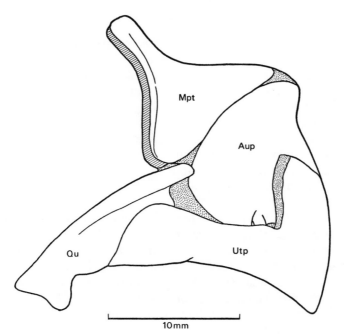

Figure 28. *Ctenurella gardineri* sp. nov. Restoration of right upper jaw in lateral view. After P57637.

The mesial surface of the right autopalatine in the holotype is only partially preserved, but clearly shown are two prominent articular areas (Fig. 27D), both developed as roughened bony surfaces rather than openings in the perichondral lining as previously described for ptyctodontids (Ørvig, 1960: pl. 29, fig. 1; Stensiö, 1969: figs 151, 153). The elongate anterior area ($aorb_1$) extends upwards and backwards from the anteroventral corner and is connected by a thickened neck of bone (*n*) to the posterior area ($aorb_2$). Both surfaces are supported by thickened perichondral bone and are elevated above the surrounding surface.

The remaining features of the mesial surface have been reconstructed on other evidence. The right autopalatine in posterior view shows a strong fold in the mesial perichondral lining which presumably extended anterodorsally across the mesial surface as a broad groove (*gr*, Fig. 27D). In ventral view the left autopalatine shows another narrow ridge bearing a roughened surface passing upwards and forwards from the posteroventral margin (*aorb*, Fig. 27D), which probably flanked this groove. Finally, the mesial surface of the left autopalatine exhibits a low depression (*mdep*, Fig. 23) which must have been situated anterodorsal to the main articular areas (Fig. 27D).

The anterior part of the ventral margin and the concave anterior margin are the only closed margins on the autopalatine. The anterior margin is rounded, whilst the ventral has a thin truncated edge with a roughened texture. In particular, the anteroventral corner of the autopalatine is completely ossified, and shows no sign of the cartilaginous process which Stensiö (1963: fig. 125) restored to lie in the groove crossing the inner surface of the upper toothplate in *Rhynchodus*. This last groove, however, is well developed in *Ctenurella gardineri* (see below).

Toothplates

The toothplates are well preserved in the holotype (Fig. 23, Plate 4B), P50906 and P50908. As previously described (Ørvig, 1960) they resemble *Rhynchodus* in having trenchant biting edges developed for shearing, in contrast to the crushing tritorial surfaces of *Ptyctodus*. Both upper and lower plates are curved with concave lateral (labial) and convex mesial (lingual) surfaces.

The upper plate is wedge-shaped in cross section as in *Rhamphodopsis* (Watson, 1938: 399). Its deeply grooved dorsal surface is flanked by mesial and lateral lamellae between which the spongy inner bone is exposed. The lateral lamella is produced anteriorly into a conspicious dorsal process behind which its dorsal edge is first gently concave then gently convex. Posteriorly the lateral lamella is distinctly lower than the mesial, which curves upwards in a broadly-convex dorsal margin. Anteriorly the mesial lamella drops below the level of the anterior process and it is cut by a deep groove just inside the anterior margin of the tooth-plate. This groove extends almost to the anteroventral corner of the biting edge, which is produced downwards to form a slight beak. Behind this beak the biting edge may show a lateral wear surface sloping downwards and inwards to a sharp edge on the lingual side, as in *Rhamphodopsis* (Watson, 1938: 399) and *Ctenurella gladbachensis* (Ørvig, 1960: 316). Behind this cutting edge there is a low rounded cusp.

The lower toothplate (Plate 2 D, E) has a concave ventral surface which attached to the meckelian cartilage and is flanked by mesial and lateral lamellae. They show the opposite development to the upper toothplate, with the lateral lamella more extensive posteriorly and the mesial more extensive anteriorly. The biting surface comprises a long, low, rounded posterior cusp and a smaller, narrow anterior cusp, with a wear surface developed on the inner side in the anterior half of the plate. The lower toothplate in P50906 is remarkably narrow compared to other specimens, suggesting some variability in toothplate form within the species.

Hyoid and branchial arches

Pharyngo-, epi- and ceratohyal elements were identified in *Rhamphodopsis* by Watson (1938), although we have no confidence in these determinations (see also Miles, 1967a: 105). Recent accounts of the visceral arch skeleton in ptyctodontids (Stensiö, 1969: fig. 152) include a ventral 'symplectic' element, presumably based on Watson's ceratohyal (1938: fig. 1), lying against the articular bone. As stated above, an element of the hyoid arch may have occupied this position although its homology and function are not known. The large element tentatively identified by Ørvig (1960: pl. 29, fig. 1) as a ceratohyal is more probably the basal of the pectoral fin. Also figured by Stensiö (1969: fig. 164) is a bone interpreted as the hyomandibula of '*Rhamphodontus tetrodon*'. Stensiö's reconstruction is presumably based on the specimen named *Rhynchodontus eximius* by Jaekel (1919: fig. 16; also Gross, 1933: fig. 16). No corresponding element has been seen in the Gogo material, and this bone may possibly be an elongated metapterygoid process or a quadrate (Miles, 1971b: 197).

Various other rod-like elements occur in the material in addition to the rostral processes described above. A short bone, expanded at one end to form a short lateral process, has been recognized in two specimens preserved near the ventral edge of the marginal plate. It may be a separate epihyal element, since it seems unlikely that the epihyal was fixed to the submarginal plate as it is in arthrodires (Miles, 1971b; Goujet, 1972). Other elongated elements in the material, expanded at one or both ends, may be from the branchial arches or the opercular skeleton.

Jaw restoration and comparisons

Several problems are posed in an attempt to reconstruct the jaw apparatus. It is clear that the right autopalatine is only slightly rotated from its correct position in the holotype in relation to the upper toothplate (Fig. 23), but its ventral surface is somewhat narrower than the broad, deeply concave dorsal surface of the toothplate, and the attachment must have been padded out by connective tissue. In its restored position the autopalatine shows no close connection to the anterior end of the quadrate and there must have been a bridging cartilaginous area. The same probably applied to the fit of the metapterygoid to the other two elements, making their precise relationships difficult to determine.

If it is assumed that the metapterygoid has been displaced upwards and forwards (i.e. rotated out of position in the same direction as the autopalatine), it may be restored as in Fig. 28 with the open, concave anteroventral margin of its mesial lamina (Fig. 26) placed against the open, convex posterodorsal margin of the autopalatine. Its position may be further fixed by aligning the fold axis of its concave lateral surface with that of the lateral surface of the autopalatine. In this position, the metapterygoid process and the two main orbital articulations on the autopalatine are also brought into line. This reconstruction is unsatisfactory in the poor fit of the metapterygoid against adjacent bones anterodorsally and posteroventrally, and it must be assumed that these areas were completed in cartilage (Fig. 28). The open lateral lamina of the metapterygoid is also difficult to explain. It is about as wide as the lateral flange of the quadrate, but there is no obvious connection between these two structures, which are separated by an angle of about 90° in their reconstructed positions.

This reconstruction differs in several ways from previous accounts. In *Ctenurella gladbachensis*, Ørvig (1960: 321) identified a dorsal 'peg-like process' and a concave 'broad shelf-like margin' as the dorsal extremities of the autopalatine. These he tentatively restored anteroventral to the orbit (1960: fig. 4B) with the process articulating against a postulated antorbital endocranial process. He likened this process to the dorsal angle of the ossification supporting the upper toothplate of

Rhynchodus (Jaekel, 1929: fig. 37; Gross, 1933: pl. 7, fig. 3). However, Stensiö (1963: fig. 124) has interpreted this ossification in *Rhynchodus* as comprising both autopalatine and metapterygoid components, the latter including the dorsal process which makes an 'otic' articulation with the endocranium behind the orbit. A hypothetical preorbital process is restored by Stensiö 'on the basis of the condition in *Ctenurella*'.

This interpretation is in fact an alternative which Ørvig (1960: 321) considered to be unlikely on the basis of the Gladbach material then available. As has been pointed out (Miles, 1967a: 104), Stensiö's interpretation of *Rhynchodus* requires a very oblique position for the upper toothplate, and postulates separate autopalatine and metapterygoid components for which there is no evidence.

Ørvig's original interpretation of *Ctenurella* was supported by his second restoration (1962: fig. 2A, B) in which a metapterygoid with otic process was shown, in addition to the orbital process of the autopalatine, with the autopalatine itself considerably reduced in height. It should be noted, however, that the evidence for this new restoration has not been published.

In the light of the new material described above, the following points may be made on Ørvig's interpretation:

(a) The metapterygoid of this account (Fig. 26) is remarkably similar in appearance to the dorsal process of the autopalatine in the holotype of *Ctenurella gladbachensis* (Ørvig, 1960: pl. 26, fig. 1). The 'peg-like process' and 'broad shelf-like margin' correspond closely to the metapterygoid process and lateral lamina of our account. (That the lateral margin of this lamina is also open in *C. gladbachensis* is clear from Ørvig's figure.)

(b) Since the autopalatine and metapterygoid are undoubtedly two discrete ossifications in *C. gardineri*, Ørvig's first interpretation (1960: 321) can be discounted, whilst his main objection to the alternative interpretation (the oblique position of the toothplates) no longer applies for the same reason—the metapterygoid can be given an oblique orientation, as proposed above, without affecting the position of the toothplates.

This leaves the metapterygoid of Ørvig's second restoration unexplained. If it is the element originally figured as the quadrate (1960: pl. 26, fig. 2), as its shape suggests (1962: fig. 2B), then the original identification may be correct, although it differs in a number of ways from the quadrate described above. Another possibility, since an additional element has evidently been identified by Ørvig as the quadrate, is that it corresponds to the ethmoidal endocranial ossification described above.

We are still left without a satisfactory explanation of the palatoquadrate in *Rhynchodus*. From the figures published by Gross (1933: pl. 7, figs 3, 4) it is clear that his specimen includes at least autopalatine and quadrate components. It can be compared with the above reconstruction of *Ctenurella gardineri* only by assuming that the metapterygoid is absent, and that the autopalatine differs in its greater height, the more anterior position of the articular areas and the deep concavity on the lateral surface.

As a consequence of the above, there is no longer any evidence for a preorbital process of the ptyctodontid palatoquadrate. It has been pointed out that the metapterygoid and autopalatine do not form a close fit anterodorsally in *Ctenurella gardineri*, and the region may have been filled with cartilage. But it seems most unlikely that a cartilaginous preorbital process was developed at this point to support an otherwise well-ossified palatoquadrate.

It would appear, therefore, that the main attachment between the braincase and the autopalatine was through the two orbital articulations ($aorb_1$, $aorb_2$, Fig. 27D) on the inner autopalatine surface. These correspond reasonably well to the two articular areas in the Wildungen specimens (Stensiö, 1969; figs 151, 153) and *Ctenurella gladbachensis* (Ørvig, 1960: pl. 29, fig. 1). It may be assumed for the present that the second articular surface attached to the similarly shaped articular area on the ethmodial ossification of the endocranium (see above). The

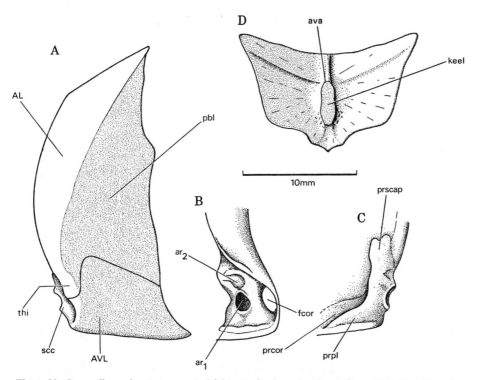

Figure 29. *Ctenurella gardineri* sp. nov. A, Right anterior lateral and anterior ventrolateral plates in anterior view; B, scapulocoracoid in lateral and C, posterior view; D, median dorsal plate in visceral view. After various specimens; scapulocoracoid based mainly on P50906 and P57665.

third, posterior articular area noted by Ørvig has a different shape and position from the posteroventral ridge (*aorb*, Fig. 27D) in *C. gardineri*, but this region is not well shown in the Gogo material. Stensiö has considered the corresponding articulation in '*Rhynchognathus*' to be an otic articulation but this is unlikely. An otic articulation with the anterior postorbital process has been assumed to occur in other placoderms (Stensiö, 1963, 1969) but it now seems that this articulation was for the hyomandibula (Miles, 1971b: 199; Goujet, 1972), and that there was only an orbital connection between palatoquadrate and endocranium, as in *Ctenurella*.

Ørvig (1962: 54) likened the metapterygoid of *Ctenurella* to the otic process of embryonic *Callorhynchus* and this comparison can still be made. However, the supposed articular structures on the otic process of *Ctenurella* are poorly developed, suggesting a weak or ligamentous attachment in contrast to the firmly fused condition in holocephalans. The ptyctodontid otic process is best regarded as an unique specialization related to the highly differentiated jaw apparatus.

Ørvig (1960: 317) concluded, contrary to earlier opinion (e.g. Dean, 1906; Watson, 1938), that the palatoquadrates in ptyctodontids must have been separated throughout their height, and that the upper toothplates could not have formed a median symphysis. In *Ctenurella gardineri* the presumed orbital attachment surface for the palatoquadrate (*apq*, Fig. 21) faces anterolaterally, and the

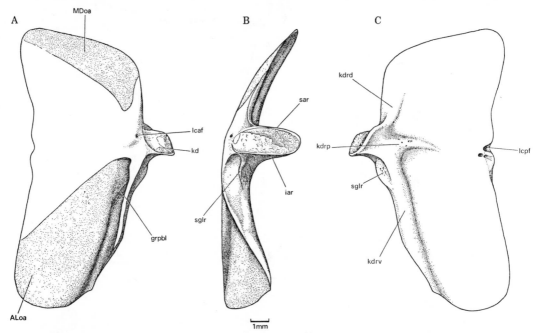

Figure 30. *Ctenurella gardineri* sp. nov. Right anterior dorsolateral plate in A, lateral, B, anterior and C, mesial view. After P57665.

resulting oblique orientation of the palatoquadrate and upper toothplate would seem to bring their anterior margins into close proximity with those of their antimeres. In addition, the anteroventral corner of the autopalatine does not form a ventral process to occupy the groove crossing the mesial surface of the upper toothplate, this groove having been interpreted by earlier workers as a symphysial area between the toothplates. On present evidence, therefore, the possibility of an anterior symphysis between these elements cannot be entirely excluded. The lower toothplate in both P50908 and P57637 is slightly longer than the upper, but the difference in length is too small to necessitate an appreciable gap between the anterior ends of the upper toothplates of each side; there is no evidence for the existence of small anterior upper toothplates in this position (cf. Ørvig, 1960; Miles, 1967a).

Trunk-armour

The trunk-shield of *Ctenurella gardineri* is essentially the same as that of *C. gladbachensis* (Ørvig, 1962: fig. 2A). It is high and short and comprises median dorsal and median ventral plates, and paired anterior dorsolateral, anterior lateral and anterior ventrolateral plates. As in *C. gladbachensis*, the median dorsal may exhibit a low median crest, but lacks the high dorsal spine of *Rhamphodopsis* (Miles, 1967a: fig. 12). Spinal plates are absent. Ørvig concluded that *Ctenurella* lacked a median ventral. This bone has, however, been identified in three Gogo specimens (P50909, P57637, P57665). It is generally similar to the median ventral of *Rhamphodopsis* (Miles, 1967a: fig. 13) with extensive overlap areas for the anterior ventrolaterals separated by a narrow median ornamented strip (Fig. 31).

The anterior dorsolateral bears a glenoid condyle as in phlyctaenioid arthrodires (see below). It also carries the main lateral-line canal to the trunk, with the canal passing within the bone between an anterior foramen situated on the external surface and a posterior foramen lying in a notch just inside the posterior margin (*lcaf*, *lcpf*, Fig. 30). There is some evidence of a ventral branch leaving the main canal near its posterior exit (Fig. 30C). No other lateral-lines are developed on the trunk plates.

The high, narrow anterior lateral has a well-developed postbranchial lamina (*pbl*, Fig. 29A), a feature apparently characteristic of ptyctodontids. The anterior ventrolateral has many of the characters of the interolateral in arthrodires (see footnote, p. 151). Its ascending postbranchial lamina is highest laterally, but does not extend to the edge of the pectoral emargination from which it is separated by the slightly thickened border of the lateral lamina (*thi*, Fig. 29A). The proportions of the ventral lamina are not well shown in the available material; this lamina is apparently longer than it is broad in *Ctenurella gladbachensis*, but broader than it is long in *Rhamphodopsis* and *Rhynchodus* (Ørvig, 1960: 324).

The Gogo material in general confirms Ørvig's reconstruction (1962: fig. 2A) of the trunk-armour of *Ctenurella*, although the shape of the overlap areas on the anterior dorsolateral plate and our restoration of the synarcual (see below) suggest that the median dorsal was more steeply inclined in profile than previously assumed. There are three aspects of the trunk skeleton that require special comment, including the scapulocoracoid and pectoral fin skeleton.

Visceral surface of median dorsal

In *Ctenurella gladbachensis*, Ørvig (1960: 324) described paired vertical laminae projecting downwards from the internal surface of the median dorsal plate, which he interpreted as perichondral bone lining the dorsal crest of a synarcual fused to the dermal bone. In this respect *Ctenurella* appeared to differ from both *Chelyophorus* and *Rhamphodopsis*, in which a median ventral keel similar to that of brachythoracids is present (Obruchev, 1964: pl. 3, fig. 5; Miles, 1967a: 107).

In *Ctenurella gardineri* a stout median ventral keel is also developed (Figs 29D and 34). Its base is attached over the length of the median dorsal and it reaches its maximum depth at about the middle of the plate. In lateral view the keel exhibits a concave anterior and a convex posterior margin separated by a distinct anteroventral angle (*ava*). The anterior margin is narrow whilst the posterior is broad and flat with a roughened surface of cancellous bone, and was presumably in contact with another skeletal element (Fig. 34). In one example (P50910) this surface is distinctly bilobed, and in several the cancellous filling has been partly lost to reveal a hollow interior, which in life must have been unossified. It is therefore probable that the keel in *C. gladbachensis* was similarly developed, but either partially unossified or poorly preserved. Behind the keel is a pit-like depression, opening posteriorly and roofed over by the low median dorsal crest near the posterior border.

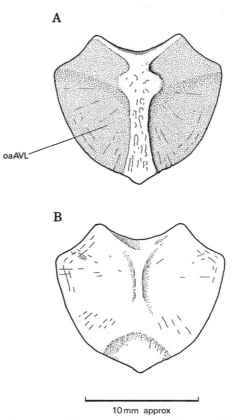

A

B

oaAVL

10 mm approx

Figure 31. *Ctenurella gardineri* sp. nov. Median ventral plate in A, ventral and B, dorsal view. After P50909 and P57667.

Glenoid condyle and dermal neck-joint

The glenoid condyle is strongly developed as a mesial projection from the inner surface of the anterior dorsolateral plate (Fig. 30). Its thickened base extends posteriorly, dorsally and ventrally in the form of broad ridges on the inner surface (*kdrp*, *d*, *v*, Fig. 30C). The dorsal and ventral ridges follow the anterior margin of the plate, and there are corresponding grooves on the external surface; the ventral groove (*grpbl*) received the postbranchial lamina of the anterior lateral plate and is deep and strongly developed. A vertical ridge projects anteriorly from beneath the condyle (*sglr*) to separate two depressions. It occupies the same position relative to the condyle as the subglenoid process of brachythoracid arthrodires.

The condyle itself (*kd*) has a smooth, rounded posterior surface and a smooth, flat ventral surface. The roughened articular surface faces anterodorsally. In anterior view the articular area is clearly defined by supra-articular and infra-articular ridges (*sar, iar*) which are continuations of the smooth posterior surface layer and appear to extend beneath the mass of cancellous bone. Thus they probably represent the rim of an anteriorly-facing concavity in the compact bone layer backing the condyle; similar concavities in *Rhynchodus tetrodon* have been interpreted as shallow articular fossae (Stensiö, 1959: fig. 74). The cancellous bone filling this concavity formed the core and the preserved articular surface of the condyle.

In the figured example the articular surface is concave dorsomesially and convex laterally, but in other specimens it may be entirely flat or somewhat concave. The functional joint surface in life must have comprised a smooth layer of connective tissue. Assuming similar development of this structure in different individuals, it seems that the condyle must have borne a convex rather than concave articular surface. In this feature, therefore, as well as in the development of internal ridges, the postbranchial groove, the subglenoid 'ridge', and overall morphology, the anterior dorsolateral of *Ctenurella* may be compared with that of various arthrodires with a highly-developed dermal neck-joint, e.g. *Dunkleosteus* (Heintz, 1932: fig. 46), *Holonema* (Miles, 1971b: fig. 68) and *Buchanosteus* (White & Toombs, 1972: pl. 7).

The condyle worked against the articular process of the paranuchal, which on its posterior surface carries a slightly elevated articular area of complementary shape (Fig. 18). Irregular dorsal and ventral ridges make this surface slightly concave, but a well-developed articular fossa of brachythoracid type is not developed, and the articular process does not generally compare closely with the brachythoracid condition. Functionally, these differences may be due to the near-vertical lateral wall of the head, the relatively large size of the articulation and the incompleteness of supporting ossifications in comparison with brachythoracids. Phylogenetically they may be due to the independent origin of the joint in arthrodires and ptyctodontids. A para-articular process is not developed, but, as noted above, the posterolateral process of the paranuchal (*oaM*, Fig. 18), which has a corresponding position, may have projected posteriorly from beneath the marginal. The posterior edge of the skull, whether formed by the paranuchal alone or by this plate plus the overlapping marginal, must have stopped against the subglenoid ridge of the anterior dorsolateral plate with the head maximally depressed, but an appreciable sliding articulation with the subglenoid ridge could not have been present.

The nature of the dermal neck-joint in other ptyctodontids is difficult to determine. The glenoid condyle seems generally to be present on the anterior dorsolateral plate but has been variously reported as either transversely or longitudinally oriented. An anterior dorsolateral has been described from Wildungen with an anteriorly directed process, whilst in *Rhamphodopsis* it was said to be developed as 'a short shelf with a slightly hollowed upper surface' (Miles, 1967a: figs 8, 9). Ørvig (1960: 315) has implied that in *Ctenurella gladbachensis* the condyle has a similar longitudinal orientation, but his published figures (1960: pl. 29, fig. 1) certainly suggest close similarity to *C. gardineri*. The same applies to the condyle in *Chelyophorus* (Obruchev, 1964: pl. 3, fig. 4).

In *Rhamphodopsis*, *Ctenurella gladbachensis* and *Rhynchodus tetrodon:* Gross the paranuchal shows no more than a simple thickening on the visceral surface for the reception of the articular condyle, and on this evidence and the supposed orientation of the condyle it has been suggested (Ørvig, 1960: 315; Miles, 1967a: 107) that ptyctodontids have a dermal joint of petalichthyid type (Ørvig, 1957). On the evidence provided by *C. gardineri*, and since a glenoid condyle of similar structure is developed in several genera*, we now suggest that a paranuchal articular process as described above is typical of the ptyctodontid dermal neck-joint. This structure would be easily broken off in compressed material leaving only the thickened base on the inner surface of the paranuchal, the preserved condition in *Rhamphodopsis* and *C. gladbachensis*.

In view of the specialized jaw apparatus and deep, flexible gill-cover it is unlikely that the ptyctodontid joint was important in feeding or respiration. Ørvig (1960: 315) suggested that the dermal posterior processes formed by the unornamented flanges behind the paranuchal and central plates may have projected beneath the median dorsal and acted to limit ventral rotation of the head relative to the trunk. This is perhaps a specialization in the neck-joint characteristic of ptyctodontids.

Scapulocoracoid and pectoral fin

Parts of a perichondrally-ossified scapulocoracoid are preserved in all specimens showing anterior lateral and anterior ventrolateral plates. It lies inside the antero-ventral angle of the trunk-armour, and comprises a posterolateral process (*prpl*, Fig. 29) attached to the inner surface of the anterior ventrolateral along the lateral edge of its ventral lamina; a dorsal scapular process (*prscap*) attached to the inner surface of the anterior lateral along the junction of its lateral and post-branchial laminae; and a mesial coracoid process (*prcor*) attached to the inner surface of the anterior ventrolateral along the junction of its postbranchial and ventral laminae. The ventral and mesial limits of this last process are not shown in the specimens.

The main body of the scapulocoracoid is high, narrow and rostrocaudally short, extending upwards from the posterolateral process to the scapular process to enclose a large coracoid fenestra (*fcor*), bounded posteriorly by the body of the scapulocoracoid and anteriorly by the inner surface of the anterior lateral plate. In this respect the fenestra differs from that of *Rhynchodus* (Stensiö, 1959: fig. 75) which is completely surrounded by perichondral bone. On the lateral surface behind the coracoid fenestra all specimens show a large circular foramen with a rimmed margin opening into the interperichondral space of the scapulocoracoid (ar_1). The posterior face of the scapulocoracoid is not well exposed but one specimen (P50906) shows a circular depression in the perichondral surface (ar_2), lying posterodorsal to this foramen and bounded by irregular dorsal and ventral ridges. The ridges continue posterodorsally but are not well exposed. All other

---

*Mark-Kurik (1974: fig. 9a, b) has described a perfectly good ptyctodontid paranuchal, with a glenoid condyle of this type, as the 'anterior dorsolateral plate of the arctolepid (?) arthrodire'.

specimens show a smooth perichondral surface behind the foramen; most of the scapulocoracoid surface is apparently visible in these specimens although the complete posterior extent of the bone is not shown.

These structures presumably represent articular areas for basals of the pectoral fin, which are not normally preserved in position in the Gogo material. Two specimens, however, show a single perichondrally-ossified element lying adjacent to the scapulocoracoid. In P50910 this element is a flat tube widening distally and of simple construction. In P57637 the element in question has a blunt lateral process projecting from the middle of its length, beside which is a small articular facet for contact with an adjacent basal element. A similar articulation between the two posterior basal elements of *Ctenurella gladbachensis* is suggested by the published figures (Ørvig, 1960: pl. 26, figs 1, 3; pl. 28, fig. 1). Neither example in the Gogo material exhibits the rugose perichondral surface seen in *C. gladbachensis* (Ørvig, 1960: 325).

According to Ørvig, *Ctenurella* had three basals in the pectoral fin attached to a row of three articular areas running forward and downward across the postero-lateral side of the scapulocoracoid. Published figures of the holotype (Ørvig, 1960: pl. 26, figs 1, 3) suggest a scapulocoracoid very similar to that of *C. gardineri*, and show a single, deep depression in the position of the circular foramen described above. Only 'mesopterygial' and 'metapterygial' elements are preserved in this specimen, and the 'propterygial' element identified in another specimen (pl. 28, fig. 1) is far removed from the scapulocoracoid. In *Rhamphodopsis* traces of the pectoral fin have been observed in a single specimen (Miles, 1967a: 109) and here again no more than two elements can be positively identified. Taking into consideration the condition in *C. gardineri*, the weight of published evidence supports the existence of only two pectoral basals in ptyctodontids.

Stensiö (1959: 210) regarded the high, short ptyctodontid scapulocoracoid as highly specialized compared to the condition in other placoderms. It lacks the long prepectoral portion seen in arthrodires and palaeacanthaspidoids, whilst the coracoid fenestra, presumably developed for deeper insertion of the ventral fin musculature does not occur in other placoderms, although *Palaeacanthaspis* shows a muscle fossa in an equivalent position relative to the articular crest (Stensiö, 1959: fig. 69A). Stensiö's suggestion (1959: 16) that the pectoral fin in ptyctodontids slopes backwards and downwards as in *Palaeacanthaspis*, is shown to be incorrect by Ørvig's account (1960) and the present material.

Ornament

The dermal ornament in *Ctenurella gardineri* comprises reticulating ridges enclosing deep pits of variable shape, and is fairly similar to that of *Rhamphodopsis* (Watson, 1938: 400) although the pits tend to be smaller and more regular in shape. The ridges tend to be finer than in *C. gladbachensis*. On the marginal a distinct area of regular ridges is always developed running subparallel to the posteroventral margin (Plate 3B). The submarginal has a reduced ornament of longitudinal ridges and grooves. The postbranchial ornament is only well shown on the left side of P50910 but is evidently similar to that described above for *Camp-bellodus* (Plates 2F and 4A). It is less extensive, being restricted to the ventral half of the postbranchial lamina of the anterior lateral.

Axial skeleton, synarcual and median ossifications

As described and figured by Ørvig (1960: 332), the axial skeleton in *Ctenurella* is made up of paired, perichondrally-ossified neural and haemal arches, lying above and below an unconstricted notochord. Each neural element comprises a main body with a dorsoventrally concave mesial surface, and a short dorsal spine sloping posteriorly. The elements of each side partially enclose a longitudinal neural canal and the left and right dorsal spines were probably in contact distally to form a second dorsal space. A median dorsal ligament may have passed through this space as suggested for other placoderms (Stensiö, 1963: fig. 5; Miles & Westoll, 1968: 447).

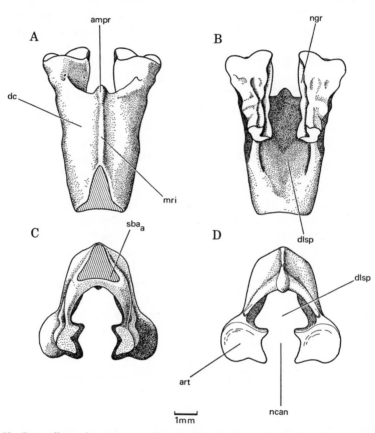

Figure 32. *Ctenurella gardineri* sp. nov. Synarcual in A, dorsal, B, ventral, C, posterior and D, anterior view. After P57665.

The haemal arch is similarly constructed to the neural, with a concave mesial face for the dorsal aorta (Ørvig, 1960: pl. 27, fig. 2) and with a ventral spine sloping posteriorly. The partial enclosure of a haemal canal in *Ctenurella* has not been observed in other placoderms. According to Stensiö (1963: fig. 5), the dorsal aorta in an undetermined pachyosteomorph lay in a haemal groove, whilst Miles & Westoll (1968: fig. 46c) assumed the dorsal aorta of *Coccosteus* to be enclosed by the distal contact of the ventral spines.

Scattered neural and haemal elements commonly occur in the Gogo material, but only one specimen (P50910; Plate 3D) shows some elements in approximately their correct positions.

Synarcual

Ørvig assumed that the anterior axial elements in *Ctenurella* were fused to form a synarcual articulating with the occipital region of the endocranium. In *C. gardineri* the synarcual (Figs 32 and 33A) is clearly shown in most specimens and its structure can be described in detail. It is formed by the fusion of the first four neural arches. These increase in size towards the anterior, and the large first element has a concave anterior face (*art*) which received the glenoid process of the endocranial occipital bone (*glpr*, Fig. 20). The fused elements of each side show a grooved mesial surface (*ngr*, Fig. 32) which partially enclosed the neural canal, but it

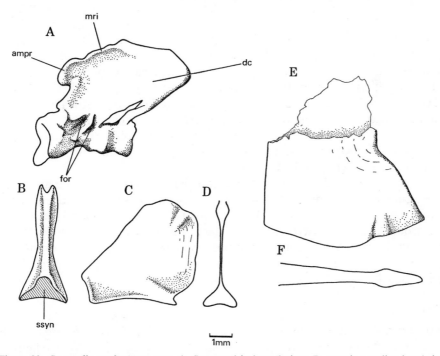

Figure 33. *Ctenurella gardineri* sp. nov. A, Synarcual in lateral view; B, anterior median basals in anterior and C, lateral view; D, posterior profile of C; E, possible posterior median basal in lateral view and F, posterior profile of E. All after P57665.

seems that the notochord lay entirely ventral to the synarcual. In Recent elasmobranchs and holocephalans, and various fossil chimaeroids (Patterson, 1965), the notochord is enclosed in the synarcual, as is also the case in the few placoderms in which the synarcual is known (Stensiö, 1963: figs 7, 8, 93, 96, 98). *Ctenurella* thus differs in this respect from all other forms in which a synarcual is developed.

In lateral view (Fig. 33A) the synarcual is short and high, with a heavily-developed dorsal crest (*dc*). It is now clear (see below) that this crest was not rigidly connected to the median dorsal plate, as was suggested by Ørvig (1960:

324–32). It is formed by the fusion of the dorsal spines into a perichondrally-ossified block, triangular in cross-section, with a concave ventral surface and a dorsal surface elevated into an irregular median ridge (*mri*, Figs 32 and 33). The ridge projects anteriorly as a blunt process (*ampr*).

The posterior element in the synarcual retains a recognizable dorsal spine, expanded dorsally to merge with the dorsal crest. In the more anterior elements, the spines are almost completely fused together and only show on the outer surface as irregular ridges running posterodorsally. The spaces between the first three spines are reduced to small foramina (*for*, Fig. 33A), through which the spinal nerve roots must have passed.

Posterodorsally the crest presents an open triangular area (*sba$_a$*, Fig. 32C). The considerable space (*dlsp*) enclosed ventrally by the sloping lateral walls corresponds to the dorsal ligament space between the more posterior neural spines.

Other median ossifications

Ørvig (1960: 329) recognized a large, perichondrally-lined anterior basal plate lying behind the median dorsal in *Ctenurella gladbachensis*, followed in one specimen by two smaller posterior plates. Two such ossifications are preserved in P50910, in which the vertebral column is relatively undisturbed (Plate 3D). The external shape of the large bone (*ba$_p$*) is not well shown, but it lies behind and presumably fitted against a smaller, triangular anterior element (*ba$_a$*). The latter is displaced to the side of the keel of the median dorsal plate. A similar triangular bone has recently been figured by Ørvig (1971: fig. 5A), and another is preserved detached in P57665 (Fig. 33B, C, D).

As oriented in Fig. 33, the anterior median element has a complete anterodorsal margin, truncated anteroventrally to form an open triangular surface (*ssyn*). The posterior margin is open and the lower half of the bone is laterally expanded to give a broad, gently concave ventral surface. P57665 has also yielded a second, larger element, which is incompletely preserved (Fig. 33E, F). Its orientation is unknown but its straight edge may be ventral, as is the case with the large posterior bone in P50910 (*ba$_p$*, Plate 3D).

Median fins

Two dorsal fins have been described in *Ctenurella* and *Rhamphodopsis* (Ørvig, 1960; Miles, 1967a). Scattered radials, some bifurcated distally as figured by Ørvig (1960: pl. 27, fig. 2) occur in the Gogo material, but nothing can be added to previous descriptions.

Relationship between vertebral column and trunk-armour

A tentative reconstruction of the trunk plates, synarcual and associated elements is given in Fig. 34. It is based mainly on specimen P57665, which has yielded detached median dorsal and anterior dorsolateral plates, a complete synarcual and two anterior basals.

The dermal plates have been drawn to scale and graphically reconstructed, and the synarcual placed in position so that its articular facets lie in the transverse axis of the glenoid condyles on the anterior dorsolaterals. The alignment of posterior neural arches is based on P50906, which shows the fifth element in position against the synarcual.

It has been assumed that the anteroventral triangular area on the anterior median basal (*ssyn*, Fig. 33B) fitted against the posterodorsal triangular area on the synarcual (*sba_a*, Fig. 32). This position is not entirely satisfactory since the thin, rounded anterodorsal edge of the anterior basal lies against the flat endochondral surface of the median dorsal keel. In P50910, however, the anterior median basal is thickened with a flat endochondral surface along this edge, as would be expected with such a contact. The significance of this variation between the two specimens is unknown. A larger posterior element, possibly that shown in Fig. 33E, F, occupies a position behind the median dorsal, and on the basis of Ørvig's descriptions there may have been additional posterior elements not shown in the figures. The restoration suggests that the median dorsal was more steeply inclined relative to the vertebral column than previously assumed (cf. Ørvig, 1960, 1962).

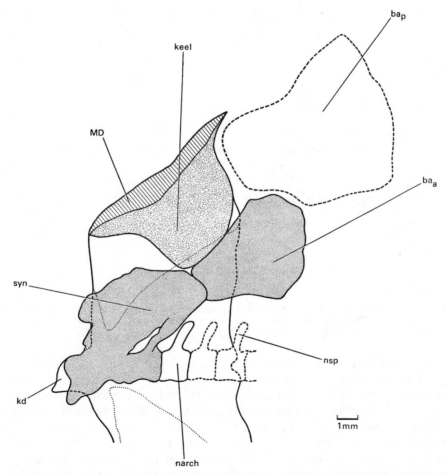

Figure 34. *Ctenurella gardineri* sp. nov. Reconstruction to show trunk-shield with synarcual, median basals and vertebral column in left lateral view. The left side of the trunk shield has been removed.

Pelvic fin and claspers

Sexual dimorphism in the pelvic fins of ptyctodontids was first noted by Watson (1938). It is best known in *Rhamphodopsis* (Miles, 1967a: 109), in which each fin has a flat endoskeletal plate and an exoskeletal covering of overlapping scales in females, the scales modified into two pairs of denticulated plates in males. As interpreted by Miles, one pair were carried on the fin and the other borne by a pelvic clasper ventral to the fin, this ventral position resembling elasmobranchs rather than holocephalans (Miles, 1967a: 113). In addition, in both sexes, small, paired, elongate plates lay transversely in front of the fin, in males supported by a small endoskeletal element. In *Ctenurella*, Ørvig (1960: 327) described a similar

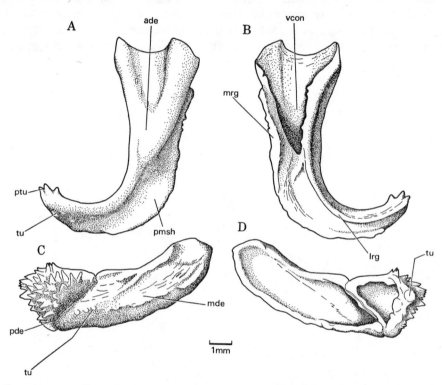

Figure 35. *Ctenurella gardineri* sp. nov. Dermal elements of pelvic claspers in A, C, external view and B, D, internal view. Anterior element in A, ventral and B, visceral view, middle and posterior elements in C, ventral and D, visceral view. P57665.

small (endoskeletal) element in front of an elongate 'basipterygium', the latter being larger and narrower in some specimens (presumably males). Paired dermal spines ornamented with pointed or hooked tubercles were said to occur 'in the region of the pelvic fins' in one specimen. The fins themselves are unknown in *Ctenurella*, since even in females they apparently lack the covering of scales seen in *Rhamphodopsis*. The paired dermal spines, since they occurred 'in front of, and clearly separated from' the (presumed) position of the fin in this specimen were interpreted as prepelvic claspers (tenacula). Later, Ørvig (1962: 56) interpreted as a copulatory organ another paired element ornamented with hooked tubercles and extending posteriorly between the basipterygia. The whole arrangement was considered to be highly reminiscent of Recent holocephalans.

Two Gogo specimens (P50906, P57665 presumably the only males) show three. paired dermal elements which seem to correspond to the two posterior pairs described in males of *Rhamphodopsis*. The absence of a scale covering in the fins of either sex in *Ctenurella* (Ørvig, 1960: 327) is corroborated by the material. Although the morphology of the three dermal elements is well known (Fig. 35), other ossifications of the pelvic girdle and fins could not be isolated and a complete understanding of the relationships of various components is not yet possible.

As oriented in Fig. 35, the anterior element (*ade*) has a hook-like shape, curving laterally to a posterior termination bearing three pointed tubercles (*ptu*). There are a few, scattered, low tubercles on the adjacent posterior surface (*tu*) but the bone is otherwise unornamented. The anterior end is expanded, with steep lateral walls enclosing a deep ventral concavity (*vcon*). The concave lateral margin is steep over its whole length, with a distinct rim in ventral view (*lrg*). The anterior part of the mesial margin is rounded and deep, extending ventrally as an irregular ridge (*mrg*). The convex posteromesial border takes the form of a low shelf (*pmsh*).

The two posterior elements fit closely together as shown in Fig. 35C, D. The more elongated element (*mde*) is slightly curved with a convex outer surface showing a few irregular ridges and occasional tubercles (*tu*). The inner (dorsal) surface is deeply concave with a high rimmed margin. The posterior element (*pde*) has a concave ventral surface densely ornamented with pointed curved tubercles (Plate 5A). As in the previous element it tends to form a hollow shell around a deeply concave inner surface, in this case rimmed by an irregular margin bearing a few, large, rounded tubercles (*tu*, Fig. 35D).

In both specimens there is evidence of an additional, small paired element, ornamented with hooked denticles, which probably corresponds to the prepelvic clasper described by Ørvig (1960: 327) in *Ctenurella gladbachensis*. In neither case is this element well seen but P50906 shows it to be about the same size as the posterior dermal element just described, from which it can be distinguished by its smooth, slightly-convex inner surface. Further close comparisons with the pelvic region of *C. gladbachensis* cannot be made on the basis of published figures. The dermal element determined by Ørvig as a basal plate (1960: pl. 27, fig. 1, *ba. pelv*) might correspond to the elongated dermal element described above, which has a similar shape. In *C. gladbachensis* this element is associated with another, bearing hooked tubercles (Ørvig, 1962: 56), although the last element has not been figured.

The males of *Rhamphodopsis* were clearly similar to *Ctenurella gardineri*. The bone showing hooked tubercles (Miles, 1967a: pl. 2, fig. 3) closely resembles the small posterior element described above (*pde*, Fig. 35); these tubercles were said to be on the lower surface and mesial edge of the exoskeletal plate of the pelvic fin (*pv. ex*, Miles, 1967a; fig. 16). This element, however, probably corresponds to the anterior element in *C. gardineri*. The coarser, pointed tubercles on the clasper of *Rhamphodopsis* (Miles, 1967a: pl. 4, fig. 4, pl. 6, fig. 3) are similar to, but more numerous than, the three tubercles on this anterior element, and either the ornamentation was reversed in *Rhamphodopsis*, or displacement of the small posterior element, not previously recognized as a discrete bone, has resulted in some misinterpretation.

On the basis of preserved positions in *Rhamphodopsis*, it was previously concluded (Miles, 1967a: 113) that the dermal clasper element lay ventral to the exoskeleton of the pectoral fin. We now reach a similar conclusion based on the structure of the elements described above. The similar curvature of the posterior unit (*pde+mde*) and the anterior element (*ade*) shows that they were closely associated in life. The posterior unit could conceivably have fitted inside the concave lateral border of the anterior element, but it seems more likely that it lay posteromesially, and was movable over the posterior mesial shelf of the anterior element. The ventral concavities in each were presumably occupied by cartilaginous supporting elements. If the anterior element corresponds to that of *Rhamphodopsis*, it could not have lain embedded in the pelvic fin as previously proposed. Presumably the hooked end was free, and it is possible that both parts were movable dermal elements of the clasper itself, lying separate and ventral to the pelvic fin.

ACKNOWLEDGEMENTS

We wish to thank Miss Kim Dennis for help with illustrations and work on the SEM, and Drs Colin Patterson, R. H. Denison and Alex Ritchie for reading and commenting on the manuscript of Part I. The photographs were taken by Mr T. W. Parmenter. One of us (GCY) also acknowledges receipt of an Australian Government postgraduate scholarship 1974–1976, during which time this work was undertaken.

REFERENCES

BEER, G. R. de., 1937. *The development of the vertebrate skull:* xxiv+552 pp. Oxford: University Press.
BRUNTON, C. H. C., MILES, R. S., & ROLFE, W. D. I., 1969. Gogo expedition 1967. *Proc. geol. Soc. 1655;* 80–3.
DEAN, B., 1906. Chimaeroid fishes and their development. *Publs Carnegie Instn, 32:* 1–172.
DENISON, R. H., 1958. Early Devonian fishes from Utah. 3. Arthrodira. *Fieldiana, Geol, 11:* 461–551.
DENISON, R. H., 1975. Evolution and classification of Placoderm fishes. *Breviora, 432:* 1–24.
EASTMAN, C. R., 1908. Devonian fishes of Iowa. *Iowa geol. Surv., 18:* 29–386.
GARDINER, B. G. & MILES, R. S., 1975. Devonian fishes of the Gogo formation, Western Australia. *Colloques int. Cent. natn. Rech. scient., 218:* 73–9.
GOUJET, D., 1972. Nouvelles observations sur la joue d'*Arctolepis* (Eastman) et d'autres *Dolichothoraci. Annls Paléont., 58:* 3–11.
GOUJET, D., 1973. *Sigaspis,* un nouvel arthrodire du Dévonien Inférieur du Spitsbergen. *Palaeontographica, 143A:* 73–88.
GROSS, W., 1932. Die Arthrodira Wildungens. *Geol. paläeont. Abh., 19:* 5–61.
GROSS, W., 1933. Die Wirbeltiere des rheinischen Devons. *Abh. preuss. geol. Landesanst., 154:* 1–83.
GROSS, W., 1937. Die Wirbeltiere des rheinischen Devons. *Abh. preuss. geol. Landesanst., 176:* 5–83.
GROSS, W., 1958. Über die älteste Arthrodiren-Gattung. *Notizbl. hess. Landesamt. Bodenforsch. Wiesbaden, 86:* 7–30.
GROSS, W., 1959. Arthrodiran aus dem Obersilur der Prager Mulde. *Palaeontographica, 113A:* 1–35.
GROSS, W., 1961. *Lunaspis broilii* und *Lunaspis heroldi* aus dem Hunsrückshiefer (Unterdevon, Rheinland). *Notizbl. hess. Landesamt. Bodenforsch. Wiesbaden, 89:* 17–43.
GROSS, W., 1962. Neuuntersuchung der Stensiöellida (Arthrodira, Unterdevon). *Notizbl. hess. Landesamt. Bodenforsch. Wiesbaden, 90:* 48–86.
GROSS, W., 1963. *Gemuendina stuertzi* Traquair, Neuuntersuchung. *Notizbl. hess. Landesamt. Bodenforsch. Wiesbaden, 91:* 36–73.
GROSS, W., 1967. Über das Gebiss der Acanthodier und Placodermen. *J. Linn. Soc., 47:* 121–30.
HEINTZ, A., 1932. The structure of *Dinichthys*. A contribution to our knowledge of Arthrodira. *Am. Mus. nat. Hist., Bashford Dean Mem. 4:* 115–224.
HEINTZ, A., 1962. New investigation on the structure of *Arctolepis* from the Devonian of Spitsbergen. *Norsk. Polarinst. Årb., 1961:* 23–40.

HEMPEL, C. G., 1966. *Philosophy of Natural Science:* x+116 pp. Englewood Cliffs, N.J.: Prentice-Hall.

HUSSAKOF, L. & BRYANT, W. L., 1918. Catalog of the fossil fishes in the Museum of the Buffalo Society of Natural Sciences. *Bull. Buffalo Soc. nat. Sci., 12:* 5–198.

JAEKEL, O., 1903. Über *Ramphodus* nov. gen., einen neuen devonischen Holocephalen von Wildungen. *Sber. Ges. naturf. Freunde Berl., 8:* 383–93.

JAEKEL, O., 1906. Einige Beiträge zur Morphologie der ältesten Wirbeltiere. *Sber. Ges. naturf. Freunde Berl., 7:* 180–9.

JAEKEL, O., 1919. Die Mundbildung der Placodermen. *Sber. Ges. naturf. Freunde Berl., 3:* 73–110.

JAEKEL, O., 1929. Die Morphogenie der ältesten Wirbeltiere. *Monogrn Geol. Palaeont, 3:* 1–198.

JARDINE, N., 1969. The observational and theoretical components of homology; a study based on the morphology of the dermal skull-roofs of rhipidistian fishes. *Biol. J. Linn. Soc., 1:* 327–61.

JARVIK, E., 1960. *Théories de l'évolution des vertébrés:* 1–104. Paris: Masson & Cie.

MARK-KURIK, E., 1973a. *Kimaspis*, a new palaeacanthaspid from the Early Devonian of Central Asia. *Eesti NSV Tead. Akad. Toim, 22/4:* 322–30.

MARK-KURIK, E., 1973b. *Actinolepis* (Arthrodira) from the Middle Devonian of Estonia. *Palaeontographica, 143A:* 89–108.

MARK-KURIK, E., 1974. Discovery of new Devonian fish localities in the Soviet Arctic. *Eesti NSV Tead. Akad. Toim, 23/4:* 332–5.

MILES, R. S., 1967a. Observations on the ptyctodont fish, *Rhamphodopsis* Watson. *J. Linn. Soc. (Zool.), 47:* 99–120.

MILES, R. S., 1967b. The cervical joint and some aspects of the origin of the Placodermi. *Colloques int. Cent. natn. Rech. scient., 163:* 49–71.

MILES, R. S., 1969. Features of placoderm diversification and the evolution of the arthrodire feeding mechanism. *Trans. R. Soc. Edinb., 68:* 123–70.

MILES, R. S., 1971a. Paleozoic fish. *McGraw-Hill Yearbook of Sci. & Tech., 1971:* 312–4.

MILES, R. S., 1971b. The Holonematidae (placoderm fishes), a review based on new specimens of *Holonema* from the Upper Devonian of Western Australia. *Phil. Trans. R. Soc., 263:* 101–234.

MILES, R. S., 1973a. An actinolepid arthrodire from the lower Devonian Peel Sound formation, Prince of Wales Island. *Palaeontographica, 143A:* 109–18.

MILES, R. S., 1973b. Relationships of acanthodians. In P. H. Greenwood, R. S. Miles & C. Patterson (Eds), *Interrelationships of fishes:* 63–103. London: Academic Press.

MILES, R. S., 1973c. Articulated acanthodian fishes from the Old Red Sandstone of England, with a review of the structure and evolution of the acanthodian shoulder-girdle. *Bull. Br. Mus. nat. Hist. (Geol.), 24:* 115–213.

MILES, R. S., 1975. The relationships of the Dipnoi. *Colloques int. Cent. natn. Rech. scient., 218:* 133–48.

MILES, R. S. & WESTOLL, T. S., 1968. The placoderm fish *Coccosteus cuspidatus* Miller ex Agassiz from the Middle Old Red Sandstone of Scotland. Part I. Descriptive morphology. *Trans. R. Soc. Edinb., 67:* 373–476.

MOY-THOMAS, J. A. & MILES, R. S., 1971. *Palaeozoic fishes*, 2nd ed.: viii+259 pp. London: Chapman & Hall.

NELSON, G. J., 1970. Pharyngeal denticles (placoid scales) of sharks, with notes on the dermal skeleton of vertebrates. *Am. Mus. Novit., 2415:* 1–26.

NEWBERRY, J. S., 1873. *Rhynchodus secans. Rep. geol. Surv. Ohio, 1:* 310–11.

OBRUCHEV. D. V., 1964. Class Placodermi. In D. Obruchev (Ed.), *Fundamentals of palaeontology, 11:* 1–159. Moscow. In Russian. (English translation, 1967—Israel program for scientific translations: Jerusalem, 168–261).

ØRVIG, T., 1957. Notes on some Paleozoic lower vertebrates from Spitsbergen and North America. *Norsk. geol. Tidsskr., 37:* 285–353.

ØRVIG, T., 1960. New finds of acanthodians, arthrodires, crossopterygians, ganoids and dipnoans in the Upper Middle Devonian Calcareous Flags (Oberer Plattenkalk) of the Bergisch-Paffrath Trough (Part I). *Paläont. Z., 34:* 295–335.

ØRVIG, T., 1962. Y a-t-il une relation directe entre les arthrodires ptyctodontides et les holocephales? *Colloques int. Cent. natn. Rech. scient., 104:* 49–61.

ØRVIG, T., 1971. Comments on the lateral line system of some brachythoracid and ptyctodontid arthrodires. *Zoologica Scr., 1:* 5–35.

ØRVIG, T., 1975. Description, with special reference to the dermal skeleton, of a new radotinid arthrodire from the Gedinnian of Arctic Canada. *Collogues int. Cent. natn. Rech. scient., 218:* 41–71.

PANDER, C. H., 1858. *Über die Ctenodipterinen des devonischen Systems:* viii+64 pp. St. Petersburg.

PATTERSON, C., 1965. The phylogeny of the chimaeroids. *Phil. Trans. R. Soc. (B), 249:* 101–219.

RITCHIE, A., 1973. *Wuttagoonaspis* gen. nov., an unusual arthrodire from the Devonian of Western New South Wales, Australia. *Palaeontographica, 143A:* 58–72.

STAHL, B. S., 1967. Morphology and relations of the Holocephali with special reference to the venous system. *Bull. Mus. comp. Zool. Harv., 135:* 141–213.

STENSIÖ, E. A., 1944. Contributions to the knowledge of the vertebrate fauna of the Silurian and Devonian of Western Podolia. II. Notes on two arthrodires from the Downtonian of Podolia. *Ark. Zool., 35:* 1–83.

STENSIÖ, E. A., 1950. La cavité labyrinthique, l'ossification sclérotique et l'orbite de *Jagorina. Colloques int. Cent. natn. Rech. scient., 21:* 9–41.

STENSIÖ, E. A., 1959. On the pectoral fin and shoulder girdle of the arthrodires. *K. svenska VetenskAkad. Handl.,* (4) *8:* 1–229.

STENSIÖ, E. A., 1963. Anatomical studies on the arthrodiran head. Part I. Preface, geological and geographical distribution, the organisation of the head in the Dolichotharaci, Coccosteomorphi and Pachyosteomorphi. Taxonomic appendix. *K. svenska VetenskAkad. Handl.,* (4) *9:* 1–419.

STENSIÖ, E. A., 1969. Elasmobranchiomorphi Placodermata Arthrodires. In J. Piveteau (Ed.), *Traité de paléontologie, 4:* 71–692. Paris: Masson.

TOOMBS, H. A. & RIXON, A. E., 1959. The use of acids in the preparation of vertebrate fossils. *Curator, 2:* 304–12.

WATSON, D. M. S., 1934. The interpretation of arthrodires. *Proc. zool. Soc. Lond., 3:* 437–64.

WATSON, D. M. S., 1938. On *Rhamphodopsis*, a ptyctodont from the Middle Old Red Sandstone of Scotland. *Trans. R. Soc. Edinb., 59:* 397–410.

WESTOLL, T. S., 1962. Ptyctodontid fishes and the ancestry of Holocephali. *Nature, Lond., 194:* 949–52.

WESTOLL, T. S., 1967. *Radotina* and other tesserate fishes. *J. Linn. Soc. (Zool.), 47:* 83–98.

WHITE, E. I., 1952. Australian arthrodires. *Bull. Br. Mus. nat. Hist. (Geol.), 1:* 249–304.

WHITE, E. I., 1968. Devonian fishes of the Mawson-Mulock area, Victoria Land, Antarctica. *Scient. Rep. transantarct. Exped., 16:* 1–26.

WHITE, E. I. & TOOMBS, H. A., 1972. The buchanosteid arthrodires of Australia. *Bull. Br. Mus. nat. Hist. (Geol.), 22:* 379–419.

WOODWARD, A. S., 1891. *Catalogue of the fossil fishes in the British Museum (Natural History), 11:* xliv+ 567 pp. London: Br. Mus. (Nat. Hist.).

WOODWARD, A. S., 1941. The head shield of a new macropetalichthyid (*Notopetalichthys hillsi*, gen. et sp. nov.) from the Middle Devonian of Australia. *Ann. Mag. nat. Hist., 8:* 91–6.

ZANGERL, R. & CASE, G. R., 1973. Iniopterygia, a new order of chondrichthyan fishes from the Pennsylvanian of North America. *Fieldiana, Geol. Mem., 6:* 1–66.

ABBREVIATIONS USED IN FIGURES

ADL	anterior dorsolateral plate	C	central plate
AL	anterior lateral plate	c_1, c_2, c_3	canals traversing lateral wall of occipital ossification
ALoa	overlap area for anterior lateral plate		
Art	articular	c_4, c_5, c_6	canals traversing orbital ossification
Aup	autopalatine	c_7, c_8	canals traversing ethmoidal ossification
AVL	anterior ventrolateral plate	cfAVL	contact face for anterior ventrolateral plate
aa	anterior angle of postorbital plate		
ade	anterior dermal element in pelvic clasper	ch	groove on inner paranuchal surface leading to postorbital fenestra
ampr	anterior median process of synarcual		
aoa	antorbital angle of preorbital plate	csc	central sensory canal
aorb	possible attachment area of autopalatine to endocranium	cspio	canal, probably for spino-occipital nerves
$aorb_1$	anterior area for attachment of autopalatine to endocranium	cuso	cutaneous sensory pit
		cv	cranial cavity
$aorb_2$	posterior area for attachment of autopalatine to endocranium	da	dorsal angle of postorbital plate
		dc	dorsal crest of synarcual
apq	attachment area for palatoquadrate	ddpr	depression for dorsal surface of upper toothplate
ar_1, ar_2	articular areas on scapulocoracoid for basals of the pectoral fin		
		dep	depression on visceral surface of preorbital plate
art	articular facet for glenoid process		
arta	articular area on paranuchal articular process	depr	depression on mesial surface of metapterygoid
arthy	articular area for hyoid arch element	dl	dorsal lamina of paranuchal articular process
artmd	articular surface for mandibular joint		
artpr	paranuchal articular process	dlsp	dorsal ligament space
ava	anteroventral angle of keel on median dorsal plate	dma	dorsomesial angle of orbital ossification
		dmad	depression on lateral surface of autopalatine, probably for adductor mandibulae muscles
ba_a	anterior median basal		
ba_p	posterior median basal		

dpr	dorsal process of upper toothplate	mr	mesial ridge of quadrate and articular
emb	ventral embayment on ethmoidal ossification	mrg	mesial ridge on anterior dermal element of pelvic clasper
est	eye-stalk attachment area	mri	median dorsal ridge on synarcual
ethoss	ethmoidal ossification	Nu	nuchal plate
fcor	coracoid fenestra	n	neck of bone connecting attachment areas on mesial surface of autopalatine
fepto	postorbital fenestra in skull-roof		
fl	flange on anterior margin of submarginal plate	nal	anterolateral notch in preorbital plate
flc	foramen in paranuchal for main lateral-line canal	narch	neural arch
		ncan	neural canal
flcv	floor of cranial cavity	ngr	neural groove
fm	foramen magnum	nP	notch in preorbital for pineal plate
fo	fold in posterior internal lamina of marginal plate	nPM	notch in ventral margin of marginal plate, probably for postmarginal plate
for	foramina for spinal nerves	nsp	neural spine
fpfc	foramen for 'profundus' sensory canal	oaAL	overlap area for anterolateral plate
glpr	glenoid process on occipital ossification	oaAVL	overlap area for anterior ventrolateral plate
gr	grooves on the mesial surface of the metapterygoid and autopalatine	oaM	overlap area for marginal plate
		oaPrO	overlap area for preorbital plate
grdf	groove on dorsal spine for anterior edge of first dorsal fin	occpr	occipital process of endocranium
		ocoss	occipital ossification of endocranium
grm	mesial groove on upper toothplate	om	orbital margin of preorbital and postorbital plates
grpbl	groove for the postbranchial lamina of the anterior lateral plate		
		ooss	orbital ossification of endocranium
grra	groove probably for the radix aorta	optm	optic margin of orbital ossification
gruo	shallow groove delimiting ridged area on external surface of marginal plate	otcap	assumed position of labyrinth cavity
		P	pineal plate
iar	infra-articular ridge	PM	postmarginal plate
ioc	infraorbital sensory canal	PN	postnasal plate
iocsb	suborbital branch of infraorbital sensory canal	PNu	paranuchal plate
		PrO	preorbital plate
kd	glenoid condyle of anterior dorsolateral plate	PSO	postsuborbital plate
		PtO	postorbital plate
kdrd	dorsal ridge on visceral surface of anterior dorsolateral plate	pbl	postbranchial lamina of anterolateral plate
kdrp	posterior ridge on visceral surface of anterior dorsolateral plate	pde	posterior dermal element of pelvic clasper
kdrv	ventral ridge on visceral surface of anterior dorsolateral plate	pfc	'profundus' sensory canal
		pln	posterolateral notch in posto plate
keel	ventral keel on median dorsal plate	pmc	postmarginal sensory canalrbital
Ltp	lower toothplate	pmsh	posteromesial shelf
la₃	posterior lamina on inner surface of marginal plate	pol	postoccular division of mesial lamina on inner surface of marginal plate
lc	main lateral-line canal on head	pp	posterior pitline
lcaf	anterior foramen for lateral-line canal	ppr	posterior process of skull-roof
lcpf	posterior foramen for lateral-line canal	pr₁, pr₂, pr₃	ventral projections on autopalatine
lfl	lateral flange of quadrate	prcor	coracoid process of scapulocoracoid
lla	lateral lamina of metapterygoid	prmpt	metapterygoid process
lr	lateral ridge of quadrate and articular	prpl	posterolateral process of scapulocoracoid
lrg	lateral ridge on anterior dermal element of pelvic clasper		
		prscap	scapular process of scapulocoracoid
M	marginal plate	psoc	postsuborbital sensory canal
MD	median dorsal plate	ptoa	postorbital angle of preorbital plate
MDoa	overlap area of median dorsal plate	ptr	protruberance on mesial surface of metapterygoid
Mpt	metapterygoid		
mde	middle dermal element of pelvic clasper	ptu	pointed tubercles on anterior dermal element of pelvic clasper
mdep	depression on mesial surface of auto-palatine		
		Qu	quadrate
ml	marginal lamina on inner surface of marginal plate	R	rostral plate
		rend	roof of endocranium
mla	mesial lamina of the metapterygoid	ri	ridge on mesial surface of metapterygoid
mp	middle pitline		

riaM	ridged area on external surface of marginal plate	sorc	supraoral sensory canal
rom	ridge inside orbital margin of postorbital plate	sos	possible suborbital shelf of endocranium
rp	rostral process	sov	supraorbital vault
rpro	preorbital ridge	sp	spine of spinal plate
rpto	postorbital ridge	ssyn	surface which probably abutted against synarcual
rsuo	supraorbital ridge	syn	synarcual
SM	submarginal plate	th	thickening on inner surface of sub-marginal plate
SO	suborbital plate		
sar	supra-articular ridge	thi	thickening on lateral margin of anterior lateral plate
sba$_a$	surface which probably abutted against the anterior median basal	thPNu	paranuchal thickening
scc	scapulocoracoid	tu	tubercle
sglr	subglenoid process	Utp	upper toothplate
sh	main shaft of quadrate	vcon	ventral concavity in anterior dermal element of pelvic clasper
she	shelf forming anteroventral margin of autopalatine	vgr	groove crossing ventral surface of autopalatine
sl	slit in lateral wall of occipital ossification	vl	ventral lamina of paranuchal articular process
soc	supraorbital sensory canal		
socl	lateral branch of supraorbital sensory canal	vn	ventral notch in autopalatine
		vpr	ventral process on ethmoidal ossification
socm	mesial branch of supraorbital sensory canal	vsend	ventral surface of endocranium
		X	bone of uncertain origin
sol	subocular division of mesial lamina on inner surface of marginal plate		

EXPLANATION OF PLATES

PLATE 1
Campbellodus decipiens gen. et sp. nov.
A, B.   Left upper toothplate in lateral and ventral views. P50907 ($\times$2).
C.   Right preorbital plate in dorsal view. Holotype, P50905 ($\times$3).
D, E.   Right lower toothplate in dorsal and lateral views. Holotype, P50905 ($\times$2).

PLATE 2
A.   *Campbellodus decipiens* gen. et sp. nov. Right upper toothplate in mesial view. Holotype, P50905 ($\times$2).
B.   *Campbellodus decipiens* gen. et sp. nov. Right autopalatine in lateral view. Holotype, P509905 ($\times$2).
C.   *Ctenurella gardineri* sp. nov. Left preorbital plate in dorsal view. P57665 ($\times$6).
D, E.   *Ctenurel la gardineri* sp. nov. Right lower toothplate in dorsal and lateral views. Holotype, P57637 ($\times$3).
F.   *Campbellodus decipiens* gen. et sp. nov. Right postbranchial lamina of anterior lateral plate. Holotype, P50905 ($\times$2).

PLATE 3
Ctenurella gardineri sp. nov.
A.   Synarcual in dorsal view. P57665 ($\times$5).
B.   Left marginal plate in lateral view. Holotype, P57637 ($\times$8).
C.   Right anterior dorsolateral plate in visceral view. P57665 ($\times$3).
D.   Anterior vertebral elements, basal plates, median dorsal and right anterior dorsolateral plate. P50910 ($\times$3).

PLATE 4
A.   *Campbellodus decipiens* gen. et sp. nov. SEM micrograph of dermal denticle from postbranchial lamina of trunk-shield. Holotype, P50905 ($\times$50).
B.   *Ctenurella gardineri* sp. nov. Head skeleton, use Fig. 23 as key-diagram. Holotype, P57637 ($\times$3).

PLATE 5
Ctenurella gardineri sp. nov.
A.   SEM micrograph of posterior element of pelvic clasper complex of bones. P57665 ($\times$50).
B.   Skeleton of head and trunk-shield as preserved. P50908 ($\times$3).

Plate 1

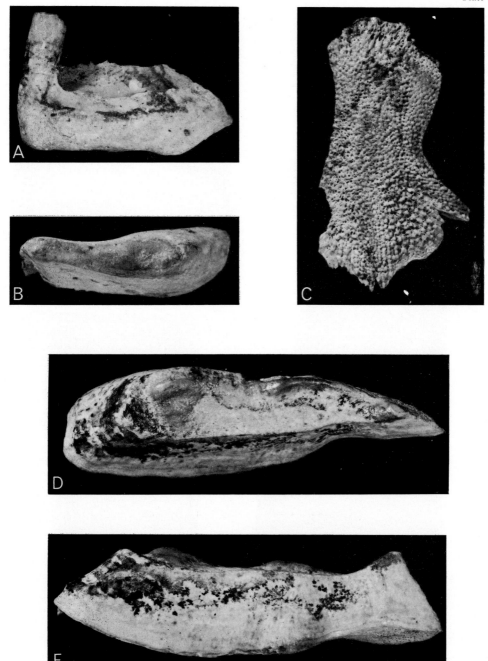

R. S. MILES AND G. C. YOUNG

Plate 2

R. S. MILES AND G. C. YOUNG

Plate 3

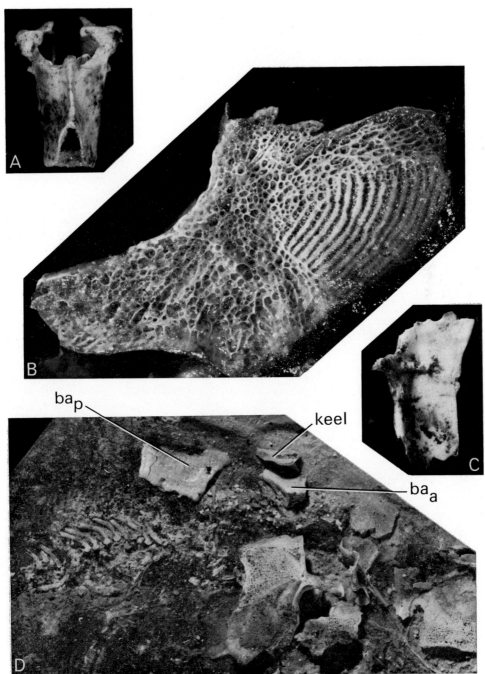

R. S. MILES AND G. C. YOUNG

Plate 4

R. S. MILES AND G. C. YOUNG

Plate 5

R. S. MILES AND G. C. YOUNG

The systematic position of acanthodian fishes

ERIK JARVIK

Section of Palaeozoology, Swedish Museum of Natural History, Stockholm, Sweden

Based in the main on an analysis of casts of *Acanthodes bronni* Agassiz it is shown that acanthodians essentially agree with extant sharks, especially with notidanids and squaloids. Some new morphological data are offered which, together with what was previously known, clearly suggest that acanthodians and selachians share a common ancestry. Hence acanthodians should be regarded as elasmobranchs. Their systematic position within this taxon is discussed.

CONTENTS

INTRODUCTION

The Gnathostomata or jawed vertebrates are generally divided into two principal groups: the Chondrichthyes and the Osteichthyes. These terms are reminiscences of the old, but long since relinquished idea that Recent elasmobranchs (sharks

199

and rays) and holocephalians are more primitive than bony fish because they have a cartilaginous endoskeleton and an unconsolidated exoskeleton of placoid scales in the skin. As is well known there is a general trend in vertebrates to a retrogressive development of the skeleton. Terms which refer to the degree of ossification should therefore be avoided, and for this reason, and because it is inappropriate to speak of the armoured placoderms as 'cartilaginous fishes', the term Elasmobranchiomorphi was introduced as a substitute for Chondrichthyes (Jarvik, 1955). In this unit the elasmobranchs and their presumed relatives (holocephalians, placoderms, acanthodians) were included. The term Osteichthyes was replaced by Teleostomi, a term coined by Bonaparte in 1837 and since the appearance of Owen's *Anatomy of vertebrates* in 1866 used by anatomists for fishes exhibiting a terminal mouth as well as an outer dental arcade including maxillary, premaxillary and dentary.

The distinction between elasmobranchiomorphs and teleostomes caused a dilemma as to the position of the Dipnoi, whereas in 1955 there was no reason to doubt that the acanthodians are elasmobranchiomorphs. The fact is that the acanthodians up to that time were generally considered to be related to elasmobranchs (by Dean, Goodrich, Gross, Holmgren, Jaekel, Reis, Romer, Säve-Söderbergh, Stensiö, Woodward, and others) or they were classified together with the Placodermi as Aphetohyoidea (Watson, 1937). More recently, however, Miles (1964, 1965, 1966, 1968, 1970, 1973a, b) and Miles & Moy-Thomas (1971) have made detailed studies of acanthodians and have added considerably to our knowledge of that group. In agreement with Nielsen (1949: 68–70, 109) Miles came to the conclusion that they are more closely related to teleostomes, in particular palaeoniscids, than to elasmobranchiomorphs and after having tested his results with the aid of Hennig's principles he placed the acanthodians as a subgroup of the Teleostomi. This view has been accepted by some writers (Romer, 1968; Halstead, 1969; Gardiner, 1973; see also Schaeffer, 1968, 1969; Heyler 1969a, b). But it has also met opposition (Nelson, 1968, 1969; Ørvig, 1972, 1973; Gross, 1973) and with the feeling that the problem was unsolved some writers (Jarvik, 1968; Ørvig, 1973) have placed acanthodians in a separate group between elasmobranchiomorphs and teleostomes.

Miles based his conclusions to a large extent on the endoskeleton of the head of the Permian form *Acanthodes bronni* Agassiz studied chiefly on casts in flexible rubber compounds. Duplicates belonging to this institute of most of these casts have been used in the subsequent account and are the basis of the new restorations presented in this paper. The most important of these casts are: P.1728 (damaged guttapercha cast of specimen figures by Reis, 1895: pl. 5, fig. 2); P.6210a/b (Museum Berlin 3A, Miles, 1968: fig. 3A; MB 3B, Miles, 1973a: pl. 2A); P.6211a/b (MB 4A, B); P.6214a/b (MB 7A, B); P.6215a/b (MB 8A, B); P.6218a/b (MB 11A, Miles, 1973a: pl. 6; MB 11B, Miles, 1973a: pl. 3B); P.6219a/b (MB 13A, Miles, 1973a: pl. 5B; MB 13B); P.6220a/b (MB 14A, Miles, 1968: fig. 1C, 1973a: pl. 4B; MB 14B); P.1621a/b (MB 16A, B); P.1622b (MB 17B, Miles, 1968: fig. 2); P.6223b (MB 18B); P.6228a/b (Senckenb. Museum, P.139a/b); P.6235a/b (Bonn, unregistered, a, Nelson, 1969: pl. 91; Miles, 1973a: pl. 4A; b, Miles, 1973a: pl. 5A). For comparative purposes skulls of *Squalus* and *Scymnorhinus* and other material of Recent sharks have been used.

DESCRIPTIONS AND DISCUSSIONS
The presumed spiracular gill slit

Miles (1964, 1965, 1973a) found the main support for teleostome, in particular palaeoniscid affinities, in the posterior part of anterior ventral endoskeletal ossification ('basisphenoid') of *Acanthodes* which he compared with the parasphenoid, a dermal bone, in palaeoniscids. In *Acanthodes* (Figs 2A, 3A and 15) this posterior part of the ossification in question shows three main structures: a median basal, or hypophysial, fenestra; a paired lateral process, interpreted as the basipterygoid process; and a groove-like depression referred to as the spiracular groove and said to correspond to the groove for the spiracular gill tube in palaeoniscids (Fig. 1C, D). In Recent fish the spiracular tube (Figs 1B and 9B), which

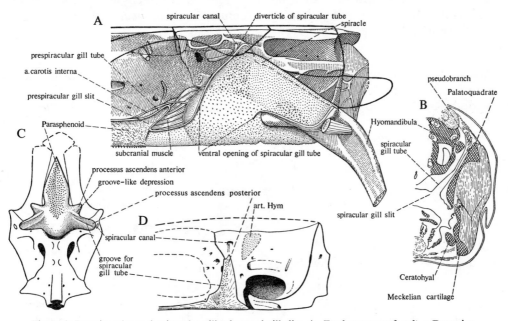

Figure 1. Prespiracular and spiracular gill tubes and gill slits. A. *Eusthenopteron foordi*, a Devonian osteolepiform. Restoration of prespiracular and spiracular gill tubes with certain adjoining structures. Lateral walls of gill tubes removed. B. *Heptanchus cinereus*, a Recent notidanid shark, larva 67 mm. Transverse section in the area of the right spiracular tube. From Luther, 1909a: fig. 8. C. *Kansasiella eatoni*, a Carboniferous palaeoniscid. Parasphenoid in relation to neural endocranium in ventral view. From Poplin, 1974: fig. 9. D. *Pteronisculus cicatrosus*, a Triassic palaeoniscid. Neural endocranium with parasphenoid in lateral view. From Jarvik, 1954: fig. 30D.

encloses the vestigial dorsal part of the spiracular gill slit (between the hyoid and mandibular arches), always opens into the pharynx. This internal opening is generally wide, and most often the spiracular tube decreases in width upwards and is closed dorsally (as was the prespiracular tube in porolepiforms and osteolepiforms, Fig. 1A) or opens outwards by a spiracle (Figs 9B and 11C; in *Eusthenopteron* unusually large). The groove in palaeoniscids lodging the spiracular tube (Fig. 1C, D) is situated on the outside of the strong processus ascendens posterior of the parasphenoid, which in turn covers the lateral commissure (suprapharyngohyal) of the neural endocranium and dorsally reaches the spiracular canal. In

Acanthodes the part of the neural endocranium that carries the presumed spiracular groove cannot possibly be the lateral commissure or any other part of the hyoid arch; and besides there is neither a spiracular canal in the endocranium nor a parasphenoid. The groove in *Acanthodes* is in fact more suggestive of the groove-like depression in palaeoniscids (Fig. 1C) which separates the anterior and posterior ascending processes of the parasphenoid than of the true spiracular groove; and like that depression it is narrow ventromedially and widens in the dorsolateral direction. The fact that the depression in *Acanthodes* so to speak widens in the wrong direction shows that it cannot be developed for a gill tube at all; and it must not be confused with the sometimes toothed groove for the prespiracular gill tube (Fig. 1A) in osteolepiforms and porolepiforms (Jarvik, 1954, 1972) which decreases in width upwards. In order to offer a more likely explanation of the groove-like depression and adjoining structures in *Acanthodes* let us turn to a comparison with sharks.

Palatobasal process

In sharks, as in acanthodians, the parasphenoid is lacking and also in sharks there is a basal (hypophysial) fenestra. This opening, which transmits the bucco-hypophysial duct and the internal carotids, is in the Devonian shark *Cladodus*

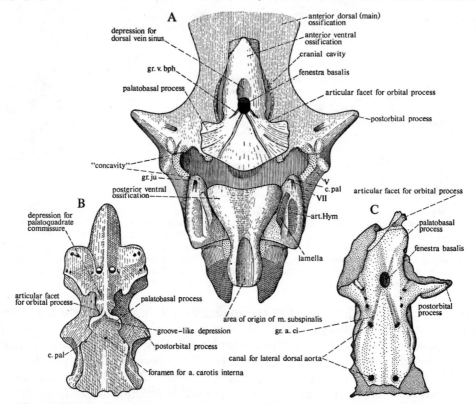

Figure 2. Neural endocrania in ventral view. A. *Acanthodes bronni*, a Permian acanthodian. Restoration of ossified parts after latex casts (cf. Figs 3, 12 and 15). B. *Squalus acanthias*, a Recent squaloid shark. From Marinelli & Strenger, 1959: fig. 213. C. *Cladodus wildungensis*, a Devonian palaeoselachian. From Gross, 1937: fig. 5.

(Fig. 2C) developed much as in *Acanthodes*. In Recent sharks (Holmgren, 1940, 1941, 1942; El-Toubi, 1949) it is generally large in larval stages, but becomes closed in the adult except for openings for the internal carotids and, rarely (*Etmopterus*, Holmgren, 1941: fig. 20; *Pristiurus*, Meurling, 1967: 11), the bucco-hypophysial duct. Lateral to the hypophysial opening, or in a corresponding position, in the area of the so called basal angle (see El-Toubi, 1949), the neural endocranium in notidanid and squaloid sharks (Figs 2B, 3B and 13C) projects into a palatobasal shelf or process. This structure, which, like the orbital process, seems to arise in the interarcual ligament connecting the palatoquadrate with the trabecula (Holmgren, 1943: 56–59; Jarvik, 1954: 81) and in sharks has a varying position (in *Cladodus*, Figs 2C and 4D, and *Chlamydoselachus*, Fig. 3C, it is far

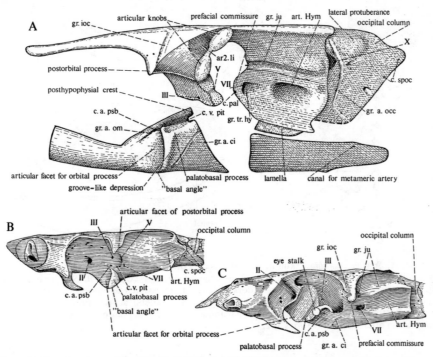

Figure 3. Neural endocrania in lateral view. A. *Acanthodes bronni*, restoration of dorsal and ventral ossifications in their presumed position in relation to each other. Ca 2/1. Mainly after casts P.6210b, P.6220a, P.6235a/b. B. *Heptanchus maculatus*, from Daniel, 1934: fig. 47. C. *Chlamydoselachus anguineus*, from Allis, 1923: fig. 8.

forwards in the orbit), must not be confused with the basipterygoid process (see also El-Toubi, 1949: 262). In several squaloids (e.g. *Squalus*, *Scymnorhinus*) the palatobasal process resembles, as to its position and configuration, the presumed basipterygoid process in *Acanthodes*. Like that process it is delimited postero-medially by a groove-like depression which widens in a dorsolateral direction and obviously corresponds to the 'spiracular groove' in *Acanthodes*. The palatobasal process, which projects ventrally and forms the 'basal angle', has a smooth surface; it lies close to the inner side of the palatoquadrate but is separated from it by connective tissue and there is no basal articulation either in the embryo or in the adult (Holmgren, 1942: 138). In *Acanthodes*, too, the presumed basipterygoid process is smooth; it lies close to but does not articulate with the palatoquadrate, and ventrally it projects slightly forming a kind of 'basal angle'. Accordingly the presumed basipterygoid process in *Acanthodes* must be a palatobasal process homologous with that in sharks.

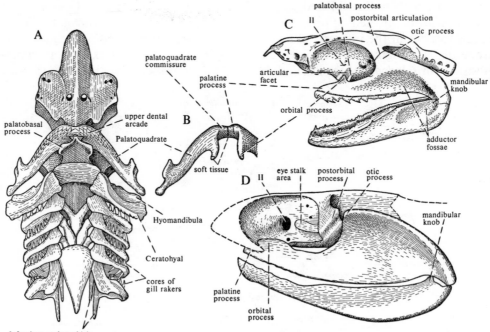

Figure 4. A. *Squalus acanthias*, skull in ventral aspect. Lower jaws, labial cartilages and branchial rays omitted. Compilation after Marinelli & Strenger, 1959: figs 213, 217. B. *Squalus acanthias*, left and part of right palatoquadrate in dorsal view. After dissected specimen. C. Neural endocranium, palatoquadrate, and lower jaws of *Heptanchus* in lateral aspect. From Gaupp, 1913: fig. 1. D. Restoration of neural endocranium, palatoquadrate and lower jaw of *Cladodus wildungensis*. Compilation after Gross, 1938: figs 1, 2.

Orbital process and orbital articulation

The view that the presumed basipterygoid process in *Acanthodes* is a palatobasal process is supported by the following conditions. In *Acanthodes* the palatoquadrate (Figs 6 and 8) projects into a strong ascending process which is received by a distinct articular facet (Fig. 3A) on the lateral side of the neural endocranium, close in front of the palatobasal process. This ascending process as to structure, position and mode of articulation agrees strikingly with the orbital process (Fig. 4A–C) of those sharks (notidanids, squaloids) in which the articular facet of the neural endocranium (Figs 3B, 4C and 13C) and the orbital articulation are in the posterior part of the orbit and, as claimed by Holmgren (1942), it is no doubt homologous with that process. This conclusion is further supported by the fact that the orbital process in *Acanthodes* (Fig. 6) presents a prominent articular ridge on its medial side, exactly as in the said sharks (Figs 4B and 7A; Gegenbaur,

1872; Luther, 1909b). Moreover, it ends with an unfinished surface suggesting that its tip was formed by soft tissue, as in *Squalus* (Fig. 4B).

Palatine process and palatoquadrate commissure

A characteristic of the palatoquadrate in sharks is that anteriorly, in front of the orbital process, it features a palatine process which bends medially and often meets its antimere to form a palatoquadrate commissure (Fig. 4A–C). Watson and Miles deny the existence of such a commissure in acanthodians. However, as is well shown in one of the casts (P.6221a), the palatine process (Figs 6 and 8) is longer than shown by Miles. Moreover, it bends medially in its anterior part in exactly the same way as does the palatine process in *Squalus* (Fig. 4A, B) and, as in that form, it ends anteromedially with an unfinished surface. These conditions show that a palatoquadrate commissure was present in *Acanthodes*, and that the palatine process was separated from its antimere only by a thin median layer of connective tissue (or cartilage), as in *Squalus*.

The presence (P.6221a) of a small depressed area (ar. lab. cart, Fig. 6B) in the proper place on the outside of the palatine process and a groove leading backwards from that area to the margin of the upper jaw indicate that the acanthodians possessed an anterior labial cartilage situated as in embryonic sharks (Holmgren, 1940: 106, 120, 125, figs 68, 81, 89, 95; Stensiö, 1963: fig. 79).

Dental arcades and position of mouth

That a palatoquadrate commissure similar to that in sharks was present in acanthodians is clearly shown also by a specimen of *Ptomacanthus* figured by Miles (1973b: pl. 6). This specimen (Fig. 5A; see also Woodward, 1915: fig. 1) presents a shark-like upper dental arcade and, no doubt, the arched tooth whorls which make up this arcade were carried by the palatine processes of the palatoquadrates. It is also an important fact that this arcade is clearly situated on the ventral side of the snout, apparently rather far behind its tip. This means that the snout was longer than usually restored. It also implies that the mouth was subterminal, as in sharks (Fig. 5B). Another remarkable feature is that the nasal openings were situated on the ventral side of the snout, close to the margin of the mouth, as sometimes in sharks.

In several acanthodians tooth whorls are present also in the lower jaw, and sometimes there is a median symphysial tooth whorl (Watson, 1937: 118; Miles, 1966; Ørvig, 1973) as, e.g. in *Chlamydoselachus* and *Heptanchus* (Gudger & Smith, 1933). The tooth whorls and other tooth-bearing jaw bones in acanthodians are carried by the palatoquadrates and the meckelian elements, and correspond to the (upper and lower) inner dental arcades in teleostomes. Maxillary, premaxillary and dentary, which in teleostomes form the outer dental arcades, are lacking, as in elasmobranchiomorphs and dipnoans.

Because acanthodians have a subterminal mouth and lack an outer dental arcade they cannot be classified as teleostomes. They are no doubt elasmobranchiomorphs, and the detailed agreement in the anterior part of the palatoquadrate and the adjoining part of the neural endocranium shows that they are in some way related to sharks. This conclusion is strongly supported by the structure of the posterior part of the palatoquadrate and numerous other data.

Postorbital joint and the double jaw joint

Behind the orbital process the palatoquadrate in *Acanthodes* (Fig. 6) expands
into a strong otic process, very much as in notidanid sharks (Figs 4C and 7A;
Gegenbaur, 1872; Fürbringer, 1897; Goodrich, 1909; Luther, 1909a; Daniel,

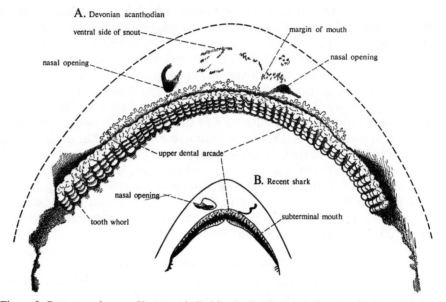

A. Devonian acanthodian

ventral side of snout

margin of mouth

nasal opening

nasal opening

upper dental arcade

B. Recent shark

nasal opening

subterminal mouth

tooth whorl

Figure 5. Representations to illustrate similarities in development of upper dental arcade, and
position of nasal openings and mouth between an early Devonian acanthodian, *Ptomacanthus
anglicus* (drawing after photograph published by Miles, 1973b: pl. 6) and a Recent shark, *Cephaloscyllium ventriosum* (from Garman, 1913: pl. 9).

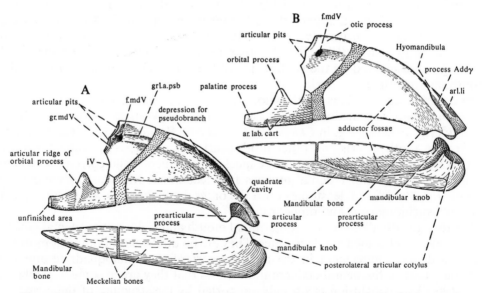

B

f.mdV

otic process

articular pits

Hyomandibula

orbital process

process Addγ

arl.li

A

articular pits

f.mdV

grl.a.psb

palatine process

gr. mdV

depression for
pseudobranch

articular ridge of
orbital process

iV

ar. lab. cart

adductor fossae

quadrate
cavity

Mandibular bone

mandibular knob

unfinished area

prearticular
process

articular
process

prearticular
process

Mandibular
bone

mandibular knob

Meckelian bones

posterolateral articular cotylus

Figure 6. *Acanthodes bronni*. Restorations of palatoquadrate, lower jaw, and (in B only) hyomandibula. A. Right side in medial and B, left side in lateral aspect. Mainly after casts P.6221a/b, P.6222b.

1934; Edgeworth, 1935); and as in *Heptanchus* and certain Palaeozoic sharks this process articulates with the posterior side of the postorbital process (postorbital joint). However, in contrast to *Heptanchus* there are two partly separate articular knobs on the neural endocranium and two corresponding articular pits on the palatoquadrate. Moreover, the otic process is pierced by a foramen (f.mdV) which, as suggested by Miles, probably transmitted the r.mandibularis trigemini or parts of that nerve. Deep fossae for the m.adductor mandibulae similar to those in notidanids and *Chlamydoselachus* (Figs 4C and 9A) are present on the external sides of the palatoquadrate and the meckelian element (lower jaw), but of greater importance is that the jaw articulation is double (Fig. 6) and developed very much as that of many sharks (Fig. 7). This characteristic double jaw articulation

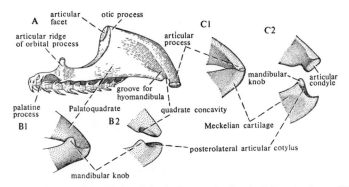

Figure 7. Palatoquadrate and double jaw joint in Recent sharks. A. Palatoquadrate of *Heptanchus cinereus* in medial aspect. After Luther, 1909 b: fig. 6; with certain details after Gegenbaur, 1872: pl. 20, fig. 3. B, C. *Chlamydoselachus anguineus*, posterior parts of palatoquadrate and meckelian cartilage in their natural position (B1, C1) and disarticulated at the jaw joint (B2, C2): B1, B2, medial and C1, C2, lateral aspects. From Allis, 1923: figs 26–29.

(Gegenbaur, 1872: 192; as to terminology see Hotton, 1952) includes a lateral joint between the condyle of the articular process of the palatoquadrate and the posterolateral articular cotylus of the lower jaw, and a medial joint between the mandibular knob and the quadrate concavity of the palatoquadrate.

Hyomandibula

In *Acanthodes* the hyomandibula (Figs 6B and 8), which, as in sharks, probably articulated with the neural endocranium ventral to the jugular vein (art. Hym, Fig. 3A), is a long rod without a canal for the truncus hyomandibularis and on the whole is suggestive of the hyomandibula in notidanids (Gegenbaur, 1872: pls 10, 15; Ruge, 1897: fig. 12; Goodrich, 1909: fig. 59) and *Chlamydoselachus* (Fig. 9; Allis, 1923). It generally includes a short dorsal and a long main ossification, but sometimes a ventral part may be separate (Miles, 1973a, 'accessory element'). This subdivision may be due to retrogressive development of the skeleton, but it may be of interest to note that the hyomandibula in sharks (Holmgren, 1940: 155, 1943: 75) is also composed of three parts: a small dorsal (suprapharyngohyal), a long middle (epihyal), and a small ventral (derived from mandibular branchial rays).

Spiracular gill tube and pseudobranch

As to the position, in acanthodians, of the hyomandibula in relation to the palatoquadrate, two fundamentally different opinions have been expressed. According to Watson (1937) there was a full-sized spiracular gill slit not closed even in the area of the jaw joint, whereas Miles (1973a: 94) claims that the hyomandibula occupied the groove-like depression on the inner side of the palatoquadrate (Fig. 6A) and that the spiracular tube was farther forwards and was closed dorsally. It is true that the hyomandibula in notidanids (Gegenbaur 1872: 167, pl. 15) lies in a groove on the inner side of the palatoquadrate, but this groove (Fig. 7A) is deepest at the jaw joint and housed only the ventral part of the hyomandibula. The dorsal part of the hyomandibula lies medial to the palatoquadrate and is separated from it by the spiracular tube (Fig. 1B). The groove-like depression

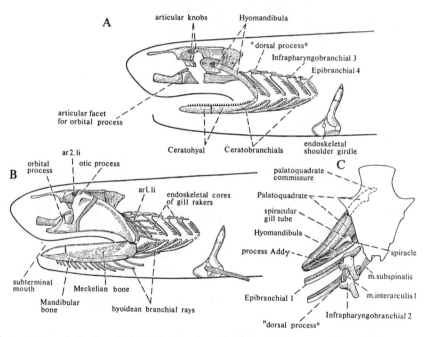

Figure 8. *Acanthodes bronni*. A, B. Restoration of skull without squamation in lateral aspect: A, with palatoquadrate and lower jaw removed. Certain details after Watson, 1937: fig. 18, and Miles, 1973b: fig. 19. C. Restoration of palatoquadrate, dorsal parts of hyoid arch and foremost two branchial arches, with spiracular tube and certain muscles. Left side, dorsal view. The neural endocranium is shown in outline.

in *Acanthodes* is, in contrast, situated well above the jaw joint and is bounded posterodorsally by a prominent edge. If the hyomandibula in *Acanthodes* occupied this depression it would (because of the edge) be expected that it had remained in this position in the fossils. But this is normally not so. In the numerous casts at my disposal (see also Miles, 1968: fig. 1; 1973a: pls 2A, 4A, 5A, 7; Nelson, 1969: pl. 91) the principal ossification of the hyomandibula, as seen in lateral view, is hidden by the palatoquadrate in its anterodorsal part or lies mainly or wholly behind that structure. Only in one cast (P.6214b) is it found in the depression (cf. Miles, 1973a: 94; Nelson, 1969: 523). Judging from these and other conditions it seems likely that the hyomandibula in *Acanthodes* was situated somewhat as in *Chlamydoselachus* (Fig. 9A). This is supported by the fact that the ventral part of the hyomandibula in *Acanthodes* (Fig. 6B) presents a crest-like process suggestive of, and most likely homologous with, the process *Addy* of Allis (1923) in *Chlamydoselachus* (Fig. 9A, B). The posterodorsal surface of this process is rough (ar1.li); moreover there is a rough and also slightly depressed area (ar2.li, Figs 3A and 8B) on the postorbital process. These conditions indicate that a ligament corresponding to that which in *Chlamydoselachus* (Fig. 9B) runs from the process *Addy* to the postorbital process was present also in *Acanthodes*. Most likely this ligament in

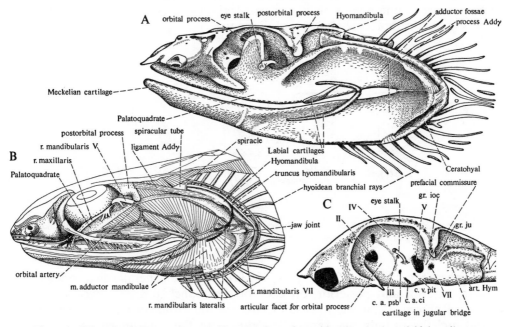

Figure 9. *Chlamydoselachus anguineus*. A. Neural endocranium with palatoquadrate, labial cartilages, lower jaw and hyoid arch in lateral aspect. From Allis, 1923: fig. 7. B. Lateral view of head showing spiracular tube, mandibular muscles and certain other soft parts. From Allis, 1923: fig. 55. C. Neural endocranium of embryo, 127 mm, in lateral view. From Holmgren, 1941: fig. 8.

Acanthodes also passed outside the spiracular tube which, as in sharks (Figs 1B and 9B), was probably situated between the dorsal part of the hyomandibula and the palatoquadrate and opened outwards by a spiracle (Fig. 8C).

Comparisons with sharks also suggest that the groove-like depression on the inner side of the palatoquadrate in *Acanthodes* lodged a well developed pseudo-branch (cf. *Ischnacanthus;* Watson, 1937: pl. 9, fig. 3). In sharks this structure has a corresponding position (Allis, 1923; Smith, 1937; Marinelli & Strenger, 1959) and in notidanids (Fig. 1B; Gegenbaur, 1872: 198; Virchow, 1890) it is large and lies close to, and is protected by, the palatoquadrate (cf. Gegenbaur, 1872: pl. 15, fig. 1).

Gill rakers

Watson's main argument for his opinion that acanthodians have a full-sized spiracular gill slit was that the hyoid arch in *Acanthodes* carries gill rakers. The gill rakers of Recent fish (and larval amphibians) are a series of elevations of varying structure, shape, and size which guard the pharyngeal openings of the gill slits and act as a straining apparatus (for review see Stadtmüller, 1924; see also Daniel, 1934; Nelson, 1968, 1969; Miles, 1973a). In certain sharks (e.g. *Squalus*, Fig. 10), as well as in holocephalians and dipnoans, the gill rakers are supported by an internal rod of cartilage. This is of particular interest because the preserved parts of the gill rakers in acanthodians are probably endoskeletal (Ørvig, 1973: 146), not exoskeletal as has generally been assumed. Accordingly the structures hitherto termed gill rakers in acanthodians are most likely endoskeletal cores homologous with the cartilaginous rods which support the gill rakers in sharks, holocephalians, and dipnoans.

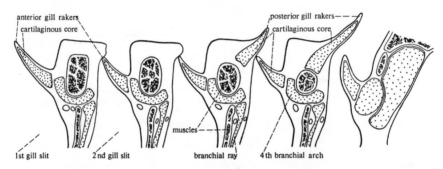

Figure 10. *Squalus sucklii*. Section through branchial arches to show branchial rays and gill rakers with supporting cartilaginous cores. From Daniel, 1934: fig. 147.

In fishes (Stadtmüller, 1924; Miles, 1973a) there are generally two series of gill rakers, anterior and posterior, on each of the branchial arches except the last. As is well shown, e.g. in specimens of *Chimaera* and *Protopterus* at my disposal, the hyoid arch bears only one series. Together with the anterior series of the first branchial arch this series guards the entrance to the first posthyoidean gill slit and is, accordingly, a posterior series. It is therefore hard to believe that the single series of the hyoid arch in *Acanthodes* (Fig. 8A), in contrast to Recent fish, should be an anterior series, as it is claimed to be by Miles (1968: 114) and Nelson (1968: 140). In one of the casts (P.6218a) the endoskeletal cores of the gill rakers belonging to the ceratohyal are directed medially, rather than laterally towards the lower jaw, a condition which indicates that they guarded the first gill slit and represent a posterior series. Most likely this is true also of the rakers of the hyomandibula, the cores of which are shorter than shown by Watson (1937: fig. 18), directed inwards and therefore hardly visible if the hyomandibula is viewed in lateral aspect. In consequence, the view that acanthodians had a full-sized spiracular gill slit lacks evidence. No doubt this slit was closed in the area of the jaw joint where the r.mandibularis VII and the r.mandibularis lateralis, as in other fishes (Fig. 9B), passed from the hyoid to the mandibular arch.

In *Acanthodes* (Miles, 1968: 114) two series of gill rakers are found on each of the foremost four posthyoidean branchial arches in juvenile individuals only. In adults the posterior series is always lacking on the foremost three arches. This is of some interest since the posterior series is lacking on the first and second branchial arches in *Squalus* (Fig. 10).

Visceral arches

Other striking similarities to sharks (Figs 4A and 13A) are that the visceral arches in acanthodians (Fig. 8) are posterior in position, and that the infrapharyngobranchials are directed backwards (Nelson, 1968; Miles, 1973a). The so called dorsal process of the latter elements (as to musculature see Fig. 8C) seems to be directed posterolaterally rather than upwards (Miles, 1973a: fig. 16F), as is the corresponding process in sharks (Wells, 1917: pl. 2; Garman, 1913: pl. 62; Marinelli & Strenger, 1959: fig. 216).

Gill covers and branchial rays

As to existing interpretations of gill covers in acanthodians, Miles (1973a: 98) remarks that they can only be accepted with reservations in most instances. They are fairly well known in *Climatius* (Fig. 11A; Miles, 1973b). This form has a large but rather short hyoidean and four posthyoidean gill covers, each projecting backward outside a gill slit, somewhat as in *Chlamydoselachus* (Fig. 11C; Gudger & Smith, 1933; Gudger, 1940). The hyoidean gill cover in *Climatius* is supported by five elongated elements which no doubt are dermal bones (Miles, 1973b: fig. 7) and, therefore, may be regarded as branchiostegal rays. Whether the long and slender rods in *Acanthodes* which hitherto have been interpreted as branchiostegal rays (Jarvik, 1963; Miles, 1973a) are exoskeletal formations is, however, doubtful. Comparison with *Chlamydoselachus* (Fig. 9A, B) and other sharks suggests that they are probably hyoidean branchial rays (Fig. 8B), and thus endoskeletal formations. Judging from a specimen figured by Miles (1966: fig. 4) the posthyoidean branchial arches also carried endoskeletal branchial rays. In this connection it may be mentioned that well-developed hyoidean branchial rays (or cartilages) are characteristic of accepted elasmobranchiomorphs (elasmobranchs, holocephalians, placoderms) and dipnoans. In teleostomes these rays are generally much modified, forming for example, the opercular process of the hyomandibula.

Sensory lines

The sensory line system of the head (Fig. 11) in acanthodians is of 'nearly a perfect selachian type' (Holmgren & Pehrson, 1949: 257). Note the similar course of the infraorbital canal in relation to the postorbital process in *Acanthodes* (gr.ioc, Figs 3A and 12A) and sharks (Figs 9 and 13D). In acanthodians (Miles, 1973b: fig. 2), as in sharks (Gudger & Smith, 1933: fig. 18), the sensory canal of the body runs between rows of slightly modified scales. Posteriorly, at least sometimes (cf. Miles, 1970: 356), the main sensory line seems to reach, or almost reach, the posterior margin of the hypochordal lobe of the caudal fin as it does in *Squalus*.

Fins

The fins in acanthodians are aplesodic, as in sharks (Jarvik, 1959; Miles, 1970). The endoskeleton of the pectoral fin in *Acanthodes* is tribasal (Miles, 1973b: fig. 20) which, according to Schaeffer (1967), is characteristic of 'hybodont- and modern-level' sharks. The preserved parts of the fin webs are completely covered with small scales very much as in Recent elasmobranchs and dipnoans. Inside these fin scales, which must not be confused with the lepidotrichia of teleostomes, there are ceratotrichia-like dermal fin rays (Watson, 1937; Jarvik, 1959; Miles, 1970); and

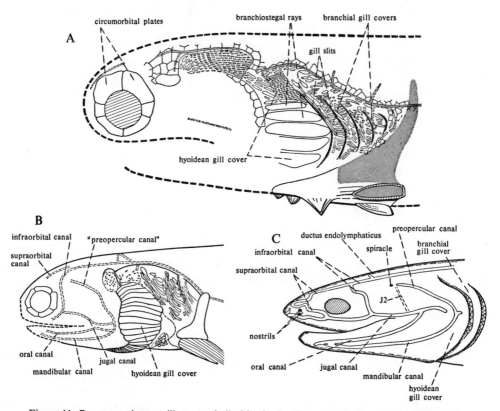

Figure 11. Representations to illustrate similarities in development of gill covers and lateral line system of head between early Devonian acanthodians and a Recent shark. A. *Climatius reticulatus*, from Miles, 1973b: fig. 10. B. *Euthacanthus macnicoli*, from Watson, 1937: fig. 3. C. *Chlamydoselachus anguineus*, mainly after Allis, 1923. The jugal 2 line (J2) of Holmgren & Pehrson (1949) in *Chlamydoselachus* is possibly homologous with the 'preopercular' line in *Euthacanthus* in which case the 'preopercular' line in *Chlamydoselachus* may be represented in acanthodians by Watson's 'opercular canal' (1937: fig. 20).

very likely these rays extended a little beyond the margin of the scale-covered part of the fin web, forming a narrow fringe, as is the case in Recent elasmobranchs and dipnoans. No ceratotrichia have been observed in the caudal fin, which, mainly for this reason it appears, has been claimed to be palaeoniscid-like (Heyler, 1969a, b). However, since ceratotrichia have been found both in the pectoral and dorsal fins (Miles, 1970) it is likely that they were present also in the caudal, and most probably they ran parallel with the rows of fin scales as they do (judged from the direction of the spines of the placoid scales) in Recent sharks. As to its shape, the caudal fin in acanthodians is suggestive of that in certain sharks (Garman, 1913; Gudger & Smith, 1933; Gudger, 1940) and, as in sharks (and many other fishes), there may be an axial lobe.

Fin spines are most often lacking in Recent sharks but may be present in front of the dorsal fins as, for example, in *Squalus*. In view of this condition the fact that spines in acanthodians are also developed in the ventrolateral fin fold cannot be attributed any great importance.

Histology

Acanthodians differ fundamentally from teleostomes in the histological structure and growth of the teeth (Ørvig, 1973; Gross, 1973) and besides, with regard to the structure of the scales and spines they agree more with elasmobranchs than with teleostomes (Gross, 1947, 1957, 1973; Ørvig, 1972).

Remarks on the neural endocranium

The neural endocranium of *Acanthodes* is imperfectly known. The ethmoidal region is not preserved and considerable parts of the postethmoid division were formed by cartilage or were membranous. Moreover, the four principal ossifications which are preserved can be studied only on latex casts and their position in relation to each other is uncertain. However, in spite of these unfortunate conditions it has been possible to extract a considerable number of data (Figs 2A, 3A, 12 and 15). As already described the anterior ventral ossification presents a structure homologous with the palatobasal process in sharks, and in front of that there is a distinct articular facet for the orbital process of the palatoquadrate, very similar in development and position, to that in notidanid and squaloid sharks. Moreover, the principal dorsal ossification presents a postorbital process similar to that in sharks, articulating, as in certain Palaeozoic sharks and notidanids, with the otic process of the palatoquadrate. As we now shall see, the neural endocranium of *Acanthodes* is also strikingly shark-like in other respects, as far as its structure can be made out with a fair amount of certainty.

Cranial roof

The dorsal side of the endoskeletal cranial roof in *Acanthodes* (Fig. 12A; P.1728, P.6210a) presents—besides grooves for sensory canals and two median ridges—a fontanelle, which, as to its position and shape, is suggestive of the fontanelle in larval stages of *Squalus* (Fig. 13B; closed in adults, Fig. 13D). Behind the fontanelle in *Acanthodes* follows a large depression whose position corresponds to that of a depression in *Squalus* (Fig. 13D). In the posterior part of the depression, close in front of the posterior median ridge, there is a paired pit which, although no foramina are discernible in its bottom, is very probably the endolymphatic fossa (cf. Fig. 13D). That this is so is supported by cast P.6210b which shows the ventral side of this part of the cranial roof. In this cast the large depression appears as an elevation (Fig. 12B) on the posterior part of which there are a paired foramen (c.d.end) and a groove running from the foramen in the

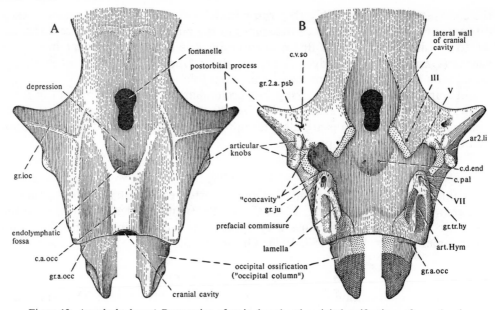

Figure 12. *Acanthodes bronni*. Restoration of main dorsal and occipital ossifications of neural endocranium in A, dorsal, and B, ventral aspects. Mainly after casts P.1728, P.6210 a/b.

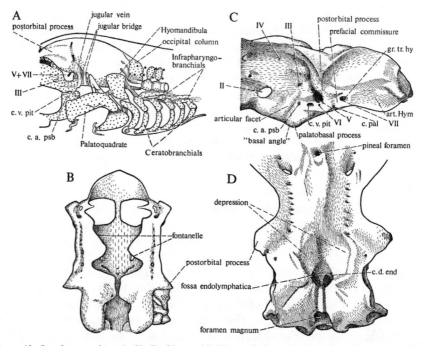

Figure 13. *Squalus acanthias*. A. Skull of larva, 35–37 mm, in lateral aspect. Anterior parts omitted. From Jollie, 1971: fig. 5. B. Neural endocranium of larva, 48 mm, in dorsal aspect. From Holmgren, 1940: fig. 69. C, D. Skull of adult in lateral and dorsal aspects. Anterior part omitted. Drawings after dissected specimen.

posterolateral direction. This foramen almost certainly transmitted the ductus endolymphaticus and, if so, it follows that this duct in *Acanthodes* ran forwards in its proximal part as it does in sharks (Retzius, 1881: pls 18, 20; El-Toubi, 1949: 249, fig. 8). Within the endolymphatic fossa the duct may have turned backward in the way characteristic of sharks (see also Goodey, 1910); however, no external opening for the duct has been observed in the dermal cranial roof (cf. Miles, 1973b: 128, fig. 5).

Lateral dorsal aorta and carotis interna

In Recent sharks (Fig. 14A, B; Allis, 1912, 1923; O'Donoghue & Abbott, 1928; Corrington, 1930; Daniel, 1934; El-Toubi, 1949) the lateral dorsal aortae run forwards close below the posterior part of the ventral side of the neural endocranium, giving off one or more metameric arteries before they join the efferent

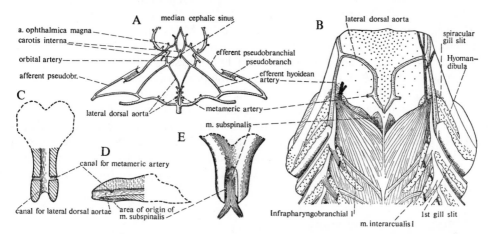

Figure 14. A. *Heptanchus maculatus*, diagram of arterial system in endocranial base. From Daniel, 1934: fig. 152 (simplified). B. *Chlamydoselachus anguineus*, posterior part of neural endocranium and adjoining parts of visceral endoskeleton with muscles and arteries in ventral aspect. From Allis, 1923: fig. 56. C-E. *Acanthodes bronni*, posterior part of posterior ventral ossification: C, dorsal view with upper part cut away and, D, sagittal section. From Miles, 1970: fig. 6. E. Ventral view after cast P. 6228a. Subspinalis muscle tentatively restored.

hyoidean arteries. They may diverge forwards or they may run parallel for a considerable distance, as in *Heptanchus* (Fig. 14A). In this form they lie close together and are enveloped in a common sheet (Allis, 1912), a condition which indicates that they were originally enclosed in the endocranial base as they were in Palaeozoic sharks (Fig. 2C). This is of interest because the posterior ventral ossification in *Acanthodes* (Fig. 14C, D) shows a median canal ('notochordal canal', Miles, 1970, 1973a) giving off a paired lateral canal. It seems likely that the median canal housed the lateral aortae, and that the paired canal transmitted a metameric artery. It is possible that the latter artery passed upwards in the groove (gr.a.occ, Figs 2A, 3A and 12; casts P.6235a/b) along the anterior margin of the occipital ossification and corresponds to the occipital artery in *Squalus* (Coles, 1928: 107; O'Donoghue & Abbott, 1928: fig. 6). After giving off a posterior branch the presumed occipital artery in *Acanthodes* crossed the lateral protuberance of the occipital ossification (this protuberance is compared by Miles with the para-occipital, or craniospinal, process in actinopterygians; see Holmgren & Stensiö, 1936: figs 320, 334; Nielsen, 1949: figs 2–4; Poplin, 1974: figs 12, 13), and then continued to the cranial roof where a branch may have passed through the canal *c.a.occ* (Fig. 12A). The course of the lateral dorsal aortae next in front of the median canal is not evidenced by the fossils but conceivably they diverged as in sharks (Fig. 14A), joined the efferent hyoidean artery, and gave off the orbital artery (external carotid). Farther forwards the lateral dorsal aortae, or internal carotids (Fig. 16D), ran forwards and downwards in a groove *gr.a.ci* (Figs 2A and 15B; cf. *Chlamydoselachus*, Fig. 3C) on the anterior ventral ossification, and, as in *Cladodus* (Fig. 2C), they probably converged towards the basal fenestra. At the posterior part of that fenestra they probably passed upwards and most likely fused into a short median vessel, the cephalic sinus, as they do in *Heptanchus*, *Squalus*, and many other sharks (Fig. 14A; Gegenbaur, 1872; Allis, 1912; Corrington, 1930; Meurling, 1967). It seems likely that this median vessel entered the median canal (c.mcs, Fig. 15A) described by Miles (1973a, 'buccohypophysial foramen') in the posterior wall of the basal fenestra, and that the internal carotids soon separated and passed into the fossa hypophyseos (Fig. 16D) by the paired canal (c.a.ci, Fig. 15A, C) which opens into the posterolateral part of that fossa. At this place the internal carotid received the efferent pseudobranchial artery which passed into the fossa by a canal (c.a.psb, Fig. 3A) seen to enter the lateral side of the wall of the cranial cavity, lateral to the foramen for the internal carotid (casts P.6218b, P.6235a). From the lateral opening of the canal *c.a.psb* a groove (gr.a.om, Figs 3A and 15C; Miles, 1973a: 84, fig. 9, alg), no doubt for the a.ophthalmica magna, runs forwards and downwards to the lateral margin of the anterior ventral ossification.

Efferent pseudobranchial artery

Provided the pseudobranch was situated in the depression on the inner side of the palatoquadrate (Fig. 6A), the efferent pseudobranchial artery probably ran forwards in the groove *gr1.a.psb* leading to the area of the foramen (f.mdV) in the otic process. In the area of that foramen it probably turned downwards following the course of the groove *gr.mdV* to continue forwards through the groove *gr2.a.psb* (Figs 12B and 15) on the ventral side of the postorbital process (in *Squalus* the artery pierces the orbital cartilage, O'Donoghue & Abbot, 1928: 842). In the orbit, the artery probably ran in an anteromedial direction to the groove *gr3.a.psb* (Fig. 15C; casts P.6218b, P.6235a) on the anteromedial part of the dorsal surface of the palatobasal process and, after giving off the a.ophthalmica magna as described above, it entered the cranial cavity through the canal *c.a.psb*, passing, as in sharks, dorsal to the trabecula (de Beer, 1924, 1937).

Jugular vein and hyomandibular articular area

The jugular vein in *Acanthodes* no doubt passed backwards fairly high up on the lateral side of the otic region in a distinct groove (gr.ju, Figs 2A, 3A and 12B; casts P.6210b, P.6220a, P.6223b, P.6235a). This groove runs from the ventromedial part of the 'concavity' which in the opinion of Miles (1968: 125, 1972a: 78–9) housed the dorsal end of the closed spiracular tube to the posterior end of the main dorsal ossification. Ventrally that groove is bounded by a rather sharp ridge conceivably housing the external semicircular canal. Ventral to that ridge is an elongate depression (art.Hym) which obviously is the articular area for the hyomandibula. Accordingly the hyomandibula articulated with the neural endocranium ventral to the jugular vein, that is, as in sharks. The main part of the 'concavity' lacks periosteal lining, and if we imagine that this unfinished area was continued ventrally by cartilage the cartilage would form a jugular bridge outside the jugular vein corresponding to that in Recent sharks (Figs 9C and 13A; lateral commissure, Holmgren, 1940, 1941; mandibular commissure, Jollie, 1971). In front of the presumed jugular bridge the jugular vein must turn rather sharply downwards in order to reach the orbit, as it obviously does in *Chlamydoselachus* (see gr.ju, Figs 3C and 9C). Since the canal (c.v.so, Figs 12B and 15A) which pierces the postorbital process is directed towards the area where the jugular vein probably passed, it is possible that this canal transmitted a tributary to the jugular

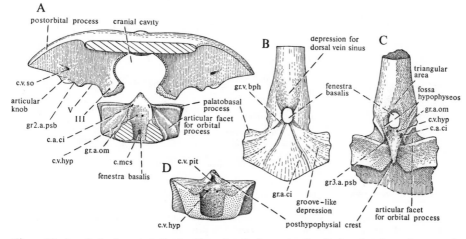

Figure 15. *Acanthodes bronni.* A. Restoration of posterior parts of main dorsal and anterior ventral ossifications of neural endocranium in anterior aspect. B–D. Anterior ventral ossification in ventral, dorsal and posterior aspects. Mainly after casts P.6210b, P.6215a, P.6235a.

vein corresponding to the supraorbital vein in *Chlamydoselachus* (Allis, 1923: 203; cf. Holmgren, 1942: 170–2). In the orbit, the jugular vein received the pituitary (interorbital) vein which, passing the notch (c.v.pit, Figs 3A and 15D) on the rear side of the posthypophysial crest ('basisphenoid pillar', Miles, 1973a), traversed the cranial base behind the fossa hypophyseos, as in sharks (Figs 3B, 9C and 13A, C; canal C of Gegenbaur, 1872). The notch for the pituitary vein has the character of a groove, and from that groove a paired depression leads to the aperture of a paired canal (c.v.hyp, Fig. 15D; casts P.6220b, P.6235a) which runs forwards and appears to open in the fossa hypophyseos, close anteromedially to the canals for the internal carotids (Fig. 15A, C). This paired canal probably transmitted a vein (Fig. 16B) corresponding to the median hypophysial vein in sharks (Fig. 16A) which may sometimes be paired (Meurling, 1967: 59).

Fossa hypophyseos

The fossa hypophyseos in *Acanthodes* (Fig. 16B) was probably bounded posteriorly by the posthypophysial crest (cf. *Heptanchus*; Gegenbaur, 1872: pl. 4, fig. 1); and anteriorly it probably extended across the dorsal opening of the basal fenestra (cf. El-Toubi, 1949: fig. 5) to the flat triangular area in front of that fenestra. This flat area formed the floor of the anterior part of the fossa, which was thus fairly large and situated as in sharks (Fig. 16A; Gegenbaur, 1872; Wells, 1917: pl. 1, fig. 4; Holmgren & Stensiö, 1936: fig. 238). Laterally it was bounded largely by cartilaginous or membranous walls which, in front of the fossa, united into an interorbital septum, as in *Scymnorhinus* and other squaloid sharks with large eyes (Holmgren, 1942: 141). The hypophysis contained in the fossa was probably drained by a paired hypophysial vein which emptied into the transverse

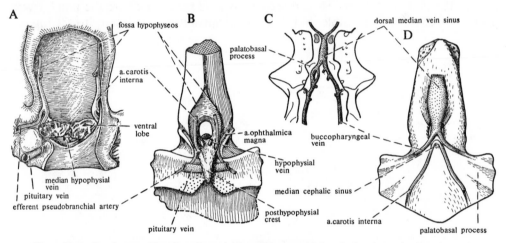

Figure 16. A. *Squalus acanthias*, fossa hypophyseos with ventral lobe of adenohypophysis and vessels in dorsal view. From Meurling, 1967: fig. 9. B. *Acanthodes bronni*, restoration of fossa hypophyseos with vessels in dorsal view. C. *Squalus sucklii*, neural endocranium in ventral view with dorsal median vein sinus and buccopharyngeal veins. From Daniel & Bennett, 1931: fig. 2. D. *Acanthodes bronni*, restoration of anterior ventral ossification in ventral view with dorsal median vein sinus, buccopharyngeal veins and arteries.

pituitary vein. After receiving the efferent pseudobranchial artery, the paired internal carotid entered the posterolateral part of the fossa conceivably to continue forwards and upwards close inside its lateral wall. In all these respects the fossa hypophyseos agrees with that in sharks, and probably the adenohypophysis was shark-like (see Meurling, 1967), possibly with a ventral lobe in the posterodorsal part of the basal fenestra.

Cranial nerves

As evidenced by a groove in the posterior part of the main dorsal ossification (cast P.6210b) the n.vagus emerged from the cranium in a posterolateral direction. The external opening of the vagus canal (X, Fig. 3A) is bounded by a notch in the anterior margin of the occipital ossification. No canal for the n.glossopharyngeus has been observed .The truncus hyomandibularis probably ran backwards towards the articular area for the hyomandibula in a groove (gr.tr.hy, Figs 2A and 3A) situated ventral to the anterior part of the jugular groove. The groove for the truncus hyomandibularis deepens forwards and is continued by the facialis canal (VII) which is bounded anteriorly by a bridge of bone (cast P.6210b). This bridge, together with the cartilage which seems to have continued it forwards to the trigeminus foramen (V), may represent the prefacial commissure (cf. Figs 3C, 9C and 13; de Beer, 1937: 390). At the anteroventral corner of that bridge lies the aperture of the palatine canal (c.pal.) A little further forwards, in front of the unossified area, the ventral side of the postorbital process presents a notch (V, Figs 2A, 3A, 12B and 15A) which, upon comparison with *Squalus* (Fig. 13), may be interpreted as a part of the trigeminus foramen. After emergence from that foramen a portion of the n.trigeminus, including at least the r.maxillaris probably passed out into the orbit via a shallow notch (iV, Fig. 6) in the palatoquadrate (cf. Fig. 9B; Luther, 1909a). However, a considerable portion of the nerve, probably the main part of the r.mandibularis, seems to have coursed upwards in a groove (gr.mdV) on the inner side of the palatoquadrate to the foramen (f.mdV) in the optic process and via that foramen to the external side of the palatoquadrate.

A little in front of the trigeminus notch is a small foramen (III, Figs 12B and 15A) which, as suggested by Miles, probably transmitted the n.oculomotorius. Ventromedial to that foramen, backwards to the area medial to the trigeminus notch, periosteal lining is lacking. It is possible that the eye stalk (if present) arose from the posterior part of this unfinished area (cf. Figs 3C and 9A, C).

Dorsal vein sinus and epibranchial muscles

It now remains to consider the two median depressions on the ventral side of the neural endocranium in *Acanthodes* (Fig. 2A), one on the anterior ventral ossification, in front of the basal fenestra, and one on the posterior part of the posterior ventral ossification.

In the roof of the mouth cavity in *Squalus sucklii*, close underneath the neural endocranium, Daniel & Bennett (1931) discovered a large median vein sinus (Fig. 16C). The sinus extends from the area above the palatoquadrate commissure (cf. Figs 2B and 4A) to the transverse level of the anterior part of the palatobasal process, where it divides into buccopharyngeal veins. The anterior median depression in *Acanthodes* (Fig. 15B) has a varying extent (Miles, 1973a: fig. 8) but at any rate in some specimens it reaches the area dorsal to the palatoquadrate commissure. Posteriorly it extends to the basal fenestra, from the margin of which a paired, rather wide groove (gr.v.bph) emerges in a posterolateral direction. These conditions strongly indicate that the anterior ventral depression in *Acanthodes* was occupied by a dorsal, median vein sinus (Fig. 16D) and that the paired groove was developed for a paired buccopharyngeal vein homologous with that in *S. sucklii*.

A unique common feature of selachians and holocephalians is the presence of epibranchial spinal muscles (Fürbringer, 1897: 407; Edgeworth, 1935: 189; Nishi, 1938: 368). One of these muscles is the subspinalis (Fig. 14B), which arises on the ventral side of the most posterior part of the neural endocranium and is inserted into the first and second infrapharyngobranchials. In *Heptanchus* the two muscles at their origin lie close together and form an unpaired median muscle. This median muscle portion lies close ventral to the dorsal lateral aortae which in this region are close together and enveloped in a common sheath (Allis, 1912: 487). In the corresponding place in *Acanthodes*, close ventral to the canal for the lateral dorsal aortae, lies the posterior median depression which extends backwards to the posterior end of the neural endocranium (Figs 2A and 14D, E). As is readily seen, this depression is most likely the area of origin of the subspinalis muscle. This view is supported by the fact that the distal end of infrapharyngobranchial 1 (cast P.6210b) presents a small but distinct area, so situated that, judging from conditions in *Heptanchus* (Davidson, 1918: fig. 3), it is most probably the area of insertion of the portion of the m.subspinalis running to the first branchial arch (Figs 8C and 14E). Larger and also well delimited areas without periosteal lining, one in front of and another behind the 'dorsal process' of the same element, indicate that dorsal interarcuales muscles (interbasales, Fürbringer; interpharyngobranchiales, Edgeworth), belonging to the category of epibranchial muscles, were also present in *Acanthodes* (Fig. 8C) as in sharks (Fig. 14B; Fürbringer, 1897; Davidson, 1918; Allis, 1923).

SUMMARY AND CONCLUSIONS

As we have seen, *Acanthodes* agrees in practically all of the numerous characters discussed above with elasmobranchiomorphs, in particular with sharks. No characters indicating relationship with any teleostome have been found, and acanthodians are hence to be classified as elasmobranchiomorphs. However, they obviously differ considerably from placoderms and holocephalians; and also from dipnoans which in several respects are elasmobranchiomorph-like. Because of the many resemblances to sharks they are to be referred to the Elasmobranchii which provisionally may be subdivided as follows (cf. Holmgren, 1940; Compagno, 1973).

1. Acanthodii
2. Selachii (sharks)
      Palaeoselachii (Palaeozoic sharks)
      Euselachii
          Squalimorphii
          Galeomorphii
3. Batoidei (rays)

The acanthodians have been shown to agree with Recent squalimorphs in the first place in the following respects (some characters are shared with other selachians, with batoids and with elasmobranchiomorphs in general): (1) the absence of an outer dental arcade; (2) the fact that the mouth was subterminal; (3) the similar development of the palatoquadrate, including: the presence of a palatoquadrate commissure, the presence of an orbital process articulating with the neural endocranium in the posterior part of the orbit, close in front of the palatobasal process, the presence of an otic process articulating, as in notidanids (and several palaeoselachians), with the postorbital process, and the presence of a double jaw joint; (4) the facts that the hyomandibula is a slender rod similar to that of notidanids and *Chlamydoselachus*, and that it is composed of three parts; (5) the fact that the efferent pseudobranchial artery entered the neural endocranium to join the internal carotid dorsal to the trabecula; (6) the fact that the hyomandibula articulated with the neural endocranium ventral to the jugular vein; (7) the presence of several gill covers behind the hyoidean gill cover; (8) the fact that the gill arches are posterior in position; (9) the facts that the infrapharyngobranchials are backwardly directed, and do not articulate with the neural endocranium; (10) the fact that the gill rakers were supported by endoskeletal cores; (11) the fact that the hyoidean branchial rays are well developed; (12) the facts that the fins are almost completely covered by fin scales, and that the dermal fin rays are ceratotrichia-like; (13) the fact that the caudal fin is shark-like; (14) the presence of three proximal elements in the endoskeleton of the pectoral fin.

In addition to these important and indisputable resemblances to squalimorph sharks there are many other similarities which, however, because of the imperfections of the fossil material and because our knowledge of the anatomy of Recent sharks is still imperfect, are somewhat less well documented. Among these resemblances may be mentioned: (15) the similarities in the lateral line system; (16) the presence of an endolymphatic fossa in the cranial roof; (17) the presence of subspinalis and interarcuales muscles; (18) the presence of a dorsal median vein sinus and buccopharyngeal veins; (19) the facts that the fossa hypophyseos was developed as in sharks, and that the vessels associated with the hypophysis ran more or less as in sharks; (20) the fact that the observed openings for nerves and vessels on the lateral side of the neural endocranium are situated very much as, for example, in *Squalus*.

These numerous common features and the impressive similarities in the histological structure of teeth, scales, and spines, can only be taken to mean that acanthodians and sharks are closely akin. However, according to the methods for establishing relationship which I have found it necessary to use (Jarvik, 1968: 499–501, 1972: 178–81), embryological data must also be in agreement with the suggested relationship. Sharks are certainly not descendants of acanthodians, but there are certain features in their ontogeny which support the view that they share a common ancestry with acanthodians. This applies to the fact that the hyomandibula arises from three components, and to the closure of the dorsal fontanelle and the basal fenestra in the ontogeny of sharks, but is perhaps most distinct in the occipital region. In *Acanthodes* this region, represented by the occipital ossification and the posterior part of the posterior ventral ossification, is fairly long. The occipital ossification is suggestive of, and certainly homologous with, the occipital column in embryonic sharks (Holmgren, 1940) and, like the latter, it is separated from the otic region by a gap and is pierced by the roots of spino-occipital nerves. In later ontogenetic stages—in connection with its incorporation into the neural endocranium—the occipital column becomes much shortened but is still rather long in notidanids and *Chlamydoselachus* (Fig. 3B, C) which are considered to be the most primitive extant sharks. As indicated by Holmgren (1942: 136), these conditions imply a recapitulation of the phyletic development.

Another case of recapitulation concerns the ventrolateral fin fold. The well-known presence of intermediate spines in the position of the ventrolateral fin fold in acanthodians and the great length of the pelvic fins in certain forms (Jarvik, 1965: 149; Miles, 1970: 350) show that the ventrolateral fin fold was well developed in these fishes. The fact that a ventrolateral fin fold with a complete series of muscle buds in the area between the pectoral and pelvic fins is also present in the ontogenetic stages of sharks (Jarvik, 1965: 146) must mean that sharks originated from forms which also had a well-developed ventrolateral fin fold.

All the facts presented in this article show unequivocally that acanthodians are closely related to sharks and, although the endoskeleton is partly ossified, they are to be included in the Elasmobranchii. As to the classification of the elasmobranchs, opinions differ widely and there are almost as many suggestions as there are writers on the subject (see Holmgren, 1941; Compagno, 1973). One major problem concerns the position of the rays (Batoidei) which, by most writers, are thought to be descendants of sharks. Another view as to the origin of rays was set forth by Holmgren who (1940) studied the ontogenetic development of the skull in both sharks and rays and, in addition (1941) made a thorough comparative anatomical analysis of the skull of the adult in numerous forms. He found profound differences between the two groups and concluded (1940: 259) that 'the disagreement is really so great that one is bound to assume a diphyletic origin of sharks and rays'. This view, disputed by Compagno (1973) and others, has unexpectedly come to the fore by the reinterpretation of the acanthodians given above. Acanthodians and sharks share several important characters which are lacking in rays, such as the orbital process and orbital articulation, the palatobasal process, epibranchial muscles, the dorsal vein sinus, aplesodic fins etc., and it may be asked if sharks are not more closely related to acanthodians than to rays. In view of the fact that the palatobasal process and the orbital articulation in acanthodians are located in the posterior part of the orbit, it is even tempting to suggest that acanthodians stand closer to the notidanid and squaloid sharks in which the conditions are exactly the same, than to palaeoselachians, galeomorphs, and *Chlamydoselachus* in which the palatobasal process and the orbital articulation are found in the anterior part of the orbit. I visualize that these tentative suggestions may appear too radical to many readers, but one thing is clear: the remarkable agreement found to exist between acanthodians and sharks, in particular notidanids and squaloids, must not be disregarded in future discussions of the intricate and still unsolved question of elasmobranch interrelationships.

REFERENCES

ALLIS, E. P. Jr., 1912. The branchial, pseudobranchial and carotid arteries in *Heptanchus* (*Notidanus*) *cinereus*. *Anat.Anz.*, *41:* 478–92.

ALLIS, E. P. Jr., 1923. The cranial anatomy of *Chlamydoselachus anguineus*. *Acta zool.*, *Stockh.*, *4:* 123–221.

de BEER, G. R., 1924. Studies on the vertebrate head. Part 1. Fish. *Q.Jl. microsc. Sci.*, *68:* 287–341.

de BEER, G. R. 1937. *The development of the vertebrate skull:* xxiv+552 pp. Oxford University Press.

COLES, E. M., 1928. The segmental arteries in *Squalus sucklii*. *Univ. Calif. Publs Zool.*, *31:* 93–110.

COMPAGNO, L. J. V., 1973. Interrelationships of living elasmobranchs. In P. H. Greenwood *et al.* (Eds), *Interrelationships of fishes:* 15–61. London: Academic Press.

CORRINGTON, J. D., 1930. Morphology of the anterior arteries of sharks. *Acta zool.*, *Stockh.*, *11:* 185–261.

DANIEL, J. F., 1934. *The elasmobranch fishes:* xi+332 pp. University of California Press.

DANIEL, J. F. & BENNETT, L. H., 1931. Veins in the roof of the buccopharyngeal cavity of *Squalus sucklii*. *Univ. Calif. Publs Zool.*, *37:* 35–40.

DAVIDSON, P., 1918. The musculature of *Heptanchus maculatus*. *Univ. Calif. Publs Zool.*, *18:* 151–70.

EDGEWORTH, F. H., 1935. *The cranial muscles of vertebrates:* 300 pp. Cambridge: Cambridge University Press.

EL-TOUBI, M. R., 1949. The development of the chondrocranium of the spiny dogfish, *Acanthias vulgaris* (*Squalus acanthias*). *J. Morph.*, *84:* 227–79.

FÜRBRINGER, M., 1897. Ueber die spino-occipitalen Nerven der Selachier und Holocephalen und ihre vergleichende Morphologie. In *Festschrift zum siebenzigsten Geburtstage von Carl Gegenbaur, 3:* 349–788. Leipzig: Wilhelm Engelmann.

GARDINER, B. G., 1973. Interrelationships of teleostomes. In P. H. Greenwood *et al.* (Eds), *Interrelationships of fishes:* 105–35. London: Academic Press.

GARMAN, S., 1913. The Plagiostomia. *Mem. Mus. comp. Zool. Harv., 36:* xiv+515 pp.

GAUPP, E., 1913. Die Reichertsche Theorie. *Arch. Anat. Physiol., 1912,* Suppl.: 1–416.

GEGENBAUR, C., 1872. *Untersuchungen zur vergleichenden Anatomie der Wirbelthiere:* x+316 pp. Leipzig: Wilhelm Engelmann.

GOODEY, T., 1910. A contribution to the skeletal anatomy of the frilled shark, *Chlamydoselachus anguineus* Gar. *Proc. zool. Soc. Lond., 1910:* 540–71.

GOODRICH, E. S., 1909. Cyclostomes and fishes. In E. R. Lankester (Ed.), *A treatise on zoology, 9:* 518 pp. Vertebrata Craniata. London: A. & C. Black.

GROSS, W., 1937. Das Kopfskelett von *Cladodus wildungensis, 1.* Endocranium und Palatoquadratum. *Senckenbergiana, 19:* 80–107.

GROSS, W., 1938. Das Kopfskelett von *Cladodus wildungensis, 2.* Der Kieferbogen. *Senckenbergiana, 20:* 123–45.

GROSS, W., 1947. Die Agnathen und Acanthodier des obersilurischen Beyrichienkalkes. *Palaeontographica, 46A:* 91–161.

GROSS, W., 1957. Mundzähne und Hautzähne der Acanthodier und Arthrodiren. *Palaeontographica, 109A:* 1–40.

GROSS, W., 1973. Kleinschuppen, Flossenstacheln und Zähne von Fischen aus europäischen und nordamerikanischen Bonebeds des Devons. *Palaeontographica, 142A:* 51–155.

GUDGER, E. W., 1940. The breeding habits, reproductive organs and external embryonic development of *Chlamydoselachus*, based on notes and drawings by Bashford Dean. *The Bashford Dean Memorial Volume: Archaic fishes,* Art. 7: 523–633. New York: Am. Mus. Nat. Hist.

GUDGER, E. W. & SMITH, B. G., 1933. The natural history of the frilled shark *Chlamydoselachus anguineus. The Bashford Dean Memorial Volume: Archaic fishes,* Art. 5: 245–319. New York: Am. Mus. Nat. Hist.

HALSTEAD, L. B., 1969. *The pattern of vertebrate evolution:* xii+209 pp. Edinburgh: Oliver & Boyd.

HEYLER, D., 1969a. *Vertébrés de l'Autunien de France. Cah. Paléont.:* 1–255. Cent. natn. Rech. scient., Paris.

HEYLER, D., 1969b. Acanthodii. In J. Piveteau (Ed.), *Traité de Paléontologie,* Tome 4,2: 21–70. Paris: Masson et Cie.

HOLMGREN, N., 1940. Studies on the head in fishes. Embryological, morphological and phylogenetical researches. Part I. *Acta zool., Stockh. 21:* 51–267.

HOLMGREN, N., 1941. Studies on the head in fishes. Part II. *Acta zool., Stockh., 22:* 1–100.

HOLMGREN, N., 1942. Studies on the head of fishes. Part III. *Acta zool., Stockh., 23:* 129–261.

HOLMGREN, N., 1943. Studies on the head of fishes. Part IV. *Acta zool., Stockh., 24:* 1–188.

HOLMGREN, N. & PEHRSON, T., 1949. The sensory lines in fishes and amphibians. *Acta zool., Stockh., 30:* 249–314.

HOLMGREN, N. & STENSIÖ, E. A., 1936. Kranium und Visceralskelett der Akranier, Cyclostomen und Fische. In L. Bolk *et al.* (Eds), *Handbuch der vergleichenden Anatomie der Wirbeltiere, 4:* 233–500. Berlin & Wien: Urban & Schwarzenberg.

HOTTON, N., 3rd., 1952. Jaws and teeth of American xenacanth sharks. *J. Paleont., 26:* 489–500.

JARVIK, E., 1954. On the visceral skeleton in *Eusthenopteron* with a discussion of the parasphenoid and palatoquadrate in fishes. *K. svenska VetenskAkad. Handl., 4(5):* 1–104.

JARVIK, E., 1955. The oldest tetrapods and their forerunners. *Sci. Monthly, 80:* 141–54.

JARVIK, E., 1959. Dermal fin-rays and Holmgren's principle of delamination. *K. svenska VetenskAkad. Handl., 4(6):* 1–51.

JARVIK, E., 1963. The composition of the intermandibular division of the head in fish and tetrapods and the diphyletic origin of the tetrapod tongue. *K. svenska VetenskAkad. Handl., 4(9):* 1–74.

JARVIK, E., 1965. On the origin of girdles and paired fins. *Israel J. Zool., 14:* 141–72.

JARVIK, E., 1968. Aspects of vertebrate phylogeny. *Nobel Symposium, 4:* 497–527.

JARVIK, E., 1972. Middle and Upper Devonian porolepiforms from East Greenland with special reference to *Glyptolepis groenlandica* n.sp. And a discussion on the structure of the head in the Porolepiformes. *Meddr Grønland, 187:* 1–307.

JOLLIE, M., 1971. Some developmental aspects of the head skeleton of the 35–37 mm *Squalus acanthias* foetus. *J. Morph., 133:* 17–40.

LUTHER, A., 1909a. Untersuchungen über die vom N. trigeminus innervierte Muskulatur der Selachier (Haie und Rochen). *Acta Soc. Sci. fenn., 36:* 1–168.

LUTHER, A., 1909b. Beiträge zur Kenntnis von Muskulatur und Skelett des Kopfes des Haies *Stegostoma tigrinum* Gm. und der Holocephalen. *Acta Soc. Sci. fenn., 37:* 1–60.

MARINELLI, W. & STRENGER, A., 1959. *Vergleichende Anatomie und Morphologie der Wirbeltiere, 3:* 173–308. Wien: Franz Deuticke.

MEURLING, P., 1967. The vascularization of the pituitary in elasmobranchs. *Sarsia, 28:* 1–104.

MILES, R. S., 1964. A reinterpretation of the visceral skeleton of *Acanthodes. Nature, Lond., 204:* 457–9.

MILES, R. S., 1965. Some features of the cranial morphology of acanthodians and the relationships of the Acanthodii. *Acta zool., Stockh., 46:* 233–55.

MILES, R. S., 1966. The acanthodian fishes of the Devonian Plattenkalk of the Paffrath Trough in the Rhineland, with an appendix containing a classification of the Acanthodii and a revision of the genus *Homalacanthus. Ark. Zool., 2*(18): 147–94.

MILES, R. S., 1968. Jaw articulation and suspension in *Acanthodes* and their significance. *Nobel Symposium, 4:* 109–27.

MILES, R. S., 1970. Remarks on the vertebral column and caudal fin of acanthodian fishes. *Lethaia (Oslo), 3:* 343–62.

MILES, R. S., 1973a. Relationships of Acanthodians. In P. H. Greenwood *et al.* (Eds), *Interrelationships of fishes:* 63–103, London: Academic Press.

MILES, R. S., 1973b. Articulated acanthodian fishes from the Old Red Sandstone of England, with a review of the structure and evolution of the acanthodian shoulder-girdle. *Bull. Br. Mus. nat. Hist. (Geol.), 24:* 111–213.

MILES, R. S. & MOY-THOMAS, J. A., 1971. *Palaeozoic fishes,* 2nd ed.: xi+259 pp. London: Chapman & Hall.

NELSON, G. J., 1968. Gill-arch structure in *Acanthodes. Nobel Symposium, 4:* 128–43.

NELSON, G. J., 1969. Origin and diversification of teleostean fishes. *Ann. N.Y. Acad. Sci., 167:* 18–30.

NIELSEN, E., 1949. Studies on Triassic fishes from East Greenland. II. *Australosomus* and *Birgeria. Meddr Grønland, 146:* 1–309.

NISHI, S., 1938. Muskeln des Rumpfes. In L. Bolk *et al.* (Eds), *Handbuch der vergleichenden Anatomie der Wirbeltiere,* 5: 351–446. Berlin & Wien: Urban & Schwarzenberg.

O'DONOGHUE, C. H. & ABBOTT, E., 1928. The blood vascular system of the spiny dogfish, *Squalus acanthias* Linné, and *Squalus sucklii* Gill. *Trans. R. Soc. Edinb., 55:* 823–90.

ØRVIG, T., 1972. The latero-sensory component of the dermal skeleton in lower vertebrates and its phyletic significance. *Zool. Scripta, 1:* 139–55.

ØRVIG, T., 1973. Acanthodian dentition and its bearing on the relationship of the group. *Palaeontographica, 143A:* 119–50.

POPLIN, C., 1974. *Étude de quelques Paléoniscidés Pennsylvaniens du Kansas. Cah. Paléont.:* 1–151. Cent. natn. Rech. scient., Paris.

REIS, O. M., 1895. Illustrationen zur Kenntnis des Skeletts von *Acanthodes bronni* Agassiz. *Abh. senckenb. naturforsch. Ges., 19:* 49–64.

RETZIUS, G., 1881. *Das Gehörorgan der Wirbelthiere. 1. Das Gehörorgan der Fische und Amphibien:* xi+222 pp. Stockholm: Sampson & Wallin.

ROMER, A. S., 1968. *Notes and comments on vertebrate paleontology:* 304 pp. Chicago & London: University of Chicago Press.

RUGE, G., 1897. Ueber das peripherische Gebiet des Nervus Facialis bei Wirbelthieren. In *Festschrift zum siebenzigsten Geburtstage von Carl Gegenbaur,* 3: 193–348. Leipzig: Wilhelm Engelmann.

SCHAEFFER, B., 1967. Comments on elasmobranch evolution. In P. W. Gilbert *et al.* (Eds), *Sharks, skates and rays:* 3–35. Baltimore: Johns Hopkins Press.

SCHAEFFER, B., 1968. The origin and basic radiation of the Osteichthyes. *Nobel Symposium, 4:* 207–22.

SCHAEFFER, B., 1969. Adaptive radiation of the fishes and the fish-amphibian transition. *Ann. N.Y. Acad. Sci., 167:* 5–17.

SMITH, B. G., 1937. The anatomy of the frilled shark *Chlamydoselachus anguineus. The Bashford Dean Memorial Volume: Archaic fishes,* Art. 6: 331–505. New York: Am. Mus. Nat. Hist.

STADTMÜLLER, F., 1924. Über Entwicklung und Bau der papillenförmigen Erhebungen (Filterfortsätze) auf den Branchialbogen der Salamandridenlarven. *Z. Morph. Anthrop., 24:* 125–56.

STENSIÖ, E. A., 1963. Anatomical studies on the arthrodiran head. *I.K. svenska VetenskAkad Handl., 4*(9): 1–419.

VIRCHOW, H., 1890. Über Spritzlochkieme der Selachier. *Arch. Anat. Physiol., 1890:* 177–82.

WATSON, D. M. S., 1937. The acanthodian fishes. *Phil. Trans. R. Soc., (B)* 228: 39–146.

WELLS, G. A., 1917. The skull of *Acanthias vulgaris. J. Morph., 28:* 417–43.

WOODWARD, A. S., 1915. The use of fossil fishes in stratigraphical geology. *Q. Jl geol. Soc. Lond., 71:* LXII–LXXV.

ABBREVIATIONS USED IN FIGURES

ar.lab.cart	area of attachment and groove for labial cartilage	gr.a.ci	groove for a.carotis interna
art. hym.	articular area for hyomandibula	gr.a.occ	groove for occipital artery
ar1.li	area of hyomandibula for attachment of ligament Adγ	gr.a.om	groove for a.ophthalmica magna
		gr.ioc	groove of postorbital process for infra-orbital sensory canal
ar2.li	area of postorbital process for attachment of ligament Adγ	gr.ju	groove for jugular vein
c.a.ci	canal for a.carotis interna	gr.mdV	groove for r.mandibularis trigemini
c.a.occ	canal for occipital artery	gr.tr.hy	groove for truncus hyomandibularis
c.a.psb	canal for efferent pseudobranchial artery	gr.v.bph	groove for buccopharyngeal vein
c.d.end	canal for ductus endolymphaticus	gr1.psb	groove of palatoquadrate for efferent pseudobranchial artery
c.mcs	canal for median cephalic sinus	gr2.psb	groove of postorbital process for efferent pseudobranchial artery
c.pal	canal for r.palatinus VII		
c.spoc	canals for spino-occipital nerves	gr3.psb	groove of palatobasal process for efferent pseudobranchial artery
c.v.hyp	canal for hypophysial vein		
c.v.so	canal for supraorbital vein	iV	notch for n.trigeminus
c.v.pit	canal or notch for pituitary (interorbital) vein	II	canal for nervus opticus
		III-X	canals or other passages for cranial nerves
f.mdV	foramen for r.mandibularis trigemini		

The homologies of ventral cranial fissures in osteichthyans

BRIAN G. GARDINER AND ALAN W. H. BARTRAM*

Queen Elizabeth College, London

Descriptions are given of the ventral cranial fissure, parasphenoid, hypophysial recess, blood supply to the head and the origin of the posterior eye muscles in two new Devonian palaeonisciforms from Western Australia, *Mimia toombsi* gen. et sp. nov. and *Moythomasia durgaringa* sp. nov. The homologies of the ventral cranial fissure are discussed and it is concluded that this fissure represents the gap between ossifications in the trabecular-polar bar and the basal plate in both palaeonisciforms and crossopterygians.

CONTENTS

INTRODUCTION

The presence of a ventral fissure, cartilage-filled in life, is a remarkable feature of the braincase of the most primitive actinopterygians known, the Palaeonisciformes. Patterson (1975) has explained how this fissure is homologous with the suture between the prootic and basioccipital bones in more advanced actinopterygians.

In some palaeonisciforms, for example *Pteronisculus* (Nielsen, 1942), *Kentuckia* (Rayner, 1951) and '*Ambodipia*' (Beltan, 1968), the ventral cranial fissure is

---

*Present address: Institut de Paléontologie, 8, Rue de Buffon, Paris.

confluent with the vestibular fontanelles and thus the endocranium contains two median ossifications in the adult. In others, such as *Boreosomus* (Nielsen, 1942) and *Kansasiella* (Poplin, 1974), the fissure is further forward, and the adult endocranium forms a single ossification. The two Devonian forms partially described in this paper belong to the latter category. However they differ from all other palaeonisciforms in that the ventral fissure passes into the postero-ventral corners of the orbits, and in having no myodome. In the discussion, attention is drawn to the primitiveness of this state of affairs.

Also in the discussion, consideration is given to the homologies that may exist between the ventral cranial fissure of actinopterygians and the fissures found in the skulls of the closest relatives of this group, the crossopterygians. As is well-known, crossopterygians are characterized by an intracranial joint which passes right through the braincase. In addition, in the rhipidistian *Eusthenopteron* and in coelacanthiforms, further ventral fissures are present behind this joint.

DESCRIPTION
Order PALAEONISCIFORMES
Family STEGOTRACHELIDAE Gardiner, 1963

Genus *Mimia* nov.

Diagnosis (provisional). Stegotrachelid palaeonisciform fishes in which: the ventral cranial fissure passes into the rear of the orbit; a pair of orbitonasal arteries passed into the orbit immediately lateral to the ventral cranial fissure; the transverse pituitary vein passed through a distinct foramen; the parasphenoid has neither basipterygoid nor ascending processes.

Type species. Mimia toombsi sp. nov.

Remarks. The generic name is derived from the aboriginal term for the fairy-like people called the Mimis whose homes are under the great rocks of the Arnhem Land plateau. Mimis are believed to be able to enter clefts in the rocks and then close these fissures behind themselve s.

Mimia toombsi sp. nov.
(Figs 1, 2, 3, 4, 5, 6 and 8A)

1970: Devonian stegotrachelid; Gardiner: 285, fig. 3.
1971: Stegotrachelid palaeoniscoid; Gardiner, in Moy-Thomas & Miles: fig. 5. 6.
1973: Gogo palaeoniscid 'A'; Gardiner: 106, figs 1, 2, 6, 8 and 9.
1973: Gogo palaeoniscid 'B'; Gardiner: figs 3 and 4.
1976: Gogo palaeoniscid; Gardiner & Miles: fig. 2.

Diagnosis. A *Mimia* with a spiracular slit between the dermopterotic and dermosphenotic and with scales ornamented with short, smooth ridges of ganoine which terminate posteriorly in sharp points.

Holotype. Western Australian Museum 70.4.245; partly disarticulated specimen wanting fins, in counterpart, from the Upper Devonian, Gogo Shales, Gogo Station (H.A.T. 67/80, see Miles, 1971), Fitzroy Crossing, W. Australia.

Remarks. The specimen is named in honour of Mr H. A. Toombs, late of the British Museum, who collected the original material on which the species is based.

Ventral fissure and associated structures

The ventral fissure (fissura oticalis ventralis of Nielsen, 1942, 1949), cartilage-filled in life, lies in the floor of the neurocranium and passes up immediately behind the foramen for the pituitary vein into the postero-ventral corner of the orbit (fv, Figs 1 to 4) anterior to the foramen for the abducens nerve (6, Figs 3 and 6). The fissure separates the basioccipital from the basisphenoid (ossifications deduced from Patterson, 1975) medially and the prootics dorso-laterally. From a broad base the fissure tapers dorsally almost to a point to open into the front of the notochordal canal just behind the 'prootic' bridge (prb, Fig. 4) and just in front of the partition between the ear and brain cavity* (z, Fig. 4). From the position

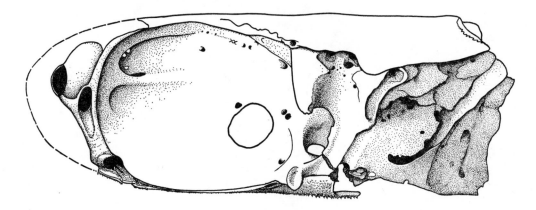

A   2mm

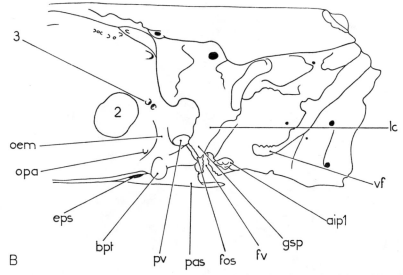

B

Figure 1. *Mimia toombsi* gen. et sp. nov. Lateral view of braincase and associated dermal bones. The individual skull roofing bones and snout bones are not shown. From BMNH P56498.

*Erroneously called zygal by Gardiner, 1973.

of the ventral fissure it is clear that the 'prootic' bridge is ossified by the basisphenoid, not the prootic as in higher actinopterygians. Laterally, at the level of
the presumed junction between the basioccipital and prootics, the ventral fissure
gives way on either side to a large foramen for the orbitonasal artery (ona, Figs
2, 3, 5 and 6).

That this foramen transmitted the orbitonasal artery (and not the internal
carotids as erroneously suggested by Gardiner, 1973, ic, Fig. 3) is assumed from

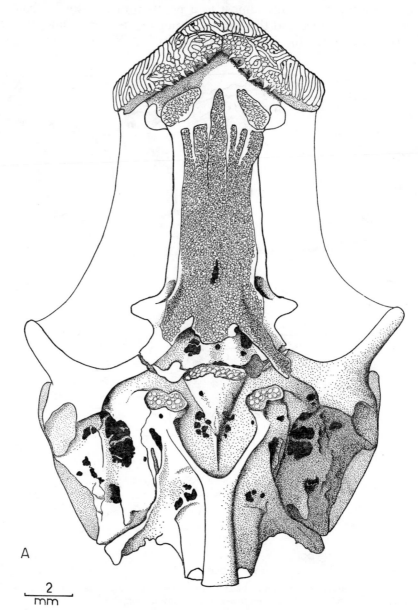

Figure 2. *Mimia toombsi* gen. et sp. nov. Ventral view of braincase and associated dermal bones.
Foramina in the orbits have been omitted. From BMNH P53247.

de Beer's (1926, 1927, 1937) descriptions of the course of the former in the development of *Lepisosteus* and *Salmo*. In both of these fishes the orbitonasal artery passes up through the cartilaginous neurocranial floor via the palatine foramen.

We have observed a similar relationship between the orbitonasal artery and the palatine nerve in sections of young *Elops* where again the artery passes up through a cartilaginous floor into the myodome, but in adult *Elops* and *Salmo* the orbitonasal artery is given off from the internal carotid as the latter passes across the floor of the myodome, much as in *Hepsetus* (Bertmar, 1962). Confirmation in

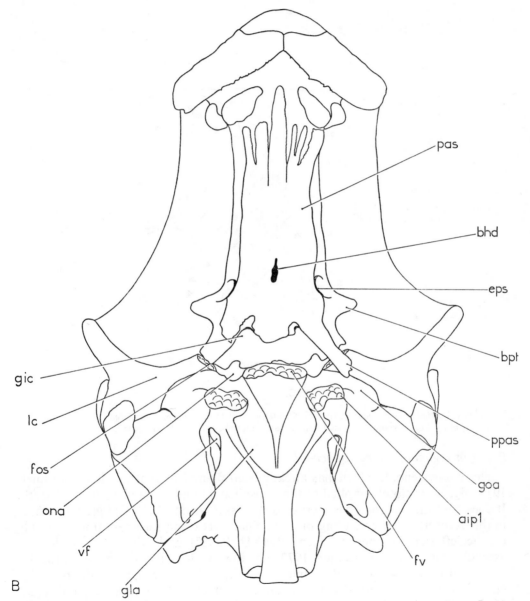

B

Figure 2B.

Mimia is provided by the presumed path of the palatine nerve, which enters the basisphenoid just anterior and immediately lateral to the orbitonasal foramen (7 pal, Fig. 6).

Lateral again to the orbitonasal foramen is another cartilage-filled fissure separating the prootics from the basisphenoid, the otico-sphenoid fissure (fos, Figs 1, 2 and 3). This fissure defines the ventral limit of the stout, long lateral commissures (1c, Figs 1 and 3), although in some specimens (Fig. 3) the fissure has been obliterated by bone growth on one side.

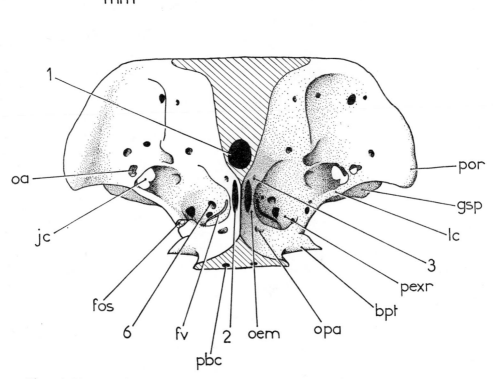

Figure 3. *Mimia toombsi* gen. et sp. nov. Orbital view of braincase which has been cut through in front of the buccohypophysial canal (cut surfaces hatched). From BMNH P56483.

Posterior region of the parasphenoid

The parasphenoid, which has already been figured in ventral view (Gardiner, 1973: fig. 3), is closely applied to the basisphenoid and covered with small teeth. It does not form a dermal basipterygoid process (Gardiner's, 1973: fig. 3 is incorrect in this respect) or an ascending process. The short posterior projection seen only on the left side in one specimen (BMNH P5327) and which was called an ascending process by Gardiner, 1973 (ap, Fig. 3) does not lie in the spiracular canal, but rather projects below and behind it, in the same plane as the posterior end of the parasphenoid. It is best regarded as a spur of the parasphenoid (ppas, Fig. 2). Posteriorly the parasphenoid is cruciate with two short postero-laterally directed arms which pass up towards the otico-sphenoid fissure. Medially a short posterior extension ends at the level of the ventral fissure or occasionally a short distance in front of it. Posteriorly this medial portion is not applied to the basisphenoid. The two grooves (gic, Fig. 2) which delimit the posterolateral arms

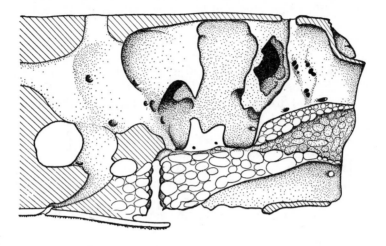

A   2mm

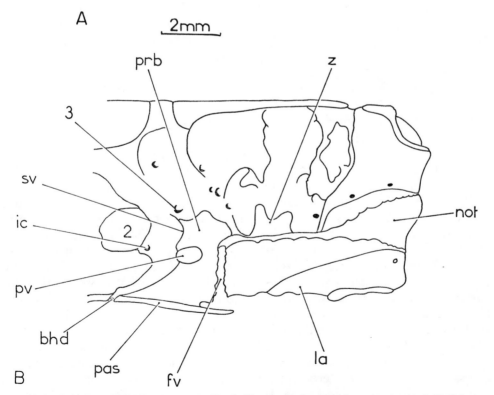

Figure 4. *Mimia toombsi* gen. et sp. nov. Vertical longitudinal section through posterior half of the braincase and parasphenoid (cut cartilage surfaces hatched). From BMNH P53234.

medially overlie the point of entrance of the carotid arteries into the back of the basisphenoid and possibly served as the point of insertion of the subcephalic muscles (the foremost trunk myomere, see Nelson, 1970). The dorsal surface of the parasphenoid is produced into a small cup around the end of the bucco-hypophysial canal. From the region of the foramen for the bucco-hypophysial canal two grooves pass back towards the otico-sphenoid fissure and must have housed the spiracular diverticulum.

Parabasal canal and blood supply to head

After leaving the aortic canal the paths of the lateral aortae can easily be recognized from the grooves on the basioccipital (gla, Fig. 2). The lateral aortae passed forwards medial to the articulatory facet for the first infra-pharyngo-branchial (aipl, Figs 1 and 2) and a notch and well-marked groove along the posterior margin of the lateral commissure (goa, Figs 1 and 2) served to transmit the orbital artery into the jugular canal. After traversing the jugular canal the orbital artery entered the orbit through its own foramen (oa, Fig. 3). From the junction with the orbital artery the carotids passed in between the back of the parasphenoid and basisphenoid to enter the parabasal canal. Just before the carotids passed below the basisphenoid they gave off a dorsal branch, the orbito-nasal artery (see above), which passed up between the ventral fissure medially and the otico-sphenoid fissure laterally through a well-marked foramen.

The parabasal canal (pbc, Figs 5 and 6) runs in the ventral surface of the basisphenoid and in places is floored by the parasphenoid. Just in front of the basipterygoid process a short canal runs laterally from the parabasal canal to open above the edge of the parasphenoid; this carried the efferent pseudobranchial artery (eps, Figs 1, 2, 5 and 6). At the junction of the efferent pseudobranchial artery with the carotid the latter divided into two. The inner branch is the cephalic artery (ic, Figs 4, 5 and 6) which passed up through a vertical canal in the anterior basisphenoid pillar to enter the cranial cavity through the pituitary foramen at the level of the pituitary vein. The other branch, the ophthalmic artery (opa, Figs 1, 3, 5 and 6), turned upwards and forwards into the floor of the orbit. The chief variations seen in the arrangement of these vessels occur after the anastomosis of the efferent pseudobranchial artery with the carotid. Thus the cephalic artery may briefly run in a groove in the lateral wall of the basisphenoid pillar and this same artery may occasionally give rise to an anterior branch (bic, Fig. 6) which runs antero-ventrally towards the snout in a short groove in the dorsal surface of the basisphenoid before passing down into the palatine canal to augment or maybe replace the palatine artery. The palatine canal in front of the division of the carotid into the cephalic and ophthalmic arteries is somewhat narrower than the parabasal canal but still must have served to transmit both the palatine artery and nerve towards the snout. The whole of this anterior extension of the palatine canal runs between the parasphenoid and the endochondral basisphenoid.

The only other structure associated with the parabasal canal is the palatine branch of the 7th nerve (see above). The palatine nerve passed antero-ventrally across the floor of the postero-ventral corner of orbit to enter the basisphenoid in front of the otico-sphenoid fissure. It then almost immediately entered the para-basal canal (7 pal, Figs 5 and 6) to run forward alongside the internal carotid.

Hypophysial recess

Although the complete extent of the basisphenoid is unclear its posterior and dorsal limits can be deduced with some confidence. The basisphenoid region consists of a hollow vertical pillar which flares dorsally to join the orbital surface at the level of the oculomotor foramen. At this point the basisphenoid forms the lateral and posterior margins of the pituitary fossa but does not separate it from the optic fenestra. Thus the large hypophysial recess is open both dorsally and anteriorly (into the optic foramen) and forms a large cone-shaped intramural

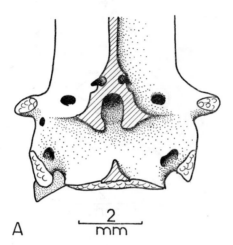

A

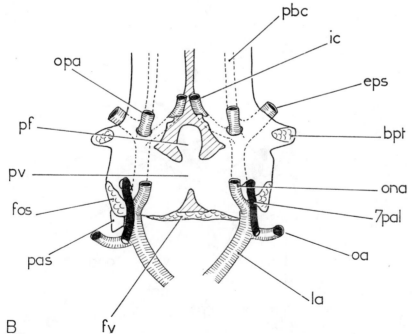

B

Figure 5. *Mimia toombsi* gen. et sp. nov. Posterior basisphenoid region of braincase cut horizontally at level of pituitary vein, in dorsal view (cut surfaces hatched). From BMNH P5324.

space (Fig. 4). Ventrally, in the foot of the pillar, this recess constricts into a
narrow bucco-hypophysial canal which curves antero-ventrally through the
stoutest point of the basisphenoid (the so-called basisphenoid bolster) and through
the parasphenoid to open in the roof of the mouth just in front of the level of the
efferent pseudobranchial foramina (bhd, Figs 1 and 4). Immediately behind the
hypophysial recess is another intramural space across which ran the hypophysial
or pituitary vein (pv, Figs 5 and 6). This second cavity is contiguous with the

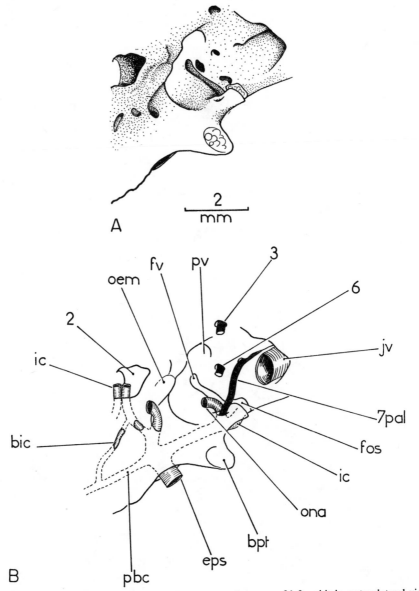

Figure 6. *Mimia toombsi* gen. et sp. nov. Posteroventral corner of left orbit in anterolateral view.
From BMNH P56501.

hypophysial recess and communicates with it anteriorly beneath the saccus vasculosus. The cavity is perhaps larger than is necessary to accommodate the pituitary vein and appears to be an intramural space which has become cut off from the more anterior intramural cavity housing the pituitary body. Viewed from above this cavity is widest anteriorly and tapers to a point posteriorly, behind the pituitary vein. The basisphenoid forms the walls around the pituitary vein, and just posterior to this vein expands into a short, stout posterior pillar which dorsally gives way to the 'prootic' bridge.

The cavity for the saccus vasculosus (sv, Fig. 4) is cup-shaped and lies antero-dorsal to the pituitary vein in the anterior surface of the 'prootic' bridge. Ventrally it is notched in the midline and from the notch a median canal passes upwards within the 'prootic' bridge to re-open dorsally within the saccus cavity. Because of the proximity of the ventral opening to the pituitary vein foramen it must have served for the passage of the v. sacci vasculosi draining the saccus vasculosus. The area available for occupation by the saccus vasculosus is much smaller than that in *Kentuckia* (Rayner, 1951) and considerably smaller than *Kansasiella* (Poplin, 1974).

Origin of posterior eye muscles

Returning now to the prominent anterior basisphenoid pillar, ventrally this pillar contributes to a stout transverse bolster which runs out towards the basipterygoid process, posterior to the foramen for the ophthalmic artery. On the lateral wall of the pillar, dorsal to the foramen for the ophthalmic artery, is a pronounced muscle scar (oem, Figs 3 and 6). This scar has the appearance of a cup-shaped depression divided into two or sometimes three components by prominent ridges. There can be little doubt that at least three of the recti muscles must have taken their origin from this point. In *Acipenser* (Marinelli & Strenger, 1973) the four recti muscles attach to the cartilaginous interorbital region and similarly in *Polypterus* (Allis, 1922) these same muscles originate on the interorbital portion of the sphenoid bone. In *Amia* on the other hand only the superior, internal and inferior recti muscles originate on the basisphenoid (presphenoid) and this they do on the anterolateral face of the transverse bolster (Allis, 1919), while the external rectus enters the myodome. There can be no doubt that at least three of the rectus muscles originated on the basisphenoid pillar in *Mimia* and that their position was more medial than in *Amia*. As for the position of the fourth (external) rectus muscle, we can only speculate; either it originated with the others, or it originated in the posteroventral floor of the orbit below the adbucens foramen and above the ventral fissure (presumably a precondition for the subsequent evolution of the actinopterygian myodome). However there is no clear cut depression in this area for it and thus not even a hint of an incipient myodome.

Genus Moythomasia Gross, 1950

1942: \Aldingeria; Gross: 431.

Diagnosis. See Gross, 1942: 430 and add: Ventral cranial fissure passes into the rear of the orbit; parasphenoid with an ascending process but no basipterygoid process.

Type species. Moythomasia perforata (Gross).

Moythomasia durgaringa sp. nov.

(Figs 7 and 8B)

1973: Gogo palaeoniscid 'B'; Gardiner, figs 5 and 7 only.

Diagnosis. Scales similar to the type-species but with up to 15 serrations posteriorly. Infraorbital 2 enlarged, over half as wide as deep. Skull roofing bones ornamented with tubercles and elongate ridges of ganoine.

Holotype. Western Australian Museum, 70.4.244; partly disarticulated head and body in counterpart from the Upper Devonian, Gogo Shales, Gogo Station (H.A.T. 67, see Miles, 1971), Fitzroy Crossing, W. Australia.

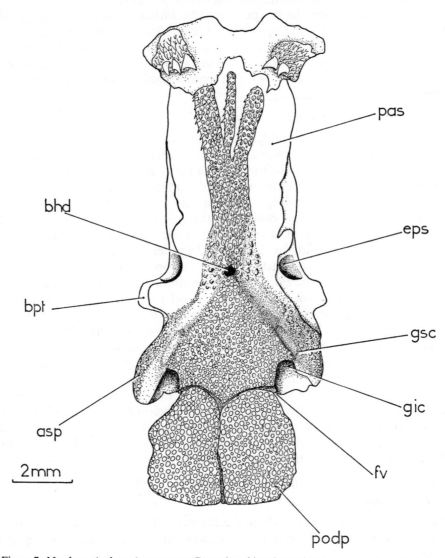

Figure 7. *Moythomasia durgaringa* sp. nov. Parasphenoid and associated structures in ventral view. From BMNH P53221.

Remarks. The species name comes from the aboriginal term for a carved ceremonial spear barbed on one side only and used as a gift to the dead in mortuary rituals. In common with other species of *Moythomasia*, on the skull roofing bones the individual ridges of ganoine appear serrated or barbed along their edges, presenting an almost herring-bone effect.

Ventral fissure and associated structures

A detailed description of this and other regions will be given elsewhere. Here only important differences from *Mimia* are dealt with.

The cartilage-filled ventral fissure occupies a similar position to that described in *Mimia* although it is not quite as extensive dorsomedially. The otico-sphenoid

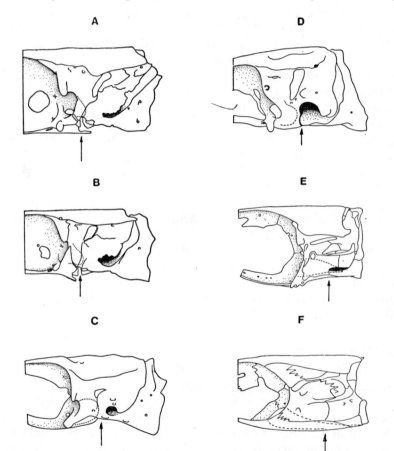

Figure 8. A. *Mimia toombsi.* No myodome; ventral fissure far forward; no ascending process. B. *Moythomasia durgaringa.* No myodome; ventral fissure far forward; ascending process present. C. *Kansasiella eatoni.* (Parasphenoid omitted) Myodome unpaired and of moderate size; ventral fissure just in front of vestibular fontanelle. D. *Pteronisculus stensioi.* (Parasphenoid omitted) Myodome large, unpaired; ventral fissure confluent with vestibular fontanelle. E. *Pholidophorus bechei.* Myodome extends into basioccipital; ventral fissure (basioccipital-prootic suture) confluent with vestibular fontanelle. F. *Elops saurus.* Myodome passes to back of skull on underside of basioccipital.
C, After Poplin, 1974; D, after Nielsen, 1942; E, after Patterson, 1975; F, after Forey, 1973.

fissure is closed anteriorly where the ascending process bridges it (Gardiner, 1973: fig. 5), but posteriorly it is open for a short distance (Fig. 8B). Thus in orbital view there is no evidence of the otico-sphenoid fissure. The lateral commissure itself occupies a more anterior position than in *Mimia* and obscures the pituitary vein when viewed laterally (Fig. 8B), it extends further forwards than in *Mimia* and is continuous with the lateral ridge at the base of the basisphenoid bolster. These differences suggest that the junction between the prechordal and chordal parts of the skull is considerably stronger in *Moythomasia* than in *Mimia*.

Posterior region of parasphenoid

The parasphenoid is closely applied to the basisphenoid but its toothed area is less extensive than in *Mimia*. In front of the opening for the bucco-hypophysial canal only the central third of the bone is toothed; laterally it is smooth and does not quite reach the margins of the basisphenoid. In one specimen (BMNH P53221) and on one side only (Fig. 7), the parasphenoid continues out for a short distance onto the basipterygoid process, but behind this region distinct ascending processes are always present. These processes not only extend onto the lateral commissure but are also toothed, and grooved by the spiracular diverticulum (gsc, Fig. 7). The grooves end anteriorly just behind the bucco-hypophysial foramen. The ascending process terminates on the lateral commissure at the level of the bottom of the jugular canal. Therefore although dermal basipterygoid processes are absent, in contrast with *Mimia* ascending processes are well developed. Posteriorly the parasphenoid has a distinct toothed triangular portion which terminates in a shallow ledge. During acid development of specimen BMNH P53221, we found two loose tooth plates in the mouth cavity. These plates not only fitted together exactly, but in turn matched the precise overlap on the back of the parasphenoid. They thus rested on that shallow ledge on the back of the parasphenoid mentioned above. These pharyngeal toothplates, which are similar to the parotic plates described by Jarvik (1954) in *Eusthenopteron*, will be described more fully elsewhere.

Origin of posterior eye muscles

On the basisphenoid bolster dorsal to the ophthalmic artery and behind the groove for the cephalic artery, in an identical position to *Mimia*, are muscle scars in the form of two distinct cups one above the other. As in *Mimia* these must have served for the origin of the rectus muscles. Whether or not the external rectus inserted here or in the back of the orbit ventrolateral to the abducens could not be determined. Certainly there is no well-defined depression in the area of the abducens which might be interpreted as an incipient myodome. Nevertheless it is tempting to reconstruct the external rectus in *Moythomasia* as originating on the base of the prootic rather than the basisphenoid since in this genus as we have seen the basisphenoid junctions with the prootics laterally (otico-sphenoid fissure) have already fused anteriorly. We have also seen that the lateral commissures are more angled forwards than in *Mimia* and are supported by the ascending processes of the parasphenoid. Also, the junction between the prechordal parts of the skull are strengthened. The greater forward inclination of the lateral com-missures in *Moythomasia* appears to have influenced the position of the trigemino-facialis chamber. This now lies in the mouth of the jugular canal (in *Mimia* it is anterior to the jugular canal mouth, see Fig. 3), and may also be correlated with a change in insertion of the external rectus muscles.

DISCUSSION

Mimia and *Moythomasia* differ from all other palaeonisciforms in possessing a ventral otic fissure which passes into the rear of the orbit, in having a pair of otico-sphenoid fissures, and in not having a myodome. *Mimia* alone differs from all other palaeonisciforms in lacking an ascending process of the parasphenoid. *Moythomasia* is the only palaeonisciform known to possess a pair of pharyngeal toothplates immediately posterior to the parasphenoid.

These facts pose three major problems. Is the ventral fissure of *Mimia* and *Moythomasia* homologous with that of other palaeonisciforms? What are the homologies between the ventral fissure of *Mimia* and *Moythomasia* and the fissures found in the braincase of crossopterygians? And what homologies can be drawn between the parasphenoid of palaeonisciforms and that of crossopterygians?

The first problem arises since the position of the ventral fissure varies among palaeonisciforms. At one extreme, in *Pteronisculus stensioi* (Nielsen, 1942), the fissure passes through the base of the otic region and runs into the vestibular fontanelles. At the other extreme, in *Mimia* and *Moythomasia*, the fissure is far forward and opens into the orbit. The position of the fissure in other palaeonisci-forms varies between these two extremes. However in all palaeonisciforms the ventral fissure has the following topographical relationships:—

(1) It is immediately anterior to the tip of the notochord.
(2) It is anterior to the partition between the ear and brain cavity.
(3) It is approximately at the level at which the lateral aortae give rise to the orbital arteries.
(4) It is immediately posterior to the region where the lateral commissures join the base of the skull.
(5) It is at the level of the posterior end of the parasphenoid.

Thus the ventral fissure appears to be homologous throughout the palaeonisci-forms. Which then is the primitive position of the fissure, anterior or posterior? The fissure in *Mimia* bears several resemblances to that between the parachordal and trabecular/polar cartilages in embryo fishes, for example *Acipenser*, *Lepisosteus*, *Exocoetus* (de Beer, 1937: pl. 30, fig. 1; pl. 38, fig. 1; pl. 50, fig. 3):—

(1) The fissure lies close behind the transverse pituitary vein, and anterior to the tip of the notochord.
(2) The fissure lies close behind, or mesial to, the points of entry of the palatine nerve into the base of the skull and the orbitonasal arteries into the orbit. These points lie close to the parachordal-trabecular junction in embryos (e.g. *Acipenser*, Holmgren, 1943: fig. 27; *Lepisosteus*, de Beer, 1937: pls 38, 39).
(3) The fissure is slightly posterior to a pair of lateral (otico-sphenoid) fissures which persist between the lateral commissures and the basisphenoid region of the skull. Referring to Holmgren's (1943: fig. 30) and Bjerring's (1968:

fig. 5; 1972: fig. 2) reconstructions of the embryonic cranium of *Amia*, it is clear that the lateral fissure of *Mimia* corresponds to the suture between the lateral commissure and the polar cartilage-trabecular bar of the embryo.

Thus the ventral fissure of *Mimia* must represent that between the chordal and prechordal skeleton of the embryo, and the anterior position of the fissure is primitive for actinopterygians. An explanation to account for the rearward position of the fissure in other palaeonisciforms has been briefly mentioned by one of us (Gardiner, 1973). Palaeonisciforms other than *Mimia* and *Moythomasia* possess a myodome for the pair of external rectus eye muscles, and the larger this muscle canal is, the further back lies the fissure (Fig. 8). Thus the size of the myodome appears to affect the position of the fissure, from which it is always separated by a thin wall of bone.

The second and third problems concern a study by Jarvik (1954) which deals with the braincase and visceral skeleton of the rhipidistian crossopterygian *Eusthenopteron*. For comparison with actinopterygians he took *Pteronisculus*, the best-known palaeonisciform at that time, from the detailed study of Nielsen (1942), and which Jarvik believed to be typical of the last group. He found in *Eusthenopteron* a cartilage-filled fissure linking the vestibular fontanelles, which he homologized with the ventral otic fissure of *Pteronisculus*, in an apparently similar position. Unfortunately it seems that *Pteronisculus* is not typical of palaeonisciforms in respect of the position of the ventral fissure. On the contrary it displays a specialized condition in which the ventral fissure has been 'pushed back' by the myodome, as argued above. Thus it is more likely that the ventral fissure of palaeonisciforms corresponds with the ventral part of the intracranial joint of crossopterygians, and that both of these represent the gap between the parachordal and trabecular regions of the embryonic braincase. It follows then that the homology between the ventral fissure of palaeonisciforms and that of *Eusthenopteron* (the 'fissura oticalis ventralis anterior' of Jarvik, 1972: fig. 99) is spurious.

Jarvik's view that the intracranial joint of crossopterygians and the ventral fissure of palaeonisciforms were not homologous led him to the conclusion that the parasphenoid in the two groups could not be strictly homologous either. The posterior edge of the parasphenoid of *Eusthenopteron* lies at the anterior limit of the intracranial joint. Between this edge and the fissure linking the vestibular fontanelles lies a pair of large parotic toothplates and a pair of smaller subotic toothplates. Jarvik argued that these plates must have been incorporated into the actinopterygian parasphenoid. Further, the ascending process of the actinopterygian parasphenoid formed from the fusion of the many small spiracular dental plates which lie dorsolateral to the parotic plates of *Eusthenopteron*. This hypothesis is greatly weakened since:—

(1) The ventral fissure of palaeonisciforms is homologous with the ventral part of the intracranial joint of crossopterygians, as argued above. Thus in both groups the hind edge of the parasphenoid lies at the same level.

(2) *Moythomasia* possesses both an ascending process and pharyngeal toothplates. The latter lie behind the ventral fissure in the same way that the parotic plates of *Eusthenopteron* lie behind the intracranial joint.

Pharyngeal toothplates have only been found in one specimen of one species of palaeonisciform. They have not been found in others either because they are absent or because, since they form a loose suture with the parasphenoid, they have fallen away. A re-examination of the hind border of palaeonisciform parasphenoids may reveal the paired embayments and overlapped region which would indicate the presence of these plates. However in many palaeonisciforms the hind edge of the parasphenoid is convex. In these cases the plates were absent, or may have been separated from the parasphenoid in life, as they are in *Eusthenopteron*. A living chondrostean, *Polyodon*, also has pharyngeal toothplates, in approximately the same position (Nelson, 1969: 522; tph, pl. 84, fig. 3). In this fish, however, the parasphenoid extends to the occiput and thus passes above them.

If the parasphenoid is homologous throughout osteichthyans, the paired grooves on either side of the bucco-hypophysial duct are surely homologous too. Jarvik held that they housed a prespiracular diverticulum in *Eusthenopteron*, but they were probably associated with the spiracular duct, as they are in palaeonisciforms. In *Eusthenopteron* the groove runs up an ascending process of the parasphenoid which Jarvik has homologized with the dermal basipterygoid process of actinopterygians. But since it carries the spiracular groove it is better homologized with the ascending process of actinopterygians, although in *Eusthenopteron* it also gives support to the endoskeletal basipterygoid process. Evidence from *Mimia*, however, suggests that the ascending process in actinopterygians and crossopterygians arose in parallel and may not be strictly homologous. *Mimia* has no ascending process, and this may be due to the presence, presumably primitive, of the otico-sphenoid fissure. It has been suggested (Jarvik, 1954: 68; Beltan, 1968: 107) that cartilage-filled fissures may inhibit the growth of dermal bone over them. Hence the fact that in palaeonisciforms the parasphenoid does not cross the ventral fissure; only in higher actinopterygians does the parasphenoid extend more caudally (Patterson, 1975). In a similar way, the otico-sphenoid fissure may inhibit the formation of an ascending process. Thus primitively, actinopterygians may have lacked the process, as in *Mimia*. When the fissure had closed the ascending process could form along the spiracular groove in order, probably, to strengthen the lateral commissure against the stresses caused by the presence of a myodome. If this view is correct, then the ascending process of *Eusthenopteron* is not strictly homologous with that of actinopterygians, since the former had not formed in connection with the lateral commissure and had no strengthening role.

In short, the differences between the braincase and mouth roofing bones of the two major groups of bony fishes are less radical than have hitherto been supposed. Also, the structure of the palaeonisciform braincase cannot be used to support a theory of the vertebral composition of the skull as Jarvik (1972) and Bjerring (1972) have done.

REFERENCES

ALLIS, E. P., 1919. The myodome and trigemino-facialis chamber of fishes and the corresponding cavities in higher vertebrates. *J. Morph.*, *32:* 207–326.

ALLIS, E. P., 1922. The cranial anatomy of *Polypterus*, with special reference to *Polypterus birchir. J. Anat.*, *56:* 189–294.

DE BEER, G. R., 1926. Studies on the vertebrate head. II. The orbito-temporal region of the skull. *Q. Jl microsc. Sci.*, *70:* 263–370.

DE BEER, G. R., 1927. The early development of the chondrocranium of *Salmo fario. Q. Jl microsc. Sci.*, *71:* 259–312.

DE BEER, G. R., 1937. *The development of the vertebrate skull:* xxiv+552 pp. Oxford: University Press.

BELTAN, L., 1968. *La faune ichthyologique de l'Eotrias du N.W. de Madagascar; le neurocrane:* 135 pp. Paris: Centre national de la recherche scientifique.

BERTMAR, G., 1962. On the ontogeny and evolution of the arterial vascular system in the head of the African characidean fish *Hepsetus odoë. Acta zool., Stockh., 43:* 255–95.

BJERRING, H. C., 1968. The second somite with special reference to the evolution of its myotomic derivatives. *Nobel Symposium, 4:* 34–357.

BJERRING, H. C., 1972. The rhinal bone and its evolutionary significance. *Zool. Scripta 1:* 193–201.

FOREY, P. L., 1973. A revision of the elopiform fishes, fossil and Recent. *Bull. Br. Mus. nat. Hist. (Geol) Suppl. 10:* 1–222.

GARDINER, B. G., 1963. Certain palaeoniscid fishes and the evolution of the snout in actinopterygians. *Bull. Br. Mus. nat. Hist. (Geol.), 8:* 255–325.

GARDINER, B. G., 1970. Osteichthyes. *McGraw-Hill Yb. Sci. Tech. 1970:* 284–6.

GARDINER, B. G., 1971. In R. S. Miles & J. A. Moy-Thomas, *Palaeozoic fishes:* xi+259 pp. London: Chapman & Hall.

GARDINER, B. G., 1973. Interrelationships of teleostomes. In P. H. Greenwood, R. S. Miles & C. Patterson (Eds), *Interrelationships of fishes:* 105–35. London: Academic Press.

GARDINER, B. G. & MILES, R. S., 1975. Devonian fishes of the Gogo Formation, Western Australia. *Colloques int. Cent. natn. Rech. scient., 218:* 73–9.

GOODRICH, E. S., 1930. *Studies on the structure and development of vertebrates:* xxx+837 pp. London: Macmillan.

GROSS, W., 1942. Die Fischfaunen des baltischen Devons und ihre biostratigraphische Bedeutung. *KorrespBl NaturfVer. Riga, 64:* 373–436.

GROSS, W., 1950. Umbenennung von *Aldingeria* Gross (Palaeoniscidae; Oberdevon) in *Moythomasia n. nom. Neues Jb. Geol. Paläont. Abh., 5:* 145.

HOLMGREN, N., 1943. Studies on the head of fishes. Part IV. General morphology of the head in fish. *Acta zool., Stockh., 24:* 1–188.

JARVIK, E., 1954. On the visceral skeleton in *Eusthenopteron,* with a discussion of the parasphenoid and palatoquadrate in fishes. *K. svenska VetenskAkad. Handl., (4)5: 1:* 1–104.

JARVIK, E., 1972. Middle and Upper Devonian Porolepiformes from East Greenland with special reference to *Glyptolepis groenlandica* n.sp. *Meddr Grønland, 187: 2:* 1–307.

MARINELLI, W. & STRENGER, A., 1973. Superklasse: Gnathostomata; Klasse Osteichthys. In *Vergleichende Anatomie und Morphologie der Wirbeltiere, 4:* 310–460. Vienna: Franz Denticke.

MILES, R. S., 1971. The Holonematidae (placoderm fishes), a review based on new specimens of *Holonema* from the Upper Devonian of Western Australia. *Phil. Trans. R. Soc. Lond. (B), 263:* 101–234.

NELSON, G. J., 1969. Gill arches and the phylogeny of fishes, with notes on the classification of vertebrates. *Bull. Am. Mus. nat. Hist., 141:* 475–552.

NELSON, G. J., 1970. Subcephalic muscles and intracranial joints of sarcopterygian and other fishes. *Copeia, 1970:* 468–71.

NIELSEN, E., 1942. Studies on Triassic fishes from East Greenland. I. *Glaucolepis* and *Boreosomus. Meddr Grønland, 138:* 1–403.

NIELSEN, E., 1949. Studies on Triassic fishes from East Greenland. II. *Australosomus* and *Birgeria. Meddr. Grønland, 146:* 1–309.

PATTERSON, C., 1975. The braincase of pholidophorid and leptolepid fishes, with a review of the actinopterygian braincase. *Phil. Trans. R. Soc. Lond. (B), 269:* 275–579.

POPLIN, C., 1974. *Étude de quelques Paléoniscidés Pennsylvaniens du Kansas:* 151 pp. Paris: Centre national de la recherche scientifique.

RAYNER, D. H., 1951. On the cranial structure of an early palaeoniscid, *Kentuckia* gen. nov. *Trans. R. Soc. Edinb., 62:* 53–83.

ABBREVIATIONS USED IN FIGURES

aip[1]	articular facet for first infrapharyngobranchial	ona	orbitonasal artery
asp	ascending process of parasphenoid	opa	ophthalmic artery
bhd	buccohypophysial canal	osm	origin of subcephalic muscle
bic	branch of internal carotid artery	pas	parasphenoid bone
bpt	basipterygoid process	pbc	parabasal canal
eps	efferent pseudobranchial artery or its foramen	pexr	possible origin of external rectus muscle
fos	fissura otico-sphenoid	pf	pituitary fossa
fv	fissura ventralis	podp	pharyngeal toothplate
gic	groove for internal carotid artery	por	postorbital process
gla	groove for lateral aorta	ppas	process of parasphenoid
goa	groove for orbital artery	prb	'prootic' bridge
gsc	groove for spiracular canal	pv	foramen for pituitary vein
ic	internal carotid artery	sv	cavity for saccus vasculosus
jc	jugular canal	vf	vestibular fontanelle
jv	jugular vein	1	foramen of olfactory tracts
la	lateral aorta	2	optic fenestra
lc	lateral commissure	3	foramen of oculomotor nerve
not	notochordal canal	6	foramen of abducens nerve
oa	foramen for orbital artery	7 pal	palatine nerve
oem	origin of posterior eye muscles		

On the individual history of cosmine and a possible electroreceptive function of the pore-canal system in fossil fishes

KEITH STEWART THOMSON

Department of Biology and Peabody Museum of Natural History, Yale University

Cosmine is a unique combination of hard tissues with the pore-canal sensory system in early fishes. Evidence is presented to show that the history of cosmine during the life of individual fishes of the family Osteolepidae (Rhipidistia) may be quite complicated. A full cosmine cover on the dermal skeleton may not be maintained at all stages. The resorption of the cosmine and presence of naked areas of spongiosa is accompanied by the development of specialized superficial structures of the dermal skeleton and probably reflects changes in environmental conditions. Comparison with superficial neuromast receptors in living fishes shows that the pore-canal system is probably electroreceptive. In the earliest fishes and the Agnatha the pore-canal system included only tonic receptors, but in the Devonian lobe-finned fishes we may identify both tonic and phasic receptors, on the basis of general morphology and size.

CONTENTS

INTRODUCTION

Only comparatively rarely does one find in a group of fossil vertebrates an organ or tissue that seems to be totally without equivalent among living forms. It is one of the pleasures of palaeontology to try to find the biological significance of such structures. One such enigmatic phenomenon is the composite set of tissues of certain groups of fossil fishes that is known as cosmine. In fact, cosmine presents a double puzzle. First it is uniquely defined, as a combination of hard tissues with a specialized sensory system. Second, the sensory system—the pore-canal system—is itself different from anything in living fishes and its function is

247

presently unknown. It is appropriate to offer a paper on cosmine in this volume of essays, because Professor Westoll was the first to see the many puzzling questions and interesting biological possibilities presented by the study of cosmine.

Cosmine occurs on the external surface of the dermal skeleton of a variety of early fossil fishes. Layers of enameloid and dentine surround the pore-canal system, the ampullae of which open to the surface through the enameloid layer. Intact areas of cosmine are histologically continuous with the underlying spongiosa of the dermal skeleton through the linking agency of a thin layer of true bone ('bone of attachment') that, as it is developmentally integrated with the dentine and enameloid, is defined as being a constituent of the cosmine (Thomson, 1975). The vascular supplies of the pore-canal system and of the dentinal pulp cavities stem from the general vascularization of the spongiosa. Thus, in thin-sections the cosmine and spongiosa seem to form one completely integrated unit. However, as is well known, the cosmine has an entirely different developmental history from that of the underlying dermal skeleton (spongiosa and isopedin). In the head, the topographic extent of the cosmine sheets usually bears no relationship to the pattern of the underlying dermal elements. Sutures between bones are completely obliterated externally by the continuous cosmine cover. Further, as Professor Westoll (1936) pointed out in his study of cosmine, the presence of such continuous sheets of cosmine must necessarily interfere with normal processes of growth in the area of the dermal skeleton. Westoll was the first to suggest that this fact is the explanation of the appearance of specimens from which the cosmine cover may be seen to be in the process of physiological resorption. He proposed that such resorption proceeds seasonally, allowing growth before a new cosmine cover is deposited.

The very existence of cosmine in fossil fishes, to say nothing of its total absence in living fishes, causes one to ask questions having to do with the fundamental biology of hard tissues. We may ask, for example: what is the developmental relationship between the pore-canal system and the dentinal and enameloid tissues; or what is the relationship between the dentine and the spongiosa? What is the biological function of the pore-canal system and what is the particular significance of its association with hard tissues? What is the phylogenetic relationship of the pore-canal system to the general seismosensory system and what is the phylogenetic significance of the hard tissues in cosmine? What are the consequences of the possession of cosmine for growth of the dermal skeleton and what are the physio-logical sequences of the periodic resorption and redeposition of hard tissue that is characteristic of cosmine?

Gross (1956, q.v. for review of the earlier histological literature) has given us a very detailed picture of the structural relationships between the hard tissues and the pore-canal system in cosmine and has portrayed in beautiful detail the arrange-ment of the pore-canal mosaics as they occur in Agnatha, Acanthodii, Dipnoi and Rhipidistia. Westoll (1936), in addition to the critically important observations noted above, also showed the difficulties in the identification of specimens showing different stages of resorption and deposition, and postulated that there must be a strong functional-environmental aspect in the biology of cosmine. Jarvik (1948, 1950) and Ørvig (1969), together with Bolaü (1951) and Bystrov (1942, 1959) have combined descriptions of details of the results of resorption with interpretations of the biological processes concerned. Thomson (1975) has developed a general model of the developmental relationships between the pore-canal system and other parts of the latero-sensory system with hard tissue formation and presented some general ideas concerning the environmental significance of cosmine metabolism.

In a detailed study of the biology of cosmine in the Lower Permian osteolepid rhipidistian *Ectosteorhachis nitidus* Cope, it was found (Thomson, 1975) that a complete covering of cosmine on the dermal skeleton occurs only in younger fishes and that in older fishes this cover breaks down, with the appearance of permanently naked areas of spongiosa and the formation of specialized remnants of cosmine on these surfaces. This conclusion runs counter to the familiar view that all cosmine-bearing fishes show a complete cosmine cover throughout the whole life history, interrupted only temporarily by each resorption between which a new complete cover is produced. The difference between these two views of the history of cosmine during the life of fishes has major implications for interpretations of the biological significance of cosmine. Further, the nature of the breakdown of the complete cosmine cover gives us insight into the developmental process involved in the control of cosmine patterns and the relationship of the cosmine to the underlying skeleton. Therefore it is important that the developmental history of cosmine be studied in detail in other fishes. In the present paper results are presented of a comparative study of cosmine in the genus of Carboniferous osteolepid rhipidistian *Megalichthys* Agassiz *in* Hibbert and some more general conclusions about the developmental history of cosmine on fishes are discussed.

In the study of *Ectosteorhachis* (Thomson, 1975) the developmental biology of cosmine is considered at length and evidence of the role of the pore-canal system in the organization of cosmine and the underlying dermal elements is used to develop a general morphogenetic model. However, it was not possible to answer fully the question of the function of the pore-canal sensory system. While knowledge of the function of the system is not absolutely necessary to the development of a model of developmental interactions between the sensory system and hard tissues, it is an important problem, solution of which may help us discover the overall significance of cosmine itself. The second portion of this paper is a discussion of the possible functional significance of the pore-canal system in fossil fishes.

This paper thus falls into two discrete parts, dealing with two important aspects of the 'cosmine problem'. Throughout, data from the osteolepid Rhipidistia will form the bulk of the evidence used. There are several reasons for this. First, the most complete suites of material are available for the Rhipidistia, many of which are well suited for the preparation of histological thin-sections showing considerable fine detail of the cosmine and dermal bone. Second, there is a possibility that the cosmine of separate groups is actually different. Certainly the developmental processes in Dipnoi (as outlined by Ørvig, 1969) are sufficiently different from those in Rhipidistia as to dictate caution in combining evidence from widely different sources in the formation of explanatory models.

THE HISTORY OF COSMINE DURING LIFE, IN OSTEOLEPID RHIPIDISTIA

The first requirement for a study of the various manifestations of cosmine in different fishes is to establish a criterion for the evaluation of the phenomena.

Westoll (1936) in referring to cosmine areas 'with rounded edges' suggests a criterion that Gross (1956) later established in full. One can distinguish two sorts of cosmine coverings in these fishes. The one is, as it were, pristine cosmine, which, whenever it was laid down, has not been modified subsequently. The margins of such cosmine are said to be *finished* and the enameloid covering laps around onto the vertical surface. A cosmine margin that has been preserved in the process of resorption, however, may be seen to be *unfinished* (see Figs 1, 2 and 3); it lacks the lappet of enameloid, and shows evidence of erosion of both dentine and enameloid. Such a criterion can be used both for the lateral margins of cosmine sheets and the margins of pores within the cosmine.

In the osteolepid species *Ectosteorhachis nitidus* a complete cover of cosmine is almost never present. Fishes are probably quite incompletely covered for the first and much of the second year of life. A complete cover builds up first in the form of patches of cosmine and by the second or third moult a complete cover is probably produced. After an unknown period of time, possibly as soon as the fourth moult, a complete cover is maintained only on the ventral parts of the head and trunk, the mandibles and the tip of the snout. Elsewhere on the head, only isolated patches of finished cosmine occur, at first with little bare space between them. On the dorsal and dorso-lateral scales the cosmine is confined to a cap over unique 'tubercles,' normally occurring one per scale and passing forward in rows that merge with the pattern of small patches of cosmine in the head. There is in addition evidence that the regular cycling of resorption and redeposition of cosmine also breaks down. It is possible that these events are associated with particular environmental factors in the life history of the fish; for instance making it physiologically impossible or selectively inadaptive to metabolize the large amounts of mineral involved in annual resorption and redeposition of a full cosmine cover.

The results of study of *Ectosteorhachis* were sufficiently interesting and sufficiently at variance with the accepted view of the life history of cosmine that it was decided to make an effort to see if similar phenomena occurred in other Rhipidistia. For it is possible that biases in collecting fossils might lead to an unnatural preponderance in collections of specimens of osteolepids bearing the beautiful full shiny cosmine cover, and that incompletely covered specimens ought to be more frequent in occurrence than has been suspected.

An osteolepid for which an abundance of well preserved material exists is the genus *Megalichthys*, of which several species have been recognized. This fish seems to be a typical osteolepid with a complete cosmine cover. However, on examination of large volumes of material, specimens with an incomplete cover come to light.

First we may recognize specimens of all species in which there is direct evidence of simple resorption of the cosmine. The surface of the dermal skeleton bears a series of patches of cosmine with surrounding naked areas. The margins of these cosmine patches are 'unfinished'; an example is shown in Fig. 1. In the study of *Ectosteorhachis*, the complete progress of resorption could not be seen clearly, although thin-sections were made showing that resorption appeared to start both at the margins of the cosmine sheets and from enlarged pores in the surface. A cavity was excavated by osteocytic activity that was largely confined to the dentine and the thin layer of bone between the dentine and the vascular bone of the spongiosa. The enameloid layer was more resistant to resorption and thus large cavities were produced that passed under the enameloid surface until the latter was sufficiently destroyed to reveal the spongiosa underneath. Evidence of an identical process is found in thin sections of *Megalichthys* dermal skeleton (cf. Figs 2 and 3). In addition, in *Megalichthys*, a consistent pattern in the course of the resorption process can be seen. At least in the skull elements, resorption seems to start wholly within the cosmine sheet, rather than at its margins. First a series of

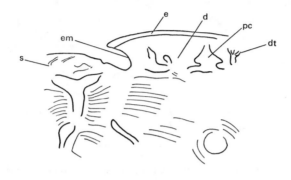

0·2mm

Figure 1. *Megalichthys macropomus* (MCZ 5143). Vertical thin-section of a scale through the margin of a portion of cosmine surface that has been under resorption. To the left, the spongiosa is exposed at the surface. The 'unfinished' nature of the margin of the cosmine is visible. d, Dentine layer; dt, dentine tubules and pulp cavity; e, enameloid layer; em, eroded margin of cosmine; pc, pore-cavity; s, spongiosa.

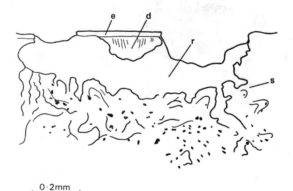

0·2mm

Figure 2. *Ectosteorhachis nitidus* (MCZ 13350). Vertical thin-section through a portion of a dermal bone showing active resorption of the cosmine. d, Dentine; e, enameloid; r, resorption cavity; s, spongiosa. (See also Thomson, 1975: fig. 23.)

pores (whether of the pore-canal system or some other pores cannot be determined) become enlarged and form the foci of resorption. These pores are arranged in radiating rows with their centre at the centre of growth of the dermal element. As the rows of pores lengthen, the more proximal pores in each row have enlarged sufficiently that the small naked patches produced begin to merge with each other, first along the rows of pores and then between adjacent rows. The result is the production of a series of irregular rows of naked areas with strips of remaining cosmine remaining between them until eventually all the cosmine is resorbed. An excellent example of this process is shown in Fig. 4.

In *Megalichthys*, exactly as in *Ectosteorhachis*, specimens showing resorption of the cosmine from the dorsal parts of the head do not show signs of any resorption

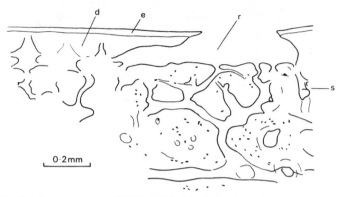

Figure 3. *Megalichthys macropomus* (MCZ 25143). Vertical thin-section through a scale showing active resorption. d, Dentine; e, enameloid; r, resorption cavity; s, spongiosa.

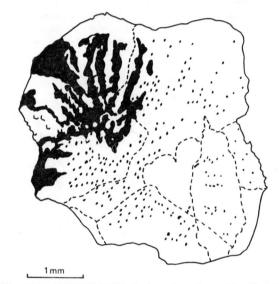

Figure 4. *Megalichthys hibberti* (BMNH P.799). Surface view of an unidentified dermal bone showing evidence of active resorption of the cosmine.

of the cosmine of the mandibles or elements of the gular series. In fact, to find osteolepid mandibles showing resorption is very rare, considerably more rare than the occurrence of resorption from dermal elements of the dorsal parts of the head One isolated mandible is known, however, (BMNH P.7852, *M. hibberti*) in which only small blebs of cosmine occur, with unfinished margins. The condition of this specimen is apparently not due to wear and thus shows that resorption of the mandibular cosmine must occur.

The process of resorption on the scales is less easy to see and it is possible that it occurs largely through encroachment from the margins. Certain specimens (e.g. BMNH P.809, *Megalichthys* sp.) show small irregular remnants of cosmine on the scales that seeem to have been produced by this process.

The above seems to be the 'normal' pattern of cosmine resorption when the whole cosmine cover is removed and replaced. There is evidence, however, that at times in the life history the cosmine cover is not fully replaced but, as in *Ectosteorhachis*, remains incomplete. The skull of the type specimen of *Megalichthys macropomus* is almost wholly naked of cosmine dorsally, with only a scattering of blebs and ridges left on the postero-lateral parts of the head (Plate 1). These isolated patches are arranged in a pattern with respect to the growth centres of the underlying dermal bone and all have unfinished margins, suggesting that they are a permanent feature. The ventral parts of the head have a complete cover of cosmine.

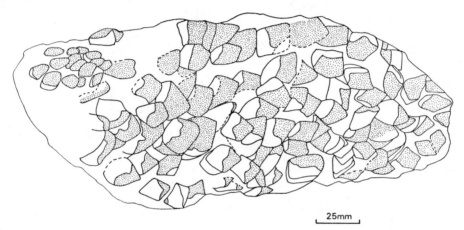

25mm

Figure 5. *Megalichthys macropomus* (MCZ 5143). Portion of the squamation showing scales in natural position. Each scale shows an incomplete covering of cosmine (stippled); see also Fig. 6.

In Figs 5 and 6, and Plate 1C, portions of the squamation of *Megalichthys macropus* are shown. Each dorsal and dorso-lateral scale only bears a partial cover of cosmine. It is incomplete on the anterior part of the scale. On each scale there is a more or less crescent-shaped area of naked spongiosa that forms the anterior margin of the exposed portion of each scale. The pattern is quite consistent and is not found in the ventral scales.

Examination of such scales in this section (Figs 7 and 8) shows that this condition is not a temporary feature of the life history of the fish. It must have persisted through a significant period of growth, for the exposed surface of spongiosa is raised significantly above the level of the cosmine. The appearance is striking in comparison with thin-sections of scales in the process of 'normal' resorption (Fig. 1). This unusual condition must have arisen through the differential growth in thickness of the exposed part of the dermal skeleton compared with that part covered with cosmine. Such differential growth must reflect growth of at least one year and possibly longer. In these specimens of *Megalichthys*, therefore we have a phenomenon comparable to, but quite different in expression from, that seen in *Ectosteorhachis*. In both, specimens may be found in which the cosmine cover of the trunk is reduced and the reduced condition is maintained over several years.

Thin-sections of *Megalichthys* scales undergoing resorption show that, as in *Ectosteorhachis*, the osteoclasts fail to modify the spongiosa proper, but they do remove a thin layer of true bone, comparable to a bone of attachment, immediately beneath the dentine layer. The extent of this layer of bone is approximately coincident with the base of the pore-canal ampullae. The result is, therefore, that

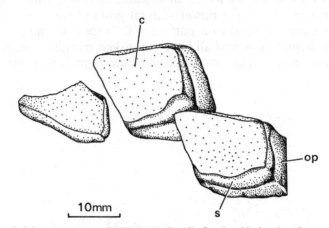

Figure 6. *Megalichthys macropomus* (MCZ 5143). Detail of scales. Notice that the cosmine does not cover the whole of the exposed portion of each scale but in front of it there is an area of overgrowing spongiosa. c, Cosmine; op, overlapped portion of scales; s, spongiosa.

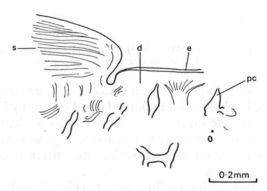

Figure 7. *Megalichthys macropomus* (MCZ 5143). Vertical thin-section through scale showing junction between cosmine-covered portion and the area of exposed spongiosa which has started to grow over the cosmine. d, Dentine; e, enameloid; pc, pore-cavity; s, spongiosa.

the pore-canal system is completely freed of surrounding hard tissue. In the scales showing a reduced cover of cosmine and a thickening of the exposed spongiosa, although the spongiosa may start to encroach over the margin of the cosmine, nonetheless, there remains a passage for the mesh-canals of the pore-canal system to turn externally and, presumably, connect with other portions of the pore-canal system buried in the external soft tissues (Fig. 8).

Although the spongiosa proper does not seem to be touched by the resorption process, it is worth noting that the uppermost layers of the isopedin (the basal lamellar bone layer) are reworked from the centre of the scale as a result of activity proceeding centripetally from the spongiosa. The result seems to be that, after a certain point in the life history, the isopedin layer (which is apparently laid down first, see Williamson, 1849) is not added to, but the scale or dermal bone grows by increase in thickness of the spongiosa partially at the expense of the isopedin. The scale or bone is often reinforced on its inner face by the addition of other bony material. Significantly, as the scale grows in thickness it seems to grow mostly at or near the centre, with the margins becoming or remaining rather thin and attenuated, as a result of an absence of growth in the spongiosa around the margins. As no new layers are added to the isopedin once its erosion from the outermost surface begins, it is not possible to age the scale by means of estimating the number of lamellae in the isopedin. A minimum age may perhaps be obtained by counting rings at the margins of the scale, where resorption apparently does not occur.

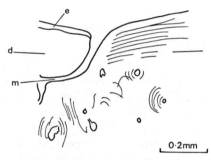

Figure 8. *Megalichthys macropomus* (MCZ 5143). Vertical thin-section through scale showing junction between the cosmine-covered portion and the area of exposed spongiosa. Note that the mesh-canal of the pore-canal system opens to the exterior at the groove between the two areas. d, Dentine; e, enameloid; m, mesh-canal; s, spongiosa.

A striking feature of many osteolepids is the consistent pattern of sculpturing that exists on the upper surface of the underlying spongiosa. In *Ectosteorhachis* such a sculptured surface is seen on cleithra, but in *Megalichthys* it seems to exist on all dermal elements, even (in the case of the material usually described as *Megalichthys intermedius*, and possibly other materials as well) under the cosmine of the scales. Sculpturing exists also in *Gyroptychius* (see Westoll, 1936), and in the form described as 'large osteolepid' (Thomson, 1976) from the Uppermost Devonian of Pennsylvania the sculpturing is very like that of *Megalichthys*. Such sculpturing presumably has a functional significance and in this connection it is interesting to note that in many respects it is identical to the sculpturing seen on the dermal bones and scales of other related forms (holoptychoid and rhizodontoid Rhipidistia, early labyrinthodont amphibians) in which there is no cosmine at all.

Westoll (1936) has documented that in other genera of osteolepids, such as *Gyroptychius*, specimens lacking cosmine may be found, but in the absence of specific evidence like that seen in *Ectosteorhachis* and *Megalichthys*, it is difficult to tell if they represent a permanent rather than temporary phase in the life history. I consider it significant, however, that in the two genera for which large suites of well preserved material are available for study, it has been possible to show that a full cosmine cover is not maintained on the dermal skeleton throughout the life history of the individual fish.

An important question concerns the very smallest growth stages. Do they show a full cosmine cover or not? Since they are growing very fast, one might presume that such small stages would be naked of cosmine, or only have an incomplete cover of discrete patches. For the Rhipidistia as a whole, however, the evidence is ambiguous. An example of the problem is the taxon *Megalichthys pygmaeus* Traquair. Specimens of these fishes range down to 120 mm in estimated total length (length of mandible 16 mm) and yet in most cases they bear a complete cover. It makes a difference, of course, whether *M. pygmaeus* is a distinct small species (in which case fish of this size might be fully grown) or whether it is a growth stage of a large species. Specimens of *Megalichthys* of the next size range (estimated length 200 mm) are very often naked of cosmine, but it is difficult to determine to what species they belong. (The taxonomy of this genus badly needs attention.)

Discussion

The data that are now becoming available concerning the developmental history of cosmine in osteolepids, like those that Ørvig (1969) has assembled on cosmine phenomena in the Devonian lungfish *Dipterus*, are of great value in studies of the general biology of cosmine. One major problem has been to determine whether there is any direct developmental relationship between the formation of the cosmine sheets and the organization of the underlying dermal skeleton, for in 'normal' fishes the complete cover of cosmine is apparently organized topographically independently of the latter. In the previous study of *Ectosteorhachis* (Thomson, 1975) I have suggested that a special relationship may exist in development between cosmine formation and the centres of growth of the underlying dermal elements. The evidence for this hypothesis depended on the distribution of the cosmine remnants (blebs and tubercles) in *Ectosteorhachis* specimens that had largely lost the cosmine. These follow the centres of growth of the individual elements and the pattern of the lateral line system. Description of the pattern of resorption on dermal elements of the head of *Megalichthys* helps confirm and extend this hypothesis. Here resorption proceeds in a pattern directly reflecting the growth centres of the underlying dermal skeleton and one may therefore suspect that the initiation of resorption is triggered or mediated in some way by the growth centre.

The incompleteness of the cosmine cover in stages of the life history of osteolepids may well be directly related to environmental conditions. It would then be expected to vary from taxon to taxon. Hypotheses concerning environmental correlations were developed in connection with my previous study of *Ectosteorhachis* (Thomson, 1975). In addition a survey of the geographic distribution of rhipidistian assemblages in the Upper Devonian of North America and Europe (Thompson,1976) shows some marked discontinuities in the distribution of cosmine-bearing Rhipidistia in comparison with those families of Rhipidistia in which cosmine is lacking. There seems to be evidence that cosmine-bearing

Rhipidistia occur in more marine and/or more riverine conditions, whereas the cosmine-free species occur in more lacustrine conditions.

There is considerable evidence that the early history of the lobe-finned fishes involved marine phases (Thomson, 1969) and it may well be that a tolerance to salt water continued in most rhipidistian fishes. *Osteolepis*, to take a well-known example, occurs in both fresh- and salt-water deposits. Any explanation of the developmental history of cosmine in particular taxa must therefore take into account the possibility that the fishes spent different parts of their life cycles in fresh- and salt-water. A fish living in salt-water has a more ample supply of minerals and other nutrients than does one in fresh-water. A working hypothesis may be presented that a full cover of cosmine is maintained equally as a store for excess calcium (see Westoll,1936), as a reserve for phosphate, and as a general protection against abrasion. The cover may be incomplete as the fish is growing. In later stages when the fish is entering reproductive condition and/or migrating to fresh-water the cosmine is partially or wholly resorbed and the mineral made available. A minimum cover in the form of discrete patches of cosmine is maintained dorsally. A fuller cover is preserved ventrally in response to a greater need for protection from abrasion. Under certain conditions (life in fresh-water?) it may have been difficult to mobilize sufficient mineral to redeposit a full cosmine cover and therefore the incomplete cover was maintained for long periods of the life history.

In connection with the above, a major question is: why is even a partial cover of cosmine maintained in the dorsal regions of the body? Where its function is probably not protective, one possibility is that it is necessary for the function of the pore-canal system. But the mosaic of horizontal mesh canals of the pore-canal system is evidently continuous outside the cosmine regions. This brings us to the second major subject of this paper.

THE FUNCTION OF THE PORE-CANAL SYSTEM

The pore-canal system of fossil fishes consists of a series of horizontal canals and ampullae connected to the surface by a set of minute pores. The canals and ampullae form a mosaic in the enclosing cosmine within which the pulp cavities of the dentinal system are regularly interspersed. The pore-canal system has a regular series of linking canals to the bases of the pulp cavities and the whole anastomosing network has a vascular supply in canals continuous with the vascular channels of the underlying spongiosa of the dermal skeleton proper.

Stensiö (1927) and Gross (1935) believed that the pore-canal system functioned in the secretion of mucus that was conducted to the external surface of the fish through the minute pore openings. They called this glandular system the 'muköses Kanalsystem' or 'Schleimkanalsystem'. However, more recently it has been agreed that the pore-canal system is a part of the laterosensory system. Such a view has been proposed by Denison (1947, 1966), Bolaü (1951) and Gross (1956) and is widely accepted. The crucial evidence is a direct continuity between the pore-canal system and the pit-line organs in fishes such as *Tremataspis* (e.g. Denison, 1947: fig. 11) and osteolepid Rhipidistia (e.g. Gross, 1956: fig. 40B). Discussion of the specific sensory function of the pore-canal system (i.e. identification of the

particular sensory modality) has been limited by our inability to identify comparable structures in living fishes. Gross (1956) attempted to compare the large
pores of certain fossil lungfish (see below) with the flask-shaped organs of Fahrenholz in *Lepidosiren* but, since the function of the latter is unknown except that they
are guessed to be chemosensory, even this limited comparison is difficult to
evaluate.

In living fishes, the existence of various puzzling neuromast receptors has been
known for a long time but only relatively recently has it been established that
many of these, particularly the ampullary canals of Lorenzini in elasmobranchs and
marine catfish and the ampullary and tuberous receptors of many bony fishes, all
serve for electroreception. In the following pages an attempt will be made to
compare the pore-canal system of fossil fishes with these modern electroreceptors.

Reconstructions of the sensory organs of the pore-canal system

In the Osteostraci and Acanthodii the pore-canal system consists of a pair of
tubes lying one above the other and separated by a perforated plate. These *upper*
and *lower mesh canals* (*Maschenkanalen*) form a polygonal or linear network in the
surface of the dermal skeleton. The upper mesh canals connect to the exterior
via short vertical canals ending at small pores. In the lobe-finned fishes the system
has been modified to a series of ampullae, the pore-cavities, usually connected only
by the upper mesh canals. There is some evidence (see Gross, 1956; Thomson,
1975) that the lower mesh canals have become condensed into the ampullary bases,
forming basal chambers, partially separated off from the main pore-cavity. Each
pore-cavity in the lobe-finned fishes opens to the exterior by its own external pore.
In all cases the whole network of canals is also interconnected to the vascular
supply of the underlying spongy bone.

In the osteostracan pore-canal system, with the double mesh canals arranged
in linked polygons, the sensory neuromasts were probably contained in the lower
mesh canal (Fig. 9A) and it seems likely that sensory hairs projected through the
perforations in the wall (*Siebplatte*) between the upper and lower canals. The upper
canal probably contained mucus (doubtfully water), as did the vertical pore canals.
It is possible that the hair endings in the mucus were contained in gelatinous
cupolae, as in the sensory hair cells of many living fishes.

In the acanthodian pore-canal system a similar arrangement to that of Osteostraci probably existed, the only major differences being that the mesh canal
system is more complicated and the main canals are arranged linearly.

In the lobe-finned fishes (Fig. 9) there are various different arrangements of the
pore-canal receptors, all departing markedly from those just given. In some
osteolepid Rhipidistia and many early Dipnoi (for example, *Osteolepis* and
Dipterus) we see first the pattern shown in Fig. 9B, where the neuromast receptors
have been confined to the basal chamber of the pore-cavity and the lower mesh
canal is reduced. It is not completely eliminated in all forms, but remnants may
persist. In these forms, the opening between the two chambers may be stellate
or in the form of a series of narrow slits. It is possible that hair processes extended
into the base of the upper pore-cavity through these openings. We have, however,
no more direct evidence of the possible presence of hair processes in these forms.
The arrangement of the receptor cells in these slits suggests a possibility for

directional response. In later osteolepid Rhipidistia, such as *Megalichthys* and *Ectosteorhachis* (e.g. Thomson, 1975) the lower mesh canal is absent and each basal chamber is larger. In all these forms there is a wide opening between the basal chamber and the pore-cavity (Fig. 9C).

In the Osteolepidae certain regions of the pore-canal mosaic include significantly larger pore-cavities that have a larger external pore. These occur in 'pore-groups' (see Jarvik, 1948, e.g. fig. 36; Bolaü, 1951: fig. 9). In all these forms, the mesh canal and the upper pore-cavity are presumed to have contained mucus.

In the holoptychoid Rhipidistia and in certain Dipnoi the pore-canal system consists exclusively of larger chambers with larger pores. The chambers are roughly cylindrical and the mesh-canal system is still present. In the dipnoan *Ganorhynchus*, the pore-cavities are extremely large and restricted to small numbers

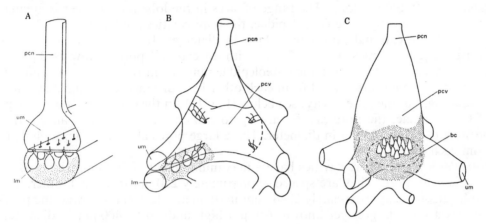

Figure 9. Reconstructions of the pore-canal system in three different types of fishes: A, an osteostracan, B, apparently the early condition for Sarcopterygii and C, a later stage for Sarcopterygii. The sensory cells are shown in white, surrounded by areas of supporting cells (stippled). bc, Basal chamber; pcn, pore-canal; pcv, pore-cavity; um, lm, upper and lower mesh canals.

in groups. There is no mesh canal system but there is a massive basal chamber. Interpretation of the soft structure is made difficult by the lack of internal differentiation of the pore-cavity. However, Gross (1956) has shown that in both *Porolepis* and *Ganorhynchus* the pore-cavity is lined with enameloid and therefore the contained soft structures must have included an epidermal lining, with the sensory cells at the base, and it is possible that the external pore was not open.

In all the cases described above, the nervous input to the sensory system came directly from the region of the underlying spongiosa, through the ramifying canals in that tissue that link to the pore-cavity bases. There is no evidence of innervation or vascularization passing directly to the upper mesh canals. In those osteolepids that show both upper and lower mesh canals there may occasionally be cross-connections between the two, the significance of which cannot be known but which might indicate the passage of blood vessels.

In the very early Dipnoan fish *Dipnorhynchus sussmilchi* the snout bears a series of branching tubular canals (rostral tubules) that have been compared (Thomson & Campbell, 1971) with the ampullary canals of elasmobranchs. I am able to offer no new observations on these structures because of the absence of available material from which to make thin-sections. However, they further indicate the diversity of the pore-canal system in lobe-finned fishes and may be related to the 'pore-group' receptors of osteolepids.

Dimensions of the pore-canal system

In making comparisons between various sensory organs in living and fossil vertebrates, the question of size is of central importance, because the function of a given system of sensory receptors will be reflected in their absolute size.

From Gross (1956) we know that in Osteostraci and Acanthodii the upper and lower mesh canals are essentially half-cylinders with a maximum diameter of approximately 50 μm. The pores range from 10–40 μm in diameter and the pore-canals are 50–100 μm long. The range of sizes in the lobe-finned fishes is greater. In the basic system of the Osteolepidae, the pore-cavities are from 150 to 250 μm high, with an external pore circa 10 μm in diameter. In some specimens Gross measured pore-cavities of only 70 μm with an external pore as low as 3 μm in diameter. The mesh canals in the Osteolepidae range from 18 to 25 μm in diameter. The pore openings are spaced from each other by a distance approximately equal to the height of the pore-cavity; see below (Fig. 13). In the case of the pore-groups of Osteolepidae, the pore-cavities are up to 300 μm in height, with an external opening of up to 125 μm in diameter. These large pores are spaced up to 400 μm from each other.

In *Porolepis*, the pore-cavities have a maximum height of 300 μm, with a 60–70 μm diameter pore. They are spaced approximately 250 μm from each other. The mesh canals are approximately 20–45 μm in diameter. In *Ganorhynchus*, the pore-cavities are the largest yet known, 650 μm high and up to 400 μm in diameter. They are spaced at intervals of approximately 500 μm.

Patterns of growth of the pore-canal system

Unfortunately, the limited fossil materials make it very difficult to determine whether the observed variation in the dimensions of the pore-canal system among given specimens reflects growth factors or whether each is constant for a given taxon regardless of size (age) of the individual. However, I believe that the question of the spacing of the pore-openings is of considerable importance as a clue to the function of the pore-canal system.

Evidence has previously been collected (see Thomson, 1975) that the pore-canal spacing in certain osteolepid Rhipidistia increases with size. The data can now be extended somewhat (Fig. 10), particularly by the addition of measurements from the very small osteolepid *Megalichthys pygmaeus*. The data seem to suggest a minimum spacing of the pore openings for osteolepids of about 100 μm. Interestingly, although we do not know the size of the individual fishes from which his materials came, in none of the more primitive fishes that Gross described (1956) was the spacing less than this.

It is also possible to confirm that the pore spacing varies systematically over the body in given taxa and that the pore spacing and the height of the pore-cavity are positively correlated. In the lobe-finned fishes one can measure the spacing between pores over the whole surface of the body in suitably preserved specimens. In Figs 11 and 12, such spacing is given for the Middle Devonian dipnoan *Dipterus valenciennesi*, and the Pennsylvanian osteolepid *Megalichthys macropomus*. In each case, there is a significant difference between the spacing on the head and on the trunk. A curious result to emerge from a review of size and spacing in the pore-canal system of fossil fishes is that there seems to be a general correlation

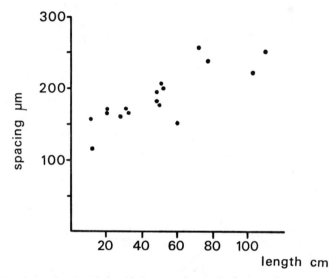

Figure 10. Graph showing the relationship between the spacing between the pore-openings and body size (length) in five species of osteolepid rhipidistian fishes (*Megalichthys pygmaeus*, *Osteolepis macrolepidotus*, *Ectosteorhachis nitidus*, *Megalichthys macropomus* and *Sterropterygion brandei*). Data from Thomson (1975) and original.

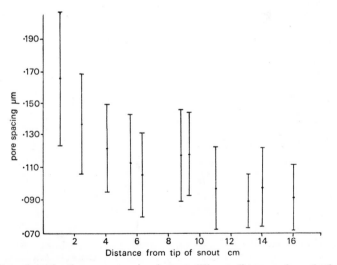

Figure 11. Plot of spacing between pore-openings at different distances along the body on the dorso-lateral part of the flank of *Dipterus valenciennesi* (Dipnoi MCZ 6635). At each position the mean spacing and standard error are shown.

between the pore spacing and the overall depth of the neuromast organ. In lobe-finned fishes the total depth is taken as the distance from the external opening to the base of the basal chamber of the pore-cavity. In those fishes in which the neuromast organs are contained in a double mesh-canal system, the overall depth measured is the distance from the surface to the base of the lower mesh-canal. A comparison of these measurements is given in Fig. 13. A similar relationship may be seen between the dimensions of the mesh-canal and the spacing of the pores. Again, the paucity of well-preserved specimens does not yet allow us to decide whether the differential spacing is a constant feature of the fish or a product of the growth process, but the limited data available suggest that the latter explanation is more probably correct.

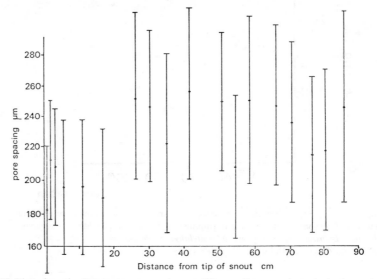

Figure 12. Plot of spacing between pore-openings at different distances along the body on the lateral part of the flank in *Megalichthys macropomus* (Rhipidistia USNM 1987). At each position the mean spacing and standard error are shown.

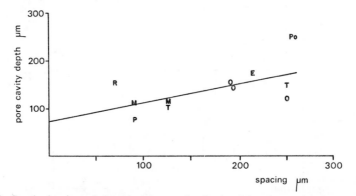

Figure 13. Graph showing relationship between the depth of the pore-cavity (see text) and the spacing between pore-openings at the surface, in the following fishes: E, *Ectosteorhachis nitidus*: M, *Megalichthys macropomus*: O, *Osteolepis* sp; P, *Poracanthodes punctatus*; PO, *Porolepis posnaniensis*: R, *Rhinodipterus secans* and T, *Tremataspis mammillata*. Data from Gross (1956) and original.

Comparison with receptors in living fishes

The most common type of neuromast receptor in the skin of living fishes is the mechanoreceptor of the lateral line organs and pit-line organs which is also present in most fishes in the form of free neuromasts in the skin of related structures such as the pit-organs of elasmobranchs. The receptor cells in such organs typically bear a large kinocilium and a number of stereocilia and, with a few notable exceptions (e.g. cochlear hair cells in the mammalian inner ear) such structures are considered typical of mechanoreceptor cells. There are two facts that lead us to conclude that the sensory function of the pore-canal system is not *typically* mechanoreceptive as in the lateral line system. First, the seismosensory lateral line and pit-line systems are well-developed in all those fishes in which the pore-canal system occurs. Secondly, there is a significant difference in absolute size between the sensory ampullae of the two systems. Therefore, the tempting possibility that the much smaller pore-canal system is but a vestigial relic of the evolutionary precursor of the more specialized lateral-line and pit-line system cannot be confirmed, and the functions of the two systems as we know them must be presumed to be different.

It is, however, nonetheless true that there is a topographic continuity of the mesh canals of the pore-canal system and the pit-line system in many fishes (see above). And both the lateral line and pore-canal systems have networks of linking canals. Furthermore, we have already noted the possibility that in Osteostraci and Acanthodii (but not the lobe-finned fishes) the sensory cells of the pore-canal system were hair cells, presumably bearing kinocilia. This is strong evidence for some sort of a mechanoreceptive function. If this is true, two additional questions remain: (a) was there some other function in addition to mechanoreception in Osteostraci and Acanthodii, and (b) is it possible that the difference in morphology of the pore-canal ampullae between the lobe-finned fishes on one hand and Osteostraci and Acanthodii on the other hand signifies a difference in function between the two types?

Much recent work has established the importance of electroreceptive neuromast organs in the skin of a wide variety of marine and fresh-water, electric and non-electric fishes (see reviews by Bennett, 1971; Fessard, 1974; and other literature cited here). These receptors are considered to be derived from typical mechano-receptive neuromasts of the seismosensory system (Lissman & Mullinger, 1968, *inter alia*) and of all the receptor organs found in the skin of fishes they bear the most obvious similarities to the pore-canal system of early fishes. At first sight, there seem to be three distinct types of electroreceptor organs in fishes: (a) the ampullary canals of Lorenzini, found in elasmobranchs, which have their receptor ampullae connected to the exterior via very long diverging jelly-filled canals, (b) the ampullary organs of teleostean fishes which consist of a flask-shaped cavity containing the receptor cells buried at the epidermal-dermal interface in the skin and opening to the surface via a short vertical canal and (c) tuberous organs, also found in teleosts, which are similar to ampullary organs but which normally lack an open canal leading to the exterior and which have the receptor cells arranged in a characteristic basal rosette. A fundamental functional difference between the tuberous organs and the two types of ampullary organs (elasmobranch and teleostean) is that the former are phasic receptors and the latter are tonic receptors.

However, the morphological distinctions between the different types are not perfectly clear. For example, the freshwater sting ray *Potamotrygon* lacks the typical ampullary canals of Lorenzini and instead bears a very large number of small ampullary receptors each with a short vertical channel (sometimes rather convoluted) to the exterior (Szabo, Kalmijn, Enger & Bullock, 1972). The marine catfish *Plotosus* (a teleostean) has long ampullary canals very similar to those of elasmobranchs (see Friedrich-Freska, 1930). The posterior tuberous organs of the weakly electric gymnotid *Hypopomus* differ from those of typical gymnotids by the presence of short horizontal canals (0·5–1·5 mm long) leading to the exterior (Szamier & Wachtel, 1970). In fact, a wide spectrum of different electroreceptors occurs in fishes, all apparently homologous with neuromast organs, each showing adaptations for a particular function, and each basically being derivable from an ampullary type of structure. A more detailed comparison of these receptors with the pore-canal system seems in order.

The structure of the ampullary canals of Lorenzini is sufficiently different from the pore-canal system, both in the arrangement of the organs and in their absolute size, that we need not spend much time on it. It is, however, worth noting that the receptor cells in at least some elasmobranch ampullae are 'hair cells' each bearing a cilium, as are the receptor cells of the catfish *Plotosus anguillaris* (Friedrich-Freska, 1930). These cilia are different from typical kinocilia and stereocilia are absent, but Waltman (1966) and Szabo (1972) postulate a mechanoreceptive function in addition to an electroreceptive function (see reviews in Murray, 1960, 1967, 1974).

Ampullary organs bearing a short vertical canal are found in many teleostean fishes (gymnotid, mormyrid and silurid), both electric and non-electric: Wachtel & Szamier (1969) show that, from their electrical properties, they are to be expected to occur more commonly in non-electric fishes and it seems to be the case that in fishes such as the glass catfish *Kryptopterus* they are extremely numerous and tuberous organs are absent, while in weakly electric fishes such as the mormyrids and gymnotids the tuberous organs are significantly more numerous than the ampullary organs. Details of the distribution of the two types of receptors may be found in Bennett, 1971; Szamier & Wachtel, 1969, 1970; Szamier & Bennett, 1974; and Szabo, 1965, 1974. For our purposes the following features of the distribution of receptors over the body are important. (1) Both ampullary and tuberous receptors are more densely distributed on the head than the trunk or fins. (2) In many fishes, the electroreceptors on the trunk are restricted to bands, zones or patches. (3) When ampullary organs alone are present, as in *Kryptopterus*, they are rather uniformly distributed. When both types of organs are present the ampullary organs are arranged in small groups. (4) In some cases, the tuberous organs are of larger size, in others the ampullary organs are larger, this presumably reflects functional differences. (5) The length of the canal leading to the ampullary organ is a function of the conductivity of the external medium. In saltwater fishes the ampullary canals are long and in freshwater fishes the ampullary canals are short. This applies to both the teleostean ampullary organs and the organs of Lorenzini (see above).

A functional explanation of this last point, based on the differences in resistivity between the tissues and the environmental medium in fresh and salt water, is given by Bennett, 1971. (The long canals of the posterior tuberous organs on *Hypopomus*, noted above, are, however, not explained by such a relationship, as this fish lives in fresh-water.) Unfortunately it has not been possible to make definite correlations between the 'depth' of the pore-canal system in fossil fishes and their environment. If such a correlation could be established, it would be good evidence for an electroreceptive function.

The overall height of the ampullary organs in all teleosts ranges between 100 and 300 µm with external pores up to 30 µm in diameter. Tuberous organs are smaller in gymnotids, but in mormyrids they are of two sizes, the larger ones being about 150 µm deep. The spacing of the ampullary receptors range from 440 µm on the head of *Electrophorus* to 2250 µm on the trunk of *Stenarchus*. The spacing of tuberous receptors ranges from *c.* 100 µm on the head of *Hypopomus* to more than 1000 µm on the trunk of *Hypopomus* and *Stenarchus*. The spacing of the 'microampullary receptors' of the freshwater sting ray *Potamotrygon* (see above) in patches may reach 500 µm ventrally near the snout but to the rear and on the back there may be as few as 5 ampullae per cm² and the patches are rare.

There is a very striking similarity in structure and topographic disposition between the pore-canal system of early fishes and the electroreceptors of fishes, particularly the bony fishes. The ampullary and tuberous receptors of bony fishes are of a similar size to the ampullae of the pore-canal system and they have a similar pattern of spacing. The ampullary organs of bony fishes and the pore-cavities of the pore-canal system in lobe-finned fishes are similar in the presence of a flask-shaped cavity with a basal neuromast receptor, a presumed jelly-filled canal leading to the surface and an external pore of similar size. The medium-sized tuberous organs of mormyrid fishes which have a double cavity also show similarity to the pore-cavity of lobe-finned fishes. A further point of similarity is that modern electroreceptors, especially in freshwater fishes, are enclosed in a material of high electrical resistivity. The resistivity of dentine is of the same order of magnitude as the measured high resistivity of the skin of mormyrids (cf. Bennett, 1971).

Only in the ampullary canals of Lorenzini (elasmobranchs and *Plotosus*) is there any form of branching canal system in living fishes but significantly, although each sensory ampulla may receive canals from a large number of separate external pore openings, there is no direct communication between adjacent ampullae. This must indicate a basic difference in function between the pore-canal system and the receptors of modern fishes, for in the former there is an external 'integration' of the incoming signal *before* it reaches the receptor neuromasts. The only other place where this is found in any fishes is in the lateral line itself.

Discussion

It is not possible to arrive at a definite answer to the question of the function of the pore-canal system. Any hypothesis must take into account the structure of the pore-canal system in early fishes, the evolutionary and presumably therefore adaptive changes in the pore-canal system among different groups of fishes, and its variation within given groups, and must be made in the light of a comparison with morphological data from living fishes. Given these requirements, the following hypothesis is developed concerning the evolution and function of the pore-canal system in early fishes.

The fact that the pore-canal system in such widely divergent groups of fishes as the Osteostraci and Acanthodii, stemming from near the bases of the two major divisions of vertebrates—agnathan and gnathostome—is essentially identical in structure, with the complicated upper and lower mesh canals separated by a common perforated wall, suggests that such a pore-canal system was a very early and basic characteristic of the vertebrates. It also suggests that the basic separation between the lateral line and pore-canal systems occurred very early. Whether the pit-line system derives from the pore-canal system or the lateral line is not clear. The contiguity of the three bespeaks their evolutionary unity (see Denison, 1966).

I suggest that the pore-canal systems in Osteostraci and the Acanthodii represent a parallel series of experiments of early fishes in the evolution of electroreceptive organs. They were derived by modification of a system of mechanoreceptors linked by mesh-canals possibly close to that from which the lateral line system itself evolved. The suggested presence of 'hair cells' within the pore-canal system of Osteostraci and Acanthodii and the strict orientation of the receptor cells in slits at the pore-cavity base in some lobe-finned fishes (e.g. Fig. 9B) seem to indicate mechanoreception in addition to electroreception. The absence of such orientation in later forms probably indicates electroreception alone.

The fundamental difference between the pore-canal system in lobe-finned fishes and Osteostraci or Acanthodii is that in the former, (1) the sensory neuromasts have become concentrated into localized ampullae, (2) a separate pore opening leads directly to each ampulla and (3) the ampullae are electrically insulated one from another except for the linking effect of the remains of the mesh canal network. This concentration of the receptor cells into ampullae is very strongly reminiscent of the arrangement of ampullary electroreceptors in modern fishes.

In all probability, the first electroreceptors in fishes were tonic receptors but in the evolution of the Dipnoi and Rhipidistia other types of receptors evolved: the rostral organs of lungfish such as *Dipnorhynchus*, the large clustered organs of *Ganorhynchus*, and the 'pore-group' organs of osteolepids. It seems likely that these larger organs were phasic receptors analogous to the tuberous receptors of mormyrids and gymotids and that their external pore-canal was plugged with epithelial tissues, as in these latter fishes. Thus, the diversification of receptor organs modified from the basic pore-canal system in lobe-finned fishes may be interpreted as an adaptive radiation of different types of phasic electroreceptors from the basic tonic ampullary pattern.

In yet later stages in the evolution of the lobe-finned fishes, and in subsequent groups of bony fishes, the electroreceptor organs, when present, were located in the soft tissues external to the dermal skeleton and no trace of them remains in any fossil forms. The question then remains, do the living Dipnoi, which lack cosmine, have electroreceptors? None have been found, probably because, to my knowledge, no-one has ever looked for them. However, the two genera *Lepidosiren* and *Protopterus* are prime candidates to possess such receptors. They dwell mostly in muddy waters with many other electric fishes, and they have reduced eyes and well-developed sensory feelers in the form of the paired fins. Perhaps the organs of Fahrenholz, with which Gross (1956) found a tentative similarity to the large pore-organs of Devonian Dipnoi, represent some kind of electroreceptor. It is an interesting subject for experimental investigation.

According to the relative development of the cosmine layer on the surface of the exoskeleton, the pore-canal system in early fishes may be enclosed in hard or soft tissues. The result is that potentially the surrounding skin can have very different electrical resistivity. Epidermal cells can be arranged to have high or low resistivity while that of dentine is probably rather more fixed and certainly cannot readily be changed once it is laid down. This is perhaps additional evidence that the patterns of cosmine cover seen in the preceding section may have a strong environmental component relating to the sensory function of the pore-canal system. The cosmine is important also as a mechanical protection of the pore-canal system, but it is probably a mistake to assume that fishes lacking a cosmine cover (either temporary or permanently) lacked the pore-canal system or some derivative of it. Therefore it is possible that electroreceptive organs are a general characteristic of fishes rather than a specialization of the more modern forms.

Finally, there is a major topographic difference between the pore-canal system of fossil fishes and the electroreceptors of modern fishes. In the latter there is no direct peripheral communication between the sensory ampullae. If the pore-canal system was an electroreceptor system the major difference from modern electroreceptor systems is that in fossil fishes the incoming signal was in some way integrated peripherally and the individual sensory ampullae potentially received a signal that contained information from more or less distant parts of the body. In modern fishes this integration is performed centrally within the nervous system (although a possibility for peripheral nervous integration exists; see Tester & Nelson, 1967). The central integration is probably the more efficient method, particularly if the range of modality and precision of the receptor systems as a whole are increased. The reduction and eventual loss of the mesh canal system thus shows the evolution of a more precise function of the system, leading to the pattern seen in living systems.

SUMMARY OF CONCLUSIONS

(1) In the life history of individual rhipidistian fishes, the covering of cosmine on the dermal skeleton may not be maintained *in toto*. Evidence from the Carboniferous osteolepid *Megalichthys*, together with previously published data, suggest that a partially naked dermal skeleton exists for considerable periods of the life history. This phenomenon probably has biological significance in terms of the history of the environments in which the fishes lived.

(2) When the full cosmine cover has broken down, the process of resorption and redeposition of the remaining cosmine may proceed out of phase. The surface of the dermal skeleton may show new specialized structures such as tubercles and regions of overgrowth of the spongiosa.

(3) Cosmine is a unique set of tissues with many functions. The hard tissues serve to give general protection and thus are always maintained in the ventral regions of the body. The hard tissues may also serve to protect the pore-canal system. However, the pore-canal system is capable of functioning while embedded merely in the soft tissue of the skin. Cosmine also acts as a store for minerals, perhaps a place of deposition of excess calcium and as a reserve for phosphate.

(4) Resorption of cosmine occurs in such a way as to indicate that the process is mediated by factors involving the centres of growth of the underlying dermal skeleton. This adds to the growing evidence that the developmental biology of cosmine, although temporally distinct from that of the dermal skeleton, may none-theless be intimately connected with it. This suggests interesting possibilities in terms of the relationship between hard tissue formation and the latero-sensory system, in general.

(5) The results of a comparison of general morphology and size relationships suggests that the pore-canal sensory system did not include typical mechano-receptive organs but was (perhaps in addition) a system of electroreceptor organs in early fishes. They were probably all tonic receptors at first, but phasic receptors may have evolved in the lobe-finned fishes by the Devonian.

(6) If this comparison is correct, electroreception is an ancient feature of all fishes.

(7) There is a wide variety of superficial receptors in lobe-finned fishes that may represent different sorts of electroreceptors.

(8) The principal difference between the electroreceptor function of the pore-canal system (if correctly identified) and that of living fishes is that in the former there is a great deal of integration of the incoming signal peripherally to the ampullary receptor sites. In the living fishes there is no system of interconnecting mesh canals and integration of the signal is more wholly nervous and central.

ACKNOWLEDGEMENTS

I am grateful to Mark Angevine for technical assistance in this study. Specimens were generously loaned by the United States National Museum, Museum of Comparative Zoology, Harvard University, and the British Museum (Natural History), for these I am grateful to Dr N. Hotton III, the late Professor A. S. Romer, and Drs R. S. Miles and P. Forey. I am also mindful of a debt to the late Professor W. Gross for valuable correspondence on the subject of cosmine and the pore-canal system. The illustrations were prepared by Linda Price Thomson and my studies have been supported by the National Science Foundation (grant GB 28823).

REFERENCES

BENNETT, M. V. L., 1971. Electroreception. In W. S. Hoar & D. J. Randall, *Fish physiology, 6*. London & New York: Academic Press.

BOLAÜ, E., 1951. Das Sinnesliniensystem der Tremataspiden und Beziehungen zu anderen Gefässsystemen des Exoskeletts. *Acta zool., stockh., 32*: 31–40.

BYSTROV, A. P., 1942. Deckknochen und Zähne der *Osteolepis* und *Dipterus. Acta zool., stockh., 23*: 63–89.

BYSTROV, A. P., 1959. The microstructure of the skeletal elements in some vertebrates of the URSS. *Acta zool., Stockh., 40:* 59–83.

DENISON, R. H., 1947. The exoskeleton of *Tremataspis. Am. J. Sci., 245:* 337–65.

DENISON, R. H., 1966. The origin of the lateral-line system. *Am. Zool., 6:* 369–70.

FESSARD, A. (Ed.), 1974. *Electroreceptors and other specialized receptors in lower vertebrates. Handbook of sensory physiology, 3*(3). Berlin, Heidlberg, New York: Springer-Verlag.

FRIEDRICH-FRESKA, H., 1930. Lorenzinische Ampullen bei dem Siluriden *Plotosus anguillaris* Bloch. *Zool. Anz., 87:* 49–66.

GROSS, W., 1935. Histologische Studien am Aussenskelett fossiler Agnathen und Fische. *Palaeontographica, 83* (Abt. A), 1–60.

GROSS, W., 1956. Über Crossopterygier und Dipnoer aus dem baltischen Oberdevon im Zusammenhang einer vergleichenden Untersuchung des Porenkanalsystems paläozoischer Agnathen und Fische. *K. svenska VetenskAkad. Handl.*, *Ser. 4*, *25:* 1–140.

JARVIK, E., 1948. On the morphology and taxonomy of the Middle Devonian osteolepid fishes of Scotland. *K. svenska VetenskAkad. Handl.*, *Ser. 3*, *25:* 1–301.

JARVIK, E., 1950. Middle Devonian vertebrates from Canning Land and Wegeners Halvø (East Greenland). Part 2, Crossopterygii. *Meddr Grønland*, *96:* 1–132.

LISSMAN, H. W. and MULLINGER, A. M., 1968. Organization of ampullary electric receptors in Gymnotidae (Pisces). *Proc. R. Soc.* (*B*), *169:* 345–78.

MURRAY, R. W., 1960. The response of the ampullae of Lorenzini to mechanical stimulation. *J. exp. Biol.*, *37:* 417–24.

MURRAY, R. W., 1967. The function of the ampullae of Lorenzini of elasmobranchs. In P. H. Cahn, *Lateral line detectors*. Bloomington: Indiana University Press.

MURRAY, R. W., 1974. The ampullae of Lorenzini. In A. Fessard, *Electroreceptors and other specialized receptors in lower vertebrates. Handbook of sensory physiology*, *3*(3). Berlin, Heidlberg, New York: Springer-Verlag.

ØRVIG, T., 1969. Cosmine and cosmine growth. *Lethaia*, *2:* 241–60.

STENSIÖ, E. A., 1927. The Downtonian and Devonian Vertebrates of Spitzbergen. Part 1, Family Cephalaspidae. *Skr. Svalbard Ishavet*, *12:* 1–391.

SZABO, T., 1965. Sense organs of the lateral line system in some electric fish of the Gymnotidae, Mormyridae and Gymnarchidae. *J. Morph.*, *117:* 229–50.

SZABO, T., 1972. Ultrastructural evidence for a mechanoreceptor function of the ampullae of Lorenzini. *J. Microsc.*, *14:* 343–50.

SZABO, T., 1974. Anatomy of the specialized lateral line organs of electroreception. In A. Fessard, *Electroreceptors and other specialized receptors in lower vertebrates. Handbook of Sensory Physiology*, *3*(3). Berlin, Heidlberg, New York: Springer-Verlag.

SZABO, T., KALMIJN, A. J., ENGER, P. S. & BULLOCK, T. H., 1972. Microampullary organs and a submandibular sense organ in the freshwater ray, *Potamotrygon. J. comp. Physiol.*, *79:* 15–27.

SZAMIER, R. B. & BENNETT, M. V. L., 1974. Special cutaneous receptor organs of fish: VII. Ampullary organs of mormyrids. *J. Morph.*, *143:* 365–84.

SZAMIER, R. B. & WACHTEL, A. W., 1969. Special cutaneous receptor organs of fish: VIII. The ampullary organs of *Eigenmannia. J. Morph.*, *119:* 261–90.

SZAMIER, R. B. & WACHTEL, A. W., 1970. Special cutaneous receptor organs of fish: VI. Ampullary and tuberous organs of *Hypopomus. J. Ultrastruct. Res.*, *30:* 450–71.

TESTER, A. L. & NELSON, G. J., 1967. Free neuromasts (pit organs) in sharks. In P. W. Gilbert, R. F. Mattewson & D. P. Ball (Eds), *Sharks, skates and rays:* 503–32. Baltimore: The Johns Hopkins Press.

THOMSON, K. S., 1969. The environment and distribution of palaeozoic sarcopterygian fishes. *Am. J. Sci.*, *267:* 457–64.

THOMSON, K. S., 1975. The biology of cosmine. *Bull. Peabody Mus. nat. Hist.*, *40:* 1–59.

THOMSON, K. S., 1976. The faunal relationships of the rhipidistian fishes (Crossopterygii) of the Catskill (Upper Devonian) of Pennsylvania. *J. Paleont.* In press.

THOMSON, K. S. & CAMPBELL, K. W., 1971. The structure and relationships of the primitive Devonian lungfish—*Dipnorhynchus sussmilchi* (Etheridge). *Bull. Peabody Mus. nat. Hist.*, *38:* 1–109.

WACHTEL, A. W. & SZAMIER, R. B., 1969. Special cutaneous receptor organs of fish: IV. Ampullary organs of the non-electric catfish, *Kryptopterus. J. Morph.*, *128:* 291–308.

WALTMAN, B., 1966. Electrical properties and fine structure of the ampullary canals of Lorenzini. *Acta physiol. scand.*, *66*, Suppl. 264: 1–60.

WESTOLL, T. S., 1936. On the structure of the dermal ethmoid shield of *Osteolepis. Geol. Mag.*, *73:* 157–71.

WILLIAMSON, W. C., 1849. On the microscopic structure of scales and teeth of some ganoid and placoid fish. *Phil. Trans. R. Soc.*, *Ser. B.*, *139:* 435–75.

EXPLANATION OF PLATE

PLATE 1

A. *Megalichthys macropomus* (Holotype, USNM 1987). Right intertemporal showing cosmine-cover reduced to small patches with finished margins. ×3.6.
B. *Megalichthys macropomus* (Holotype, USNM 1987). Left squamosal showing cosmine cover reduced to small patches arranged more or less concentrically. ×3.6.
C. *Megalichthys macropomus* (MCZ 5143). Portion of squamation showing scales with reduced cosmine cover. ×2.0.

Plate 1

(Facing page 270)

The axial skeleton of the coelacanth, *Latimeria*

S. MAHALA ANDREWS

Royal Scottish Museum, Edinburgh

A frozen specimen of the living coelacanth, *Latimeria chalumnae* Smith, dissected to show the axial skeleton, shows several interesting and previously undescribed peculiarities of structure. These are described and where possible their functional significance is discussed. The dorsal cartilages abutting against the notochord differ from the previously published description by Millot & Anthony (1956, 1958a, b) in some important respects, and invite comparison with other crossopterygians, tetrapods, lungfishes and primitive actinopterygians. The comparisons pinpoint one of the problems of homology of central elements between these groups, *viz.* whether the crossopterygian pleurocentrum is derived ontogenetically from neural arch material or perichordal tube. A speculative solution is tentatively offered.

CONTENTS

INTRODUCTION

A frozen specimen of the living coelacanth, *Latimeria chalumnae* Smith, was acquired by the Royal Scottish Museum, Edinburgh in 1969 (Plate 1A). After casting and distribution of frozen tissues to various recipients, it was dissected to show the skeleton (Plates 2 and 3; Figs 1 to 4). It was then fixed in 5% formalin for seven days and stored in 1% propylene phenoxetol; the present dark colour of the notochord and cartilage seems to be due to ageing in this solution. Millot & Anthony (1956, 1958a, b) gave a good general description of the axial skeleton of *Latimeria* from material fixed in formalin before dissection. The Royal Scottish Museum specimen agrees with their published description in most features, but it also shows several interesting and previously undescribed peculiarities of structure. These will be described here, followed by comments where possible on their functional significance.

The dorsal cartilages abutting against the notochord differ from Millot & Anthony's (*op. cit.*) description in some important respects, and invite comparison with other crossopterygians, tetrapods, lungfishes and primitive actinopterygians. The comparisons pinpoint one of the problems of homology of central elements between these groups, *viz.* whether the crossopterygian pleurocentrum is derived ontogenetically from neural arch material or perichordal tube. Most of the tetrapod terminology for vertebral elements employed by Andrews & Westoll (1970a) for crossopterygians has been used here, for reasons given in the discussion. However, the term 'arcual' and other reminiscences of Gadovian theory have been avoided as far as possible, following Williams' (1959) review of tetrapod vertebrae in which he rejected it. This work, together with Panchen's (1967) discussion of it and Schaeffer's (1967) attempt to summarize and interpret ontogenetic and phylogenetic patterns in osteichthyan vertebrae, form the background to the problem of homology discussed here.

DESCRIPTION

The axial skeleton of *Latimeria* consists of a relatively enormous fibrous tube, the notochord, with paired rows of small cartilaginous elements forming almost continuous cartilaginous tracts dorsally and ventrally, and supporting the bony neural and haemal arches and spines where present. Most of the notochordal fibres run circumferentially. Notochord, cartilages and spines are all invested in a very thick sheet of fibrous connective tissue (a small remnant is seen in Plate 2B), the inner layers of which adhere tightly around and between the arches and spines. This investment is continuous with the vertical and horizontal septa and the myosepta, and is composed of layers of differently-directed fibres.

The notochord

The structure and size of the notochord were described by Millot & Anthony (1956, 1958a, b). In the trunk region, the notochord (*n*, Fig. 1) seems to be expanded to a diameter very much greater than 'normal' for a notochordal fish, but at the level of the second dorsal and anal fins (Plate 1) it tapers rather suddenly to 'expected' proportions in the tail. This will be discussed later, under 'Aspects of function'.

A representative vertebra
(nos 52–56, Figs 1 and 3)

The vertebral elements will be described mainly from the left side, as shown in the plates and most figures, because this side has few irregularities. A representative vertebra, from the posterior trunk region, has two pairs of dorsal (*contra* Millot & Anthony, 1956, 1958a, b) and one pair of ventral cartilages abutting against the notochord. The rather rounded anterior pair of dorsal cartilages (neural arch bases, *n.a.b*, Fig. 1) supports the ossified neural arches (*n.a*) which are surmounted by a long, slender neural spine (*n.sp*). Where the neural spine joins the neural arches the bone is often expanded to form thin anterior and posterior flanges at the level of the dorsal longitudinal ligament (*d.lig*). The paired posterior dorsal cartilages (pleurocentra, *p.c*) are also rounded, and in this region they are smaller than the neural arch bases. They cover the notochord between successive neural arch bases and lie alongside the neural canal (*n.can*). Both pairs of dorsal cartilages are domed outwards to a thickness of 2–3 mm. The paired ventral cartilages (haemal arch bases, *h.a.b*), serially homologous with the more anterior intercentra (*i.c*, Fig. 2), are oval-oblong and support the ossified haemal arches (*h.a*) which often have triangular expansions at their tips. More posteriorly the latter fuse into a median haemal spine (*h.sp*, Fig. 3).

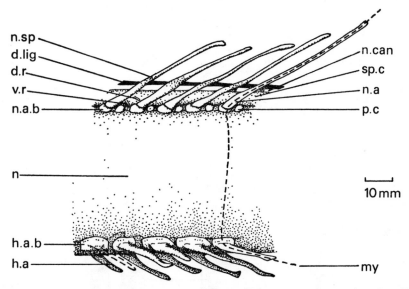

Figure 1. *Latimeria chalumnae* Smith, specimen 1969.76. Five representative vertebrae (nos 52–56) from the posterior trunk region, seen from the left side. × 0·75.

Regional variation

Regional variation is considerable. The most anterior part of the column (vertebrae nos 1–20) and the supplementary caudal lobe (vertebra no. 93 onwards) will be described later.

The neural spines maintain a more or less constant height throughout the trunk. They do not articulate with the first and second dorsal fin supports, but a single neural spine (vertebra no. 47) is shortened to accommodate the second dorsal fin support. The neural arches and neural arch bases are similar in shape and size up to the end of the trunk (Figs 2 and 3). In the main caudal region (Fig. 4, vertebrae nos 67–92) the tips of the neural spines are modified to articulate with epineural spines (*e.n.sp*) which indirectly support the bifurcating radials of the tail fin, as described by Millot & Anthony: the neural spines shorten and become more and more gently inclined, and the neural arch bases increase in size and become more oblong posteriorly. The dorsal ligament is well developed until the eighth epichordal fin-supporting vertebra (no. 75), after which its place is taken by

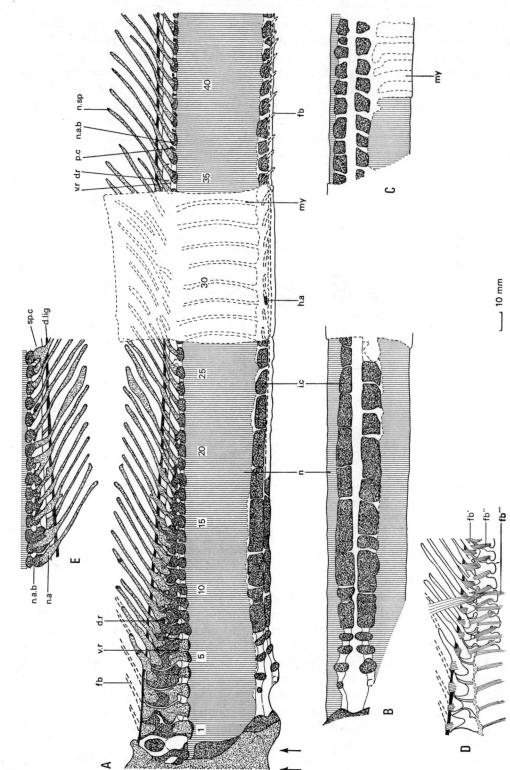

_10 mm

small ovoid chondrifications (dorsal caudal cartilages, *d.cc*, Fig. 4) separating the next ten neural arches. Behind the last of these, the neural arches are sub-parallel to the notochord and the dorsal ligament cannot be distinguished from the general fibrous investment. Pleurocentra are found dorsally throughout at least the posterior three quarters of the column (vertebra no. 36 onwards, *cf*. Millot & Anthony, 1958b: fig. 1, where they are shown only in the supplementary caudal lobe). They are at first very small (Fig. 2A) but they increase in size posteriorly, equalling the neural arch bases in size and becoming more oblong in the main caudal region, and continuing right into the supplementary caudal lobe.

Ventral to the notochord the haemal arch bases tend to increase in length anteriorly, but it is difficult to say where they cease to carry haemal arches. Ossification in the latter ceases as shown in Fig. 3, but gives way to structures of hard fibrous cartilage (*fb*) which only gradually become indistinguishable from the fibrous investment; indeed, a small haemal arch apparently of cartilage is found among them in the middle of the trunk (vertebra no. 29, Fig. 2A). *Latimeria* resembles *Glyptolepis* (Fig. 5C) in having haemal arches in the trunk region (not epi-haemal spines, Andrews & Westoll, 1970b: fig. 23). In the anterior half of the trunk of *Latimeria* the one-to-one relationship between the neural spines and the paired ventral elements (intercentra, Fig. 2) is lost, and long oblong cartilages extend uninterrupted through two, three or more segments. The breaks between successive cartilages are not symmetrical on the two sides of the body, often even where there is one pair of elements per vertebra (Fig. 2B, C, e.g. vertebra no. 37). Posteriorly, the haemal arches elongate considerably before finally fusing to a long, slender median haemal spine (*h.sp*, Fig. 3) on vertebra no. 66, a short distance before the main caudal region. Here, their tips are modified to articulate with epihaemal spines (*e.h.sp*) supporting the hypochordal lobe of the tail fin, while the haemal arch bases become smaller and more cylindrical posteriorly. There is a series of caudal cartilages (*v.cc*, Fig. 4) separating the haemal arches of vertebrae nos 73–86, very similar to those separating the neural arches but extending slightly further anteriorly and posteriorly, and being more often subdivided into two or three ovoid elements between successive arches. Lying against the notochord between the haemal arch bases in the main caudal region is another series of tiny oval cartilages (haemal supports, *h.s*, Fig. 4), often two or three per segment (see also below in connection with the supplementary caudal lobe).

The most anterior region

The most anterior region of the vertebral column is very specialized. Here the distinction between neural arches and neural arch bases remains clear, but from vertebra *c*. no. 20 forwards the neural arch bases increase in height, appearing to push the ossified neural arches further and further dorsally, together with the dorsal ligament. The cartilaginous arch bases come to surround the ventral roots of the spinal nerves (*v.r*, Fig. 2A), developing anterior and posterior processes

Figure 2. *Latimeria chalumnae* Smith, specimen 1969.76. The anterior part of the vertebral column, ×0·5. A. Seen from the left side. The two arrows indicate the region where the notochord is fused to the otico-occipital. B and C. Ventral view. D. Dorsal part of the most anterior region seen from the left side, with part of the connective tissue investment in place. E. Dorsal vertebral elements seen from the right side showing fusions. For key to shading, see Fig. 3.

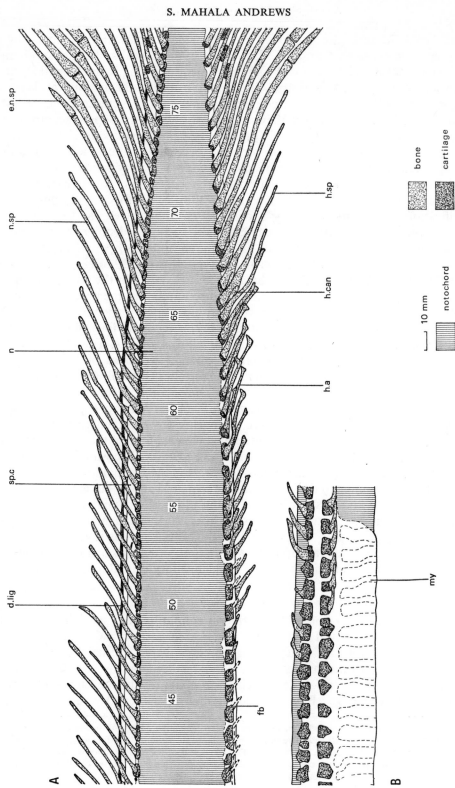

rather like zygapophyses above them. Further anteriorly still they also come to surround the dorsal roots (*d.r*) and the surfaces between successive neural arch bases become very complex. The first few neural arch bases only, completely surround the dorsal roots in front of them and have a certain amount of ossification ventrally. Neural spines are lacking on the first three vertebrae, although neural spine-shaped bundles of hard fibres (*fb*) take their place; the dorsal ligament runs along the tops of their truncated neural arches to the back of the braincase. On vertebrae nos 6, 9 and 11 (7, 9 and 11 on the right side) the neural arches bear prominent transverse processes, with which no ossified or cartilaginous skeletal elements are associated. Instead, they are buried in the thick layers of investing fibrous connective tissue, and a strong oblique bundle of hard fibres runs backwards and downwards from each of them (and equivalent spots on neural arches without processes between vertebrae nos 1 and 13) to the ventral part of the succeeding neural arch base (*fb"*, Fig. 2D). Further dorsally a parallel oblique bundle of fibres (*fb'*) runs from each posterior bony flange at the level of the dorsal ligament to the ventral end of the neural arch behind. Where the neural spines are lacking anteriorly, these bundles pass uninterrupted to the other side of the animal. These two sets of bundles gradually become indistinguishable posteriorly from the undifferentiated fibres of the innermost layer of connective tissue investment, which run obliquely between the more posterior neural arches and neural arch bases. Superficial to these two sets of bundles in the anterior region, another series of vertical bundles (*fb'''*) stands out from the surface of the connective tissue investment as buttress-like ridges between the notochord and the neural arches.

Ventral to the notochord in this anterior specialized region, it becomes more difficult to see the boundaries of the elongated cartilaginous elements passing forwards, because they appear to be separated by fibrous cartilage instead of fibres of the same kind as the connective tissue septa. However, right at the beginning of the column (vertebrae nos 3–6) there are four pairs of button-like, round or oval, beautifully domed cartilages, separated quite widely from each other in the longitudinal direction by narrow tracts of fibrous cartilage, and in the transverse direction by a dilation of the haemal canal.

The supplementary caudal lobe

The axial skeleton of the supplementary caudal lobe is also very specialized. There is a sudden marked change in the structure of the vertebral elements following vertebra *c*. no. 92 (Fig. 4). There are four pairs of pleurocentra in the next segment and about three in the succeeding one, and they suddenly increase dramatically in height. The elements of each pair come to fuse over the top of the neural canal (*n.can*) and form a laterally compressed, very short, broad, rhomboidal 'neural spine', all entirely cartilaginous. There is probably considerable variation in the number of elements in this transitional region, as Millot & Anthony (1958b: fig. 1) show one long segment containing ten or more pairs of pleurocentra, although their figure seems diagrammatically regular (cf. *ibid:* pl. LXXV). The

Figure 3. *Latimeria chalumnae* Smith, specimen 1969.76. Vertebral column of the posterior trunk region, ×0·5. A. Seen from the left side. B. Ventral view.

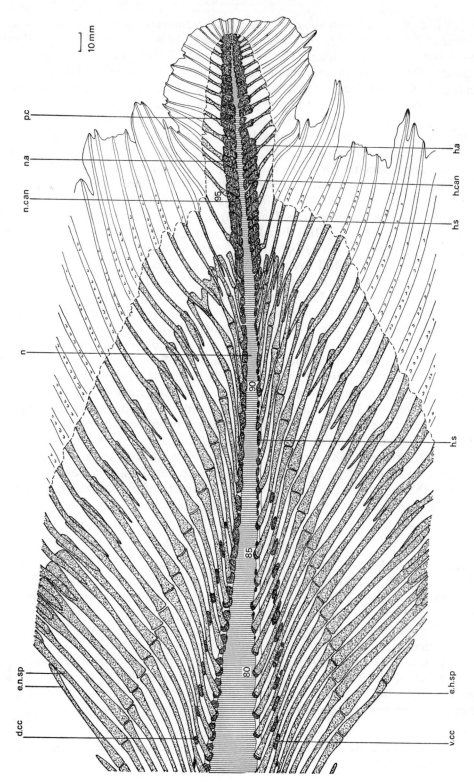

number of pleurocentra per segment within the supplementary lobe is also rather irregular, varying from none to three pairs. Between them, the neural arch bases plus neural arches are also entirely cartilaginous, and are only distinguishable from them in that they lie in the plane of a myoseptum (see Millot & Anthony, 1958b: pl. LI) and carry mostly ossified, rod-like neural spines supporting the fin-web directly. There are occasional spaces between elements at the level of the neural canal, but no spinal nerves could be found issuing through them.

Ventral to the notochord, the cartilaginous haemal supports (which appear in the main caudal region, see above) behave in a similar manner to the pleurocentra: at the base of the supplementary lobe they suddenly increase in height, come to roof over the haemal canal and increase in the number of pairs per segment. There are up to five pairs between each pair of cartilaginous haemal arch bases plus haemal arches, the latter being distinguished by their mostly ossified, rod-like haemal spines each lying in the plane of a myoseptum and supporting the fin web.

The multiplication of elements in the supplementary caudal lobe of *Latimeria* is reminiscent of diplospondyly (e.g. in the tail of *Amia*), but as the term polyspondyly is inappropriate in the absence of centra, 'polyapsidy' is suggested as a name for the condition, indicating the arch-like form of the pleurocentra and haemal supports. *Coccosteus* (Miles & Westoll, 1968: 449) may have had two sets of neural and haemal arches per segment in the tail, but as they all bear long neural spines it is more likely that each is formed at the junction of a myoseptum with the vertical septum, so that diplapsidy would not be involved.

Individual variations

Irregularities of vertebral structure are relatively common among fish (Rabinerson, 1925; Andrews & Westoll, 1970a: 282–3) and the present specimen is not unusual in showing several fusions of elements and other irregularities. Adjacent neural spines have occasionally fused together (e.g. vertebrae nos 13–14, 18–19, Fig. 2E) and this has resulted in their neural arches being irregularly distributed and right and left sides being out of step for the next part of the column, until rectified by the formation of separate half neural arches without neural spines (e.g. vertebrae nos 20, Fig. 2A and 63, Fig. 3). The forking of some of the more posterior epineural spines in the main caudal region (Fig. 4) is probably due to similar irregularities (cf. also Millot & Anthony, 1958b: pl. LXXX). On the other hand, the frequent fusion of ventral elements in the anterior trunk is apparently the normal pattern for *Latimeria*, although this region is clearly subject to a great deal of individual variation in detail, as shown by the asymmetry between right and left sides. In a similar way, pleurocentra and haemal supports seem to be normally numerous in the supplementary caudal lobe, but the exact numbers are very variable, so much so that neural and haemal spines may not fall opposite each other (vertebrae nos 94–96).

Figure 4. *Latimeria chalumnae* Smith, specimen 1969.76. Vertebral column of the tail, ×0·5. For key to shading, see Fig. 3.

DISCUSSION
Aspects of function
The notochord

Because the notochord seems so greatly expanded in the trunk region, its proportions relative to the size of the body and the length of individual vertebrae have been compared with a variety of other living and fossil notochordal fishes (Table 1).

It appears that the diameter of the notochord (N) in the trunk region of *Latimeria* is 'normal' with respect to the depth of the body (D) (i.e. its ratio D/N falls within the range of other fishes measured*). In relation to the length of the body (L) the notochord diameter in the trunk region is very large (making L/N very low), although approached by *Holophagus* and equalled by *Fleurantia**. However, it is in relation to the segment length (S), reflecting the size of the vertebral elements, that it appears so disproportionate (N/S=*c*. 5 compared to a 'normal' averaging *c*. 1·5; compare Fig. 5A with B). It is tempting to correlate this with the large number of segments in *Latimeria* (*c*. 110, nearly two thirds of which belong to the trunk): this is about twice as many as for example in *Eusthenopteron foordi* whose

Table 1. Proportions of the notochord in the trunk region of various fishes, relative to body depth, body length (snout to tail length has been taken instead of the preferable, but not always obtainable, length of the notochord) and segment length. The proportions of fossil forms (*) may be unreliable since it is possible that all the dorsal or ventral elements may be displaced together during fossilization. *Fleurantia* is particularly suspect in this regard. The letters refer to the following sources: (A) Andrews & Westoll, 1970b; (B) Goodrich, 1930; (C) Moy-Thomas & Miles, 1971; (D) Millot & Anthony, 1958b; (E) Romer, 1966; (F) Schaeffer, 1948

		N notochord diameter (mm)	S trunk segment length (mm)	L body length (mm)	D body depth (mm)	N/S	L/N	D/N
*Xenacanthus**	(C)	9·1	7·6	762	109	1·19	83·7	11·9
Acipenser	(B & original)	9·6	8·3	696	100	1·16	72·5	10·4
*Chondrosteus**	(E)	14·1	9·8	915	123	1·44	64·9	8·7
*Hypsocormus**	(E)	16·0	5·3	800	190	3·02	50·0	11·8
*Fleurantia**	(C)	8·7	2·9	250	45	3·00	28·7	5·2
Neoceratodus	(B & original)	12·0	11·0	1000	180	1·09	83·3	15·0
Protopterus	(B)	12·0	6·5	620	100	1·85	51·6	8·3
*Osteolepis**	(A)	2·5	1·5	100	13	1·66	40·0	5·1
*Megalichthys**	(A)	60·0	20·0	—	—	3·00	—	—
*Eusthenopteron**	(C)	21·0	16·3	1210	160	1·29	57·6	7·6
*Glyptolepis**	(A)	16·4	8·2	755	127	2·00	46·0	7·7
*Coelacanthus**	(C)	10·0	4·6	570	80	2·17	57·0	8·0
*Diplurus**	(F)	12·0	6·2	676	108	1·94	56·3	9·0
*Holophagus**	(E)	15·7	12·1	558	149	1·29	35·5	9·5
Latimeria	(D)	40·0	7·2	1180	272	5·55	29·5	6·8
Latimeria	(original)	48·0	10·0	1382	371	4·80	28·8	7·7

*See legend to Table 1.

total is *c.* 55. However, such a correlation could not be investigated because it was not possible to obtain reliable vertebral counts in any other genera, the tail skeletons of which are either not ossified, not preserved or not figured. *Eusthenopteron* may nevertheless be a good example for comparison, as its tail is not attenuated like that of many heterocercal and eel-like forms, although not so truncated as that of the coelacanth. Although the relationship of body size to vertebrae in *Megalichthys* is unknown, this form also has short, high trunk vertebrae, and as noted by Andrews & Westoll (1970b: 420) the most likely explanation of the size groupings of the trunk and tail vertebrae is that the notochord was expanded in the trunk region.

The expanded trunk notochord in *Latimeria* may be an adaptation for its mode of swimming. The fish seems to be mainly sedentary (Schaeffer, 1948), most of its forward motion probably being slow and produced by sculling movements of the second dorsal and anal fins or undulation of the rays of the epi- and hypochordal tail fins. Its paired fins (especially the pectorals) are probably used for manoeuvring among irregular rocks and its strong tail only put to full use for powerful acceleration in lunges after its prey. The expansion of the notochord in the trunk may stiffen the anterior part of the body to improve directional stability during acceleration. However, more work is necessary to elucidate the significance and limitations of notochord dimensions and possible reasons for numerous short and high trunk vertebrae. The fusion of the anterior intercentra to form long cartilages extending through segmental boundaries and the close articulation of the anterior neural arch bases by complex zygapophysis-like processes probably contribute to this stiffening behind the head, in particular protecting the neural and haemal canals from bending too far. The thick fibrous connective tissue investment of the vertebral column is obviously a very important part of the axial skeleton in a notochordal fish. The layers of fibres are orientated so as to resist excessive bending and torsion, and anteriorly the strong, well differentiated oblique bundles presumably brace the anterior trunk region more firmly against torsion (functioning as a geodetic framework, Parrington, 1967).

The serial relationship between ossified neural spines or haemal arches and bundles of fibres differentiated from the connective tissue investment in their places, is consistent with the formation of skeletal elements directly from connective tissue where stress requires them (Laerm, 1976 and references). Possibly the dorsal bicipital ribs of *Eusthenopteron foordi* originated in this way and ossified quite late in development, while its sacral attachment (Andrews & Westoll, 1970a) was still only a rib-shaped bundle of fibres.

The function of the anterior ventral button-like cartilages is uncertain.

The tail

Latimeria has a typical coelacanth tail. The very high oil content of the body renders the fish neutrally buoyant, and the tail is symmetrical with stout, rather widely separated bony fin rays. The dorsal and ventral series of caudal cartilages probably counteract compression forces which might otherwise affect the neural and haemal canals in the powerful main caudal region. It is the supplementary caudal lobe which makes coelacanth tails peculiar; whatever its function, it is likely to be important and closely linked with the evolution of the coelacanthiforms, since it is one of their most characteristic and constant features through at least 350 million years.

The axial structure of the supplementary caudal lobe is very different from the main caudal region: the notochord tapers at the beginning of the tail to a diameter which, in the supplementary lobe, is in normal proportion to the length of the vertebrae, these are polyapsid and mainly cartilaginous, the fin web is supported directly on the neural and haemal spines, and (Millot & Anthony, 1958a) the lepidotrichia are supple throughout their length in contrast to the rigid ossified rays of the main epi- and hypochordal lobes. The functions of these two regions thus appear to be quite different (cf. Schaeffer, 1948). The main caudal is one of the most completely ossified regions of the body and is well fitted for its strong side-strokes, while nearly all the above modifications of the supplementary lobe point to flexibility. The two regions are often quite widely separated by a gap in vertebral structure (Millot & Anthony, 1958b: fig. 1; Schaeffer, 1948: fig. 1). The body musculature continues into the supplementary lobe as a slender finger, whose very different innervation (Millot & Anthony, 1965: 78, fig. 1) is also consistent with a separate function. The presence of a strong paired ridge, formed by a thickening of the dermis and arching of the lateral line scales (Plate 1A; Millot & Anthony, 1958a: fig. 1831, and 1965: pl. VI, top right), which runs across the main caudal region to the supplementary lobe, suggests that the latter may be used to control turbulence, which may be considerable in a large fish with spiny fins suddenly accelerating. It may also act as a brake if flexed strongly sideways, fully extended, or it may move to balance the forces produced by the second dorsal and anal fins in sculling. Another possibility which presents itself is that it is a signalling device, perhaps supported by the frequency of its mutilation (Millot & Anthony, 1958b) although this may be due to damage on sharp rocks.

Evidence bearing on the homology of pleurocentra

Leaving aside the specialized anterior region and the supplementary caudal lobe, there is a very close resemblance between the neural arches, pleurocentra, haemal arches and intercentra in *Latimeria* and those known in other cross-opterygians (Andrews & Westoll, 1970a, b), and the homologies are particularly clear in comparison with *Eusthenopteron foordi* (Fig. 5A, B) in spite of the discrepancy in the size of the notochord. The main differences concern the degree of ossification, for *Eusthenopteron* has far more bone close to the notochord. However, the bony neural arches of *Eusthenopteron* must have been separated from the notochord by thin cartilaginous neural arch bases, and the ossified parts of its pleurocentra and intercentra were probably also surrounded by cartilage. The homology between the vertebral elements of *Eusthenopteron* and those of tetrapods (Fig. 5E) is generally agreed (Gregory, Rockwell & Evans, 1939; Jarvik, 1952; Panchen, 1967; Parrington, 1967; Romer, 1966; Andrews & Westoll, 1970a) and for this reason the tetrapod terminology has been applied to *Latimeria*.

A similar vertebral plan is thus found in coelacanthiforms (*Latimeria*, Fig. 5A), porolepiforms (*Glyptolepis*, Fig. 5C), earlier osteolepiforms (*Osteolepis, Eusthenopteron*, Fig. 5D, B) and probably onychodontiforms (*Onychodus*)—vertebrae in

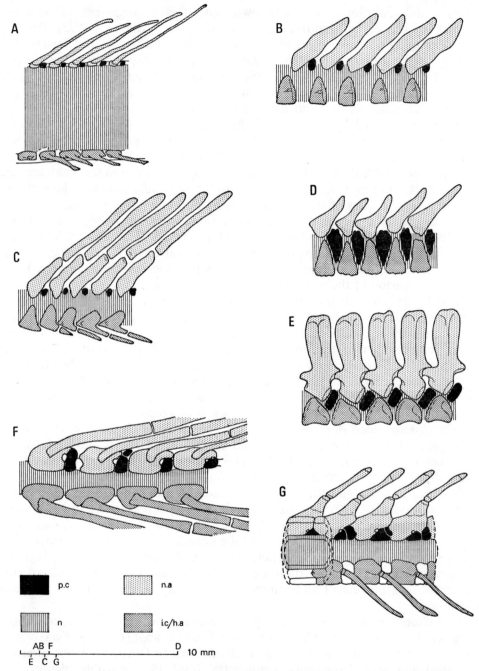

Figure 5. Vertebrae from the posterior trunk region in left lateral view. A, *Latimeria;* B, *Eusthen-opteron;* C, *Glyptolepis;* D, *Osteolepis;* E, *Eryops;* F, *Neoceratodus;* G, *Acipenser.* A and B are from specimens with very similar body length, ×0·5. C–G are drawn with the notochord approximately the same diameter as B. Dorsal ribs have been omitted from B and E. B, Redrawn from Andrews & Westoll, 1970a; C, D. redrawn from Andrews & Westoll, 1970b; E. Redrawn from Romer, 1966; F. Drawn from an unregistered specimen in the Royal Scottish Museum; G. Redrawn from Goodrich, 1930.

primitive rhizodontiforms are unknown—and this arrangement of vertebral elements is likely to be primitive for crossopterygians (Andrews, 1973). Panchen (this Volume) is probably right in considering the condition of the vertebrae in *Osteolepis* more primitive than that of *Eusthenopteron* or *Glyptolepis*. However, we cannot tell whether the known ossifications of the two latter forms were extended by cartilage as far as in *Osteolepis* or not, although it is likely at least that spaces such as those ventrally between the ossified parts of the intercentra of *Eusthenopteron* were buffered by cartilage. If the elements were so extended, the only advance of *Eusthenopteron* and *Glyptolepis* over the more primitive condition was regression of ossification, and no freeing of the sides of the notochord would have been involved. In contrast to this, *Latimeria* is advanced both in the small amount of ossification in its vertebral column, and in the negligible extent of the vertebral elements over the sides of the column. Nearly all known fossil coelacanths were also advanced in the regression of ossification, which probably accounts for elements like their pleurocentra not being preserved. The way in which the vertebral elements are confined to dorsal and ventral tracts in *Latimeria*, leaving the sides of the notochord completely free, may be a specialization connected with the great expansion of the notochord in the trunk region.

When compared with the lungfish *Neoceratodus* (Fig. 5F) the homologies of the cartilaginous vertebral elements also seem very clear. In particular, the elements called 'interdorsals' by Goodrich (1930) occupy the same relative position and have similar relationships to the neural arches, notochord and nerve roots as the pleurocentra in *Latimeria* and other crossopterygians. In *Neoceratodus* the trunk is without pleurocentra, the neural arch bases taking up all the space instead, and in *Latimeria* the pleurocentra do not persist in the anterior trunk, possibly because of the close articulation of the neural arch bases in this region.

In the chondrostean *Acipenser* (Fig. 5G), the vertebral elements cover a much greater proportion of the notochord and the whole of the neural and haemal canals, and, apart from the neural spines and ventral ribs, they are all cartilaginous. In this respect they resemble space-filling elements of the shark kind (Williams, 1959), which may account for the greater subdivision of 'interdorsals' and 'interventrals' (Goodrich, 1930). In spite of this it still seems reasonable to homologize the two small elements ('interdorsals', Goodrich, 1930; 'intercalaries', Schaeffer, 1967) between the neural arch bases and the notochord, with the crossopterygian pleurocentrum. In *Polyodon* (Schaeffer, 1967: fig. 3A), although the cartilaginous elements are space-filling, they do not extend far laterally over the notochord, and correspond in arrangement and numbers very closely to the plan seen in crossopterygians.

Fossil chondrosteans and especially holosteans show very diverse conditions of the vertebrae (Schaeffer, 1967) but in some forms at least (*Caturus, Eurycormus*) the centra appear to be composed in a similar manner to *Osteolepis*, from 'alternating dorsal and ventral crescentic hemicentra' (*ibid.* 192) (possibly pleurocentra and intercentra extending to cover the sides of the notochord?). However, all the elements called 'intercalaries' by Schaeffer may not be homologous. Those of *Amia*, which he suggested as possible homologues of the pleurocentra of *Eusthenopteron*, are *embryonic* elements (and may not therefore be strictly comparable with adult structures in fossils) associated with the ventral ends of the neural arches behind them, and in the trunk behaving during development as 'cushions' between these and the bony centra. It seems more likely that they are equivalents of neural arch bases (*contra* Andrews & Westoll, 1970a: 273) than equivalents of pleurocentra, although in view of the diversity of holostean centrum formation, the latter remains a possibility. The homologies of the various actinopterygian and dipnoan centra are beyond the scope of this paper.

Up to this point the homologies discussed have all been of the anatomical or phylogenetic kinds (Panchen, this Volume). The comparisons clearly suggest that tetrapod pleurocentra were derived from the similarly named elements in cross-opterygians (probably from an *Osteolepis*-like form with alternating thin crescentic intercentra and pleurocentra almost completely covering the notochord, *ibid.*). The comparisons also show that these elements ('pleurocentra') are quite widely distributed among primitive (notochordal) bony fish. However, considerations of serial and probable ontogenetic homologies in *Latimeria* lead to conflicting conclusions.

In *Latimeria* the distinction between bony spines (neural and haemal spines and their arches) and the dorsal (neural arch bases and pleurocentra) and ventral (intercentra or haemal arch bases and haemal supports) rows of cartilages is particularly clear. The neural arch bases, however, are continuous with the cartilaginous core (*knorpelkern*) of the neural arches and spines, and in other crossopterygians (e.g. *Eusthenopteron*) there is ossification in the pleurocentrum, so that neural arches and pleurocentra seem likely to form one longitudinal series. In the supplementary caudal lobe, moreover, the serial homologues of the pleurocentra actually roof over the neural canal, and apart from lacking a bony neural spine are indistinguishable from the true neural arches. In spite of Schaeffer's (1967: 191) statement of caution that 'details of the ontogeny can rarely be inferred from the adult condition', it is very difficult to envisage the development of this region in terms other than subdivision of continuous dorsal and ventral tracts of cartilage according to local stresses associated with the position of the myosepta and the need for increased flexibility. In other words, the pleurocentra and neural arch bases/neural arches must apparently have a common ontogenetic origin. This conflicts with what we know of the ontogeny of the tetrapod pleuro-centrum. In amniotes this develops in the perichordal tube and the neural arches develop separately (Williams, 1959). In modern amphibians (*ibid.*) the homology of the centrum as a pleurocentrum is less certain, but it too forms perichordally while the neural arches have a separate origin. We must therefore ask whether the ancestral crossopterygian pleurocentra were ontogenetically derived from neural arch material along with the neural arches (in which case the pleurocentrum must somehow have changed during evolution to develop in the perichordal tube) or whether some other explanation is possible. It is not a problem of whether tetrapods have an autocentrum or a centrum formed by outgrowth from the 'arch bases', i.e. of the significance of the neurocentral suture (Williams, 1959: 19), since even in crossopterygians the pleurocentrum is distinct and separate from the neural arch and its base; indeed, the constancy of the crossopterygian vertebral pattern is remarkable.

The embryology of living notochordal fishes may help to answer this question, although little is known of it. In *Protopterus annectens*, Mookerjee, Ganguly &

Brahma (1954) have described paired dorsal and ventral cartilages in the peri-chordal tube, separated from the neural arch cartilages by a zone of perichordal tissue. This form, however, has thin ring-like centra in the adult, into which the perichordal cartilages later become incorporated. *Protopterus dolloi* and *Neocera-todus* do not have centra in the adult, but their neural and haemal arch bases ('basidorsals' and 'basiventrals') are quite extensive over the surface of the notochord. The ontogeny of these cartilages has not been described, but they are probably likewise perichordal (Schaeffer, 1967: 190). In *Acipenser* the neural arches and 'intercalaries' develop together in a dorsal tract condensed from sclerotomic mesenchyme previously spread over the notochord and neural tube (*ibid.*). These scattered facts suggest that crossopterygian pleurocentra and neural arch bases/neural arches may have developed together along the junction between the circum-neural tissue and the perichordal tube, the neural arches forming in the positions of the myosepta and the pleurocentra between them. The former would then be free to develop upwards and the latter downwards (as in *Osteolepis*?) and the centres of origin of the two could eventually separate, probably at the amphibian stage. The recently discovered advanced embryos of *Latimeria* may be expected to throw some light on this.

SUMMARY

Some previously unknown features, shown by a new specimen of the axial skeleton of *Latimeria*, are described. The functional significance of some of its peculiarities of structure is discussed, in particular the great expansion of the notochord in the trunk region and the strange supplementary caudal lobe of the tail.

The dorsal cartilages abutting against the notochord are compared with other crossopterygians, tetrapods, lungfishes and primitive actinopterygians. The comparisons point clearly to anatomical and phylogenetic homology of the elements here called pleurocentra in all these groups, but raise the problem of whether fish and tetrapod pleurocentra are ontogenetically homologous. A tenta-tive solution to this problem is suggested.

The vertebral column of *Latimeria* presents an abundance of problems, the majority still not fully solved, for example:

(a) What is the reason for such short, high vertebrae and/or such an expanded trunk notochord?

(b) What is the function of the anterior button-like cartilages, ventral to the notochord?

(c) How does *Latimeria* really use the supplementary caudal lobe?

(d) Do the pleurocentra and neural arches develop from neural arch material or perichordal tube, or the junction between the two?

(e) Are the neural and haemal arches and spines derived ontogenetically from material from two adjacent segments (resegmentation) or intrasegmentally?

Perhaps because the pattern may be primitive for bony fishes, vertebral structure throws no light on the interrelationships of crossopterygian groups and at present there is no satisfactory way of classifying them. The groups previously associated as Rhipidistia were united only by shared primitive characters, but are divided by numerous specializations, so that 'Rhipidistia' cannot be upheld as a natural taxon. The scheme of relationships proposed by Andrews (1973) may also be unacceptable, but the difficult problem of how these groups *are* related remains with us.

ACKNOWLEDGEMENTS

I should like to thank my mother, Mrs M. B. Andrews, for her skilful help in the tedious work of clearing away the investing fibres from the neural and haemal spines. I am most grateful to Dr A. L. Panchen and Dr C. D. Waterston for helpful discussion and especially to the former for allowing me to quote his unpublished material. I also wish to thank Lt. Col. and Mrs G. E. Aldridge for encouragement and help with translation, and Mr R. C. M. Thomson and Mr K. B. Smith for photography. Finally, I wish to acknowledge my debt of gratitude to Professor T. S. Westoll for stimulating and guiding my interest in fossil fishes.

REFERENCES

ANDREWS, S. M., 1973. Interrelationships of crossopterygians. In P. H. Greenwood *et al.* (Eds), *Interrelationships of fishes:* 137–77. London: Academic Press.

ANDREWS, S. M. & WESTOLL, T. S., 1970a. The postcranial skeleton of *Eusthenopteron foordi* Whiteaves. *Trans. R. Soc. Edinb.*, *68*(9): 207–329

ANDREWS, S. M. & WESTOLL, T. S., 1970b. The postcranial skeleton of rhipidistian fishes excluding *Eusthenopteron*. *Trans. R. Soc. Edinb.*, *68*(12): 391–489.

GOODRICH, E. S., 1930. *Studies on the structure and development of vertebrates.* London: Macmillan.

GREGORY, W. K., ROCKWELL, H. & EVANS, F. G., 1939. Structure of the vertebral column in *Eusthenopteron foordi* Whiteaves. *J. Paleont.*, *13:* 126–9.

JARVIK, E., 1952. On the fish-like tail in the ichthyostegid stegocephalians. *Meddr Grønland, 114*(12): 1–90.

LAERM, J., 1976. The development, function and design of amphicoelous vertebrae in teleost fishes. *Zool. J. Linn. Soc.*, *58*(3): 237–54.

MILES, R. S. & WESTOLL, T. S., 1968. The placoderm fish *Coccosteus cuspidatus* Miller ex Agassiz from the Middle Old Red Sandstone of Scotland. Part I. Descriptive morphology. *Trans. R. Soc. Edinb.*, *67*(9): 373–476.

MILLOT, J. & ANTHONY, J., 1956. Considérations préliminaires sur le squelette axial et le système nerveux central de *Latimeria chalumnae* Smith. *Mem. Inst. scient. Madagascar (Ser. A), 11:* 167–87.

MILLOT, J. & ANTHONY, J., 1958a. *Latimeria chalumnae*, dernier des crossoptérygiens. In P. Grassé (Ed.), *Traité de Zoologie, 13*(3): 2553–97. Paris: Masson.

MILLOT, J. & ANTHONY, J., 1958b. *Anatomie de* Latimeria chalumnae. *Tome I. Squelette, muscles et formations de soutien.* Paris: C.N.R.S.

MILLOT, J. & ANTHONY, J., 1965. *Anatomie de* Latimeria chalumnae. *Tome II. Système nerveux et organes des sens.* Paris: C.N.R.S.

MOOKERJEE, H. K., GANGULY, D. N. & BRAHMA, S. K., 1954. On the development of the centrum and its arches in the Dipnoi *Protopterus annectens. Anat. Anz., 100:* 217–30.

MOY-THOMAS, J. A. & MILES, R. S., 1971. *Palaeozoic fishes.* London: Chapman & Hall.

PANCHEN, A. L., 1967. The homologies of the labyrinthodont centrum. *Evolution, N.Y., 21*(1): 24–33.

PARRINGTON, F. R., 1967. The vertebrae of early tetrapods. *Colloques int. Cent. natn. Rech. scient., 163:* 269–79.

RABINERSON, A., 1925. Beiträge zur vergleichende Anatomie der Wirbelsäule der Knorpelfische. *Anat. Anz., 59:* 433–54, 481–95, 513–22, 560–5; *60:* 354–60.

ROMER, A. S., 1966. *Vertebrate paleontology,* 3rd ed. Chicago: University Press.

SCHAEFFER, B., 1948. A study of *Diplurus longicaudatus* with notes on the body form and locomotion of the Coelacanthini. *Am. Mus. Novit., 1378:* 1–32.

SCHAEFFER, B., 1967. Osteichthyan vertebrae. *Zool. J. Linn. Soc., 47*(311): 185–95.

WILLIAMS, E. E., 1959. Gadow's arcualia and the development of tetrapod vertebrae. *Q. Rev. Biol., 34*(1): 1–32.

ABBREVIATIONS USED IN FIGURES

d.cc	dorsal caudal cartilage	i.c	intercentrum
d.lig	dorsal longitudinal ligament	my	course of myoseptum
d.r	dorsal root of spinal nerve	n	notochord
e.h.sp	epihaemal spine	n.a	neural arch
e.n.sp	epineural spine	n.a.b	neural arch base (anterior dorsal cartilage)
fb	bundles of hard fibres in connective tissue investment	n.can	neural canal
		n.sp	neural spine
h.a	haemal arch	p.c	pleurocentrum (posterior dorsal cartilage)
h.a.b	haemal arch base	sp.c	spinal cord, covered by perimeningeal adipose tissue
h.can	haemal canal		
h.s	haemal support	v.cc	ventral caudal cartilage
h.sp	haemal spine	v.r	ventral root of spinal nerve

EXPLANATION OF PLATES

PLATE 1

Latimeria chalumnae Smith. Royal Scottish Museum specimen 1969.76, a male weighing 32 kg, caught off Vanamboini, Grand Comore, on 24 March 1969.

A. Entire specimen after thawing. $\times 0\cdot15$.

B. Dissected axial skeleton. $\times 0\cdot15$.

PLATE 2

Latimeria chalumnae Smith. Same specimen as Plate 1.

A. Anterior region of axial skeleton. $\times 0\cdot5$.

B. Axial skeleton of posterior trunk region. $\times 0\cdot5$.

PLATE 3

Latimeria chalumnae Smith. Same specimen as Plate 1.

A. Axial skeleton of base of tail. $\times 0\cdot5$.

B. Axial skeleton of tail region. $\times 0\cdot5$.

Plate 1

S. MAHALA ANDREWS

Plate 2

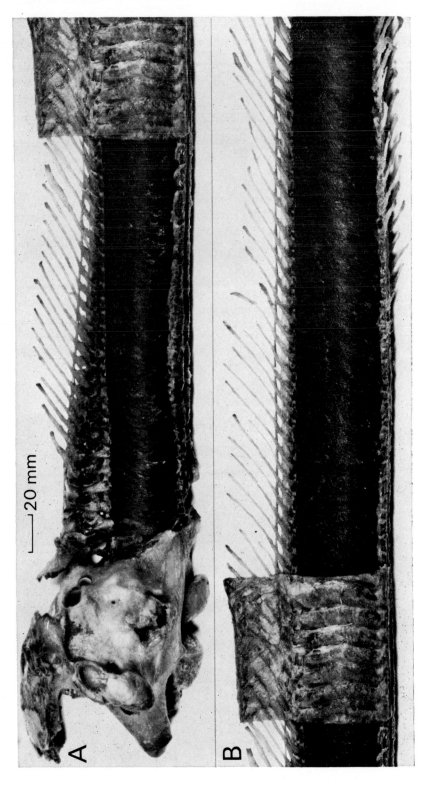

S. MAHALA ANDREWS

Plate 3

20 mm

A

B

S. MAHALA ANDREWS

The origin and early evolution of tetrapod vertebrae

A. L. PANCHEN

Department of Zoology, University of Newcastle upon Tyne

Vertebral structure has been used in the classification of the fossil stegocephalian amphibia since the latter part of the last century and their division, on this basis, into Labyrinthodontia and Lepospondyli has persisted to the present day. Theories of vertebral homology have also been used extensively in the reconstruction of the phylogeny of primitive tetrapods and to corroborate their ancestry amongst crossopterygian fish.

The temnospondylous labyrinthodonts may plausibly be linked with the crossopterygians in this way, but with *Osteolepis* rather than the popular *Eusthenopteron*. The vertebrae of the former may also be primitive for crossopterygians.

However, the vertebrae of *Ichthyostega*, often regarded as a link between those of fish and tetrapod, are anomalous, as are those of several early temnospondyls. Nor do the vertebrae of other Palaeozoic tetrapods suggest a common ancestry with temnospondyls.

A functionally diplospondylous vertebra, of gastrocentrous type, was probably the common ancestral condition of at least the unrelated batrachosaur labyrinthodonts, microsaurs and reptiles, in apparent conflict with the fish-temnospondyl link.

No other evidence unites batrachosaurs with temnospondyls, or microsaurs with other 'lepospondyls', nor does vertebral structure illuminate the interrelationships or antecedents of the living amphibian orders.

CONTENTS

INTRODUCTION

The structure of the vertebrae has long been used as one of the principal characters (often *the* principal character) defining the major taxa of Palaeozoic and Triassic amphibia. Because of this it has also been used in reconstructions of the interrelations of those taxa and thus of the phylogeny of early tetrapods.

These extinct orders of Amphibia (the 'Stegocephalia') bear no obvious close relationship to the three extant orders (the Lissamphibia: *sensu* Parsons & Williams, 1962, 1963) and the candidature of any stegocephalian group as that ancestral to the reptiles is subject to damaging criticism (Panchen, 1972, 1975).

289

Until relatively recently, however, it was assumed that vertebral structure corroborated the division of stegocephalians into two well-defined taxa, the Labyrinthodontia and the Lepospondyli, which, together with the Lissamphibia, constitute the three subclasses of the class Amphibia (Romer, 1966).

It was also assumed that while the status of the Lepospondyli as a natural group was regarded by some as unsatisfactory, that of the Labyrinthodontia was well established. Within the Labyrinthodontia the structure of the compound vertebrae appeared to reinforce the division of the subclass into natural groups and to parallel well-defined evolutionary trends within those groups. Furthermore there appeared to be a strong case for maintaining that the vertebrae of labyrinthodonts could be extrapolated back to a single ancestral condition and that the ancestral condition was closely paralleled in the crossopterygian fishes, which are the agreed ancestral group for all tetrapods.

In the light of recent discoveries all these assumptions may be called into question. Particularly the story of the early evolution of tetrapod vertebrae is certainly less simple and less certain than it appeared to be even less than ten years ago, and I am very glad of the opportunity to attempt a re-appraisal.

TAXONOMIC HISTORY

By the end of the last century the practice of classifying the stegocephalians on the basis of their vertebral structure was well established. An influential example of such a classification is that of Zittel (1895) set out below:

> CLASS AMPHIBIA
> ORDER STEGOCEPHALI
> Suborder PHYLLOSPONDYLI
>   Family Branchiosauridae
> Suborder LEPOSPONDYLI
>   Family Microsauridae
>   Family Aïstopodidae
> Suborder TEMNOSPONDYLI
> Suborder STEREOSPONDYLI
>   Family Gastrolepidotidae
>   Family Labyrinthodontidae
> ORDER COECILIAE
> ORDER URODELA
> ORDER ANURA

In later versions of this classification the Temnospondyli were subdivided into two sections, Cope's Embolomeri (excluding *Anthracosaurus*, which joined *Loxomma* in the family Gastrolepidotidae) and Rhachitomi (Cope, 1880, 1882).

Zittel's classification of his order Stegocephali, based primarily on vertebral structure (as indicated by the termination '-*spondyli*') yields a series of groups all of which, while not necessarily generally accepted, are in current use. The single exception is the family Gastrolepidotidae: *Anthracosaurus* is now recognized as one of the Embolomeri (Panchen, 1977) and *Loxomma* and related genera are generally regarded as Rhachitomi.

The Embolomeri, Rhachitomi and Stereospondyli of Cope and Zittel comprise the Labyrinthodontia of current usage. The term Labyrinthodontia was used in the classification drawn up by a committee of the British Association in 1874 (Miall, 1875) in the same sense as Zittel's Stegocephali, and this was endorsed by Nicholson & Lydekker (1889). However, the latter authors also used the term 'Labyrinthodontia Vera' in the current sense of Labyrinthodontia and this current usage was finally established by Watson in an authoritative classification of pre-Jurassic tetrapods (Watson, 1917).

Zittel's suborder Lepospondyli corresponds exactly to the current subclass Lepospondyli. Today it is generally divided into three orders, Microsauria, Nectridea and Aïstopoda as proposed by the Miall Committee. Zittel on the other hand placed the Nectridea within the Microsauria.

The suborder Phyllospondyli was based on the pioneer work of Credner (1891) on a series of small stegocephalians from the Lower Permian of central Europe. These were forms with poorly ossified vertebrae consisting of leaf-like neural arches with little or no apparent ossification of the centrum, which was restored by Credner as a curious ventral hemicylinder.

The Lepospondyli were united by the possession of a single spool-like centrum closely sutured or fused to the neural arch, while the Temnospondyli (in Zittel's sense) were large stegocephalians with compound centra consisting of an anterior intercentrum and a posterior pleurocentrum, the latter often divided bilaterally. It was the realization that the Stereospondyli represented the culmination of the trends, particularly in vertebral evolution, shown by the rhachitomous Temnospondyli, that led to the general acceptance of the Labyrinthodontia as a natural group.

This realization was due to the work of D. M. S. Watson, which ushered in the era of phylogenetic vertebrate palaeontology. In a study of rhachitomous and stereospondylous labyrinthodonts, Watson (1919) was able to show a series of morphological trends in skull structure extending in time from the Lower Permian to the Upper Trias. This led him to regard the Rhachitomi and Stereospondyli as grades, the former having given rise to the latter, perhaps many times. Several authors had previously suggested the derivation of the stereospondylous vertebra from the rhachitomous condition by expansion of the crescentic intercentrum to a disc and loss of the pleurocentrum. Watson established beyond doubt that rhachitomous and stereospondylous labyrinthodonts belonged in a single taxon.

Watson extrapolated the trends he had shown in Permian and Triassic labyrinthodonts to show the primitive nature of the British Carboniferous forms (Watson, 1926, 1929). Of these the anthracosaurs were known to have embolomerous vertebrae, while the postcranial skeleton of the commoner loxommatids was unknown. Watson therefore assumed, not unreasonably, that the loxommatids were embolomerous. His classification of labyrinthodonts thus contained three grades characteristic of Carboniferous, Permian and Triassic and termed Embolomeri, Rachitomi (*sic*) and Stereospondyli respectively.

In describing the Carboniferous amphibia of Scotland, Watson (1929) gave an account of two non-labyrinthodont species, *Adelogyrinus* and *Dolichopareias* from the Lower Carboniferous Oil Shale group. These forms appeared to be aberrant early microsaurs, but Watson erected a new order Adelospondyli for them outside the Lepospondyli on the basis of a vertebral structure consisting of a single centrum sutured rather than fused to the neural arch. It was subsequently shown, however, that this character was shared by many microsaurs and even occurred in some but not all vertebrae within a single vertebral column (Steen, 1938) and was thus unreliable taxonomically (Westoll, 1942; Romer, 1950).

Later Watson (1940) concluded that *Eugyrinus* from the British Coal Measures was characterized by precocious development of the trends he had demonstrated in the rhachitomous and stereospondylous labyrinthodonts, and was a primitive phyllospondyl. He also concluded that *Eugyrinus* was related to '*Protobatrachus*', recently described as an anuran ancestor (Piveteau, 1937), by way of *Amphibamus* and similar forms from the Middle Pennsylvanian of Illinois. Thus the Phyllospondyli were regarded as the group from which the frogs arose.

Meanwhile Romer (1930, 1933) and Steen (1931, 1934), influenced by Watson's opinion, added numbers of small Carboniferous rhachitomes with poorly known vertebral structure to Credner's 'branchiosaurs' in the order Phyllospondyli and Romer (1933) even included the recently described ichthyostegids from the Devonian or Lower Carboniferous of East Greenland, whose vertebral structure was then unknown. However, Steen (1937, 1938) was later to have doubts as to the validity of the order and Romer (1939) finally and correctly concluded that the phyllospondyls were a collection of small and in some cases larval or neotenous early labyrinthodonts.

Paralleling attempts at unravelling the phylogeny and the classification of early tetrapods, using the structure of their vertebrae, were speculations as to the ontogeny and phylogeny of the vertebrae themselves. Perhaps the greatest influence on such speculations were the theories of Gadow (1896, 1933). Gadow & Abbott (1895) used the developmental stages of elasmobranch vertebrae as an archetype for the ontogeny and phylogeny of the vertebrae of all craniates. In the particular elasmobranchs which they described the vertebrae were formed from four pairs of 'arcualia'. Gadow thus attempted to reconstruct the ontogeny of tetrapod vertebrae in terms of the ossification of these cartilaginous arches.

The living Urodela and possibly the Apoda were known to have centra formed by ossification in the notochordal sheath with little or no preformation of the centrum in cartilage, so that Gadow's theories were clearly inapplicable here. Romer (1945) assumed that the spool-like centra of the Lepospondyls were similarly formed and thus included the Urodela and Apoda, together with the fossil forms, in the subclass Lepospondyli.

The labyrinthodonts, on the other hand, were assumed to have vertebrae formed in the Gadovian manner and were thus placed, together with the frogs (following Watson), in a second subclass the Apsidospondyli.

In his masterly review of the labyrinthodonts themselves, Romer (1947) supported a dichotomous division of the Labyrinthodontia. It was generally accepted, following Watson's classic studies, that the anthracosaurs of the Coal Measures and early Permian were related to reptiles and particularly to *Seymouria* and its allies, considered by Watson to be reptiles. The contemporary loxommatids on the other hand had a skull structure which placed them with the rhachitomes.

Säve-Söderbergh (1934) had already produced a dichotomous classification of all vertebrates in which the two were widely separated on this basis. Romer proposed a similar separation within the labyrinthodonts. Zittel's term Temnospondyli was revived for the ichthyostegids, rhachitomes and stereospondyls, while the use of Säve-Söderbergh's term Anthracosauria (Anthracosauroideae: Watson, 1929) was expanded to include the Seymouriamorpha. The term Embolomeri was then available to characterize the anthracosaurs in Watson's sense.

However, an important review on the development of tetrapod vertebrae by Williams (1959) demonstrated the inapplicability of Gadow's theories to amphibian ontogeny. Williams claimed to show from published accounts that the centrum of all tetrapods is formed from the combination of the caudal half-sclerotome of one segment with the cranial half-sclerotome of the segment behind it. It is thus intersegmental in ontogeny as well as position.

In most tetrapods, but not the Urodela and Anura, the separation of the half-sclerotomes of a single segment is clearly marked by the development of a sclerocoel. Thus resegmentation of the sclerotome halves in the latter groups is less certain. However, nothing corresponding to Gadow's arcualia can be seen in any tetrapod.

The precursor of the centrum is a perichordal tube of mesenchyme which first condenses in the region of the sclerocoel to give a perichordal ring which gives rise to an intervertebral disc. From this disc develops the intercentrum present in the tails of many reptiles, while in the trunk region it forms the intervertebral joint and in mammals contributes to the meniscus and the epiphyses of adjacent centra.

In most tetrapods the rest of the perichordal tube chondrifies and is subsequently ossified to form the definitive centrum, but in the urodeles and apoda ossification was thought to be direct without any preformation in cartilage.

Consequently in Romer's (1945) sense all tetrapods must probably be regarded as 'lepospondyl'. Thus, as a result of Williams' review and a study by Parsons & Williams (1962, 1963) of the common features of the living orders of amphibia, Romer subjected his major groupings of the Class Amphibia to a radical revision in the next and final edition of his text (Romer, 1966).

Parsons' & Williams' thesis was that the three extant orders of amphibia should be regarded as belonging to a single natural group characterized by pedicellate teeth in which crown and root (pedical) were separated by connective tissue. As a name for the proposed taxon they revived Gadow's use of the term Lissamphibia.

Thus the revival of the concept of the Lissamphibia countered not only Romer's (1945) acceptance of the diphyletic origin of the modern orders at the primitive tetrapod level, but the more extreme theory of Jarvik (1972 and references therein) that the Anura on one hand and the Urodela and Apoda on the other were descended from different groups of crossopterygian fish. An even more extreme theory, now generally regarded as discredited, was diphyletic origin from the Crossopterygii and Dipnoi, as originally proposed by Holmgren (1933, 1949 etc.) and supported by Säve-Söderbergh (1934) in his dichotomous classification.

Romer's 1966 classification of Amphibia is set out in outline below:

CLASS AMPHIBIA
SUBCLASS LABYRINTHODONTIA
    Order ICHTHYOSTEGALIA

Order TEMNOSPONDYLI
Suborder Rhachitomi
Suborder Stereospondyli
Suborder Plagiosauria
Order ANTHRACOSAURIA
Suborder Schizomeri
Suborder Diplomeri
Suborder Embolomeri
Suborder Seymouriamorpha
SUBCLASS LEPOSPONDYLI
Order NECTRIDEA
Order AÏSTOPODA
Order MICROSAURIA
SUBCLASS LISSAMPHIBIA
SUPERORDER SALIENTIA
Order PROANURA
Order ANURA
SUPERORDER CAUDATA
Order URODELA
Order APODA

The suborder Plagiosauria is characterized by a unique type of vertebral column with single elongate, platycoelous centra and 'intervertebral' neural arches, whose transverse plane alternates with that of the centra (Fig. 3D). The anomalous vertebrae, together with the form of the dermal girdle, led Nilsson (1939, 1946) to assign them a taxonomic position outside the Labyrinthodontia. However, their cranial characters, apart from a quite extraordinary flattening in the typical Triassic forms, may be considered labyrinthodont. In 1959 I proposed their retention within the Labyrinthodontia as an order equivalent in rank to the Temnospondyli and Anthracosauria (*sensu* Romer) while he has reduced them to subordinal rank, an arrangement with which I am content.

The Seymouriamorpha are now generally regarded as Amphibia as a result of Romer's review and of Špinar's (1952) demonstration of larval stages in the seymouriamorph *Discosauriscus.*

The vertebrae of seymouriamorphs are closely comparable to those of early reptiles. The pleurocentrum, as in embolomeres, is the principal central element, but whereas the intercentrum in the latter is a complete but thinner disc, that in the seymouriamorphs is a crescentic wedge, as in rhachitomes, and is considerably smaller than the pleurocentrum. No taxonomic term has been proposed to include the species bearing this type of vertebra, probably because they were until relatively recently regarded as cotylosaurian reptiles. The descriptive term 'gastrocentrous', rescued from Gadow's vertebral nomenclature, is normally used.

Recent discoveries and redescription of the critical material has necessitated a complete revision of the classification of the Anthracosauria (*sensu* Romer) (Panchen, 1970, 1975). The suborder Schizomeri was based on the association of the post-cranial skeleton of the early temnospondyl *Pholidogaster* and the skull of a primitive anthracosaur (*Eoherpeton*, Panchen, 1975), while the suborder

Diplomeri resulted from the similar confusion of an embolomere *Diplovertebron* from Bohemia with the very reptiliomorph *Gephyrostegus* from the same locality and horizon and often the same block (Carroll, 1970; Panchen, 1970). Thus these two suborders lapse. However, Carroll's work on *Gephyrostegus* and related forms (Carroll, 1969a, 1970) shows that they represent a well-defined taxon distinct from both embolomeres and seymouriamorphs. In addition the description of primitive Lower Carboniferous anthracosaurs (Romer, 1970; Hotton, 1970; Panchen, 1975) made necessary the creation of a new taxon, the Herpetospondyli.

I have long favoured the separation of the Seymouriamorpha from the anthracosaurs in Watson's sense and have thus adopted Efremov's (1946) term Batrachosauria to replace Anthracosauria (*sensu* Romer) and thus constitute one of the orders into which the labyrinthodonts are divided:

Order BATRACHOSAURIA Efremov
Suborder Anthracosauria Säve-Söderbergh
Infraorder Herpetospondyli Panchen
Infraorder Embolomeri Cope
Infraorder Gephyrostegoidea Carroll
Suborder Seymouriamorpha Watson

THE VERTEBRAE OF CROSSOPTERYGIANS

It is generally accepted that the crossopterygian fishes, in the sense in which that term is normally used, are the group which gave rise to all the tetrapods. Andrews (1973) has discussed the interrelationships and classification of the Crossopterygii and the classification, definition and content of the constituent subdivisions as she sets them out are accepted here.

The majority opinion is that the Osteolepidida of the superorder Quadrostia Andrews (Osteolepiformes of Jarvik) is the subgroup from which all tetrapods are probably to be derived.

Although Jarvik's diphyletic theories of the origin of tetrapods are not generally accepted, his excellent accounts of the anatomy, and particularly the cranial anatomy, of 'Rhipidistia' (osteolepiforms and porolepiforms) are an essential basis for comparison of these fishes with early tetrapods. The rhipidistian which Jarvik has described in the greatest detail is the osteolepiform *Eusthenopteron foordi* Whiteaves. Although probably too late in geological time (Frasnian, early Upper Devonian; Westoll, 1949) it appears to be anatomically an almost perfect tetrapod ancestor. It is not surprising therefore that the vertebrae of *Eusthenopteron* have frequently been used as a point of departure when considering the origin of tetrapod vertebrae.

The vertebrae of *Eusthenopteron* have been described by Gregory, Rockwell & Evans (1939), Jarvik (1952) and Andrews & Westoll (1970a). All three accounts agree that an individual vertebra consists of three distinct elements: a neural arch; a principal centrum in the form of a thin hoop or crescent of bone, incomplete dorsally and often separated into halves ventrally; and a bilateral pair of small nodules of bone situated posteriorly at the vertical level of the base of the neural arch (Fig. 1A).

Gregory *et al.* and Andrews & Westoll refer to the principal centrum as the intercentrum and the paired nodules as pleurocentra, thus homologizing them with the components of the labyrinthodont centrum. Andrews & Westoll comment particularly on the resemblence of the *Eusthenopteron* vertebra to that of rhachitomous labyrinthodonts. In this respect they support the theory of Romer (1947, 1964) that the rhachitomous vertebra is primitive for labyrinthodonts (and thus ancestral for the amniotes assumed to be derived from them). Schemes for the homology of the parts of the labyrinthodont vertebra have also assumed this theory (Williams, 1959; Panchen, 1967).

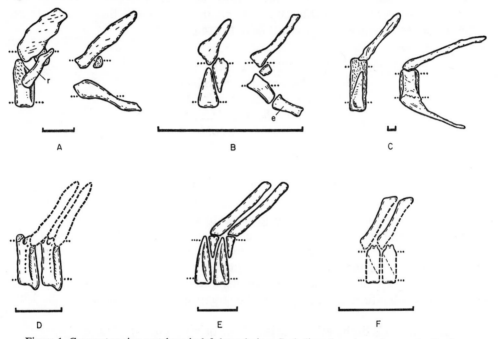

Figure 1. Crossopterygian vertebrae in left lateral view. Scale lines 1 cm in each case. A. *Eusthenopteron foordi*, mid-trunk and mid-caudal. B. *Osteolepis panderi*, mid-trunk and anterior caudal. C. *Megalichthys hibberti* and *M.* sp., trunk and anterior caudal resp. D. *Ectosteorhachis nitidus*, 2 trunk. E. *Lohsania utahensis*, 2 trunk. F. *Thursius pholidotus*, 2 trunk.
   A–C, Redrawn after Andrews & Westoll; D–E, redrawn after Thomson & Vaughn; F, redrawn after A. & W. but orientation of centra reversed (see text). Maximum diameter of restored notochord indicated in each case: r, dorsal rib; e, epihaemal spine.

Jarvik, on the other hand, is cautious about the identity of the principal centrum and refers to it as the 'ventral vertebral arch'. He interprets the whole vertebra in Gadovian terms, referring to the 'pleurocentrum' as the 'interdorsal' and suggesting that the ventral arch is formed from the fusion of the basiventral of one segment with the interventral of the next succeeding segment. He thus assumes the type of resegmentation of centra in ontogeny seen in amniotes and possibly all tetrapods (and incidentally allows one to infer the homology of the *Eusthenopteron* ventral arch with the single centrum of amniotes!).

The evidence given by Jarvik for the intersegmental position of the ventral vertebral arch is the presence of a vertically directed groove in the posterior part of the lateral face, interpreted as for the segmental artery. The presence of this groove is confirmed in trunk and anterior caudal vertebrae by Andrews & Westoll, who give the same interpretation.

Jarvik's basis for the interpretation of the paired nodules as interdorsals is his report of the presence of 'two more or less distinct notches, one antero-dorsal and one postero-dorsal' interpreted as for the ventral and dorsal roots of the spinal nerve. The position of the nodule would then correspond to that of the inter-dorsal in elasmobranchs (Goodrich, 1930: fig. 18).

However, Andrews & Westoll were unable to find any trace of these notches in their extensive and beautifully prepared material. They also point out that the 'pleurocentra' in their natural orientation lie below the level of the neural canal and are thus associated with the notochord rather than the nerve cord. This is corroborated by a specimen in which two adjacent neural arches have fused and there is a foramen, presumably for the spinal nerve, at a much higher level in the plane of fusion.

Other interpretations of the 'pleurocentra' have been suggested. I suggested (Panchen, in Andrews & Westoll, 1970a) that they might be interpreted as separately ossified transverse processes, as the tubercular head of the characteristic short 'bladeless' dorsal ribs appeared to articulate with the 'pleurocentra', but the authors disagreed because the 'pleurocentra' extended into the caudal region beyond the most posterior preserved rib.

Schaeffer (1967) and Thomson & Vaughn (1968) have suggested that the 'pleurocentra' are to be homologized with intercalaries of the type seen in *Amia*, but this suggestion was made before Andrews' and Westoll's redescription and is not tenable if their assertion of the relationship of the pleurocentra to the notochord is correct.

Other important features of the vertebrae of *Eusthenopteron* must be noted. Firstly, Andrews & Westoll note that there appears to be no firm articulation, and certainly not a suture, between the neural arch and the principal centrum, although the latter is usually unfinished dorsally and thus may have extended further round the notochord in cartilage. Secondly, undulations in the anterior and posterior margins of the neural arch at the level of the top of the neural canal are described by Jarvik as incipient zygapophyses, and correspond with the margins of adjacent vertebrae. They bear no distinct facets, however, and may not have been in contact. Thirdly, backwardly directed haemal spines are born directly on the principal centrum and in the 'sacral' region coexist with ribs on the same vertebrae, thus confirming that the latter are dorsal ribs.

Finally it must be emphasized that in *Eusthenopteron*, as in most other 'rhipidistians' in which the vertebrae have been described, there appears to have been no significant constriction of the notochord by the ossified centrum. In every case the centrum must have provided reinforcement in the notochordal sheath in the plane of the myoseptum but the notochord must still have been the compression member in the vertebral column. In *Eusthenopteron* there is also considerable separation between successive centra and bending and torsional stresses in the plane of the axis must also have been born to a considerable extent by the notochord and its sheath.

Other crossopterygian vertebrae have been described by Andrews & Westoll

(1970b) and by Thomson & Vaughn (1968). *Tristichopterus*, a form closely related to *Eusthenopteron*, appears to have very similar vertebrae (A. & W.).

Eusthenopteron and *Tristichopterus* are placed together in the family Eusthenopteridae of the order Osteolepidida. They represent advanced members of that order. Vertebrae have been described in a number of forms in the more primitive family Osteolepididae of the same order.

Osteolepis itself has vertebrae which are much more fully ossified than those of *Eusthenopteron* (Fig. 1B). Vertebrae are described from *O. panderi* (Pander) and are probably similar in other species (A. & W.). The centrum in the trunk region takes the form of two crescentic wedges, the first antero-ventral one occupies the position of the principal centrum in *Eusthenopteron* and is surely homologous with it. The second is almost as large and is postero-dorsal, so that in the trunk region the notochord was surrounded by alternating dorsal and ventral three-quarter hoops of bone, with alternating oblique boundaries between them in side view.

Once again the two components are homologized by Andrews & Westoll with the intercentrum and pleurocentrum respectively of labyrinthodont vertebrae and it is difficult to disagree. Anteriorly both intercentrum and pleurocentrum are consolidated in the midline, but in the midtrunk region the intercentrum is split and paired pleurocentral blocks are quite widely separated. In the tail region the pleurocentra are considerably reduced, disappearing behind the anal fin, and the dorsal extension of the intercentrum is somewhat smaller. As in *Eusthenopteron* the intercentrum or principal centrum bears the haemal arch.

No groove for the segmental artery is reported and ossified ribs, either dorsal or pleural, were also absent. Dorsally the trunk pleurocentra are indented on each side, presumably for the spinal nerve. They seem to have equally developed facets for the preceding and succeeding neural arches.

Osteolepis is a Middle Devonian osteolepidid and the contemporary *Gyroptychius* probably had similar vertebrae. Those of the later, Carboniferous and Permian, members of the family are puzzling but the Middle Devonian *Thursius* may show an intermediate condition (A. & W.).

Best known are the vertebrae of *Megalichthys* from the Carboniferous (Fig. 1C), which are common as isolated specimens in the Coal Measures. In the trunk region there is normally a single annular centrum, but anteriorly these may be split into bilateral halves. They show no sign of the course of the segmental artery or nerve or of any division into 'intercentrum' and 'pleurocentrum'.

The neural arches are very slender and usually separate from the centra in the trunk region: in the tail, however, they are firmly sutured or fused. The haemal arches extend back from directly below the centrum.

The vertebrae of two Permian forms, *Ectosteorhachis* and *Lohsania*, are described by Thomson & Vaughn. Only the centra of *Ectosteorhachis* are described but paired facets, probably for neural arches, are preserved postero-dorsally. Below them are possible rib facets. The neural arches may also have contacted a facet on the centrum behind: Thomson & Vaughn give a reconstruction on this basis which gives a considerable separation between successive centra.

Apart from the presence of rib facets, the centra of *Ectosteorhachis* differ from those of *Megalichthys* in showing short grooves dorsally, anterior to the posterior neural arch facets. These are interpreted as for the spinal nerves. There is also a strongly marked ridge suggesting the vertical course of the myoseptum (Fig. 1D). In *Megalichthys* some centra show a postero-ventral triangular area of periosteal bone on either side, but are of unfinished bone dorsal to this.

It is probable that the trunk neural arches of *Megalichthys* were similarly orientated to those of *Ectosteorhachis*, but in the tail region they originate directly over the centrum. In *Megalichthys* with its extremely slender neural arches it is probable that successive trunk centra were in contact.

Lohsania is separated taxonomically from *Ectosteorhachis* almost entirely on the structure of the vertebrae, there being two separate elements to each trunk centrum. The 'principal centrum' is also annular but tapers uniformly towards the dorsum in lateral view. It bears an ill-defined lateral ridge, possibly for the myoseptum. The second element is described by Thomson & Vaughn as the accessory centrum. It is a shallow crescentic wedge above the notochord, seen in side view as an isosceles triangle with its apex directed downwards. Dorsally it is ridged for the articulation of the neural arch (Fig. 1E). Thus *Lohsania* seems to have had a vertebral column, of which the centra formed a series, of alternating dorsal and ventral wedges, as in *Osteolepis*. However, there was a much greater disparity in size between dorsal and ventral elements.

Thomson & Vaughn interpret the vertebrae of *Lohsania* in Gadovian terms. The accessory centrum and its neural arch are considered to be induced by the interdorsal and basidorsal respectively and to be associated in the same vertebra with the principal centrum *behind*. The latter is thought to be formed by basiventral posteriorly and interventral anteriorly. Thus playing by Gadow's rules they assume that there was no resegmentation in the vertebrae of *Lohsania*. They then extend this interpretation to the single centrum of *Ectosteorhachis* (and presumably *Megalichthys*) assuming that it was formed by the fusion of the interdorsal ($\equiv$ accessory centrum) to the front of the principal centrum. This interpretation is further extended to *Eusthenopteron*, so that while they agree with Jarvik's interpretation of the accessory element as the interdorsal, they associate it with the 'ventral vertebral arch' behind. Thus they deny in the same Gadovian terms the resegmentation postulated by Jarvik.

Andrews & Westoll on the other hand interpret the accessory central elements of *Eusthenopteron*, *Osteolepis* and *Lohsania* as homologous, occupying a postero-dorsal position within the compound centrum and also homologous with the pleurocentra of tetrapods. I agree with A. & W's interpretation of the homology of the accessory centra with one another and with their views on their relative position for the following reasons:

(1) In the holospondylous forms (*Megalichthys*, *Ectosteorhachis*) the neural arch extends backwards from the postero-dorsal 'corners' of the centrum. In the other three fish the accessory centrum is situated either directly below the ventral end of the neural arch pedicel or closely associated with it. This is the condition of the pleurocentrum in temnospondylous labyrinthodonts.

(2) In *Eusthenopteron* a trunk rib extends from the capitular head on the principal centrum to the tubercular head directly lateral to the accessory centrum, suggesting that they all pertain to the same vertebral unit.

(3) In all rhipidistians in which caudal vertebrae have been described the haemal arches extend back from directly below the principal centrum just as from the intercentrum of temnospondylous labyrinthodonts.

These points also convince me of the correctness of Andrews' and Westoll's identification of the principal centrum as the intercentrum and the accessory element(s) as the pleurocentrum. *However, at this stage the homologies implied can only be suggested with respect to the corresponding elements in the vertebrae of temnospondylous labyrinthodonts* (see below).

Unfortunately Andrews & Westoll weaken their case somewhat by assuming that the holospondylous centra of *Megalichthys* and *Ectosteorhachis* are formed by the fusion of the pleurocentrum with the intercentrum *behind* it, i.e. across the vertebral boundary. This gives essentially the same homologies as Thomson & Vaughn's position, if speculations about resegmentation and arcualia are ignored.

As evidence for the mode of formation of the holospondylous centrum Andrews & Westoll quote the closer association of pleurocentrum with intercentrum behind in *Lohsania*, the unfinished area antero-dorsally on the *Megalichthys* centrum and the vertebrae of *Thursius* referred to above.

Of the few isolated central elements of *Thursius* known, some are intercentra and pleurocentra of the *Osteolepis* type, but a total of three centra are complete rings. The presence of the dorsal notches for the spinal nerves together with the ventral paramedian ridges (which occur commonly in early tetrapods as well as rhipidistians) corroborates the idea that the rings are formed by fusion of inter-centrum and pleurocentrum. There does not seem, however, to be any evidence of fore-and-aft orientation, so that there is no intrinsic evidence of the direction in which the fusion has occurred (Fig. 1F).

Thus the only evidence for Andrews' and Westoll's theory is the form of the surface of the *Megalichthys* centrum, but the antero-dorsal unfinished area might well represent an expansion of the intercentrum into a previously unossified area as in the intercentra of stereospondylous labyrinthodonts.

Other rhipidistian vertebrae have been described by Andrews & Westoll (1970b) in *Rhizodopsis* (Osteolepidida, Rhizodopsidae), *Rhizodus* and *Strepsodus* (Quad-rostia, Rhizodontida, Rhizodontidae) and the porolepiform *Glyptolepis* (Binostia, Holoptychiida, Holoptychiidae).

Those of *Rhizodopsis* are not well known but appear, at least in the posterior trunk and anterior tail region, to have had annular centra. Neural arches, haemal arches and ribs are not described. In *Rhizodus* the position is reversed: a series of neural arches is known as are poorly preserved haemal spines, but, despite the enormous length of *R. hibberti* (*c.* 6–7 m) from which the vertebrae are known and the close relationship to *Strepsodus*, no centra are preserved. It seems improbable that they were unossified.

The centra of *Strepsodus* are annular but differ from those of other rhipidistia in being considerably longer and more massive, so that they constrict the notochordal canal considerably.

The vertebrae of *Glyptolepis*, on the other hand, did not apparently constrict the notochord and in this and other respects resemble those of *Eusthenopteron*. The intercentra are similar but even less extensive and with considerable regional variation in height. The paired pleurocentra are minute and the notochord must have had a broad lateral exposure. An odd feature is the very anterior presence of epineural spines above the neural spines. The epineurals commence at about the fifteenth vertebra, very little behind the base of the pectoral fin. No ribs are reported.

The similarity of the vertebrae between *Eusthenopteron* and *Glyptolepis*, one osteolepiform and one porolepiform, suggested to Andrews & Westoll that this was the primitive condition for rhipidistians. I think, however, that another interpretation is possible.

Early members of the family Osteolepididae, notably *Osteolepis* itself, are generally regarded as primitive osteolepiforms, characterized by heavy ossification,

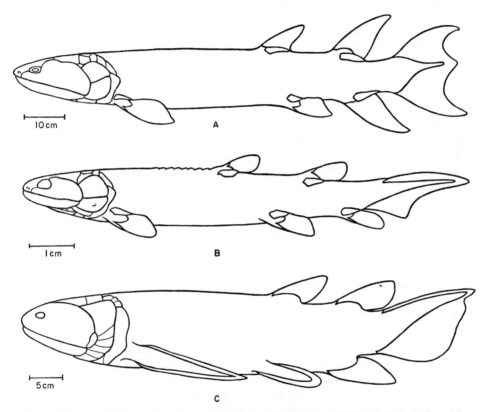

Figure 2. Restored body outline of crossopterygians, after Andrews & Westoll. Scale indicated in each case. A, *Eusthenopteron foordi*; B, *Osteolepis panderi*; C, *Glyptolepis paucidens*.

the presence of cosmine on scales and dermal bones and a heterocercal tail, as well as the relatively early geological horizon. *Eusthenopteron* on the other hand is advanced in the reduction of cosmine, the symmetrical tail (Fig. 2) and probably in the presence of dorsal ribs (Andrews, 1973).

It seems reasonable to suggest, therefore, that the less complete ossification of its vertebrae represents a phylogenetic reduction from an *Osteolepis*-like ancestor.

In later Osteolepididae (*Megalichthys, Ectosteorhachis, Lohsania*), however, there are various degrees of consolidation into ring centra. This is paralleled in

Rhizodopsis and reaches its maximum development in *Strepsodus* (and one may surmise in *Rhizodus*).

Glyptolepis is the only porolepiform from which the vertebrae are described, but it is an advanced porolepiform of the family Holoptychiidae. It therefore seems reasonable to suggest that the form of its vertebrae represents a parallel reduction in ossification to that in *Eusthenopteron* and not a shared primitive condition. A bold prediction, which would corroborate this hypothesis, is that the vertebrae of *Porolepis* when described will resemble those of *Osteolepis*!

The above hypotheses would gain credibility if some functional explanation could be given for the evolutionary trends suggested: I think it can.

Parrington (1967) compared the vertebrae of rhachitomous labyrinthodonts to a geodetic framework of two pairs of girders wound onto a cylindrical form, with the girders at about 45° to the axis of the cylinder. This arrangement is a very effective one for resisting flexure and torque. In the rhachitomous vertebra, however, Parrington suggested that it was a 'negative' of such an arrangement, with intercentra and pleurocentra corresponding to the spaces and gaps revealing unconstricted notochord corresponding to the girders. Such an arrangement he saw as permitting limited torque and flexure more easily than simple cylindrical centra. He was able to illustrate this with respect to torque by simple mechanical models.

Parrington did not consider the effect of the articulation of successive neural arches on his model, but if the model is applied to the vertebrae of *Osteolepis*, in which zygapophyses are not developed, it seems more apt than its application to tetrapods.

The vertebrae of *Osteolepis* may be compared to a geodetic framework in which the 'negative girders' (i.e. the junctions between the components) are at a greater angle to the axis than 45°. This situation also prevails in primitive temnospondyls (see below). Thus the *Osteolepis* vertebra lies between Parrington's two models giving more resistance to torque than the 45° model but less than the model with cylindrical centra.

This correlates well with the fact that *Osteolepis*, like most primitive Osteichthyes, has a heterocercal tail. A minimum resistance to twisting of the axis, probably even in the trunk, is necessary for the heterocercal tail to produce its lifting function (Simons, 1970) and Andrews & Westoll (1970b) note that the junction angle in the centra approaches 45° behind the pelvic fin.

In later members of the family Osteolepididae, however, the tail approaches a symmetrical form (Andrews & Westoll, 1970b); torsion is therefore unnecessary for tail function and unless controlled would be deleterious for swimming. The later members of the family are also considerably larger than *Osteolepis*.

O. panderi varies between 72–135 mm, *O. macrolepidotus* 125–305 mm, (Jarvik, 1948) while *Megalichthys* reaches a length of about 2 m.

Thus in a fish of *Megalichthys* size the development of ring centra to resist torque might become a necessity if torsion of the axis was no longer necessary. This would be more emphatically the case in the enormous *Strepsodus* and *Rhizodus* of the family Rhizodontidae.

The tail of *Eusthenopteron* has also attained external symmetry but the whole fish was of modest size, reaching some 1·5 m only in the very largest specimens

(Andrews & Westoll, 1970a). Controlled torsion would no longer have been necessary for tail function and two factors would have allowed reduction in vertebral ossification. The first was the limited size, but, more importantly, the second was the origin of dorsal ribs.

The development of dorsal ribs was, as Westoll (1962) has pointed out, one of the critical factors in conversion of the vertebral column of fish from what is principally a compression member to the girder of tetrapods. In *Eusthenopteron* on the other hand, ribs and vertebrae may have braced the myosepta in what was probably a relatively rigid trunk with a thickly sheathed and thus relatively inflexible notochord (Andrews & Westoll, 1970a). Only the absence of ribs or the need of a combination of flexibility and strength in a larger fish would require the development of ring centra.

In *Glyptolepis* ribs are absent and the vertebrae are in other respects similar to those of *Eusthenopteron*. In this case, however, it is generally agreed that one is dealing with a relatively sluggish animal very similar in habit to the living dipnoan *Epiceratodus* (Andrews & Westoll, 1970b). Both have a virtually unconstricted notochord and while *Epiceratodus* has pleural ribs in the trunk, torsion in *Glyptolepis* was probably controlled by the extension forward in the trunk of epineural and epihaemal spines. Thus reduction of ossification of the vertebrae from the presumed ancestral condition would carry no disadvantages.

To summarize the present section: the vertebrae of *Eusthenopteron* need not, of necessity, be regarded as representing a condition ancestral to that of tetrapods and those of *Osteolepis* are more likely candidates. *Eusthenopteron*, however, does show the potentiality for the development of dorsal ribs, so important to tetrapods. Further, the variety of vertebrae within 'rhipidistia' is so great that, without functional considerations, almost any theory of the origin of tetrapod vertebrae could be corroborated by a well chosen rhipidistian.

Lastly, it is possible to see, in the development of ring centra in later Osteolepididae from the *Osteolepis* condition, a remarkable parallel with the origin of the stereospondylous condition from that in rhachitomous temnospondyls (see below). The massive vertebrae of *Strepsodus*, assuming an ancestral *Osteolepis*-like condition, considerably strengthen the parallel.

THE VERTEBRAE OF PRIMITIVE TETRAPODS

Description of the vertebrae in ichthyostegids is based on very limited material (Jarvik, 1952). Following Jarvik the general tendency has been to interpret them as being closely similar to those of *Eusthenopteron* (e.g. Romer, 1964; Panchen, 1967; Thomson & Bossy, 1970), but I now doubt whether this conclusion is tenable.

As described by Jarvik, each vertebra consists, as in *Eusthenopteron*, of a neural arch, a hoop-like 'ventral vertebral arch' and paired 'interdorsals'. Important differences are documented, however. Firstly, successive neural arches are articulated together by well-developed zygapophyses of tetrapod type. Secondly, each neural arch bears a well-defined facet for the tubercular head of the rib, terminating a short transverse process.

Information on the principal centrum ('ventral vertebral arch') seems less secure. In the trunk and anterior tail regions it appears, unlike that of *Eusthenopteron*, to be always entire ventrally rather than split into two halves in some cases. It also appears to have an articular facet on each side for the neural arch, in contrast to *Eusthenopteron*, and to bear a well-defined facet for the capitulum of the rib.

As a basis for discussion, however, I want to suggest that, despite attempts to postulate an *Eusthenopteron-Ichthyostega*-temnospondyl series in vertebral evolution, with the intercentrum as the principal central element, the centrum of *Ichthyostega* might be better interpreted as homologous in all four senses (anatomical, ontogenetic, serial and phylogenetic: see below) with the pleurocentrum of batrachosaurs and reptiles. My reasons, mostly negative, are as follows:

(1) The centrum is complete and longest ventrally and appears similar to that of primitive anthracosaurs such as *Proterogyrinus* ('*Mauchchunkia*' Hotton, 1970).

(2) In all labyrinthodonts it is the pleurocentrum, when present, which has the persistent articulation with the neural arch, as the centrum of *Ichthyostega* may have.

(3) The centra of *Ichthyostega* show no trace of a groove for the segmental artery in contrast to the intercentra of *Eusthenopteron* and other rhipidistians as well as the embolomerous anthracosaur *Eogyrinus* (Panchen, 1966, 1967).

(4) While Jarvik restores the capitular facet in a postero-lateral position on the centrum, as in the intercentra of labyrinthodonts, it is perhaps improbable but not impossible that he has reversed the orientation of the centrum. The specimen on which this data is based (Jarvik, 1952: fig. 15) lacks enough of the neural arch to adjudicate. It should be noted, however, that reversal of the centrum gives a more normal, oblique orientation of the line between the rib-heads (Fig. 3A).

(5) Haemal arches borne on low crescentic intercentra are present in the tail of *Ichthyostega* but it is not clear that the latter are serially homologous with the trunk centra. Indeed, the specimen referred to above (No. 4) bears three centra which are complete hoops and totally unlike the shallow crescents of the caudal intercentra, yet the specimen is referred by Jarvik to the anterior caudal region. *Peltobatrachus* (Panchen, 1959) provides a cautionary parallel in this respect.

Even more doubt must surround Jarvik's 'interdorsals'. As in *Eusthenopteron* he describes a notch for the dorsal root of the spinal nerve and 'conceivably' one for the ventral root but the evidence seems meagre. If the notches are present and if the interdorsals are situated relative to the neural arches in the position restored by Jarvik (i.e. at the level of the neural canal, Fig. 3A), they cannot be interpreted as pleurocentra and must be intercalaries as suggested by Thomson & Bossy.

Finally the ribs of *Ichthyostega* are quite extraordinarily broad as emphasized by Jarvik and particularly well shown in a specimen originally figured by Stensiö (1931: pl. 36). The supporting function of the ribs of tetrapods has been repeatedly emphasized and in *Ichthyostega* there is little doubt that the limited ossification of the centra and the unconstricted notochord are correlated with this (Panchen, 1966; Thomson & Bossy, 1970).

Jarvik has always emphasized the distinctness of the ichthyostegids from the labyrinthodonts proper and this is endorsed by Thomson & Bossy. The latter authors further note the distinctness of the vertebrae from those of labyrinthodonts. If they are equally different from those of any rhipidistians the ichthyostegids are removed from an ancestral condition in any speculation about the origin of tetrapod vertebrae.

Turning to the temnospondyl labyrinthodonts, the evidence from another well-known early group appears, surprisingly, to be of little value. When Romer (1947) pointed out that the loxommatids of the Carboniferous were to be included in the Temnospondyli on the structure of the skull, he suggested not only that they were not embolomerous (contra Watson: see above), but that they would prove to be

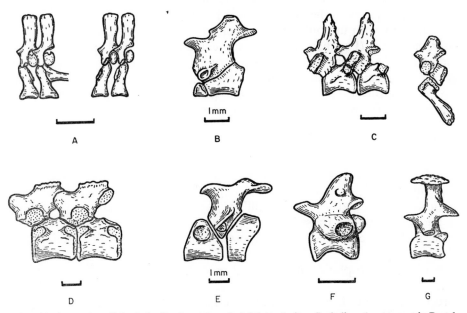

Figure 3. Anomalous 'labyrinthodont' vertebrae in left lateral view. Scale lines 1 cm except in B and E. A. *Ichthyostega* sp., 2 anterior caudals after Jarvik, and some redrawn with orientation of centra reversed (see text). B. *Doleserpeton annectens*, single trunk, after Bolt. C. *Peltobatrachus pustulatus*, 2 trunk and single mid-caudal. D. A plagiosaur, 2 trunk after Nilsson. E. *Tersomius* cf. *T. texensis*, trunk, after Daly. F. *Parioxys bolli*, immediate postsacral (with maximum development of intercentrum), after Carrol. G. ?*Fayella chickashaensis*, trunk, after Olson.

rhachitomous. The brief description by Baird (1957) of neural arches and intercentra in a loxommatid was generally accepted as a triumphant vindication of Romer's prediction. The material has recently been redescribed and figured by Beaumont (1977) from casts supplied by Dr Baird and, while confirming that the loxommatids (or at least *Megalocephalus*) were not embolomerous, it certainly does not establish that they were rhachitomous. The only central elements preserved are very low, apparently poorly ossified intercentra, which seem more similar to those of early anthracosaurs than to those of typical rhachitomes such as *Eryops*, or primitive forms such as *Neldasaurus* and *Pholidogaster* (see below).

Other temnospondyls, as defined by Romer (1947, 1966) are known to have vertebrae which are neither rhachitomous nor stereospondylous. Notably this is the case within the superfamily Dissorophoidea, taxonomically members of the Rhachitomi.

The Lower Permian *Doleserpeton* Bolt (1969) has gastrocentrous vertebrae (Fig. 3B) with the pleurocentrum very much the principal central element and the intercentrum reduced to a degree comparable with that of seymouriamorphs and early reptiles.

It should also be noted that the vertebrae of *Doleserpeton* provide a very good antecedent condition for that found in the Upper Permian terrestrial plagiosaur *Peltobatrachus* and the later typical Triassic plagiosaurs (Fig. 3C, D).

I originally argued (Panchen, 1959) that the single trunk centrum of plagiosaurs (including *Peltobatrachus*) was the pleurocentrum, largely on the evidence of its primary association with the neural arch in front in *Peltobatrachus* and its apparent serial homology with undoubted pleurocentra in the tail of that animal.

Later, however, (Panchen, 1967) on the evidence of the rib articulation which extends from the neural arch to the centrum in front of it in *Peltobatrachus*, I suggested an effectively intrasegmental centrum. This would have been formed by the union of the cranial half-sclerotome with the caudal half-sclerotome behind and thus across the normal vertebral boundary.

Romer (1968), however, would not accept the plagiosaur centrum as anything other than 'super-stereospondylous', but did at the same time point out that the group might well have arisen from a dissorophoid rhachitome. This suggestion has much greater force with the discovery of *Doleserpeton*. It was recently worked out in some detail in an unpublished paper by Miss Christine Janis (personal communication) who also drew in Bolt's suggestion of the relationship of *Doleserpeton* and Lissamphibia (based primarily on vertebral structure and the common possession of pedicellate teeth) to propose a common ancestry for plagiosaurs and lissamphibians.

Most other dissorophoids have normal rhachitomous vertebrae but two notable exceptions should be mentioned. Daly (1973) assigns rich material of a small dissorophid from the Lower Permian of Oklahoma to *Tersomius* cf. *texenis*. The vertebrae bear large intercentra of rhachitomous type, but the pleurocentra are large and, although paired, meet in the midline ventrally (Fig. 3E).

Another possible dissorophoid is represented by material referred by Olson (1972) to the dissorophid *Fayella chickashaensis* Olson. The referred vertebrae consist of hollow cylinders apparently fused to the neural arches. The centra invite interpretation as pleurocentra, but in two instances small pieces of bone are situated between adjacent centra in a dorsal position and suggest rhachitomous pleurocentra to Olson.

Olson proffers two hypotheses, firstly that the attribution is correct, that the bone nodules are pleurocentra and that the holospondylous elements are intercentra ('hypocentra'). In this case the animal is indeed 'super-stereospondylous' and the dissorophoids present a spectrum from gastrocentrous (*Doleserpeton*) through Daly's specimen to rhachitomous and on via *Parioxys* (Carroll, 1964; Moustafa, 1955) (if indeed this is dissorophoid) to stereospondylous and, in the fusion of intercentrum and arch, beyond (Fig. 3).

The second hypothesis is that the attribution is wrong, that the new material represents a seymouriamorph and that the holospondylous elements are pleurocentra. Olson favours the first interpretation because of the lack of small ventral intercentral wedges and the 'tenuous evidence' of the dorsal bony elements.

Holmes & Carroll (1976) describe a small temnospondyl from Scotland as *Caerorhachis*. It is assumed to be from the Namurian of Midlothian and in most respects the skull resembles that of the primitive edopoid *Dendrerpeton*. Significantly, however, it lacks an otic notch.

Nevertheless the vertebrae resemble those of primitive anthracosaurs with hoop-like pleurocentra as the dominant central elements and small crescentic intercentra. The vertebral count appears to be 31 presacral vertebrae, longer than the 23 or 24 of *Ichthyostega* but similar to that of osteolepiforms.

The cases reviewed above emphasize the appalling (to the taxonomist) plasticity of early temnospondyl vertebrae. Turning, however, to the orthodox temnospondyls, the evolutionary trends from the vertebrae of Lower Permian rhachitomes, such as *Eryops*, through 'neorhachitomes' to the few truly stereospondylous Triassic genera (metoposaurs, *Mastodonsaurus*) remain as secure as when they were first demonstrated by Watson (1919).

However, as Thomson & Bossy have emphasized, there are, apart from the loxommatids and dissorophoids, two early offshoots from the presumed 'central rhachitome' stock. These are the colosteids, represented first in the Viséan by *Pholidogaster* and *Greererpeton* and the trimerorhachoids, appearing first in the Westphalian as *Saurerpeton*.

Both these groups are characterized by the primitive retention of the intertemporal bone in the skull roof and, significantly, the absence, or (in some trimerorhachoids) slight development of the otic notch (Chase, 1965; Romer, 1969; Panchen, 1975).

The vertebrae of these forms might well be regarded as primitive for rhachitomes and may be typified by those of *Pholidogaster* (Romer, 1964), *Neldasaurus* (Chase, 1965) and '*Eobrachyops*' (*Isodectes*: Baird, in Welles & Estes, 1969). The centra are strikingly like those of *Osteolepis* with intercentra and pleurocentra alternating deep wedges in side view, the intercentra being complete crescents, the pleurocentra paired arcs (Fig. 4A).

However, the vertebrae of *Trimerorhachis* itself do not conform to this pattern, having massive intercentra and small pleurocentra.

The vertebrae of *Pholidogaster* impressed Romer as being so different from the normal rhachitomous type that he termed them 'schizomerous'. However, in a recent account of the vertebrae of *Eryops* based on Professor Romer's own notes, Moulton (1974) has shown convincingly that the trunk vertebrae of this archetypal rhachitome can be reconstructed with cartilaginous extensions that bring them much nearer to the putative primitive type (Fig. 4B).

The vertebrae of Batrachosauria (anthracosaurs and seymouriamorphs) are more consistent in pattern. In every case the undoubted pleurocentrum is the principal central element, is complete ventrally and is articulated or fused to the neural arch. Where the two are connected by articulation or suture the contact surface on the pleurocentrum faces antero-dorsally and frequently does not extend to the back of the pleurocentrum. Thus the pleurocentrum lies postero-ventrally to the neural arch, while the intercentrum is antero-ventral.

The batrachosaur intercentrum is always smaller in volume than the pleuro-entrum, at least in the trunk region. In almost every case it is also shorter antero-posteriorly as well as in height.

In *Proterogyrinus* ('*Mauchchunkia*' Hotton, 1970), representing the most primitive anthracosaurs (Herpetospondyli, Panchen, 1975), the pleurocentra are well ossified, but perforated for the notochord, while the intercentra are typical crescents extending little more than half-way up the height of the pleurocentra (Fig. 4C).

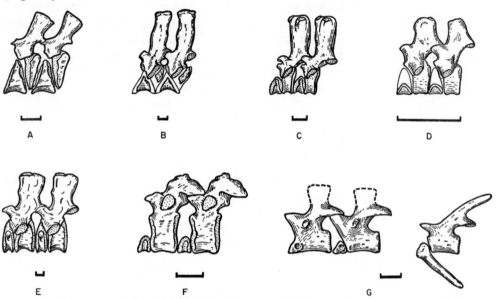

Figure 4. Temnospondyl and batrachosaur vertebrae in left lateral view. Scale lines 1 cm in each case. A. *Neldasaurus wrightae*, 2 trunk, after Chase. B. *Eryops megacephalus*, 2 trunk, after Moulton. C. *Proterogyrinus* ('*Mauchchunkia*'), 2 trunk, after Hotton. D. *Gephyrostegus bohemicus*, 2 trunk, after Carroll. E. *Eogyrinus attheyi*, 2 trunk. F. *Seymouria baylorensis*, 2 trunk, after White. G. *Kotlassia prima*, 2 trunk and caudal, after Bystrow.
    Cartilaginous extensions of centra (unshaded) as restored by authors in B and D.

The vertebrae of *Gephyrostegus* as described by Carroll (1970) are similar in outline but significantly less well ossified, the pleurocentra being hollow shells and the intercentra similarly reduced.

Carroll suggests, probably correctly, that the intercentra were extended dorsally in cartilage to fill the considerable gap above them (Fig. 4D). The difference in degree of ossification in the vertebrae of *Proterogyrinus* and *Gephyrostegus* probably relates to their difference in size (of the order of 3 : 1 for the diameter of the pleurocentrum); more significant is the difference between the vertebrae of *Gephyrostegus* and those of microsaurs and the earliest reptiles, most of which are even smaller.

The vertebrae of the Embolomeri (Fig. 4E) are easily derived from those of *Proterogyrinus* by dorsal extension of the ossification of the intercentrum to surround the notochord completely. The intercentrum is not, however, a complete disc, except perhaps in the sacral vertebra (Panchen, 1970), but is always tapered dorsally in side view. Neither is it significantly longer than that of *Proterogyrinus* or *Gephyrostegus* and is markedly shorter than the pleurocentrum in almost every case. The vertebrae of *Eogyrinus* are best known (Panchen, 1966). In this form the notochordal perforation is reduced to a small foramen through both intercentrum and pleurocentrum while in *Anthracosaurus* and '*Eobaphetes*' (cf. *Leptophractus*) it is occluded altogether (Panchen, 1977).

Successive intercentra and pleurocentra formed a series of ball-and-socket joints with intercentra convex anteriorly and posteriorly, (except in the immediate region of the notochordal canal) and fitting between the amphicoelous pleurocentra.

It is generally agreed that this functional diplospondyly is correlated with a secondary elongation of the presacral column to about 40 vertebrae and an anguilliform mode of swimming (Panchen, 1967, 1970; Parrington, 1967; Carroll, 1969a; Thomson & Bossy, 1970). However, Hotton notes that the centra of *Proterogyrinus*, with about 26 presacral vertebrae, already have an incipient ball-and-socket arrangement. Thus this arrangement was perfected in the large, long-bodied, aquatic embolomeres, from its incipient state in the Herpetospondyli.

The condition in the Herpetospondyli is reasonably certainly ancestral to that in the Seymouriamorpha, but in the latter group a new factor appears. That is the reduction, not only of the intercentra themselves in height and length, but also of the space available for their dorsal extension in cartilage (Fig. 4F, G). At the same time the neural arch and pleurocentrum become more solidly connected, emphasizing the supportive function of the column as against the locomotor one (Panchen, 1967). The poorly ossified vertebrae of *Discosauriscus* are undoubtedly correlated with its small size and partially neotenous condition (Špinar, 1952).

In *Seymouria* the intercentra are very low and only about half the length of the pleurocentra in the trunk (White, 1939). It is possible that they were chondrified to the height of the pleurocentra as a dorsally tapering space is available, but it does not seem probable. In *Kotlassia* on the other hand, despite incipient neoteny, the intercentra are even smaller relative to the pleurocentra in similar-sized vertebrae (Bystrow, 1944). Furthermore the anterior surface of the pleurocentra is so shaped that any independent dorsal extension of the intercentra seems improbable. This is further emphasized by the fusion of intercentra and pleurocentra in the mid-trunk region. In the tail region, however, functional diplospondyly was probably retained.

This reduction of the intercentra reaches an extreme condition in *Solenodonsaurus*, a batrachosaur of roughly the same size, in which successive trunk pleurocentra were in contact, allowing only for minute intercentral wedges ventrally (Carroll, 1970).

The vertebrae of *Seymouria* and *Kotlassia*, but not *Solenodonsaurus*, parallel those of *Limnoscelis*, a primitive and aberrant reptile, and *Diadectes*, of disputed affinity, in having swollen, laterally expanded neural arches with widely-spaced almost horizontal zygapophyses. Many authors have noted that vertebrae so equipped would allow horizontal flexure, but almost eliminate torsion. Parrington (1967) correlates this with a relatively small head, in contrast to the condition in terrestrial rhachitomes, as part of his thesis that controlled torsion in the latter was necessary to retain balance.

One of the most important results of studies of stegocephalians during the last decade has been the realization that the vertebrae of microsaurs are in no way fundamentally different from those of batrachosaurs or reptiles. Thus if the taxon Lepospondyli has any validity the Microsauria should not be members of it.

Steen (1938) reported free caudal haemal arches in the microsaur *Microbrachis* indicating that the centra were pleurocentra and later that there were trunk

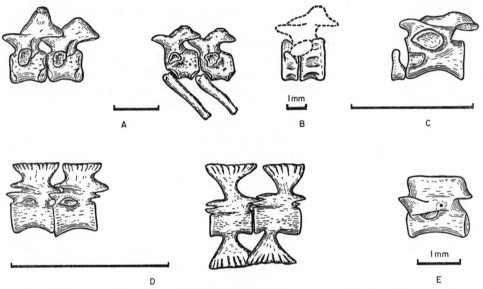

Figure 5. 'Lepospondyl' vertebrae in left lateral view. Scale lines 1 cm except in B and E. A. *Pantylus*, 2 mid-trunk, 2 mid-caudal, data after Carroll (1968). B. *Acherontiscus caledoniae*, single trunk, data after Carroll (1969b). C. *Cardiocephalus* cf. *C. sternbergi*, sacral, after Daly. D. *Urocordylus scalaris*, 2 trunk, 2 caudal, after Steen. E. *Phlegethontia* sp. trunk, after Thomson & Bossy.

intercentra (Brough & Brough—née Steen—1967). This, however, was not generally accepted (Romer, 1950; Carroll & Baird, 1968) because of the difficult nature of the *Microbrachis* material.

Carroll (1968) described undoubted haemal arches in the Lower Permian microsaur *Pantylus*, which also, being relatively large and heavily built, has swollen neural arches paralleling those of seymouriamorphs and diadectids (Fig. 5A). He thus accepted that the principal centra in microsaurs were pleurocentra.

Following Carroll's account in which he also notes the presence of haemal arches in *Lysorophus* (also Olson, 1971), regarded by some as an aberrant microsaur, other microsaurs with compound centra have been described.

Carroll (1969b) describes a microsaur, with an incomplete but probably characteristic skull but embolomerous vertebrae, as *Acherontiscus* (Fig. 5B). The single specimen probably came from the Namurian A of the Lothian Coalfield near Edinburgh and thus probably pre-dates all other microsaurs apart from *Adelogyrinus* and related (or identical? Dr S. M. Andrews—personal communication) forms from the Viséan of the same area (Brough & Brough, 1967).

Subsequently Vaughn (1972) has described a microsaur from the Missourian (early Stephanian) of Colorado (*Trihecaton*) which bears crescentic trunk intercentra. Daly (1973) has shown that the well-known microsaur *Cardiocephalus* from the Lower Permian bears trunk intercentra of distinctive shape (Fig. 5C), as well as haemal arches, and Carroll & Currie (1975) note the presence of trunk intercentra in the latest established microsaur, *Goniorhynchus stovalli* Olson from the Hennessey formation of Oklahoma.

Thus the vertebrae of microsaurs may parallel all stages in the development of vertebrae in batrachosaurs. Isolated microsaur pleurocentra and their attached neural arches are closely similar to those of captorhinomorph reptiles (e.g. Gregory, Peabody & Price, 1956). It is also noteworthy that the microsaur *Trihecaton* described by Vaughn has teeth with a relict infolding of the enamel, thus removing yet another absolute distinction between labyrinthodonts and microsaurs.

Little need be said about the vertebrae of primitive reptiles (see Romer, 1956 for a review). In primitive reptiles trunk intercentra are retained and persist in the living *Sphenodon*. Caudal intercentra are retained in most reptile groups with well-developed tails and, if present, bear haemal arches. Primitively reptile vertebrae were notochordal, again a condition retained in *Sphenodon*.

Thus there is little or no doubt of the homology of the principal centrum throughout the batrachosaurs, microsaurs and reptiles, or that of the intercentrum. Furthermore, resegmentation by fusion of sclerotome halves across the segmental boundary occurs in all amniotes which have been studied (Williams, 1959) and it is thus reasonable to assume that this was also the mode of formation of the principal centrum in batrachosaurs and microsaurs.

The amphibian groups whose vertebrae have not yet been discussed are the extant orders and the Palaeozoic Nectridea and Aïstopoda. Wake (1970) has reviewed vertebral ontogeny in the extant orders. He notes that while the middle region of the husk-like centrum of urodeles is not preformed in cartilage the ends certainly are. Similarly the centrum in caecilians (Apoda) is to some degree preformed in cartilage and thus in this respect neither is clearly distinct from that of amniotes. Also, while doubting the presence of a definitive sclerocoel in caecilians, Wake thinks it probable, as asserted by Williams, that the centrum is formed by resegmentation as in amniotes. The situation is, however, much more doubtful in urodeles.

In the Anura, as in the other two orders, no clear sclerocoel develops. However, resegmentation has been reported by Smit (1953) although other authors including Wake consider this doubtful. As in the urodeles and Apoda some of the centrum is preformed in cartilage while other regions ossify directly and are thus membrane bone in the sense of Patterson (1977, this Volume). An additional complication in Anura is the presence within the order of two patterns of centrum formation: epichordal and perichordal. In perichordal frogs, in which the centrum surrounds the notochord, much of the centrum is membrane bone, while in epichordal forms, in which the centrum ossifies only above the notochord, it is largely preformed in cartilage.

Haemal arches are developed only in the urodeles, as the only group with persistent tails. They lie directly below the caudal centra and not in an 'intercentral' position below the intervertebral discs.

This is also the position occupied by the haemal arch in the Palaeozoic Nectridea such as *Urocordylus*. The caudal vertebrae of Nectridea bear fan-shaped neural and haemal arches so that they appear almost bilaterally symmetrical about a horizontal plane (Fig. 5D).

Vaughn (1963) has described a small, apparently caudal vertebra from a coprolite of early Permian (Wolfcampian) age from New Mexico. It bears an intracentral haemal arch of urodele-like form. In this and other respects the vertebra is strikingly like that of living urodeles. He has also described a very frog-like vertebra from contemporary deposits in Utah. The latter vertebra is remarkable in lacking any notochordal perforation (Vaughn, 1965).

Whatever the taxonomic provenance of these specimens they are very different from the vertebrae of any known Lower Permian tetrapod and might conceivably extend the history of the two respective lissamphibian orders back to at least the Lower Permian.

The vertebrae of Aïstopoda are of 'lepospondyl' form with a single elongate centrum to which is fused the neural arch. In Aïstopoda the neural arch is low with little development of the spine. Normal haemal arches are absent but a pair of ridges run longitudinally below the caudal centra and thus define the haemal canal.

Baird (1964) notes that in the aïstopod family Phlegethontiidae and possibly in the other family Ophiderpetontidae the neural arch pedicels below the posterior zygapophyses and just behind the transverse processes are perforated for the spinal nerve (Fig. 5E). This character is unknown in the Nectridea but the arch pedicels are perforated for at least the dorsal root in some living urodeles (e.g. Wake & Lawson, 1973).

Lund (1976) mentions a very small specimen of *Phlegethontia linearis* in which separate intercentra and pleurocentra are visible in the neck region, divided in the plane of the intravertebral foramen. Nobody else has described such a condition in any aïstopod and conceivably it represents an ontogenic stage in the development of a holospondylous vertebra.

DISCUSSION

In discussing the homologies of vertebral centra in the various groups of early tetrapods and the related crossopterygian fish several agreed hypotheses can be disposed of fairly quickly.

Firstly it is now generally accepted that the principal centrum of batrachosaurs, microsaurs and amniotes is homologous throughout. More or less satisfactory series can be reconstructed within each group to show that the principal centrum expands at the expense of the intercentrum and in some microsaurs and all but primitive amniotes becomes the sole ossified centrum in the trunk region.

It is important, however, to realize what is implicit and even more, what is *not* implicit in the above hypothesis. Firstly I have urged (Panchen, 1967) that the single centrum of amniotes be regarded as the homologue of the whole compound centrum of labyrinthodonts. This has been accepted by Thomson & Bossy (1970) and Wake (1970), although the latter author is rightly sceptical about accepting my 1967 ontogenetic hypothesis.

Thus in one sense the single trunk centrum of a reptile or an advanced microsaur, such as *Pantylus*, is homologous with pleurocentrum *plus* intercentrum of an anthracosaur (or of their own tail regions). This is the sense in which I use the term anatomical homology. Intercentrum and pleurocentrum form a unit with the same relationship to surrounding structures (notably ribs, myosepta and segmental vessels as well as the neural arch) as the holospondylous centrum of advanced amniotes.

Anatomical homology does not imply ontogenetic homology, which asserts similar embryological derivation. However, I wish to propose as a probable (but untestable!) hypothesis that the centrum, compound or single, of batrachosaurs, microsaurs and amniotes, shows ontogenetic homology throughout; i.e. that it is produced by the process of resegmentation known in living amniotes. This hypothesis further implies that there is no necessary relationship between the division of the adult compound centrum into intercentrum and pleurocentrum and the embryological structure of half-sclerotomes and embryonic intervertebral discs (contra Williams, 1959).

However, the agreed phylogenetic homology is between the single trunk centrum of amniotes and microsaurs and the pleurocentrum of batrachosaurs, primitive microsaurs and primitive amniotes. The hypothesis in this case is that a reconstructed phylogenetic series of the pleurocentrum expanding at the expense of the intercentrum represents the actual phylogeny, or more correctly the result of a true phyletic series of ontogenies.

The term serial homology may be used in all three senses, analogous to anatomical, ontogenetic and phylogenetic homology, to assert similarity of relationship to other segmental structures within metamerically repeated segments along the body. In the review above I have used it analogously to phylogenetic homology to assert the homology of the pleurocentrum in the tail of (e.g.) *Ichthyostega* and *Peltobatrachus* with the respective undivided trunk centrum of each animal.

While suggesting that the ontogeny and phylogeny of the vertebrae and particularly the centra are similar in batrachosaurs, microsaurs and reptiles I do not regard this as evidence of any close relationship between the three groups—quite the reverse. I regard the possession of an essentially diplospondylous gastrocentrous vertebra as a primitive character of unknown taxonomic extent in early tetrapods. There is compelling reason to believe from the structure of the skull that the three groups are not closely related (Carroll & Baird, 1968; Panchen, 1972, 1975).

Further, parallel evolution of vertebral structure (and of other parts of the postcranial skeleton) is almost certainly due to similar selection pressures on relatively small terrestrial tetrapods with a common endowment of primitive characters. In this respect the persistently poorer ossification and more primitive condition of the vertebrae of the anthracosaur *Gephyrostegus*, suggested by Carroll (1969a, 1970) as a relict of reptile ancestry, makes it appear much less terrestrially adapted than its microsaur and reptile contemporaries.

I suggested in 1967 that the adaptive reason for the increasing dominance of the pleurocentrum in the batrachosaurs (other than Embolomeri) and reptiles was the increasing dominance of the load-bearing girder function of the vertebral column over the locomotor compression-member one. The pleurocentrum is always the element most closely associated with the neural arch and thus with the articulations between centra. This thesis would of course, also apply to the microsaurs.

In the classic series in temnospondyl labyrinthodonts the opposite is the case. Aquatic locomotion favoured that central element, the intercentrum, most closely associated with the myoseptum and ribs. Thus the oblique split in the centrum separating intercentrum and pleurocentrum moves postero-dorsally in phylogeny, favouring the intercentrum; until in a few stereospondyls the pleurocentrum disappears even as a hypothetical cartilage element (Watson, 1919; Panchen, 1967).

This is essentially the hypothesis that I put forward in 1967, the important difference being that there is now no palaeontological evidence of backward convergence between the batrachosaur ('anthracosaur') line on one hand and the temnospondyl one on the other.

There is a parallel lack of any shared derived characters in the skull; every character normally used to diagnose the subclass Labyrinthodontia of current classification is primitive for tetrapods. Notably this is the case for the eponymous tooth structure, the pattern of dermal bones of the skull roof (at least in temnospondyls) and the compound centra of the vertebrae.

The only shared derived character is the characteristic otic notch. However, the notchless condition in trimerorhachoids, colosteids and the gastrocentrous Namurian temnospondyl *Caerorhachis* described by Holmes & Carroll (1976) suggests that the otic notch has arisen within the temnospondyls, conceivably more than once. Thus it would represent a common response in the unrelated temnospondyls and batrachosaurs to problems of hearing in large primitive terrestrial tetrapods (Panchen, 1972).

The vast majority of temnospondyls have rhachitomous vertebrae and the primitive type of rhachitomous vertebra ('schizomerous') present in colosteids, some trimerorhachoids and some edopoids (e.g. *Dendrerpeton*: see Carroll, 1967) is easily derived from that of the crossopterygian *Osteolepis*.

If this is accepted as evidence of a true ancestor-descendant relationship, however, the rhachitomous vertebra must be regarded as primitive for temnospondyls. This implies either that it was also primitive for all tetrapods, in a fish or early tetrapod ancestor, or that the vertebrae of *Ichthyostega*, batrachosaurs, 'lepospondyls' and reptiles were derived from some other crossopterygian fish pattern (or patterns).

If the latter is the case there is no necessary phylogenetic homology between the intercentrum of temnospondyls and that of batrachosaurs *et al.*, or between the paired pleurocentra of temnospondyls and the single pleurocentrum of the latter groups. Nor can ontogenetic homology be asserted. The only reasonable hypothesis would be a limited anatomical homology based on relative position (as Wake, 1970, has pointed out) and relationship to the neural arch.

Turning to the Nectridea, Aïstopoda and extant amphibia, it is probable, as Williams suggested, that their single holospondylous centrum is phylogenetically as well as anatomically and ontogenetically homologous with that of amniotes, i.e. that it is a pleurocentrum.

The position of the haemal arches in Nectridea and urodeles might be regarded as evidence against this. However, haemal arches become intravertebral in the tails of some lizards and all snakes.

Thus, were it not for the *Osteolepis*—primitive temnospondyl link, my inclination would be to say that the 'normal' condition for early tetrapods was to have vertebrae in which the pleurocentrum is dominant and is closely associated by articulation, suture or fusion, with the neural arch. The ancestral condition on this hypothesis would be a *functionally* embolomerous one in which ball-and-socket diplospondyly existed between the large pleurocentrum and the smaller, partly cartilaginous intercentrum.

This comes very close indeed to Watson's classic theory of the primitive embolomerous vertebra. However, a true embolomerous vertebra, derived from the primitive condition, only evolved in the Embolomeri in association with extreme elongation of the trunk and, in many cases, large size (Panchen, 1966).

If the primitive embolomerous or gastrocentrous vertebra is accepted, the presence of aberrant vertebrae in some temnospondyls (*Caerorhachis*, some dissorophoids, all plagiosaurs) might be regarded as corroborating evidence, with *Caerorhachis* representing a primitive condition. However, Holmes & Carroll (1976) in describing *Caerorhachis*, despite making some of the points put forward in this discussion, finally conclude that the axis was functionally notochordal and thus that the pattern of vertebral ossification was of no great taxonomic or functional significance.

The common possession of a gastrocentrous/holospondylous centrum was used to unite batrachosaurs and 'lepospondyls' by Brough & Brough (1967) as class 'Eoreptilia' in a scheme of classification in which 'Eobatrachia' (Ichthyostegalia and temnospondyls), Amphibia (the Lissamphibia) and Reptilia were also separate classes. Microsaur—batrachosaur—reptile relationship, implicit in this scheme, has already been rejected, but Thomson & Bossy (1970) incline towards an association of the Nectridea and Aïstopoda with the batrachosaurs, largely on skull characters. Lund (1976), apparently influenced by Thomson & Bossy, actually places the two lepospondyl groups within his *subclass* Anthracosauria but neither presents the framework within which this subclass exists nor diagnoses it.

Wake (1970) concludes that nothing in the vertebral ontogeny of the three living orders of Amphibia either supports or refutes their association as Lissamphibia. However, Carroll & Currie (1975) have recently suggested the origin of the Apoda from microsaurs in contradistinction to the origin of Anura and Urodela from a common, probably dissorophoid, ancestor.

Taxonomic conclusions from this review are mostly negative. The Labyrinthodontia are not a natural group but are at least diphyletic, with temnospondyls and batrachosaurs probably unrelated and ichthyostegids close to neither. Microsaurs are not lepospondyls, but the Nectridea and Aïstopoda may still belong together. Nothing can be said about the origin and interrelationships of the extant orders from their vertebral structure.

ACKNOWLEDGEMENTS

It is pleasant and appropriate to be able to acknowledge the stimulus to my speculations on vertebrae of a discussion with Professor Stanley Westoll many years ago on the function of tetrapod ribs. I am also glad to be able to recall being shown the vertebral column of *Osteolepis* long before it was described.

I am grateful to Dr Eileen Beaumont, Dr Robert L. Carroll and Mr Robert Holmes, Miss Christine Janis and Dr Richard Lund for permitting me to cite their unpublished work.

As with all my previous papers the manuscript was typed by my wife, Rosemary Panchen.

REFERENCES

ANDREWS, S. M., 1973. Interrelationships of crossopterygians. In P. H. Greenwood *et al.* (Eds), *Interrelationships of fishes:* 138–77. London: Academic Press.

ANDREWS, S. M. & WESTOLL, T. S., 1970a. The postcranial skeleton of *Eusthenopteron foordi* Whiteaves. *Trans. R. Soc. Edinb.*, *68:* 207–329.

ANDREWS, S. M. & WESTOLL, T. S., 1970b. The postcranial skeleton of rhipidistians excluding *Eusthenopteron*. *Trans. R. Soc. Edinb.*, *68:* 391–489.

BAIRD, D., 1957. Rhachitomous vertebrae in the loxommid amphibian *Megalocephalus*. *Bull. geol. Soc. Am.*, *68:* 1698.

BAIRD, D., 1964. The aïstopod amphibians surveyed. *Breviora*, No. 206: 1–17.

BEAUMONT, E. H., 1977. Cranial morphology of the Loxommatidae (Amphibia: Labyrinthodontia). *Phil. Trans. R. Soc.* (B)

BOLT, J. R., 1969. Lissamphibian origins: possible protolissamphibian from the Lower Permian of Oklahoma. *Science, N.Y. 166:* 888–91.

BROUGH, M. C. & BROUGH, J., 1967. Studies on early tetrapods. I. The lower Carboniferous microsaurs. II. *Microbrachis*, the type microsaur. III. The genus *Gephyrostegus*. *Phil. Trans. R. Soc.* (B), *252:* 107–65.

BYSTROW, A. P., 1944. *Kotlassia prima* Amalitzky. *Bull. geol. Soc. Am.*, *55:* 379–416.

CARROLL, R. L., 1964. The relationships of the rhachitomous amphibian *Parioxys*. *Am. Mus. Novit.*, No. 2167: 1–11.

CARROLL, R. L., 1967. Labyrinthodonts from the Joggins Formation. *J. Paleont.*, *41:* 111–42.

CARROLL, R. L., 1968. The postcranial skeleton of the Permian microsaur *Pantylus*. *Can. J. Zool,.* *46:* 1175–92.

CARROLL, R. L., 1969a. Problems of the origin of reptiles. *Biol. Rev.*, *44:* 393–432.

CARROLL, R. L., 1969b. A new family of Carboniferous amphibians. *Palaeontology*, *12:* 537–48.

CARROLL, R. L., 1970. The ancestry of reptiles. *Phil. Trans. R. Soc.* (B), *257:* 267–308.

CARROLL, R. L. & BAIRD, D., 1968. The Carboniferous amphibian *Tuditanus* (*Eosauravus*) and the distinction between microsaurs and reptiles. *Am. Mus. Novit.*, No. 2337: 1–50.

CARROLL, R. L. & CURRIE, P. J., 1975. Microsaurs as possible apodan ancestors. *Zool. J. Linn. Soc.*, *57:* 229–47.

CHASE, J. N., 1965. *Neldasaurus wrightae*, a new rhachitomous labyrinthodont from the Texas Lower Permian. *Bull. Mus. comp. Zool. Harv.*, *133:* 153–225.

COPE, E. D., 1880. Extinct Batrachia. *Am. Nat.*, *14:* 609–10.

COPE, E. D., 1882. The rhachitomous Stegocephali. *Am. Nat.*, *16:* 334–5.

CREDNER, H., 1891. Die Urvierfüssler (Eotetrapoda) des sächsischen Rotliegenden. *Naturw. Wschr.*, *5:* 471–5; 483–4; 491–7; 507–9.

DALY, E., 1973. A Lower Permian vertebrate fauna from southern Oklahoma. *J. Paleont.*, *47:* 562–89.

EFREMOV, J. A., 1946. On the subclass Batrachosauria, a group of forms intermediate between amphibians and reptiles. (In Russian, English summary). *Izv. Akad. Nauk SSSR* (*Biol.*), *6:* 616–38.

GADOW, H. F., 1896. On the evolution of the vertebral column of Amphibia and Amniota. *Phil. Trans. R. Soc.* (B) *187:* 1–57.

GADOW, H. F., 1933. *The evolution of the vertebral column.* Cambridge: University Press.

GADOW, H. F. & ABBOTT, E. C., 1895. On the evolution of the vertebral column of fishes. *Phil. Trans. R. Soc.* (B) *186:* 163–221.

GOODRICH, E. S., 1930. *Studies on the structure and development of vertebrates.* London: Macmillan.

GREGORY, J. T., PEABODY, F. E. & PRICE, L. I., 1956. Revision of the Gymnarthridae, American Permian microsaurs. *Bull. Peabody Mus.*, *10:* 1–77.

GREGORY, W. K., ROCKWELL, H. & EVANS, F. G., 1939. Structure of the vertebral column in *Eusthenopteron foordi* Whiteaves. *J. Paleont.*, *13:* 126–9.

HOLMES, R. & CARROLL, R. L., 1976. Temnospondyl amphibian from the Mississippian of Scotland. *Breviora*

HOLMGREN, N., 1933. On the origin of the tetrapod limb. *Acta zool., Stockh.*, *14:* 185–295.

HOLMGREN, N., 1949. Contributions to the question of the origin of tetrapods. *Acta zool., Stockh., 30:* 459–84.

HOTTON, N., 1970. *Mauchchunkia bassa* gen. et sp. nov. an anthracosaur (Amphibia, Labyrinthodontia) from the Upper Mississippian. *Kirtlandia*, No. 12: 1–38.

JARVIK, E., 1948. On the morphology and taxonomy of the Middle Devonian osteolepid fishes of Scotland. *K. svenska VetenskAkad. Handl., 25*(3), No. 1: 1–301.

JARVIK, E., 1952. On the fish-like tail in the ichthyostegid stegocephalians. *Meddr Grønland, 114*, No. 12: 1–90.

JARVIK, E., 1972. Middle and Upper Devonian Porolepiformes from East Greenland with special reference to *Glyptolepis groenlandica* n.sp. *Meddr Grønland, 187*, No. 2: 1–307.

LUND, R., 1976. Anatomy and relationships of the Pennsylvanian amphibian family Phlegethontiidae. *Ann. Carneg. Mus.*

MIALL, L. C., 1875. Report of the committee . . . on the structure and classification of the labyrinthodonts. *Rep. Br. Ass. Advmt Sci., 1874:* 149–92.

MOULTON, J. M., 1974. A description of the vertebral column of *Eryops* based on the notes and drawings of A. S. Romer. *Breviora*, No. 428: 1–44.

MOUSTAFA, Y. S., 1955. The skeletal structure of *Parioxys ferricolus*, Cope. *Bull. Inst. Egypte, 36:* 41–76.

NICHOLSON, H. A. & LYDEKKER, R., 1889. *A manual of palaeontology* . . . 3rd ed., Vol. *II*. London: Blackwood.

NILSSON, T., 1939. Cleithrum und Humerus der Stegocephalen und rezenten Amphibien. *Acta Univ. lund., 35*(10): 1–39.

NILSSON, T., 1946. A new find of *Gerrothorax rhaeticus* Nilsson, a plagiosaurid from the Rhaetic of Scania. *Acta Univ. lund., 42*(10): 1–42.

OLSON, E. C., 1971. A skeleton of *Lysorophus tricarinatus* (Amphibia: Lepospondyli) from the Hennessey Formation (Permian) of Oklahoma. *J. Paleont., 45:* 443–9.

OLSON, E. C., 1972. *Fayella chickashaensis*, the Dissorophoidea and the Permian terrestrial radiations. *J. Paleont., 46:* 104–14.

PANCHEN, A. L., 1959. A new armoured amphibian from the Upper Permian of East Africa. *Phil. Trans. R. Soc. (B), 242:* 207–81.

PANCHEN, A. L., 1966. The axial skeleton of the labyrinthodont *Eogyrinus attheyi. J. Zool., Lond., 150:* 199–222.

PANCHEN, A. L., 1967. The homologies of the labyrinthodont centrum. *Evolution, 21:* 24–33.

PANCHEN, A. L., 1970. *Teil 5a: Anthracosauria. Handbuch der Paläoherpetologie*. Stuttgart: Fischer.

PANCHEN, A. L., 1972. The interrelationships of the earliest tetrapods. In K. A. Joysey & T. S. Kemp (Eds), *Studies in vertebrate evolution: essays presented to Dr F. R. Parrington, FRS:* pp. 65–87. Edinburgh: Oliver & Boyd.

PANCHEN, A. L., 1975. A new genus and species of anthracosaur amphibian from the Lower Carboniferous of Scotland and the status of *Pholidogaster pisciformis* Huxley. *Phil. Trans. R. Soc. (B), 269:* 581–640.

PANCHEN, A. L., 1977. On *Anthracosaurus russelli* Huxley (Amphibia: Labyrinthodontia) and the family Anthracosauridae. *Phil. Trans. R. Soc. (B)*

PARRINGTON, F. R., 1967. The vertebrae of early tetrapods. *Colloques int. Cent. natn Rech. Sci.*, No. 163: 269–79.

PARSONS, T. S. & WILLIAMS, E. E., 1962. The teeth of Amphibia and their relation to Amphibian phylogeny *J. Morph., 110:* 375–89.

PARSONS, T. S. & WILLIAMS, E. E., 1963. The relationships of the modern Amphibia: a re-examination. *Q. Rev. Biol., 38:* 26–53.

PATTERSON, C., 1977. Cartilage bones, dermal bones and membrane bones, or the endoskeleton versus the exoskeleton. In S. M. Andrews *et al.* (Eds), *Problems of vertebrate evolution:* 77–121. London: Academic Press.

PIVETEAU, J., 1937. Un amphibien du Trias inférieur. Essai sur l'origine et l'évolution des amphibiens anoures. *Annls Paléont., 26:* 135–77.

ROMER, A. S., 1930. The Pennsylvanian tetrapods of Linton, Ohio. *Bull. Am. Mus. nat. Hist., 59:* 77–147.

ROMER, A. S., 1933. *Vertebrate paleontology*, 1st ed. Chicago: University Press.

ROMER, A. S., 1939. Notes on branchiosaurs. *Am. J. Sci., 237:* 748–61.

ROMER, A. S., 1945. *Vertebrate paleontology*, 2nd ed. Chicago: University Press.

ROMER, A. S., 1947. Review of the Labyrinthodontia. *Bull. Mus. comp. Zool. Harv., 99:* 1–368.

ROMER, A. S., 1950. The nature and relationships of the Palaeozoic microsaurs. *Am. J. Sci., 248:* 628–54.

ROMER, A. S., 1956. *Osteology of the reptiles*. Chicago: University Press.

ROMER, A. S., 1964. The skeleton of the Lower Carboniferous labyrinthodont *Pholidogaster pisciformis. Bull. Mus. comp. Zool. Harv., 131:* 129–59.

ROMER, A. S., 1966. *Vertebrate paleontology*, 3rd ed. Chicago: University Press.

ROMER, A. S., 1968. *Notes and comments on vertebrate paleontology*. Chicago: University Press.

ROMER, A. S., 1969. A temnospondylous labyrinthodont from the Lower Carboniferous. *Kirtlandia*, No. 6: 1–20.

ROMER, A. S., 1970. A new anthracosaurian labyrinthodont, *Proterogyrinus scheelei*, from the Lower Carboniferous. *Kirtlandia*, No. 10: 1–16.

SÄVE-SÖDERBERGH, G., 1934. Some points of view concerning the evolution of the vertebrates and the classification of this group. *Ark. Zool.*, *26A*, No. 17: 1–20.

SCHAEFFER, B., 1967. Osteichthyan vertebrae. In C. Patterson & P. H. Greenwood (Eds), *Fossil vertebrates. Papers presented to Dr E. I. White* (also *J. Linn. Soc. (Zool.) 47:* 185–95). London: Academic Press.

SIMONS, J. R., 1970. The direction of the thrust produced by the heterocercal tails of two dissimilar elasmobranchs: the Post Jackson shark, *Heterodontus portusjacksoni* (Meyer) and the Piked dogfish *Squalus megalops* (Macleay). *J. exp. Biol.*, *52:* 95–107.

SMIT, A. L., 1953. The ontogenesis of the vertebral column of *Xenopus laevis* (Daudin) with special reference to the segmentation of the metotic region of the skull. *Annale Univ. Stellenbosch*, *29:* 79–163.

ŠPINAR, Z. V., 1952. Revise některých moravských Diskosauriscidů (Labyrinthodontia). *Rozpr. ústred. Úst. geol.*, *15:* 1–160.

STEEN, M. C., 1931. The British Museum collection of Amphibia from the Middle Coal Measures of Linton, Ohio. *Proc. zool. Soc. Lond.*, *1930:* 849–91.

STEEN, M. C., 1934. The amphibian fauna from the South Joggins, Nova Scotia. *Proc. zool. Soc. Lond.*, *1934:* 465–504.

STEEN, M. C., 1937. On *Acanthostoma vorax* Credner. *Proc. zool. Soc. Lond.*, *1937:* 491–9.

STEEN, M. C., 1938. On the fossil Amphibia from the Gas Coal of Nýřany and other deposits in Czechoslovakia. *Proc. zool. Soc. Lond.*, *108:* 205–83.

STENSIÖ, E. A., 1931. Upper Devonian vertebrates from East Greenland collected by the Danish Greenland Expeditions in 1929 and 1930. *Meddr Grønland*, *86* (No. 1): 1–212.

THOMSON, K. S. & BOSSY, K. H., 1970. Adaptive trends and relationships in early Amphibia. *Forma Functio*, *3:* 7–31.

THOMSON, K. S. & VAUGHN, P. P., 1968. Vertebral structure in Rhipidistia (Osteichthyes, Crossopterygii) with description of a new Permian genus. *Postilla*, No. 127: 1–19.

VAUGHN, P. P., 1963. New information on the structure of Permian lepospondylous vertebrae—from an unusual source. *Bull. Sth. Calif. Acad. Sci.* 62: 150–8.

VAUGHN, P. P., 1965. Frog-like vertebrae from the Lower Permian of southeastern Utah. *Contr. Sci.*, No. 87: 1–18.

VAUGHN, P. P., 1972. More vertebrates including a new microsaur from the Upper Pennsylvanian of Central Colorado. *Contr. Sci.*, No. 223: 1–30.

WAKE, D. B., 1970. Aspects of vertebral evolution in the modern Amphibia. *Forma Functio*, *3:* 33–60.

WAKE, D. B. & LAWSON, R., 1973. Developmental and adult morphology of the vertebral column in the plethodontid salamander *Eurycea bislineata*, with comments on vertebral evolution in the Amphibia. *J. Morph.*, *139:* 251–99.

WATSON, D. M. S., 1917. A sketch classification of the pre-Jurassic tetrapod vertebrates. *Proc. zool. Soc. Lond.*, *1917:* 167–86.

WATSON, D. M. S., 1919. The structure, evolution and origin of the Amphibia. —The 'Orders' Rhachitomi and Stereospondyli. *Phil. Trans. R. Soc. (B)*, *209:* 1–73.

WATSON, D. M. S., 1926. Croonian Lecture. —The evolution and origin of the Amphibia. *Phil. Trans. R. Soc. (B)*, *214:* 189–257.

WATSON, D. M. S., 1929. The Carboniferous Amphibia of Scotland. *Palaeont. hung.*, *1:* 219–52.

WATSON, D. M. S., 1940. The origin of frogs. *Trans. R. Soc. Edinb.*, *60:* 195–231.

WELLES, S. P. & ESTES, R., 1969. *Hadrokkosaurus bradyi* from the Upper Moenkopi formation of Arizona, with a review of the brachyopid labyrinthodonts. *Univ. Calif. Publs geol. Sci.*, *84:* 1–54.

WESTOLL, T. S., 1942. Ancestry of captorhinomorph reptiles. *Nature, Lond.*, *149:* 667–8.

WESTOLL, T. S., 1949. On the evolution of the Dipnoi. In G. L. Jepsen, *et al.*, (Eds), *Genetics, paleontology and evolution:* 121–184. Princeton: University Press.

WESTOLL, T. S., 1962. Some crucial stages in the transition from Devonian fish to man. *The evolution of living organisms: Symp. R. Soc. Victoria 1959.*

WHITE, T. E., 1939. Osteology of *Seymouria baylorensis* Broili. *Bull. Mus. comp. Zool. Harv.*, *85:* 325–409.

WILLIAMS, E. E., 1959. Gadow's arcualia and the development of tetrapod vertebrae. *Q. Rev. Biol.*, *34:* 1–32.

ZITTEL, K. von., 1895. *Grundzüge der Palaeontologie (Palaeozoologie).* München & Leipzig: R. Oldenbourg.

Evolution of the pelvis in birds and dinosaurs

ALICK D. WALKER

Department of Geology, University of Newcastle upon Tyne

A consideration of some aspects of the pelvic musculature of living and fossil reptiles and birds leads to a modification of the ideas of Galton (1969) concerning the underlying causes of the backward rotation of the pubis in ornithischian dinosaurs. A key feature is believed to be the dorsal migration of the puboischiofemoralis internus in ornithischians, crocodiles and birds, leading to the weakening of the pubis as a preliminary to backward rotation. A reconstruction of part of the musculature of the pelvis in *Ornithosuchus* and theropod dinosaurs, somewhat different from that of Romer (1923c), gives a plausible explanation for the non-rotation of the pubis in theropods and suggests that it is highly unlikely that birds have originated from this group. The sequence of muscular changes leading to the backward rotation of the pubis in birds probably paralleled that of ornithischians up to a certain point, but the final cause is attributed to the modification of the hind limbs to form an arrestor mechanism upon landing from a glide during the transition from the quadrupedal to the bipedal arboreal stages. Reasons are given for supposing that the pelvis of the Jurassic crocodile *Hallopus* could have been opisthopubic.

CONTENTS

INTRODUCTION

Romer (1922, 1923a, b, c, 1927a, b, 1942) published a series of important papers on reptilian and avian pelvic muscles with particular reference to the restoration of the musculature in fossil archosaurs. These studies have been very influential and are the foundation on which any subsequent work must be based. Since then few modifications to Romer's ideas have been made, the most notable exception being that of Galton (1969) on the pelvic muscles of ornithischian dinosaurs and birds, particularly in connection with the ornithopod *Hypsilophodon*. Other studies include general papers by Colbert (1964) and Charig (1972) on the evolution of the archosaur pelvis and hind-limb, reconstruction of the pelvic musculature of the ornithomimid *Dromiceiomimus* by Russell (1972), and brief remarks on certain aspects of the musculature by Bakker (1971), and Bakker & Galton (1974). It seems to me, however, with the advantage of hindsight, that Romer placed too much reliance on the alligator as a guide to the restoration of dinosaurian muscula-ture, and not enough upon lizards and *Sphenodon*. This seems to stem in part from his early work (1922) on *Iguana*. This is not to suggest that the crocodile should be completely discounted as a source of information—far from it—but simply to sound a note of caution. Romer assumed that a dorsal migration of the puboischio-femoralis internus as in the alligator and crocodile (Gadow, 1882a: 'quadratus lumborum', 1882b) was a common archosaurian character. However, the peculi-arities of the crocodilian pubis—its rod-like form, exclusion from the acetabulum and lack of a symphysis—mark it out as very different from that of other reptiles, in particular the pubis of the Thecodontia, hence it is unwise to assume that its musculature represents a primitive archosaurian type.

As a preliminary to further discussion, it is necessary at this point to make some comments on the relationship between birds and crocodiles. As a result of work on *Sphenosuchus*, a reptile from the Upper Trias of South Africa, the conclusion has been reached (Walker, 1972, 1974) that birds and crocodiles share an immediate common ancestry in the Trias, and are not independently descended from theco-dontians. The evidence for this conclusion has yet to be published in detail, but it is based mainly on the forward position of the quadrate head (the 'proquadrate' condition) in both groups, the details of the otic capsule (elongated cochlear duct, large, laterally-placed perilymphatic sac, etc.), backward growth of the embryonic infrapolar processes in connection with the formation of 'basitemporals', and pattern of pneumatization of the skull, also initiation in *Sphenosuchus* of the specializations of the bird coracoid. *Sphenosuchus* has a 'proquadrate' streptostylic quadrate and cranial kinetic system of primitive avian type (the interpretations previously given (Walker, 1972, 1974) are incorrect in certain respects), and the quadrate is unique in known reptiles (but resembles that of birds) in having separate orbital and pterygoid processes. Thus it appears to be closer to the direct line leading to birds than to 'true' crocodiles, and it is assumed that the latter are derived from the Sphenosuchia (perhaps polyphyletically) by fixation of the quadrate and loss of kinesis. The essential characters which unite these groups originated while the ancestral stock was still in the reptilian stage, and they denote the emergence of a higher level of organization than that of the Theco-dontia, approximating to that of living crocodiles. To express these relationships it is necessary to erect a new Class, the Proquadrata, to include the Sphenosuchia, Crocodilia and Aves as subclasses.

The foregoing will make it clear that I differ fundamentally from Ostrom (1973, 1974a, b, 1975a, b, 1976) on the question of the origin of birds. I believe that birds originated from quadrupedal, arboreal ancestors in late Triassic to mid-Jurassic times, and progressed through various stages of gliding to flapping flight. It seems to me that the evolution of the bird shoulder girdle and its musculature (Walker, 1972) took longer than has previously been realized, and that the early arboreal stages of climbing and jumping were necessary in order to preadapt the shoulder musculature for its later function of moving the wing. An unpublished comparative study of the musculature of the coracoid in thecodontians and theropod and prosauropod dinosaurs leads to the conclusion that Ostrom's identification of a 'biceps tubercle' in the theropod *Deinonychus* (Ostrom, 1974b) is incorrect, and that this projection was for the origin of a tendinous portion of the coraco-branchialis, probably representing the posterior part of the coracobrachialis brevis. The homologous projection is present in the Middle Triassic thecodontian *Lago-suchus* (Bonaparte, 1975: fig. 8), and in prosauropod dinosaurs (von Huene, 1932: pls 9, 27) as well as in theropods. Its position in the former two groups (it is represented by a ridge in the thecodontians *Stagonolepis* and *Ornithosuchus* (Walker, 1961, 1964)) shows that it cannot have been for the origin of the biceps, unless one is prepared to ignore the musculature of living reptiles and simply place muscle-origins arbitrarily on the coracoid. The biceps in *Deinonychus* would almost certainly have originated from about the middle of the medial margin of the coracoid, as it normally does in reptiles (Fürbringer, 1876, 1900), and the apparent resemblance to the coracoid of *Archaeopteryx* is merely a superficial one. The often-repeated statement (Ostrom, 1973, 1974a, b; Bakker, 1975; Bakker & Galton, 1974) that the shoulder girdle of *Archaeopteryx* is reptilian and that the glenoid faces downward is incorrect; the girdle in fact is essentially avian and the glenoid faces laterally, with an anterodorsal-posteroventral orien-tation of the axis of the cavity.

THE PUBOISCHIOFEMORALIS INTERNUS

This muscle, henceforth abbreviated to 'pifi', primitively originates in reptiles on the inner or dorsal surface of the pubo-ischium and passes round the anterior border of the girdle, between the processus lateralis pubis and the anterior notch of the ilium. Normally it has two or three insertions on the femur. In lizards the pifi is usually divisible into three portions. Unfortunately the nomenclature applied to these divisions is confused, Gadow (1882b), Romer (1923b, 1942) and Galton (1969) using four different systems of numbering. In an attempt to simplify matters the divisions will here be termed pifi dorsalis, medialis and ventralis, referring to the order in which they pass round the anterior border of the girdle (Fig. 1). Ventralis is absent in crocodiles, (and also in birds) and while all three divisions are present in many lizards, they may scarcely be differentiated in some. In *Sphenodon* (Perrin, 1895; Osawa, 1898) the pifi is undivided at origin (except for a small anterior portion) and has a continuous fleshy insertion on the femur. The obturator nerve is usually taken to separate dorsalis and medialis in lizards, while it is convenient to restrict ventralis to that part which passes out below the ambiens, although this does not quite correspond to Romer's Part III (1942). (It is not feasible in fossils to attempt to distinguish between a possible portion of III *above* the ambiens and Part II (sensu Romer, 1942), hence the portion of III which runs above the ambiens in lizards will be ignored in the present work.) The apparent absence of the pifi ventralis in the specimen of *Iguana* used by Romer (1922) as one of his 'types' (although I believe it can be made out on his pl. XLIII), and its absence also in crocodiles, appears to have led him to rule out the possibility of there having been a sub-ambiens portion of the pifi in archosaurs. However, I believe that a substantial pifi ventralis was present in many fossil representa-tives of this group, and that recognition of this fact is important to the under-standing of the changes which took place in the archosaur pelvis.

The muscle is well documented in lizards by Gadow (1882b), Perrin (1892) and Rabl (1916), and has been particularly clearly illustrated by Romer himself (1942) in the case of *Lacerta*. Other genera listed by these authors include: *Iguana*, *Varanus*, *Chalcides*, *Cnemidophorus*, *Uromastix*, *Ptyodactylus* and *Chamaeleo*. *Sphenodon* (Perrin, 1895) has no division of the muscle passing out below the ambiens, but in this form the processus lateralis is very close to the ambiens origin, so that there is scarcely room for it, and from Perrin's account it seems that the ventral portion ends at the ilio-pubic ligament. The pifi ventralis is not a large muscle in lizards, but it originates from the down-turned anterior portion of the pubis, passes out *above* the processus lateralis (a point to be emphasized), and

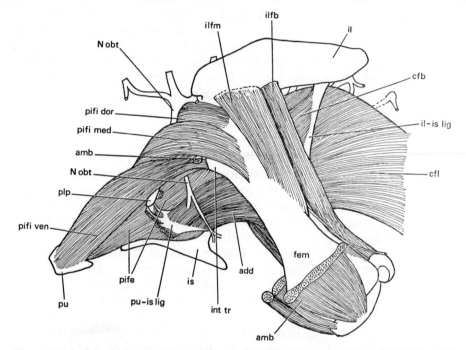

Figure 1. Part of the thigh musculature of *Lacerta*, left lateral view, after Romer (1942: Stage IV). For key to abbreviations see p. 358.

inserts on the internal trochanter of the femur along with the anterior part of the puboischiofemoralis externus (pife). Its course around the pubis is marked by a concavity of the lateral edge, between the processus lateralis and the slight projection near the acetabulum which indicates the origin of the ambiens (Fig. 2). There is thus a clear resemblance between the very rudimentary and attenuated pubic 'apron' in, for example, *Varanus*, and the much longer one found in such thecodontians as *Stagonolepis* (Fig. 2C) and *Ornithosuchus*. In these Triassic forms the projection for the ambiens origin is clearly marked (Walker, 1961, 1964), and the distal thickening indicates the attachment of the rectus abdominis, and is thus homologous with the processus lateralis of the lepidosaurian pubis. The course of pifi ventralis is indicated by a change in the nature of the lateral edge of the pubis, passing upwards, from a rough or sharp edge to a smooth, rounded one just below the ambiens projection. This is particularly clear in the case of *Ornithosuchus* (BMNH R 2410), where a slight notch passes obliquely up and back round the edge of the pubis, ascending towards the proximal end of the femur (Fig. 3). The existence of a pifi ventralis is thus well-attested in these thecodontians. Presumably it had a large area of origin on the anterior surface of the 'apron', corresponding broadly to that of the pife on the posterior side, and, tapering to a tendon, passed obliquely proximally round the lateral edge of the pubis to an insertion on the inner side of the proximal end of the femur. The muscle appears to have been developed in all thecodontians, and a measure of its importance can be gauged from the length of the 'apron' which, by and large, increased throughout the Trias.

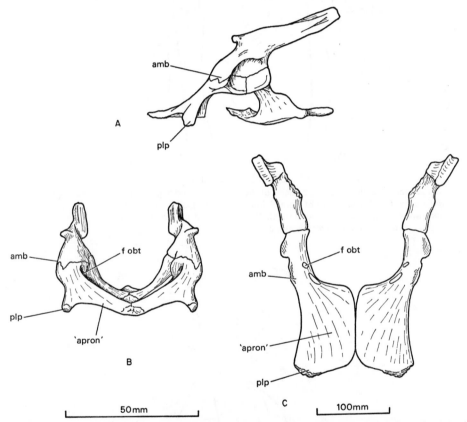

Figure 2. A, B. Left lateral and anterior views of pelvis of *Varanus*. C. Anterior view of pelvis of the Upper Triassic thecodontian *Stagonolepis*. In most lizards the ambiens arises from the pubis, but in *Varanus* and some other genera from the ilium.

A similar muscle can be inferred in the case of the saurischian dinosaurs (Fig. 4), in which there is very frequently a marked anterior concavity of the pubis, in side view, below the bulge which marks the ambiens origin. An origin for part of the pifi from this region was indicated by Gregory for *Tyrannosaurus* and *Camarasaurus* (previously unpublished drawings reproduced as Fig. 1 in Romer, 1923c). From a study of the pelvic muscles of the alligator, and consideration of the changes likely to have been brought about by the assumption of parasagittal excursion of the hind-limb, Romer (1923b, c) postulated a general dorsal migration of the origins of the muscles of the pubo-ischium in archosaurs. However, there seems to be no pressing reason why dorsal movement should predominate, and

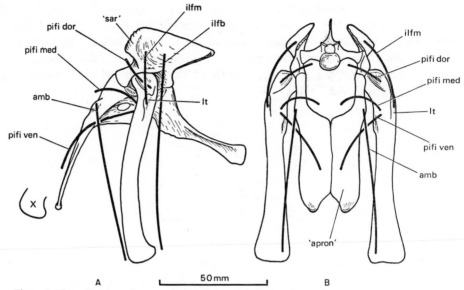

Figure 3. Lines of action of some muscles of the pelvis in the advanced Upper Triassic thecodontian *Ornithosuchus* (girdle and femur of a small individual). A, Left lateral view; B, anterior view. The femora are perhaps too near the vertical. 'X'. Distal end of femur in theoretical maximum protracted position, using pifi ventralis alone.

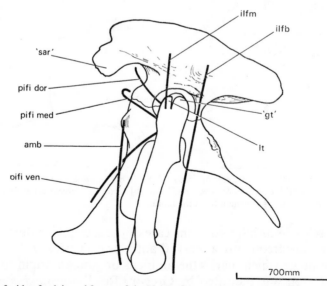

Figure 4. Left side of pelvis and femur of the Upper Cretaceous carnivorous dinosaur *Tyrannosaurus*, showing lines of action of some muscles. Data on girdle and femur from Osborn (1906, 1916). Scale approximate.

it seems to me, with the advantage of hindsight, that Romer was unduly influenced by the peculiarity of the crocodilian pifi, which has entirely left the pubis. Provided a muscle originates sufficiently far anterior or posterior to the main arc of movement of the femur, and passes obliquely to an insertion near to the proximal end, it can still bring about significant movement of the bone. Furthermore,

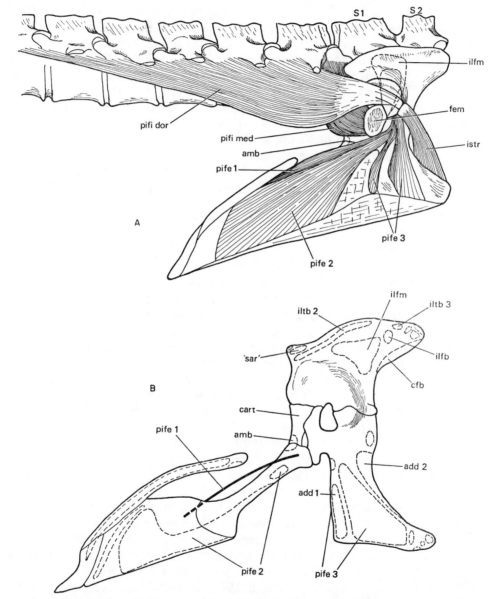

Figure 5. A. Left lateral view of the deep pelvic musculature of *Alligator mississippiensis*, after Romer (1923b: pl. 20). The femur is represented as cut half-way down. The iliofemoralis has been added in broken lines from Romer's pl. 19, fig. 2, in which it is shown in a more anterior position than in the figure below. B. Left side of pelvic girdle of *A. mississippiensis*, showing muscle origins and line of action of pife l, after Romer (1923b). Not to same scale as A.

since pifi ventralis was certainly a *protractor*, great power was not required, so that its proximal insertion was quite adequate for the work it had to do. Restoration of certain of the pelvic muscles of *Ornithosuchus* (Fig. 3A) shows that pifi ventralis alone was theoretically capable of protracting the femur slightly beyond the end of the pubis. Its effectiveness would have fallen off as the femur approached the pubis, but further protraction would have been possible by the more dorsal portions of the same muscle. Furthermore, the origin of the pifi ventralis from the anterior surface of the pubis would have made it a more effective protractor than the pife from the rear of the 'apron', since it had the advantage of the pulley-effect of its course round the anterior edge. Charig (1972) was puzzled by the problem of what he termed the 'femur knocking on the pubis'. However, this difficulty is only an apparent one, and disappears when the actual musculature is considered, as distinct from the theoretical examples discussed by Charig. Another important point is that the insertions on the femur of the protractors arising from the lower part of the pubis are so far proximal that they are always *behind* the anterior edge of the pubis, even with the femur extending directly forwards. The crocodile is an exception to this, again owing to the peculiarity of the pubis.

Russell (1972), in a reconstruction of the pelvic musculature of the ornithomimid *Dromiceiomimus*, shows a pubic head of the pife originating from the anterior surface of the pubis, as well as a pubic head of the pifi above it. This seems to be due to a misunderstanding of the state of affairs in the crocodile. In the latter, as Romer (1923b) has shown (Fig. 5), there is an additional head of the pife (Romer's pife 1), not found in *Sphenodon* or lizards, which arises from the dorsal or anterior surface of the pubis and passes smoothly round the concave lateral border of the bone to join the more normally-situated posterior portions of the same muscle. This pife 1 bears a strong superficial resemblance to the muscle which is here termed pifi ventralis. However, its obturator innervation in the crocodile, and the manner of its junction with 'pife 2', show that it is part of the *ventral* musculature, i.e. the pif externus, and it has only been enabled to spread round on to the dorsal surface precisely because the pif internus has vacated this area completely. Furthermore, the pifi ventralis passes round *above* the processus lateralis of the pubis in lizards, whereas 'pife 1' in the crocodile is *below* the equivalent of the processus lateralis which, according to Romer (1923b: 535) is quite close to the acetabulum. It is most unlikely that a 'pife 1' would have occurred in an animal at the same time as a well-developed pifi ventralis, and the existence of the muscle should be looked on as a crocodilian specialization.

In ornithischian dinosaurs Galton (1969) has made out a good case for the existence of a portion of the pifi (his 'part 2') originating from the prepubic process. It would seem unlikely that a portion of the pifi ventralis had survived the reduction of the pubis to a rod and its backward rotation in the Trias, especially when the very small size of the prepubic process in Triassic and early Jurassic ornithischians (Thulborn, 1972; Charig, 1972) is taken into account (Fig. 6). Furthermore, the pifi ventralis has also been lost in birds. The course of the postulated muscle relative to the ambiens is critical to its identification. The ambiens would have originated from the upper end of the pubis close to the acetabulum, and its initial path would have been down the inner side of the thigh;

the prepubic process in more advanced ornithischians curves somewhat laterally and the femur has an offset head, so that it seems probable that the muscle would have passed above (or lateral to) the ambiens to reach the inner side of the femur towards the proximal end, and hence was a pifi medialis, equivalent to the lower division of the muscle in crocodilians, as Galton suggested.

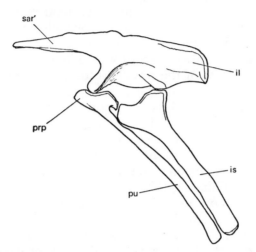

Figure 6. Left side of pelvis of the early Jurassic ornithischian dinosaur *Scelidosaurus*, after Charig (1972).

THE 'GREATER TROCHANTER'

Romer (1923c) placed the insertion of the pifi dorsalis (his 'part 2') partly on the summit of the 'greater trochanter' in *Tyrannosaurus*, and entirely on it in *Thescelosaurus* (1927b). Thus, although he actually says very little about its function, it is evident that he regarded this muscle as a retractor of the femur in dinosaurs, because of the supposed leverage of the proximally-projecting 'greater trochanter'. This assumption has been followed by Galton (1969) and others. More recently, however, Charig (1972) and I (Walker, 1972) have independently come to the conclusion that this mode of action of the pifi dorsalis is unlikely. Charig makes the point that the dinosaurian 'greater trochanter' never extends far enough proximally to afford significant leverage to the pifi, if it attached to it. Other reasons for rejecting the pifi dorsalis (or indeed any part of the pifi) as a retractor in dinosaurs are as follows:

(a) The variability of the proximal extension of the 'greater trochanter', which in fact hardly ever seems to rise above the head of the bone proper, and in certain groups (prosauropods, sauropods, stegosaurs, ankylosaurs) falls well short of it. It seems odd that the function of the muscle should thus vary from being a protractor to a retractor from group to group.

(b) There were undoubtedly more efficient retractors of the femur in dinosaurs —the caudifemorales, adductors and the posterior part of the pife which, particularly the first two, gave a much more direct backward pull on the femur. The weak contribution that the pifi dorsalis might have made, with its adverse leverage, could have added nothing significant to this retraction. This is particularly evident when one recalls that retraction is the power-stroke of the limb.

(c)   In all living reptiles the pifi mainly functions as a protractor of the femur. It also tends to abduct the femur, but there is no question of its acting as a retractor. This is still the case in the crocodile, in which both the pifi dorsalis and medialis have migrated dorsally and occupy very different origins from those of primitive reptiles. (It may be as well to point out here that in birds the pifi dorsalis has entirely disappeared (Romer, 1942: 281) and only an attenuated pifi medialis remains. The bird iliotrochanterici are derived from the iliofemoralis and are not homologous with any part of the pifi. This topic will be referred to below).

The insertion-area of the pifi dorsalis is well marked on the femora of *Ornithosuchus* (Fig. 3), where it forms a depression on the outer side of the proximal end. The position of this depression immediately proximal to the lesser trochanter allows the same basic relationship to be inferred as in a crocodile or lizard, namely that the iliofemoralis on its way to the lesser trochanter bridged over the insertion of the pifi dorsalis. A depression in the same position is seen in many theropod femora, e.g. *Allosaurus* (Gilmore, 1920: pl. 14) and *Tyrannosaurus* (Osborn, 1916). There is, in dinosaur femora in general, a depression proximally on the anterior surface medial to the lesser trochanter, and there is little doubt that this was for the insertion of the pifi dorsalis (Romer's pifi 2). It may, as Galton suggested (1969) have extended back along the side of the 'greater trochanter' via the cleft between the trochanters. This position for the insertion of the pifi dorsalis is in general agreement with the situation in the alligator, except that in dinosaurs both the pifi dorsalis and the iliofemoralis insertions have moved to a more proximal position, as a consequence of the assumption of the 'vertical' limb posture.

It might seem at first that this proximal position of insertion, partly opposite the head of the femur, would make it impossible for the pifi to protract the limb. This might have been so if the movement of the bone about its head was always that of a simple pivot about an axis through the centre of the head. However, this may not have been the case. The head would almost certainly not have been an exact fit into the acetabulum—one could hardly imagine a more perfectly engineered ball-and-socket articulation than the human hip-joint, yet the head of the femur is not a precise fit in the acetabulum and 'only small regions of contact exist in most positions' (Barnett, Davies & MacConnaill, 1961: 177). If the head of the femur is of slightly smaller radius than the socket into which it fits, and the proximal end of the bone is held against the upper part of its socket, then the effective fulcrum is transferred to a proximal transverse line, or more likely in practice a narrow band, of contact (Fig. 7). The strip of bone actually in contact with the acetabulum will change from instant to instant as the femur is protracted, but at any given moment the fulcrum will be situated at the proximal extremity of the bone. Thus any muscle or tendon which inserted distal to this contact-surface would have had a lever-arm for protraction, even if it attached directly opposite to the ball-like head. Protraction takes place during the recovery stroke, so that the hind-limb, 'travelling light', requires relatively little muscular effort to move it forward and the short lever-arm of the pifi dorsalis would have imparted a rapid, though relatively weak, forward movement to the femur. It is necessary, of course, on this assumption for the femur to be held against the upper part of its socket to counter the weight of the limb itself, but adequate muscular force was available for this in the iliofemoralis, attaching to the lesser trochanter. This was primarily a 'holding muscle', best developed in bipedal dinosaurs, preventing the body from falling over sideways when the opposite leg was raised from the ground, and it may reasonably be supposed to have held the femur of its own side firmly in place during protraction. In many dinosaurs the acetabulum is distinctly larger than the head proper and it may be that a certain amount of rolling took place between the proximal end of the bone and the upper part of the acetabulum, although this is more controversial (cf. also Charig, 1972).

The dinosaurian 'greater trochanter' thus remains something of a puzzle. Apparently no muscle inserted on it; rather, it has the appearance of an articular surface which, as far as I am aware, is always continuous with the head proper in well-preserved femora. (Thulborn describes a separate, blade-like 'greater tro-chanter' in *Fabrosaurus* (Thulborn, 1972), but this is quite unusual and appears to be due to crushing and incomplete preservation). Its uniform fore-and-aft

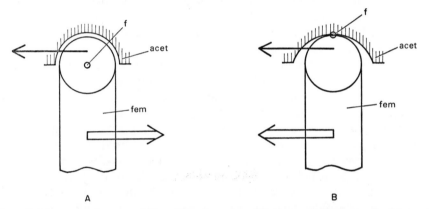

A   B

Figure 7. Diagrammatic parasagittal sections through acetabulum and proximal end of femur to illustrate the effect on the musculature of proximal transference of the fulcrum. Anterior end to left; solid arrow, a muscle inserting proximal to the apparent position of the fulcrum; open arrow, direction of movement of the femur: A, assuming purely rotatory movement about an axis through the centre of the head; B, assuming only limited contact of the proximal surface against the upper part of the acetabulum. Disparity in curvature exaggerated in B. For further explanation see text.

curvature, particularly in ornithischians, suggests a condyle rather than a muscle-attachment process. The 'greater trochanter' projects laterally from the acetabulum and does not, in any dinosaur, appear to have been entirely covered by the most lateral part of the supra-acetabular rim. The most likely explanation for its existence seems to be that put forward by Gregory & Camp (1918: 534), namely that it marks the position of a bursa, since it occurs at the spot where the ilio-femoralis and iliofibularis (which I believe was larger than as restored by Romer) turned down rather abruptly from the ilium and passed to the lesser trochanter and along the back of the thigh respectively. Owing to the more nearly parasagittal swing of the femur in dinosaurs these muscles were no longer able to pass out as directly as they had done in the Thecodontia. The iliofemoralis also passed closely over the dorsal part of the pif i as the latter approached its insertion medial to the lesser trochanter. Possibly the 'greater trochanter' articulated with a cartilaginous lateral prolongation of the acetabular lining. There is a considerable resemblance to the proximal end of the bird femur, and this is more marked in the case of the ornithischians, which usually have a saddle-shaped depression running antero-posteriorly across the proximal end of the bone, separating the head proper from a broader (from front to back) lateral convexity (Romer, 1927b; Galton, 1969, 1974) which forms the so-called 'greater trochanter'. The dinosaurs, however, have no true antitrochanter such as is present in the bird, against which the outer part of the proximal end of the femur bears, but the postulated cartilaginous extension may have taken its place. In the fowl, the antitrochanter is slow to ossify. All in all, the term 'greater trochanter' seems to be a misnomer in dinosaurs, and, whatever its true nature, the projection is better regarded as part of the head, as was pointed out by Gregory (in Gregory & Camp, 1918: 534).

THE LESSER TROCHANTER

This has been referred to already and would warrant little further discussion were it not for the fact that Bakker (1971), Bakker & Galton (1974) and Galton (1975) have departed from the traditional interpretation of the significance of this process, which has been as the insertion of the iliofemoralis. Von Huene (1908: 294 and fig. 289) was perhaps the first to indicate the attachment of the iliofemoralis here, but he also regarded it as the site of insertion of 'quadratus lumborum' (pifi dorsalis) and 'puboischiofemoralis posterior 2' (ischiotrochantericus). Both of these muscles insert proximally to the iliofemoralis in the alligator (Romer, 1923b) and in lizards (Perrin, 1892). Von Huene refers to the process as the 'trochanter major', a usage also followed by Gregory (in Gregory & Camp, 1918: 534 and pl. XLVIII), who also calls it the 'external trochanter'. Gregory also placed the insertion of the iliofemoralis on this projection. The usage of 'trochanter major' persisted for some time (e.g. Haughton, 1924; von Huene, 1932) but has gradually been superseded by 'lesser trochanter' (Gilmore, 1914, 1920), although in fact the older term is more appropriate. Gregory labels the area on which the iliofemoralis inserts in the alligator as 'trochanter major', although crocodilian femora do not normally show any distinctive features in this region. However, a femur of the Triassic stagonolepidid *Typothorax* in the American Museum of Natural History (No. 2710), which is very crocodilian in general shape, shows the insertion-area of the iliofemoralis in the crocodilian position, and this surface falls away abruptly medially and laterally and projects proximally as a raised 'tongue'. This may be the femur referred to by von Huene (1915) as having a 'trochanter major', but his figure does not show it. It is clear, however, that the beginning of a lesser trochanter is shown by this bone, elevating the iliofemoralis over pifi dorsalis, which would have inserted immediately proximal to it. A stage further than this is shown by the lesser trochanter in prosauropod dinosaurs, in which it is generally rather a low, elongated ridge, overhanging little, if at all, towards the proximal end. Its primitive nature in this group was noted by von Huene (1908). In more advanced dinosaurs the lesser trochanter projects proximally as a well-defined process, separated from the 'greater trochanter' by a cleft, and is, on the whole, directed towards the area on the ilium on which the iliofemoralis might be expected to have originated, that is, the anterior portion of the blade from above approximately the middle of the acetabulum forwards to the region of the anterior embayment. (I consider that the position of the origin of this muscle in lizards, e.g. *Iguana* (Romer, 1922: pl. XLIV) is a better guide than in crocodiles. In the latter the iliofibularis is unusually small and the iliofemoralis origin is set further back.) It is incorrect to say, as Romer does (1927b: 254) that the lesser trochanter 'points upwards and, in most positions of the limb, markedly backward'. In the ornithischians Romer shows it directed towards the area where he puts the origin of the iliotrochantericus. It is doubtful, however, whether any such muscle existed in these dinosaurs, and this area was probably taken up by the iliofemoralis, as in saurischians.

Relevant to the question of the function of the lesser trochanter in archosaurs is the situation in the avian pelvis and femur (Fig. 9). The anterior part of the iliac blade (anterior iliac fossa) and its preacetabular ventral edge in birds give rise to the iliotrochanterici, of which there are commonly three, the posterior being usually by far the largest and the medius smallest and frequently absent or fused with the others (Hudson, 1937; Hudson, Lanzillotti & Edwards, 1959; George & Berger, 1966). These muscles converge to form tendons which insert in a row down the external surface of the trochanter and trochanteric ridge near its upper edge, in the order anterior, medius and posterior passing proximally. Romer (1927b), in spite of some doubts, homologized the bird iliotrochanterici with the pifi dorsalis (his part 2), in consistency with his view (1923a) that the anterior part of the bird ilium had 'picked up' the dorsal portion of the pifi, a process which he supposed to have occurred to some extent also in the Ornithischia. At that time his study of the development of the thigh musculature of the chick (Romer, 1927a) had not yielded clear-cut evidence as to the homologies of these muscles. He therefore restored *Thescelosaurus* with an iliotrochantericus arising in the normal position of the iliofemoralis, passing somewhat backward and downward to insert on the summit of the 'greater trochanter'. In addition, on the bird analogy, he restored a rather small iliofemoralis externus arising from the region where the 'antitrochanter' occurs in more specialized ornithischians, passing forward and downward to an insertion on the lesser trochanter. These two muscles cross over each other in a very awkward-looking manner in his restoration (1927b: fig. 16).

It seems to have been generally overlooked that Romer later changed his mind concerning the homology of the iliotrochanterici. As a result of his work on the development of the thigh musculature of *Lacerta* (Romer, 1942: 280–1) he abandoned his former opinion and concluded that the dorsal part of the pifi had been completely lost in birds, the iliotrochanterici being equivalent to part of the iliofemoralis of reptiles. Thus the only remaining part of the pifi in birds would be the iliacus (iliofemoralis internus or pif i medialis or pif i 1 of Romer, 1923b), a very small muscle passing from the ilium to the medial surface of the femur above the ambiens (when the latter is present), and thus approximately in the crocodilian position. The iliofemoralis, on the other hand, would be represented by up to four muscles: the three iliotrochanterici and the iliofemoralis externus. This change was embodied in Table 2 of the third edition of 'The Vertebrate Body' (Romer, 1962: 280).

Other reasons why this later conclusion is likely to be correct are:

(a) The backward migration of the pife to form the obturator, along with the backward rotation of the pubis, is well shown in Romer's figures (1927a) of Stages II to V of the chick, and thus the phylogenetic history is recapitulated. There is no indication, in contrast, of any capture by the ilium of pifi dorsalis or dorsal migration of the latter in ontogeny; the iliofemoralis externus and iliotrochanterici simply differentiate and enlarge from the originally single 'deep dorsal mass' as the blade of the ilium grows forwards.

(b) It is difficult to visualize the capture of the pifi dorsalis by the blade of an ilium of avian type assuming, as did Romer, that dorsal migration of the pifi to a position like that of the crocodile was a common archosaurian change. In Triassic crocodiles such as *Orthosuchus* (Nash, 1975) (Fig. 10A) and *Protosuchus* (Colbert & Mook, 1951) the ilium has a narrow anterior process set high up, approximately on the same level as the transverse processes of the posterior dorsal vertebrae. Examination of a recent crocodile skeleton confirms this high position. The bird anterior iliac expansion on the other hand, does not, as Romer says (1923a: 144), lie in the region where the pifi dorsalis originates in the crocodile (the centra and undersides of the transverse processes of the last six presacral vertebrae), but is situated almost entirely *above* the transverse processes and ribs of the synsacrum. It is thus difficult to see how the ilium could have 'picked up' the dorsal part of the pifi. Had such a process begun it would appear that in early stages, owing to the width of the sacrum, the part of the pifi dorsalis captured would have been reduced in length by about half, which would be likely to have had a drastic effect on its function.

(c) The 'trochanter' of birds is in the same position as the lesser trochanter of advanced thecodontians and dinosaurs, and the iliotrochanterici and iliofemoralis externus (gluteus medius et minimus) insert on the outside of the trochanter and down the trochanteric ridge in a manner similar to that usually postulated for the iliofemoralis of dinosaurs, a relationship which has been deduced by earlier workers from the pattern of insertion of the iliofemoralis and pifi dorsalis found in modern lizards, *Sphenodon* and crocodiles. Also, in the alligator (Fig. 8E), the iliofemoralis insertion divides the proximal heads of the femorotibialis, in a similar manner to the division of the same muscle by the iliotrochanterici in birds.

Thus the evidence is self-consistent and leads to the conclusion that the lesser trochanter of archosaurs served for the insertion of the iliofemoralis.

The account of the proximal femoral trochanters of the Theropoda in Romer's 'Osteology of the Reptiles' (1956: 367) is confused and does not agree with his own earlier reconstructions of pelvic musculature (1923c). It seems clear that he did not intend to say that the iliofemoralis inserted on the 'greater trochanter' and that this projection was 'comparable to the greater trochanter of mammals', since he himself had pointed out (1927b: 257) that it is the *lesser* trochanter of dinosaurs which is homologous with the greater trochanter of mammals. These curious statements do not seem to stem from any change of view on his part and must, I think, be regarded simply as mistakes. Possibly the confusion arose from the fact that 'iliofemoralis internus' has been used by some authors, including

Romer himself (1927b: 254) for part of the reptilian pifi. It has also been commonly used for the remnant of the pifi in birds, here called medialis.

In Bakker's paper (1971: fig. 7) small diagrams of certain pelvic muscles of the dinosaurs *Stegosaurus* and *Triceratops* are shown. The iliofemoralis origin is placed just in front of the 'antitrochanter', more or less where Romer has the iliotrochantericus in *Thescelosaurus*. This seems reasonable, but the muscle was,

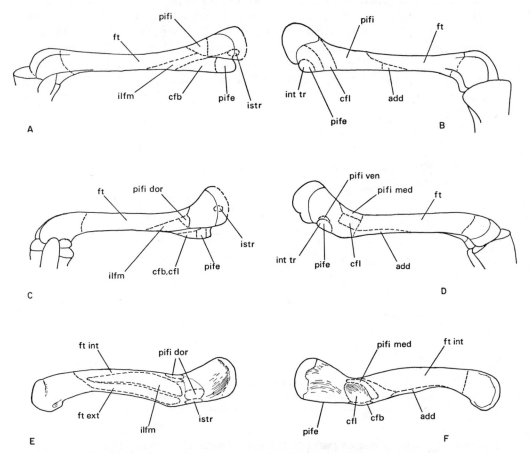

Figure 8. Muscle insertions and origins on reptilian femora. A, B. Lateral and medial views of left femur of *Sphenodon*, after Perrin (1895), C, D. Same of the lizard *Uromastix*, after Perrin (1892), E, F. Same of *Alligator mississippiensis*, after Romer (1923b). A fine dotted line has been added to E, joining the two insertions of pifi dorsalis shown by Romer, to indicate that these are parts of one tendon. In *Crocodylus acutus* the insertion is continuous (Gadow, 1882a).

I think, larger in these dinosaurs and occupied the outer surface of the ilium as far as to the anterior embayment. He puts the insertion on the side of the 'greater trochanter' (labelled 'crest for iliofemoralis'). The pifi is shown as originating partly from the anterior process of the ilium and partly from the space below this, presumably thus from the posterior presacral vertebrae as in the alligator. The insertion is placed upon the lesser trochanter, labelled 'crest for puboischio-femoralis internus'. Bakker gives no reason for this departure from the accepted interpretation, but simply states that the iliofemoralis inserts on a high external crest which, from his figure, is evidently the 'greater trochanter'. A similar arrangement is shown by Galton (1975: fig. 3G), again without explanation. As stated above, however, in lizards and crocodiles the pifi dorsalis (Romer, 1942: 'part I') inserts on the outer side of the upper surface of the femur, nearly reaching its lateral edge, proximal to the insertion of the iliofemoralis which bridges over it (Fig. 11). The same general relationship also obtains in *Sphenodon* (Fig. 8), but the pifi has a continuous insertion area and is not divided into two parts as in the alligator. There is thus no justification for placing the insertion of the iliofemoralis so far proximally and posteriorly and the comparison with mammals is, in consequence, not valid. In the case of the origin of the pifi dorsalis, it seems that Bakker may have been influenced by Romer's earlier views on the homology of the bird iliotrochanterici but, as pointed out above, Romer later gave good

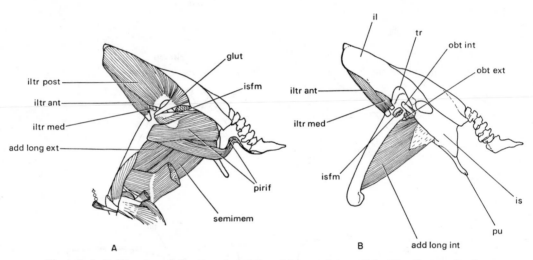

Figure 9. A, B. Two stages in the dissection of the pelvic musculature of the Blue Grouse, *Dendragapus obscurus*, left lateral views, after Hudson *et al.* (1959), to show the relationship of the iliotrochanterici and gluteus (iliofemoralis externus) to the trochanter and trochanteric ridge.

reason for regarding these as parts of the iliofemoralis. Thus the muscles from the ilium which insert on or near the upper edge of the avian trochanter are all parts of the iliofemoralis (not to be confused with the 'iliofemoralis internus', also called the iliacus, which is equivalent to the pifi medialis of reptiles). Again, as already mentioned, during the swing of the femur the lesser trochanter points on average towards the area on the ilium which would be the expected place of origin of the iliofemoralis. This is true also in the case of theropods, but it seems to me likely that the iliofemoralis was not as large as usually restored in that group, and that a substantial iliofibularis occupied a considerable area of the ilium behind it. Also, although the bird iliotrochanterici do insert in part almost at right angles to the trochanteric ridge, rotation of the femur about its long axis is an important part of their function (Hudson, 1937; Cracraft, 1971). The iliotrochantericus posterior is usually by far the largest of the three and is a strongly convergent muscle which inserts on a curved, anteriorly-convex ridge on the outer side of the trochanter. The posterior portion passes to the posterior part of the trochanter, more or less in line with the long axis of the femur. The situation in dinosaurs differs in another important respect from that in birds; in the former the lesser trochanter projects freely proximally from a base situated lower down the shaft, whereas the bird trochanter and trochanteric ridge are continuous with the side of the shaft. Under these circumstances it is reasonable to suppose that the direction in which the trochanter projects indicates the main direction of pull exerted by the muscle attached to it. A principal direction of tension at right angles to a freely-projecting process would appear to be mechanically unsound and thus unlikely to have occurred in dinosaurs.

Thus although it might appear possible to locate the iliofemoralis insertion in dinosaurs on the side of the 'greater trochanter', this would mean that the lesser trochanter would be left without a clear functional raison d'être, since the pifi dorsalis is unlikely to have inserted on it. On the other hand, the role usually allotted to it is consistent with the arrangement in *Sphenodon*, lizards and crocodiles (Fig. 8), and also with what is known at present about its phylogenetic history.

Bakker & Galton (1974: 169) attribute an additional function to the lesser trochanter—that of providing an increase in the area of origin of the femorotibialis, which they consider gave 'increased muscle-power for knee extension'. It is difficult to see that any significant increase in the power of this muscle could have been achieved in this way, since the area involved is very small compared to the total area of origin of the muscle and the elevation permitted by the ridge leading up to the trochanter is not great enough to provide any significant increase in leverage at the knee. Only the distal, declining part of the ridge would have been left free by the iliofemoralis, although even this might have been covered by it. In birds, which seem to present the closest analogy, the femorotibialis origin (Hudson, 1937; George & Berger, 1966) is divided proximally by the insertions of the tendons of the iliotrochantericus anterior and medius, which attach to the distal part of the ridge.

THE ORNITHISCHIAN PELVIS

In both birds and ornithischian dinosaurs the pubis has rotated backward to lie alongside the ischium as a slender rod. The homology of this posterior process has long been known in birds from embryological studies (Johnson, 1883), and confirmed recently in the case of the ornithischians by the discovery of primitive specimens of Upper Triassic and Lower Jurassic age (Fig. 6) in which the prepubic process is very small but the backwardly-directed true pubis is as long as the ischium (Charig, 1972).

Romer (1927b: 246) considered that 'the true pubis had at the beginning of ornithischian development been relieved of all duties except that of bearing the 'obturator' (i.e. the anterior part of the pife), and that it has shifted backward with the effect of giving that muscle a more advantageous position.' As Galton (1969: 34) points out, however, this is a result, rather than a cause, and does not itself provide an explanation for the backward rotation of the pubis, since the anterior part of the pife arising from its posterior surface was originally a protractor of the femur, but became converted to a retractor—the 'obturator'. Galton suggested that backward rotation became possible when protraction of the femur was mainly carried out by a well-developed 'sartorius' (anterior part of the iliotibialis) arising from a long, narrow anterior iliac process and passing down to the knee. According to him, this muscle functionally replaced the anterior part of the pife. The latter could thus be converted to a retractor if other selection pressures existed which favoured backward rotation. Galton points out that a posteriorly-directed pubis is advantageous to a reptile tending to bipedalism since this extends the viscera backwards and hence also the centre of gravity. There is good evidence for full bipedalism in early ornithischians (Casamiquela, 1967; Thulborn, 1972), and both Thulborn and Galton (1971, 1972) agree that this type of locomotion was probably characteristic of the earliest ornithischian stock. Charig (1972: 139) gives essentially the same explanation as Galton of the muscular changes leading to pubic rotation in ornithischians.

The weakness of Galton's hypothesis lies in the assumption, made earlier as we have seen by Romer (1923c, 1927b), that the pifi dorsalis inserts on or near the summit of the 'greater trochanter' and was a retractor of the femur. Reasons have already been given for doubting this and for suggesting that, on the contrary, the pifi dorsalis was always a protractor in dinosaurs, as in other reptiles. On this view, there was no necessity for the development of a major new protractor in ornithischian evolution. The weakening of the ornithischian pubis as a prelude to backward rotation is more plausibly to be compared with that of the crocodilian pubis. The muscle most conspicuously absent from the modern crocodilian pubis, as Romer (1923b) has shown, is the pifi, which has entirely vacated the pubis for a more dorsal position (Fig. 5). It is thus reasonable to assume that the weakening of the crocodilian pubis in the Trias was due precisely to this dorsal migration of the pifi. The pubis in Triassic crocodiles (Broom, 1904, 1927; Colbert & Mook, 1951; Nash, 1975) is a straight or gently curved slender rod less expanded distally than in post-Triassic and Recent members of the group (Fig. 10a). Since the distal expansion and its cartilaginous continuation in the alligator provide the main areas of origin of the anterior divisions of the pife, it seems likely that these muscles were less well developed in Triassic forms, and in particular, that the portion ('pife 1') which arises from the dorsal surface in living forms had not yet spread round to any significant extent to take up the space vacated by the pifi. This conclusion appears to be supported by the form of the pubis in modern crocodiles which, although sometimes appearing in illustrations to be straight, is actually concave dorsally and has a marked sinus in the lateral margin. The 'extra' head of the pife (Romer, 1923b: 'pife 1') (Figs 5 and 11) is thus enabled to pass freely back below the ambiens to the proximal end of the femur. Although it is difficult to estimate to what extent the pifi dorsalis and medialis had migrated dorsally in extinct archosaurs, we can be virtually certain that they had done so in the Triassic crocodiles (sensu stricto), since the pelvis resembles so closely those of living forms (Nash, 1975). Furthermore, the high position of the anterior iliac process provides confirmation that the pifi dorsalis passed below it, running back from the lumbar vertebrae (although Nash: fig. 23, has accidentally transposed the positions of insertion of this portion and the pifi medialis). The narrow, pointed anterior iliac process in *Orthosuchus*

(Nash, 1975) is not unlike those of early ornithischians, although not as long, and again, the high position of the process and the wide lateral exposure of the lumbar vertebrae in the latter group strongly suggest that comparable dorsal migration of the pifi had occurred. Thus it seems probable that comparison with crocodiles will provide useful indications concerning the reasons for pubic rotation in the Ornithischia. Two qualifications must be entered, however, first that the crocodilian exclusion of the pubis from the acetabulum probably took place subsequently to the loss of the pifi from it and thus does not concern the present discussion, and second, that the anterior part of the crocodilian iliac blade has degenerated since the Trias. The formation of a 'pife 1' may thus have occurred to compensate for the reduction of the 'sartorius'.

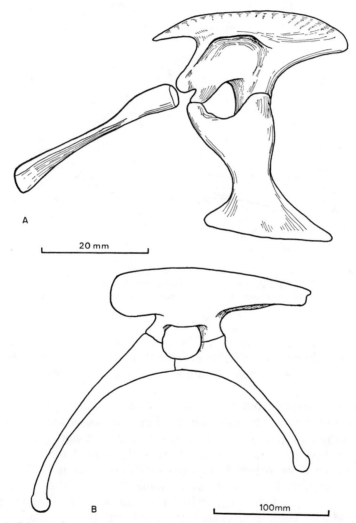

Figure 10. Left lateral views of pelves of: A, the Upper Triassic crocodile *Orthosuchus*, after Nash (1975): B, the Upper Triassic coelurosaur *Coelophysis*, after Colbert (1964). Scale in B approximate.

338 A. D. WALKER

In lizards the pifi dorsalis has an indirect course, starting from the dorsal surface of the ischium, turning through some 90° or more round the anterior border of the ilium below the anterior iliac process, and passing as a strong tendon to the external border of the proximal end of the femur. According to Snyder (1954) it produces a rapid protractive movement of the femur. In crocodiles, in contrast, it is a very long and powerful muscle, passing directly backward from the last six presacral vertebrae to a similarly-placed tendinous insertion. It thus appears even better placed to effect similar movements. In both groups the pifi medialis has a more anterior and distal, fleshy insertion. However, it is likely that the similarities have arisen independently, and that the primitive condition in each would have been like that of *Sphenodon*, with an undivided fleshy insertion

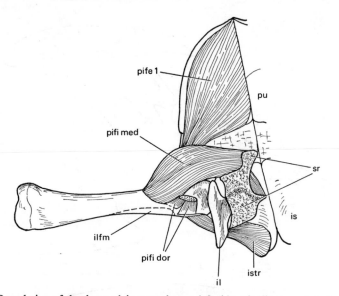

Figure 11. Dorsal view of the deep pelvic musculature, left side, of *Alligator mississippiensis*, after Romer (1923b), with vertebral column and ribs removed and sacral ribs cut. Iliofemoralis insertion added.

(apart from the ventralis division). The primitive nature of the girdle, with a plate-like pubo-ischium, in the early Triassic thecodontian *Proterosuchus* (Cruickshank, 1972) implies a correspondingly primitive musculature in the earliest archosaurs.

Since there is evidence from *Sphenosuchus* (see Introduction) that there was an arboreal stage in the evolution of crocodiles (high degree of pneumatization of the skull, modification of the coracoid, relatively large humerus, hollow bones) it does not seem unreasonable to suppose that the dorsal migration of the pifi in this group reflects an original need for rapid protraction of the hind limb in climbing and jumping. Presumably in climbing the animal would have used its limbs alternately in pairs, first the paired fore-limbs, then the paired hind-limbs so that it would have been an advantage to have been able to bring the appropriate pair of limbs forward very quickly while the other pair retained its grip. (Rapid climbing in this manner is carried out by living 'flying' squirrels, and the use of the fore-limbs as a *pair* in the arboreal bird ancestor may well have been pre-adaptive for flapping flight). The coracoid of *Sphenosuchus* shows the beginnings of specialization for rapid protraction of the humerus (Walker, 1972), and it seems probable that the unusual origin of the supracoracoideus muscle in modern crocodiles is a relic of this situation. Fürbringer (1876) has shown that this muscle originates in large part from the inner surface of the anterior expansion of the coracoid, wraps round the medial margin of this portion of the bone (and also round the biceps origin), and then passes back to its insertion at the proximal end of the deltopectoral crest. Presumably the extension of the origin on to the inner surface lengthens this portion of the muscle and thus enables the humerus to be moved through a wider arc, while essentially maintaining the direction of pull. Since crocodiles have been mainly amphibious or aquatic since the Trias, some modification of the pelvic musculature may be expected to have taken place, but the close similarity of the pelvic girdle and hind limb of Upper Triassic forms to those of living members of the group suggests that this has not been great. Nash (1975) concluded that *Orthosuchus* was less well adapted to an aquatic environment than are living crocodiles, and the forward origin of the 'sartorius' in this genus (see below) tends to bear this out. A similar conclusion was reached in the case of *Protosuchus* by Colbert & Mook (1951: 174), while Kermack (1956) described a small unnamed crocodilian from an Upper Triassic fissure-filling as being a very agile, land-living form with long limbs and a very long tail. The failure of crocodiles to rotate the pubis backward (except perhaps in the case of *Hallopus*) is thus probably due to the general change in their mode of life from cursorial-arboreal to amphibious-aquatic towards the end of the Trias. The exclusion of the pubis from the acetabulum and further elongation of the coracoid (in order to position the glenoids laterally) probably ensued as a result of this change.

To return to the ornithischian dinosaurs, a comparable specialization of the pifi dorsalis seems to have taken place as a preliminary to backward rotation of the pubis. In this case the need for rapid protraction of the femur appears to have been connected with the early development of a thoroughgoing bipedalism. Such a specialization would have been as important in a fleet, cursorial biped as in an active tree-dweller. However, Galton's argument (1969) concerning the development of a long anterior iliac process undoubtedly has validity, since it seems that the pubis did not turn back until this process had attained a considerable length—it had not turned back in the crocodile *Orthosuchus*, for example. Clearly, the 'sartorius' cannot act as an efficient femoral protractor (as distinct from a tibial extensor) until it has attained an origin well in front of the acetabulum, since in early stages its line of action is too close to the long axis of the femur. It seems, then, that the final step was the functional replacement of the other main protractor, the anterior part of the pife arising from the back of the pubis, by the 'sartorius'. The pubis, having lost all divisions of the pifi, and probably having suffered some reduction of the pubic head of the pife, would then have been functionally almost useless in its original position and could rotate backwards, permitting the posterior extension of the body cavity with the accompanying advantage of backward displacement of the centre of gravity. The pubic head of the pife would then be in a better position from which to act as a retractor—the 'obturator'.

However, it is necessary to consider the reasons why an additional protractor was required, since pifi dorsalis and medialis would seem to have been sufficient in themselves. The explanation seems to be that the pifi dorsalis is more effective in the initial half, approximately, of the recovery swing of the femur, but its effectiveness falls off after the femur has passed its lowest point. The 'sartorius', on the other hand, passes close to the axis of the femur when the latter is retracted, but the angle it makes with the femur at the knee progressively increases during protraction, increasing accordingly the moment-arm of the muscle about the acetabulum. The maximum is reached at about the probable maximum protracted position of the limb. The 'sartorius' would have been in an increasingly contracted

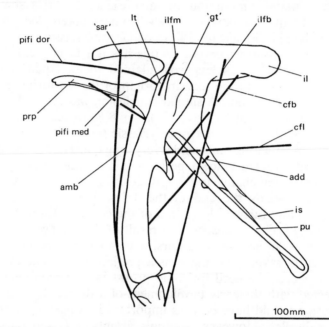

Figure 12. Lines of action of some muscles of the pelvis and femur in the Lower Cretaceous ornithopod dinosaur *Hypsilophodon*, modified from Galton (1969).

state towards the end of protraction, but the increasing moment-arm would tend to offset this. In the case of the pigeon, Cracraft (1971: 233) states that the 'sartorius' 'probably functions to protract the femur (and tibiotarsus) rapidly over a great distance'. The anterior part of the pife was not well fitted to perform a similar function in the original position of the pubis since its effectiveness decreased as the femur approached that bone. Thus it seems that the process envisaged by Galton was the 'last straw' rather than the main reason for pubic rotation, and was preceded by weakening of the pubis due to dorsal migration of the pifi.

Before leaving the subject of the ornithischian pelvis some remarks may be made about certain other aspects of hind limb musculature. Since the avian iliotrochanterici are probably derivatives of the reptilian iliofemoralis, and do not represent the pifi dorsalis, there is no reason to suppose that the latter muscle has been 'captured' by the bird ilium. It follows that there is even less reason to suppose that this change occurred in ornithischian dinosaurs, there being good evidence that the pifi dorsalis was in the crocodilian position in this group. It is thus confusing to perpetuate the restoration of an 'iliotrochantericus' as well as an 'iliofemoralis externus' on the ornithischian ilium, and it is suggested that a single iliofemoralis probably originated from the anterior part of the bone, approximately where Romer (1927b) placed the 'iliotrochantericus', but extending also a little more posteriorly, and passed directly down to the lesser trochanter (Fig. 12). There was probably a well-developed iliofibularis arising from a larger posterior area of the ilium than previously restored, beginning behind the ilio-femoralis. Such a muscle could rapidly flex the crus at the beginning of the recovery phase of the hind limb, bringing the centre of gravity proximally and thus reducing the moment of inertia about the acetabulum, as in the pigeon (Cracraft, 1971). If the limb were stabilized in the extended position by other muscles, contraction of the iliofibularis could have helped materially in retraction also. The 'antitro-chanter' of more advanced ornithischians may have served for a specialized anterior tendinous portion of the iliofibularis, since in birds the comparable muscle, the biceps femoris, originates from a narrow line on the hinder part of the ilium, immediately below the origin of the iliotibialis, and the anterior part of the origin, above the acetabulum, is tendinous (Hudson, 1937; Hudson et al., 1959). The functional significance of the ornithischian 'antitrochanter', however, requires more consideration than is possible in this paper.

Galton (1969) considered that the ornithischian prepubic process provided origin for either a pubotibialis or, more probably, part of the pifi (his 'part 2'). It seems likely, as suggested above, that the course of this muscle, if part of the pifi, was above the ambiens, and it was thus a pifi medialis. In primitive ornithischians this appears to have been approximately in the crocodilian position, that is, arising from the internal surface of the lower part of the ilium and the undersides of the sacral ribs and passing round the anterior sinus of the ilium above the small prepubis or pectineal process to reach the inner side of the femur. The fact that the 'tunnel' below the anterior iliac process tends to become reduced in extent in more advanced ornithischians, concomitantly with the anterior extension of the prepubic process, tends to support the identification of the latter as an origin for part of the pifi, which apparently extended its origin forwards to gain a more direct course to the femur. At the same time the pifi dorsalis seemingly became reduced in importance. This change appears to have been connected with increase in size and loss of bipedalism, but again, the evidence requires a more careful examination. Finally, an 'S-shaped ridge' on the outer surface of the proximal end of the femur in *Hypsilophodon*, below the 'greater trochanter', was regarded by Galton (1969, 1974: 95) as separating the insertions of pifi dorsalis (his 'pifi 1') anteriorly and 'iliotrochantericus' posteriorly. This ridge, however, well shown by Galton in a stereophotograph (1969: fig. 8A) is closely paralleled in the femur of birds (best shown by larger species), and in them the muscle inserting posteriorly is the obturator internus. The resemblance is very striking, and in view of the remarks made above about the superfluity of the 'iliotrochantericus' in ornithischians, may well be significant.

THE SAURISCHIAN PELVIS

One of the major problems in considering the musculature of extinct archosaurs is that of estimating the extent to which dorsal migration of the pifi had taken place. The presence of a sinus in the anterior margin of the ilium is not sufficient in itself to prove that this change had occurred since, as Charig (pers. comm.) points out, there is frequently an anterior iliac process in lizards, and in this group the pifi originates from the primitive position, that is the dorsal surface of the pubo-ischium. There seems no justification for the assumption that an upward and forward shift of the origin of the pifi was a common archosaurian character. This assumption was based by Romer on the situation in modern crocodiles but, as noted above, the pubis in crocodiles is very different from that of thecodontians and thus its musculature does not necessarily represent a primitive archosaurian condition. One can only be reasonably certain that the pifi dorsalis was in the crocodilian position when there is wide lateral exposure of the posterior presacral vertebrae below a narrow, dorsally-placed anterior process of the ilium. These conditions hold for primitive ornithischians and Triassic crocodiles, but not for saurischians. (The ilium in the theropod dinosaur *Compsognathus corallestris* (Bidar, Demay & Thomel, 1972: figs 10, 11) might appear to contradict the foregoing, but it seems probable that the left ilium is largely missing from the specimen and that what is seen is the inner view of the right ilium, emerging above the sacral vertebrae. Thus the iliac blade may actually be deeper than it seems.)

In advanced thecodontians and saurischians there is good reason to think that a sub-ambiens portion of the pifi, that is the pifi ventralis, was well developed; indeed, by analogy with living lizards, the great development of the pubic 'apron' in advanced thecodontians was probably largely due to the increase in size of this muscle. As already pointed out, its origin from the anterior side of the pubis gave it an advantage as a protractor over the anterior part of the pife coming from the back of the pubis. It may well have existed in the proterosuchian thecodontians, since it appears to be present in a reduced form in *Sphenodon* (Gadow, 1882b: 411; Perrin, 1895: 94), and the pubis of *Proterosuchus* (Cruickshank, 1972: fig. 8) has a sinus in the lateral border between the acetabulum and the processus lateralis ('pubic tubercle' of Cruickshank). A similar sinus is figured by Gow (1975: fig. 31) in the Upper Permian diapsid *Youngina* and the Lower Triassic *Prolacerta*, the latter regarded by Gow as a thecodontian. (The steep descent of the pubes in such thecodontians as *Stagonolepis* (Walker, 1961) is sometimes looked on as in some way connected with the turning back of these bones in ornithischians, but it is more probably related to the bulk of these armoured reptiles—the thickened distal knob of the pubis suggests that the bone functioned as a support when the belly rested on the ground).

In the saurischian dinosaurs there is nearly always a concavity in the lateral (or anterior) border of the pubis of greater or lesser extent, and this is most marked just below the ambiens projection, where the pifi ventralis would be expected to pass out (Fig. 4). The great anteroventral elongation of the pubis in the coelurosaurs (using the customary division of the Theropoda into Coelurosauria and Carnosauria for the present purpose) evidently constituted a distinct specialization of the pifi ventralis for rapid protraction of the femur (Fig. 10B).

The anterior position of origin of this muscle, its great length and its proximal insertion on the inner side of the femur would have made it an efficient agent for this purpose in the earlier stages of protraction. As the femur approached the pubis its effectiveness would have decreased, but the more dorsal portions of the same muscle would then have been well placed to continue protraction. It is not necessary to suppose that in theropods these upper portions had moved their origins to any great extent from the primitive position, though the midline fenestration of the girdle ventral to the acetabulum in more advanced members suggests that some dorsal migration had taken place. Dorsalis may thus have moved to approximately the position where medialis is located in the alligator and perhaps in addition may have arisen from the last one or two lumbar vertebrae in those forms in which the pelvis is very compressed and the sacral ribs in consequence very short. In other words, much as Romer has the origin of his 'pifi 1' in *Tyrannosaurus* (1923c).

It is hardly necessary to give many examples of the coelurosaurian specialization of the pubis, but the following may be mentioned: *Coelophysis* (Colbert, 1964) (Fig. 10B), Upper Triassic; *Compsognathus longipes* (Colbert, 1962: pl. 35), *C. corallestris* (Bidar *et al.*, 1972), Upper Jurassic, and the 'ostrich dinosaurs', the ornithomimids (Osborn, 1916; Russell, 1972), best known from the Upper Cretaceous. Although in some coelurosaurs the pubis descends more steeply, this is not an argument against the presence of the muscle, first, because it was at its most effective in the early part of the recovery phase of the limb, and second, because the attitude of the pubis depends upon the orientation of the sacrum— in some forms the sacrum may have been tilted up a little anteriorly. There is no necessity, however, for the whole vertebral column and pelvis of theropods to be tilted up, as advocated by Charig (1972: 135), since the femoral protractors would have worked perfectly well with the pelvis in the normal attitude. The '45° tilt' of the column in the restoration of *Ornithosuchus* in walking pose (Walker, 1964: fig. 10) is unlikely to have been held except for short periods, and in bipedal running the column would probably have been disposed essentially horizontally. This is not to say that these predatory forms did not use an upright stance on a 'tripod base' when surveying the surrounding country.

In the more advanced theropods the pubes converge ventrally to join at an expansion called the pubic 'foot' or 'boot'. The area available for the origin of the pifi ventralis would thus have been somewhat reduced. However, the same restriction applied to the pife originating from the posterior side of the same region, and the existence of both muscles would have rendered protraction more effective than hitherto realized. In addition, in lizards the anterior parts of the origin of the pifi (ventralis of this paper) sometimes arise in opposition from each side of a median ligament as well as from the pubes themselves (Perrin, 1895: 95). The presence of a similar arrangement in theropods would thus have permitted a greater thickness of the pifi ventralis at its origin, compensating for the convergence of the pubes. (It may be as well to point out that, although the pubes of advanced theropods appear to be rod-like in side view, there is normally a reduced 'apron' with a median symphysis starting at about one-third the way down the bone. The pubes part again more distally and finally re-unite at the 'foot' (Osborn, 1906: 293; Gilmore, 1920: pl. 11).) The median ventral aperture was presumably closed by a sheet of connective tissue completing the areas of origin of the pifi ventralis in front and the pife anterior behind. In more primitive theropods and advanced semi-bipedal thecodontians the pubic 'apron' is narrow but essentially parallel-sided.

For similar reasons to those of Charig (1972: 135) I think it is unlikely that the iliofemoralis acted to any significant extent as a femoral protractor, as Colbert (1964) suggested. As noted above, it seems to have been essentially a 'holding' muscle, attaching to the lesser trochanter. Romer (1923c) restored a very large iliofemoralis in *Tyrannosaurus*, passing from most of the concave lateral surface of the blade to the lesser trochanter and the femoral shaft more distally. However, the theropod ilium, including that of *Tyrannosaurus*, often shows a vertical ridge running up from above the middle of the acetabulum (Lambe, 1917; Osborn, 1916; Russel, 1972; Madsen, 1974). Faint indications of a similar division are present in some ilia of *Megalosaurus bucklandi*. According to Russell, the division is a ridge in tyrannosaurids, while in ornithomimids there is a shallow sulcus separating two crests. Russel therefore restored two heads of the iliofemoralis in *Dromiceiomimus*. However, for reasons already mentioned, it seems to me that Romer under-estimated the size of the iliofibularis and overestimated that of the iliofemoralis, particularly in theropods, and that it is more likely that the posterior area gave rise to a large iliofibularis. It certainly seems very improbable that there would have been an 'iliofemoralis posterior' inserting at the posterior end of the 'greater trochanter' as Russell has shown it. On the other hand, the lesser trochanter is always at the anterior side of the proximal end of the femur and is directed on the whole towards the anterior of the two areas on the theropod ilium, while reasons have been given for concluding that the iliofemoralis inserted on this process. Gregory & Camp (1918: pl. XLVI) in fact show a similar distribution of the origins of the two muscles in the coelurosaur *Ornitholestes* to that suggested here. A well-developed iliofibularis would have been advantageous to theropod dinosaurs for the same reasons as outlined above for ornithischians. The presence of the 'cnemial process' on the lateral side of the tibia in large theropods, bracing the fibula at just the spot where the iliofibularis might be expected to insert, affords further evidence for the importance of this muscle.

The pelvis of prosauropods and sauropods would appear to present few problems. In both there is a sinus, often marked, in the anterior border of the pubis below the ambiens origin, indicating the passage of pifi ventralis, and the pubes are relatively broad. There is no reason to suppose that pifi dorsalis had migrated very far, since there are ample areas of origin available on the inner side of the ilium and the sacral ribs. The lesser trochanter is degenerate in sauro-pods, in keeping with the quadrupedal locomotion of these animals; nevertheless the iliofemoralis area on the ilium appears to have been very large, extending both behind and in front of the acetabulum. Presumably therefore a large posture-maintaining iliofemoralis was present, which is not surprising in animals of this bulk, but the lack of a definite lesser trochanter suggests that speed of locomotion was a key factor. The slow, lumbering sauropods (Charig, 1972: 135) lost the lesser trochanter, although it was not very well developed in prosauropods, while the more nimble quadrupedal ceratopsians retained it. Not surprisingly the iliofibularis seems to have been very small in sauropods.

Thus there were two contrasting styles of adaptation for rapid bipedal locomotion in dinosaurs. The ornithischians attained full bipedalism early, probably in the Middle Trias, and in them the pifi dorsalis migrated dorsally and anteriorly to an origin like that in living crocodiles giving, with the 'sartorius', rapid protraction of the femur in the recovery stroke of the limb. This was accompanied by backward rotation of the pubis. In the theropod dinosaurs on the other hand, specialization proceded in a different direction, with the anteroventral extension of the pubis providing origin for an enlarged and elongated pifi ventralis, which served essentially the same purpose as the elongated pifi dorsalis in primitive ornithopods, that is, rapid protraction of the femur. Both types may have originated from forms with a relatively short pubis, since there is no necessary correspondence between the length of this bone before and after rotation. The pubis of primitive ornithopods and early birds had an important role to play as an area for the attachment of the abdominal musculature and, after rotating back, presumably tended initially to adopt the same length as the ischium since this would have maximized the posterior extent of the body cavity. Thus the length of the ischium would appear to be a principal determining factor, at least, immediately after the pubis had swung back.

It may be asked why these two styles of specialization emerged. The answer would appear to lie basically in the different feeding habits of the two groups. Ornithischians from their beginning adopted a herbivorous diet without defensive adaptations, so far as is known, in the early period of their history. (The significance of the canine-like teeth of the heterodontosaurids is uncertain (Thulborn, 1971)). Theropods, on the other hand, were carnivores at this time, as predominantly later also. Thus it seems that the small early ornithischians relied on fleetness of foot to escape from predators, which may initially have been thecodontians, while the early bipedal saurischians could 'look after themselves' and furthermore were able to prey upon a variety of more sluggish reptiles. The greater depth of the anterior iliac process in theropods, in contrast to that of primitive ornithischians, suggests that the pifi dorsalis and medialis never migrated so far dorsally and anteriorly, or became as elongated, as in the latter group. Theropods evidently retained a basically more primitive pifi throughout their history, and appear to have taken longer to acquire full bipedalism than did ornithischians. Probably the Triassic coelurosaurs were not as swift as the contemporary ornithopods.

As a postscript to this section, it may be pointed out that the specialization of the ventral part of the pifi (and also of the anterior part of the pife) in theropods, which was particularly marked in coelurosaurs, makes it very unlikely that backward rotation of the pubis would later have occurred in this group. Hence it is highly improbable that birds have originated from theropod dinosaurs, as Ostrom claims.

THE AVIAN PELVIS

It is presumed that the preliminary stages in the evolution of the bird pelvis were similar to those passed through by the ornithischian dinosaurs, namely that the pifi as a whole migrated dorsally to a position like that which it adopts in modern crocodiles. The occurrence of a small remnant of the pifi medialis in approximately the same position as in the crocodile tends to confirm that this was the case. Whether or not there is a particularly close relationship between crocodiles and birds, as I have suggested (Walker, 1972, 1974), I find it difficult to accept a purely terrestrial origin of birds from theropod dinosaurs, without an intermediate arboreal stage, as postulated by Ostrom (1973, 1974a, 1975a, b, 1976). The energy requirements for direct take-off from the ground are clearly much greater than those needed for steady-state flight, and it seems to me that long before a purely terrestrial biped had acquired the necessary muscle-power and wing area for take-off from the ground, it would have become completely inadaptive and at great risk from predators. The ability to take-off directly from the ground was probably a relatively late acquisition in bird evolution, not attained even by *Archaeopteryx* in the Upper Jurassic. It seems to me to be much more logical to visualize a progression from a quadrupedal arboreal reptilian ancestor, through a stage of gliding, to flapping flight; indeed I consider that the particular adaptations of the shoulder-girdle and its musculature which evolved in the arboreal ancestor were a necessary prerequisite for the development of efficient flapping flight. These improvements seem to have taken a long time to acquire, and were not completed in *Archaeopteryx*, in which there was still no ossified sternum and direct elevation of the wing by the supracoracoideus muscle had not yet been achieved (Walker, 1972).

Thus it is considered that dorsal migration of the pifi was a consequence of the need for rapid protraction of the femur in the arboreal ancestor. However, the later stages of modification of the pelvis and its musculature proceeded differently from those in the Ornithischia. In the birds, the pifi dorsalis has been entirely lost (Romer, 1942), and the iliofemoralis has expanded forwards and divided into a number of heads, forming the iliotrochanterici and the iliofemoralis externus. The pifi ventralis has also disappeared and the pifi is represented only by a very small medialis, or 'iliacus' (iliofemoralis internus). It seems a reasonable deduction that in the bird ancestor, the iliotrochanterici functionally replaced the pifi dorsalis, since they run back from a similar, but more dorsal origin, to a similar insertion-position on the femur. The function of the ilio-trochantericus posterior in living birds has been the subject of disagreement. Hudson (1937) considered it, like the smaller anterior and medius portions, to be a protractor and inward rotator of the femur, whereas Cracraft (1971), as a result of work on the pigeon (*Columba livia*), maintained that in the latter, and perhaps in other birds as well, the iliotrochantericus posterior is a retractor. He bases this conclusion largely on the fact that, in the pigeon, the trochanter projects proximally 1–2 mm beyond the 'neck' of the bone. However, the iliotrochantericus posterior is usually a strongly-converging muscle, running back to insert on a ridge on the side of the trochanter. This ridge varies somewhat in birds, but considering forms which are not obviously specialized for swimming or other purposes, it usually begins distally more or less parallel to the upper or anterior edge of the trochanter and curves strongly backwards proximally, often following the curve of the tro-chanter round, so that at its proximal end it faces more or less proximally. Thus the anterior part of the muscle, which is the longest, runs back approximately at right angles to the femoral axis and inserts opposite or slightly distal to the head of the bone, and the more posterior parts progressively swing round so that the shortest, most posterior portion runs down almost parallel to the femoral axis. Consequently it may not be correct to conclude from the proximal projection of the trochanter that the muscle retracts the femur, particularly if there be any validity in the suggestion that the femoral axis of rotation may be transferred during protraction to the contact between the proximal surface of the trochanter and the antitrochanter, rather than running directly through the head. Possibly the anterior part protracts and rotates the femur while the posterior part retains largely its old 'holding' function. In many birds, the crow (*Corvus corone*) for example, the trochanter does not project freely proximally, and it would seem unlikely that the function of the iliotrochantericus posterior would differ fundamentally between crow and pigeon. Whatever its function in living birds, however, the iliotrochantericus posterior in *Archaeopteryx* (or the equivalent part of an undivided iliofemoralis) could not have been a retractor, since the trochanter is *distal* to the proximal termination of the femur. Preparation of the hitherto-buried head of the right femur of the London specimen, which agrees with the Berlin specimen, makes this quite clear.

The principal reason for the loss of the pifi dorsalis in birds and its replacement by the iliotrochanterici is therefore believed to be the different attitude of the femur in this group. The avian femur is usually a relatively short bone, shorter compared to the tibia than in coelurosaurs and ornithopods, and has a smaller, more forwardly-directed arc of movement. Cracraft's stick-diagrams and data on the pigeon walking (1971) show that the femur in this bird moves very little, the angle moved through in the parasagittal plane varying from 15°–46° in thirteen locomotor cycles of different speed, with an average of 28° (no correlation was found between the size of the arc and the speed of the gait). In the slow walk, moderate walk and run, the angle between the femur and the pelvic axis varied between 46° and 83°. The anterior end of the pelvis is carried with an upward tilt of varying amount, so that the angles which the femur makes with the horizontal are somewhat less than these.

In bipedal dinosaurs, on the other hand, the femur appears to have moved through a wider arc, and to have been directed, on the whole, more downwards, perhaps so that its central position was roughly at right angles to the pelvic axis; this is believed to be the reason for the persistence of the pifi dorsalis in bipedal ornithopods and (probably in a somewhat reduced form) in other ornithischians. With the femur directed on the whole approximately at right angles to the line of the long pifi dorsalis, its function as a protractor would remain unimpaired. If, however, the femur came gradually to be directed in a shorter, more forward-lying arc, the effectiveness of the pifi dorsalis would have been correspondingly reduced. The iliofemoralis, in contrast, lying dorsal to the acetabulum, would have been in a position to exert some protractive force, as well as acting as an abductor and resister of adduction, and by extending its origin forwards this protractive component would continually have increased. In this way the pifi dorsalis in the bird ancestor could gradually have been replaced by the anterior part of the iliofemoralis, which eventually differentiated to form the iliotrochanterici. The strong forward growth of the iliofemoralis in both ontogeny and phylogeny has presumably obliterated any trace in the embryo of the original dorsal migration of the pifi.

As the iliac blade extended forwards, the origin of the 'sartorius' would have
been carried forward along with it, a process which would have had selective
value in that it increased the protractive tendency of this muscle. Thus it seems
likely that both the 'sartorius' and the iliofemoralis contributed to the regression
of the pifi dorsalis. A parallel change to that which occurred in the ornithischian
dinosaurs then probably took place. The only remaining protractor on the pubis,
the pife anterior, having been functionally replaced by the 'sartorius', the
pubis was free to rotate backwards, with the same advantages as in the early
Ornithischia. The more forwardly-directed femur would also have reduced the
effectiveness of the pife anterior as a protractor, and increased that of the
'sartorius'. Backward rotation of the pubis may thus have taken place when the
anterior iliac process was shorter than in ornithischians.

Very probably the increasing tendency of the anterior portion of the iliofemoralis
to rotate the upper side of the femur inwards, due to its insertion on the side of the
trochanter, was an important factor favouring the forward extension of its origin
(see below). It is difficult to decide whether a ridge-like lesser trochanter developed
first, with a 'holding' function for the iliofemoralis in response to the more vertical
excursion of the femur, and that this ridge was then suitably placed for the iliofemoralis to rotate the femur about its long axis, or whether there was a more
gradual change from a more crocodilian arrangement of the muscle insertions.
In the crocodile the rotatory tendency of the pifi dorsalis appears to be balanced
by the ischiotrochantericus and, probably, the anterior part of the pife; in the
bird the iliotrochanterici are opposed by the homologous muscles—the ischio-
femoralis and the obturators (the latter representing the anterior part only of the
pife, the posterior part having been lost (Romer, 1927b: 245)). This may, in
fact, be the reason why the posterior part of the pife has been lost in birds, since
it seems that the 'pife 2' (of Romer, 1923b) would be the portion likely to have
contributed the most contra-rotation, hence this portion survived as the obturators.
(Romer's 'pife 1', as already pointed out, is probably an exclusively crocodilian
speciality, and very probably a post-Triassic crocodilian speciality at that).

So far the reasons underlying the change in attitude and decrease in the arc of
movement of the femur in birds have not been considered. The key factor here is
believed to be the modification of the hind limbs to form an 'arrestor mechanism'.
At some stage in the evolution of the gliding (or gliding with weak flapping)
proavian it would have been necessary for the facility to be acquired of landing
on the hind limbs alone, in contrast to the previous use of all four limbs in arresting
movement. This must have been a critical transition in the evolution of birds,
involving the acquisition of a higher order of sensory and muscular coordination
than hitherto, with increased control over the glide-path and the ability to induce
stalling of the aerofoil at just the right moment for landing. However, there are
indications in the skull of the Upper Triassic reptile *Sphenosuchus* that the appropriate regions of the brain were already undergoing enlargement, presumably in
response to arboreal life. Thus the floccular recess is well developed and there is a
characteristically avian (though relatively small) inverted shield-shaped supraoccipital to which the epiotics are fused, indicating enlargement of the cerebellum.

The features of the bird hind limb which make it effective as an arrestor include
the relatively short femur with small arc of movement, directed forward as well as

downward; the well-developed femorotibialis muscle; and the mesotarsal ankle-joint with what I have called a 'tendon-sling tarsus' (Walker, 1972) formed from the tendons of the gastrocnemius and the digital flexors. The femur has of necessity to be relatively short in the attitude which it usually adopts in birds, otherwise when landing the moment-arm of the body's mass about the fulcrum of the knee would be excessive. This point also applies in bipedal walking, especially in large birds. According to Cracraft (1971), the muscles chiefly important in resisting protraction of the femur on landing are the adductors (superficialis and profundus) which, owing to their insertion on the distal half (or more in some cases) of the femur, have a large moment-arm about the acetabulum. The smaller arc moved through by the femur is probably a result, in part at least, of the long areas of origin and insertion of these muscles on the ischium and femur—they correspond broadly to Charig's type 'X' (1972). The femorotibialis probably assists by resisting flexion at the knee, but this point is not mentioned by Cracraft. According to him, the external part of the gastrocnemius is 'possibly the portion of the muscle that absorbs most of the shock and acts as a break (*sic*) to tarsometatarsal flexion' (presumably 'brake' is meant here) during landing (Cracraft, 1971: 236). That the gastrocnemius is one of the main shock-absorbers on landing would seem to be clear from ordinary observation of birds, and it is this shock-absorbing function which I believe has led to the formation of the mesotarsal joint in birds and bipedal dinosaurs. In the latter a shock comparable to that incurred by the bird on landing took place due to the fact that the whole weight of the body came upon *one* foot at each stride. It would have been as uncomfortable for a bipedal archosaur to have run 'flat-footed' as it is for ourselves. In agreement with Charig (1972) I think it likely that the type of tarsus with a calcaneal heel was a necessary inter-mediate step before the acquisition of full bipedalism in archosaurs. The arguments for this view cannot be put here, but it may be pointed out that a reduced or vestigial calcaneal heel has been reported in the prosauropod *Rio-jasaurus* (Bonaparte, 1971: 153), the pseudosuchian *Lagosuchus* (Bonaparte, 1975: 40) and the coelurosaur *Saltopus* (Walker, 1970: 325). Bonaparte's evidence in the case of *Lagosuchus* is particularly clear and demonstrates an intermediate stage between the crocodilian and mesotarsal types of joint.

The above points are made in order to show (a) that the nature of the ankle-joint is not necessarily a bar to the suggested crocodile-bird special relationship, (although it is not implied that the ancestral stock had a tarsus identical with that of modern crocodiles—there was a certain amount of variation in tarsal structure in the Thecodontia under the general label 'crocodilian') and (b) that similarities of femoral construction could have arisen convergently in birds and bipedal dinosaurs in response to similar, but not identical functional requirements. The rather bird-like femur in the Upper Jurassic crocodile *Hallopus* (Walker, 1970: fig. 6 and p. 355) reinforces this latter point.

Another aspect of the arrestor mechanism which has probably been influential in moulding the bird ilium and femur is the role of the iliotrochanterici in resisting outward rotation of the upper side of the femur. The slightly distal position of the trochanter in *Archaeopteryx* shows that part of the primitive function of the anterior portion of the iliofemoralis (or iliotrochanterici if differentiated) in the bird ancestor was a protractive tendency, and it has been suggested above that, in addition, inward rotation of the femur would have been favoured by a more forward position of origin of this muscle. It seems likely that, on landing, there is a tendency for the knees to spread outwards and the femora to rotate laterally. Thus it may be that the anterior extension of the iliotrochanterici evolved in order to resist this rotation, as much as actively to protract the femur. A further point concerning the avian hind limb is that the joint between the trochanter and the antitrochanter may well have arisen as an improvement in the use of the limb as an arrestor, making it possible for birds to land directly on the ground with less risk of femoral dislocation, in spite of the femur having an offset head. The position and strong buttressing of the antitrochanter tend to suggest this, coupled with its absence in *Archaeopteryx* (see below).

The timing of the events in the evolution of the bird pelvis is not easy to decide. If the indications from the crocodilian pelvis are a reliable guide, it would seem that the pifi had migrated dorsally by the Upper Trias, leading to weakening of the pubis. Unfortunately, little is known of the pelvis in the Sphenosuchia. As regards *Archaeopteryx*, I believe that Ostrom (1973, etc) has overstated the case as regards the reptilian characteristics of this form. As indicated in the Introduction the shoulder girdle (*pace* Ostrom) is essentially avian, with a typical furcula and laterally-directed glenoid. It is intended to consider these and other points in another publication, but as regards the present discussion, I do not accept that the pubis in *Archaeopteryx* was directed downwards, as suggested by Ostrom (1973) and am doubtful about the reconstruction given by Wellnhofer (1974). The Eichstätt specimen which he figures is about two-thirds the size of the Berlin specimen and somewhat more than half of the London specimen, hence there may have been cartilaginous borders to the bones around the acetabulum which would preclude an exact fit now. On the other hand, the right half of the pelvis of the Berlin example, which I have studied, is so little disturbed that it seems preferable to accept its evidence. The pubis is not damaged proximally, as Ostrom states (1973), or very little, and is only very slightly displaced laterally at the contact with the ilium. The proximal articular surface of the left pubis of the London specimen has been exposed by more recent preparation and shows a flat, almost circular facet, set obliquely to the long axis of the pubis, indicating that the latter was inclined quite strongly backward. The Berlin and London specimens are in agreement in showing an inclination of the pubis of about 135°, somewhat greater than the figure of 120° given by Wellnhofer for the Eichstätt example (Fig. 13).

As noted earlier there is no necessary correspondence in the length of the pubis before and after rotation, so that the elongated pubis of *Archaeopteryx* is not evidence of coelurosaurian ancestry. The pubis does not simply rotate back in its adult form during phylogeny; each successive stage is produced by a separate individual ontogeny, consequently it would have been perfectly possible for the pubis to have elongated gradually during phylogenetic rotation, had selection favoured such a change. *Archaeopteryx* has a pubis which is considerably longer than the ischium, but so have many other birds.

Preparation of the Berlin specimen in 1972 revealed the presence on the pubis of a short pectineal process, immediately below the contact with the ilium. That it should project from the pubis is not surprising, as the ambiens normally originates from the pubis in reptiles, hence it might be expected to have done so in an early bird. In carinate birds the pectineal process, when present, arises from the ilium, but in ratites it may be partly or entirely derived from the pubis (Bellairs & Jenkin, 1960: 259).

The femur/tibia ratio in *Archaeopteryx* is 71% in the Eichstätt example (Wellnhofer, 1974), 75% in the London specimen and 78%, in the Berlin specimen. Ratios for some modern birds which are not particularly specialized in the hind limb (although it is difficult to make a meaningful comparison) are: jackdaw (*Corvus monedula*) 58%; pigeon (*Columba livia*) 67%; Greenland falcon (*Falco rusticolus*) 80%. In the domestic fowl (*Gallus gallus*) the ratio is 70%. These birds are roughly of the same order of size as *Archaeopteryx*. Thus *Archaeopteryx* agrees more closely with birds in this respect than with coelurosaurs, in which the femur/tibia ratio ranges from 81% to 93%, the lowest value occurring in the

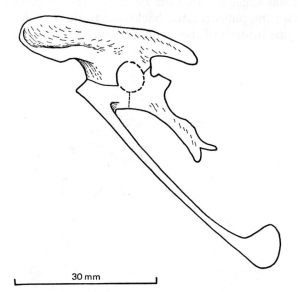

30 mm

Figure 13. Left lateral view of the pelvis of *Archaeopteryx*, based on the Berlin and London specimens. Scale as for London specimen.

highly cursorial late Cretaceous *Dromiceiomimus* (Russell, 1972). (In some of these examples it is the tibiotarsus or tibia plus astragalus that has been measured, but this does not make a great deal of difference to the ratios). In comparison, the femur/tibia ratio in the Upper Triassic ornithopod *Fabrosaurus* (Thulborn, 1972), is 81%.

Thus, by the late Jurassic, birds as represented by *Archaeopteryx* had acquired a backwardly-rotated pubis with a pectineal process, considerable forward expansion of the iliac blade, a relatively short femur, and a mesotarsal joint. The attitude of the pubis indicates that the centre of gravity had been displaced somewhat backwards, while the iliac blade development, shortened femur and mesotarsal joint denote that substantial progress had been achieved in the use of the hind limb as an arrestor during landing. Thus it seems reasonably certain that *Archaeopteryx* would have been able to land on its hind limbs alone on a branch or other eminence, although to do so on level ground would probably have been risky and difficult (except against a wind) because of its lack of an antitrochanter and its inability to reduce its speed by active wing-flapping. Paradoxically, it may have been the need to reduce speed before landing from a glide which elicited the first attempts at downbeating of the wings in the proavian, rather than the endeavour to use them as wings in the true sense. To take off, *Archaeopteryx* would only have had to launch itself out from a branch or, if on the ground, first climb a tree or other high point.

THE PELVIS OF *HALLOPUS*

The Upper Jurassic crocodile *Hallopus* was originally restored (Walker, 1970) with a pubis extending forwards and downwards, that is, 'propubic' (a term proposed by Charig (1972: 124), as being less ambiguous than 'triradiate'). Later (Walker, 1972) it was suggested that the *Hallopus* pelvis might have been 'opisthopubic', that is, with the pubis rotated backward to lie close to the ischium. Some brief remarks in justification of this proposal may perhaps be given now.

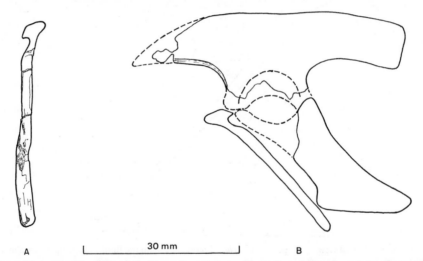

A   30 mm   B

Figure 14. A. Possible pubis of the Upper Jurassic crocodile *Hallopus*, as preserved. B. Restoration of the pelvis in left lateral view.

The possibility of the attainment of opisthopuby by crocodiles had not arisen (perhaps not surprisingly!) at the time of my 1970 paper, but the relationship now postulated between upward migration of the pifi, weakening of the pubis, and its subsequent backward rotation, suggests that it should be taken into account. The element originally identified, with considerable doubt, as the distal end of the pubis (Walker, 1970: fig. 1, bone 14) is the incomplete impression of a rather flat bone, the outline of which matches well the posteroventral corner of the ischium (bone 13) and, since the impression is incomplete at the edge which would be anterior, it is very likely to be the antimere of it (it is difficult to decide whether bone 13 is from the left or right side of the pelvis, so that a more precise identification is not possible). On the other hand, bone 24 (Fig. 14a) resembles an opisthopubic pubis in having a notch close to the (presumed) proximal end and a short forward projection on the opposite side which can be interpreted as a pectineal process. The ambiens would thus have been in the reptilian position. The bone broadens gradually distally and appears to be incomplete since it terminates at a hollow in the matrix which has been excavated to expose the proximal end of the left ulna. Bone 24 was briefly considered as a possible pubis (Walker, 1970: 332), but the slenderness of the proximal end appeared to rule out its identification as a normal reptilian pubis and the possibility of opisthopuby was not considered. In the restoration given later (1972) it was assumed that the more expanded distal end lay in the transverse plane and met its fellow in a short distal symphysis, hence it appears parallel-sided in the restoration.

Hallopus, as previously interpreted (Walker, 1970), was a small, highly specialized cursorial crocodile. Apart from *Hallopus*, archosaurs, as far as is known, did not include extreme quadrupedal cursoriality as an adaptive type, hence there are no closely-related functional analogues. The pelvis might therefore be expected to have had unusual features. One of the indications of the likelihood of rapid locomotion in *Hallopus* is the shortness of the humerus and femur compared to the distal segments of the limbs. The femur/tibia ratio is very low at 78%, the ratio humerus/radius+radiale+metacarpal III is 58%, that of femur/tibia+metatarsal III is 53%. Thus it seems likely that, as in cursorial mammals, the humerus and femur moved through rather short arcs and were directed on the whole as much backward and forward, respectively, as downward. Certainly, the limbs of *Hallopus* moved in the parasagittal plane. Thus the functional analogy in the hind-limb is with birds rather than cursorial dinosaurs, and backward rotation of the pubis may have taken place for rather similar reasons. The pifi dorsalis, on this interpretation, would have been largely replaced by the anterior extension of the iliofemoralis and the 'sartorius'. That there had been some anterior expansion of the iliofemoralis in *Hallopus* is shown not only by the anterior expansion of the iliac blade, but by the nature of the trochanter (referred to in the original paper as the 'lesser trochanter'). This is a thin ridge arising from the outer side of the shaft and dying out proximally, in a similar way to the bird or *Archaeopteryx* trochanter, rather than forming a proximally-projecting, solid process as in the lesser trochanter of dinosaurs. Other resemblances to the bird femur, previously noted (Walker, 1970: 355, 357) are (a) the presence in *Hallopus* of a proximal ventrolateral ridge on which the pife probably inserted, which appears to be homologous with the obturator ridge of the bird femur, and (b) the very weak fourth trochanter. The possible significance of these in relation to the musculature of the hind limb was dealt with in the original paper.

The anterior process of the ilium in *Hallopus* is not, as preserved, as elongated as in primitive ornithopod dinosaurs or in *Archaeopteryx*, but is somewhat more so than in the Triassic crocodile *Orthosuchus*. It is considerably deeper than in the ornithopods and *Orthosuchus*, but not as deep as in *Archaeopteryx*. The anterior end, however, is incomplete in *Hallopus* and may originally have been longer. As in bird evolution, the more forward attitude of the femur would have made the pife anterior even less effective as a protractor than in ornithopod dinosaurs. This can be seen from the skeletal restoration of *Hallopus* (Walker, 1970: fig. 10), in which the left femur is shown in front of the (propubic) pubis. The femoral attitude would also have made the 'sartorius' a more effective protractor, so that the anterior iliac process in *Hallopus* need not have been as long as in early ornithischians for backward rotation to have taken place. Taking all these factors into account, it does not seem impossible that *Hallopus* could have been opisthopubic. However, because of the small size of the proximal end, the pubis would appear to have been excluded from the acetabulum. This does not agree with the assumption made earlier that pubic exclusion in late Triassic crocodiles was correlated with the adoption of amphibious habits.

SUMMARY

(1) A substantial sub-ambiens division of the M. puboischiofemoralis internus, the pifi ventralis, was probably present in thecodontians and its increase in length is held to have been mainly responsible for the elongation of the pubic 'apron' in advanced members of this group.

(2) The dinosaurian 'greater trochanter' was not a muscle insertion-area, but an articular surface of some kind, and either lay below a bursa at the region where the iliofemoralis and iliofibularis turned downward, or articulated with a cartilaginous extension of the acetabulum. The pifi dorsalis did not insert on it and this muscle was a protractor, not a retractor, of the femur.

(3) It is possible that the fulcrum of the femur/ilium joint in dinosaurs was not through the centre of the head of the femur, but at the proximal surface of the head plus 'greater trochanter'.

(4) The lesser trochanter served for the insertion of the iliofemoralis, as has usually been accepted hitherto. The pifi dorsalis inserted proximal to it or between it and the head of the femur.

(5) Attention is called to the fact that Romer (1942) gave reasons for considering the avian iliotrochanterici to be derived from the reptilian iliofemoralis, and not from the dorsal part of the pifi, as he had suggested earlier (1927b). The pifi dorsalis has therefore almost certainly been lost by birds.

(6) Dorsal migration of the pifi in the Trias led to the reduction of the crocodilian pubis to a rod. It is therefore suggested that a similar muscular change was the prelude to backward rotation of the pubis in ornithischian dinosaurs and birds. In the ornithischians, the reason for upward migration of the muscle is believed to be the early acquisition of full bipedalism, requiring rapid protraction of the femur. In crocodiles an early arboreal stage is held to have been responsible, involving a similar functional need. The same is thought to be true of birds, but the later development of the hind limb as an arrestor upon landing was probably the basic reason for backward rotation in bird ancestry, and also led to the suppression of the pifi dorsalis. In ornithischians the improvement of the 'sartorius' as a femoral protractor was probably the 'last straw', rather than the basic reason for pubic rotation.

(7) The pifi ventralis remained large in saurischians, and the dorsalis and medialis portions probably suffered only moderate change in their positions of origin. The ventralis was particularly well developed in 'coelurosaurs', giving a comparable specialization for rapid protraction of the femur to that of the dorsalis in ornithopods. The difference in adaptive styles in these two groups was probably due to the early adoption of herbivory in ornithischians, which relied on fleetness of foot to escape from predators, whereas full bipedalism took longer to acquire in the more conservative, carnivorous theropod dinosaurs.

(8) The specialization of the pif i ventralis (and also of the anterior part of the pife) in theropods makes it unlikely that the pubis would subsequently have rotated backwards; hence it is improbable that birds have originated from this group.

(9) The pubis of *Archaeopteryx* has a short pectineal process and is thought to have been directed back at ca. 135°. *Archaeopteryx* could probably land on its hind limbs alone on a branch or other high point, but would have had difficulty in landing on level ground. Nor could it take off directly from the ground; having no efficient wing elevator muscle, and lacking an ossified sternum it could not beat its wings sufficiently rapidly or powerfully.

(10) The forward attitude of the femur in the small cursorial Jurassic crocodile *Hallopus* and its rather bird-like nature, with other factors, suggest that the pelvis may have been opisthopubic. What appears to be a suitably-shaped pubis occurs close to the ilia and ischia.

ACKNOWLEDGEMENTS

It is appropriate to pay tribute first of all to the work of the late Alfred S. Romer, without which this paper clearly could not have been written. The ideas put forward in the paper arose partly out of a discussion with Dr Alan Charig at the British Museum (Natural History), but he is not responsible for any of my errors. I am grateful to Dr Charig for reading the manuscript and for permitting additional preparation of the London specimen of *Archaeopteryx*, and to Mr Peter Whybrow for carrying it out. Mr Cyril Walker has helped me with material on many occasions in the British Museum and his assistance is appreciated. My thanks also go to Dr Hermann Jaeger, of the Museum für Naturkunde, East Berlin, for his hospitality and the preparation of the pelvis of the Berlin *Archaeopteryx*. Finally, it is a pleasure to acknowledge the continued support and encouragement given by Professor T. Stanley Westoll to my own and other people's researches on fossil vertebrates in this University.

REFERENCES

BAKKER, R. T., 1971. Dinosaur physiology and the origin of mammals. *Evolution, 25:* 636–58.

BAKKER, R. T., 1975. Dinosaur renaissance. *Scient. Am.*, April, 1975: 58–78.

BAKKER, R. T. & GALTON, P. M., 1974. Dinosaur monophyly and a new class of vertebrates. *Nature, Lond., 248:* 168–72.

BARNETT, C. H., DAVIES, D. V. & MacCONNAILL, M. A., 1961. *Synovial joints, their structure and mechanics*. London: Longmans.

BELLAIRS, A.d'A. & JENKIN, C. R., 1960. The skeleton of birds. In A. J. Marshall, (Ed.) *Biology and comparative physiology of birds, I:* 241–300. New York & London: Academic Press.

BIDAR, A., DEMAY, L. & THOMEL, G., 1972. *Compsognathus corallestris*, nouvelle espèce de dinosaurien theropode du Portlandien de Canjuers (Sud-Est de la France). *Annls Mus. Hist. nat. Nice, 1:* 9–40.

BONAPARTE, J. F., 1971. Los tetrapodos del sector superior de la formacion Los Colorados, La Rioja, Argentina (Triasico Superior). *Op. lilloana, 22:* 5–183.

BONAPARTE, J. F., 1975. Nuevos materiales de *Lagosuchus talampayensis* Romer (Thecodontia—Pseudosuchia), y su significado en el origen de los Saurischia. *Acta geol. lilloana, 13:* 5–90.

BROOM, R., 1904. On a new crocodilian genus (*Notochampsa*) from the Upper Stormberg Beds of South Africa. *Geol. Mag. (N.S. Dec. V), 1:* 582–4.

BROOM, R., 1927. On *Sphenosuchus* and the origin of the crocodiles. *Proc. zool. Soc. Lond.:* 359–70.

CASAMIQUELA, R. M., 1967. Un nuevo dinosaurio ornitisquio Triasico (*Pisanosaurus mertii*, Ornithopoda) de la formacion Ischigualasto Argentina. *Ameghiniana, 5:* 47–64.

CHARIG, A. J., 1972. The evolution of the archosaur pelvis and hind-limb: an explanation in functional terms. In K. A. Joysey & T. S. Kemp (Eds), *Studies in vertebrate evolution: essays presented to Dr F. R. Parrington, FRS.:* 121–55. Edinburgh: Oliver & Boyd.

COLBERT, E. H., 1962. *Dinosaurs: their discovery and their world.* London: Hutchinson & Co.

COLBERT, E. H., 1964. Relationships of the saurischian dinosaurs. *Am. Mus. Novit., 2181:* 1–24.

COLBERT, E. H. & MOOK, C. C., 1951. The ancestral crocodilian *Protosuchus. Bull. Am. Mus. nat. Hist., 97:* 143–82.

CRACRAFT, J., 1971. The functional morphology of the hind-limb of the domestic pigeon, *Columba livia. Bull. Am. Mus. nat. Hist., 144:* 173–268.

CRUICKSHANK, A. R. I., 1972. The proterosuchian thecodonts. In K. A. Joysey & T. S. Kemp (Eds), *Studies in vertebrate evolution: essays presented to Dr F. R. Parrington, FRS:* 89–119. Edinburgh: Oliver & Boyd.

FÜRBRINGER, M., 1876. Zur vergleichenden Anatomie der Schultermuskeln. III Teil. *Morph. Jb., 1:* 636–816.

FÜRBRINGER, M., 1900. Zur vergleichenden Anatomie des Brustschulterapparates und der Schultermuskeln. IV Teil. *Jena. Z. Naturw., 34:* 215–718.

GADOW, H., 1882a. Untersuchungen über die Bauchmuskeln der Krokodile, Eidechsen und Schildkröten. *Morph. Jb., 7:* 57–100.

GADOW, H., 1882b. Beiträge zur Myologie der hinteren Extremität der Reptilien. *Morph. Jb., 7:* 329–466.

GALTON, P. M., 1969. The pelvic musculature of the dinosaur *Hypsilophodon* (Reptilia: Ornithischia). *Postilla,* No. 131: 1–64.

GALTON, P. M., 1971. *Hypsilophodon,* the cursorial non-arboreal dinosaur. *Nature, Lond., 231:* 159–61.

GALTON, P. M., 1972. Classification and evolution of ornithopod dinosaurs. *Nature, Lond., 239:* 465–6.

GALTON, P. M., 1974. The ornithischian dinosaur *Hypsilophodon* from the Wealden of the Isle of Wight. *Bull. Br. Mus. nat. Hist. (Geol.), 25:* 1–152.

GALTON, P. M., 1975. English hypsilophodont dinosaurs (Reptilia: Ornithischia). *Palaeontology, 18:* 741–52.

GEORGE, J. C. & BERGER, A. J., 1966. *Avian myology.* London: Academic Press.

GILMORE, C. W., 1914. Osteology of the armoured Dinosauria in the United States National Museum, with special reference to the genus *Stegosaurus. Bull. U.S. natn. Mus., 89:* 1–143.

GILMORE, C. W., 1920. Osteology of the carnivorous Dinosauria in the United States National Museum, with special reference to the genera *Antrodemus (Allosaurus)* and *Ceratosaurus. Bull U.S. natn. Mus., 110:* 1–154.

GOW, C. E., 1975. The morphology and relationships of *Youngina capensis* Broom and *Prolacerta broomi* Parrington. *Palaeont. afr., 18:* 89–131.

GREGORY, W. K. & CAMP, C. L., 1918. Studies in comparative myology and osteology, No. III. *Bull. Am. Mus. nat. Hist., 38:* 447–563.

HAUGHTON, S. H., 1924. The fauna and stratigraphy of the Stormberg Series. *Ann. S. Afr. Mus., 12:* 323–497.

HUDSON, G. E., 1937. Studies on the muscles of the pelvic appendage in birds. *Am. Midl. Nat., 18:* 1–108.

HUDSON, G. E., LANZILLOTTI, P. J. & EDWARDS, G. D., 1959. Muscles of the pelvic limb in galliform birds. *Am. Midl. Nat., 61:* 1–67.

HUENE, F. von, 1908. Die Dinosaurier der Europäischen Triasformation, etc. *Geol. paläont. Abh. (Suppl.-Bd.), 1:* 1–419.

HUENE, F. von, 1915. On reptiles of the New Mexican Trias in the Cope collection. *Bull. Am. Mus. nat. Hist., 34:* 485–507.

HUENE, F. von, 1932. Die fossile Reptil-Ordnung Saurischia, ihre Entwicklung und Geschichte. *Mon. Geol. Paläont., 4*(1): 1–361.

JOHNSON, A., 1883. On the development of the pelvic girdle and skeleton of the hind limb in the chick. *Q. Jl. Microsc. Sci.* (N.S.) *23:* 399–411.

KERMACK, K. A., 1956. An ancestral crocodile from South Wales. *Proc. Linn. Soc. Lond., 166:* 1–2.

LAMBE, L. M., 1917. The Cretaceous theropodous dinosaur *Gorgosaurus. Mem. Geol. Surv. Canada, 100:* 1–84.

MADSEN, J. H. Jr., 1974. A new theropod dinosaur from the Upper Jurassic of Utah. *J. Paleont., 48:* 27–31.

NASH, D. S., 1975. The morphology and relationships of a crocodilian, *Orthosuchus stormbergi,* from the Upper Triassic of Lesotho. *Ann. S. Afr. Mus., 67:* 227–329.

OSAWA, G., 1898. Beiträge zur Anatomie der *Hatteria punctata. Arch. mikrosk. Anat. EntwMech., 51:* 481–691.

OSBORN, H. F., 1906. *Tyrannosaurus,* Upper Cretaceous carnivorous dinosaur (second communication). *Bull. Am. Mus. nat. Hist., 22:* 281–96.

OSBORN, H. F., 1916. Skeletal adaptations of *Ornitholestes, Struthiomimus, Tyrannosaurus. Bull. Am. Mus. nat. Hist., 35:* 733–71.

OSTROM, J. H., 1973. The ancestry of birds. *Nature, Lond., 242:* 136.

OSTROM, J. H., 1974a. *Archaeopteryx* and the origin of flight. *Q. Rev. Biol., 49:* 27–47.

OSTROM, J. H., 1974b. The pectoral girdle and forelimb function of *Deinonychus* (Reptilia: Saurischia): a correction. *Postilla,* No. 165: 1–11.

OSTROM, J. H., 1975a. On the origin of *Archaeopteryx* and the ancestry of birds. *Colloques int. Cent. natn. Rech. scient., 218:* 519–32.

OSTROM, J. H., 1975b. The origin of birds. *Annu. Rev. Earth & Planet Sci.*, *3:* 55–77.

OSTROM, J. H., 1976. *Archaeopteryx* and the origin of birds. *Biol. J. Linn. Soc.*, *8:* 91–182.

PERRIN, A., 1892. Contributions sur l-étude de la myologie comparée: membre postérieur chez un certain nombre de Batraciens et de Sauriens. *Bull. scient. Fr. Belg.*, *24:* 372–552.

PERRIN, A., 1895. Recherches sur les affinités zoologiques de *l'Hatteria punctata. Annls Sci. nat.* (*Zool.*), *20*(7): 33–102.

RABL, C., 1916. Ueber die Muskeln und Nerven der Extremitäten von *Iguana tuberculata* Gray. *Anat. Hefte, I Abt.*, *53:* 681–789.

ROMER, A. S., 1922. The locomotor apparatus of certain primitive and mammal-like reptiles. *Bull. Am. Mus. nat. Hist.*, *46:* 517–606.

ROMER, A. S., 1923a. The ilium in dinosaurs and birds. *Bull. Am. Mus. nat. Hist.*, *48:* 141–5.

ROMER, A. S., 1923b. Crocodilian pelvic muscles and their avian and reptilian homologues. *Bull. Am. Mus. nat. Hist.*, *48:* 533–52.

ROMER, A. S., 1923c. The pelvic musculature of saurischian dinosaurs. *Bull. Am. Mus. nat. Hist.*, *48:* 605–17.

ROMER, A. S., 1927a. The development of the thigh musculature of the chick. *J. Morph.*, *43:* 347–85.

ROMER, A. S., 1927b. The pelvic musculature of ornithischian dinosaurs. *Acta zool.*, *Stockh.*, *8:* 225–75.

ROMER, A. S., 1942. The development of tetrapod limb musculature—the thigh of *Lacerta. J. Morph.*, *71:* 251–98.

ROMER, A. S., 1956. *Osteology of the reptiles.* Chicago: University of Chicago Press.

ROMER, A. S., 1962. *The vertebrate body*, 3rd ed. Philadelphia: W. B. Saunders.

RUSSELL, D. A., 1972. Ostrich dinosaurs from the Late Cretaceous of western Canada. *Can. J. Earth Sci.*, *9:* 375–402.

SNYDER, R. C., 1954. The anatomy and function of the pelvic girdle and hindlimb in lizard locomotion. *Am. J. Anat.*, *95:* 1–46.

THULBORN, R. A., 1971. Origins and evolution of ornithischian dinosaurs. *Nature, Lond.*, *234:* 75–78.

THULBORN, R. A., 1972. The postcranial skeleton of the Triassic ornithischian dinosaur *Fabrosaurus australis. Palaeontology*, *15:* 29–60.

WALKER, A. D., 1961. Triassic reptiles from the Elgin area: *Stagonolepis, Dasygnathus* and their allies. *Phil. Trans. R. Soc.* (*B*), *244:* 103–204.

WALKER, A. D., 1964. Triassic reptiles from the Elgin area: *Ornithosuchus* and the origin of carnosaurs. *Phil. Trans. R. Soc.* (*B*), *248:* 53–134.

WALKER, A. D., 1970. A revision of the Jurassic reptile *Hallopus victor* (Marsh), with remarks on the classification of crocodiles. *Phil. Trans. R. Soc.* (*B*), *257:* 323–72.

WALKER, A. D., 1972. New light on the origin of birds and crocodiles. *Nature, Lond. 237:* 257–63.

WALKER, A. D., 1974. Evolution, organic. *McGraw-Hill Yearb. Sci. Technol.:* 177–9.

WELLNHOFER, P., 1974. Das fünfte Skelettexamplar von *Archaeopteryx.* The fifth skeletal specimen of *Archaeopteryx. Palaeontographica, Abt. A, 147:* 169–216.

ABBREVIATIONS USED IN FIGURES

acet	acetabulum	int tr	internal trochanter
add	adductor femoris	is	ischium
add long ext	adductor longus et brevis, pars externus (adductor femoris, pars)	isfm	ischiofemoralis (ischiotrochantericus)
		istr	ischiotrochantericus
add long int	adductor longus et brevis, pars internus (adductor femoris, pars)	lt	lesser trochanter
		N obt	obturator nerve
amb	ambiens, ambiens origin	obt ext	obturator externus (puboischiofemoralis externus, pars)
'apron'	descending anterior portion of pubes		
cart	cartilage in crocodilian pelvis intervening between ilium, pubis and ischium	obt int	obturator internus (puboischiofemoralis externus, pars)
cfb	caudifemoralis brevis	pife, pife 1, etc.	divisions of the puboischiofemoralis externus
cfl	caudifemoralis longus		
f	fulcrum	pif i	puboischiofemoralis internus (undivided)
fem	femur		
f obt	obturator foramen	pif i dor	puboischiofemoralis internus dorsalis
ft	femorotibialis	pif i med	puboischiofemoralis internus medialis
ft ext	femorotibialis externus	pif i ven	puboischiofemoralis internus ventralis
ft int	femorotibialis internus	pirif	piriformis (caudifemoralis longus and brevis)
glut	gluteus medius et minimus, iliofemoralis externus (iliofemoralis, pars)		
		plp	processus lateralis pubis
'gt'	'greater trochanter'	prp	prepubic process
il	ilium	pu	pubis
ilfb	iliofibularis	pu-is lig	pubo-ischiadic ligament
ilfm	iliofemoralis	'sar'	'sartorius' or iliotibialis 1 origin
il-is lig	ilio-ischiadic ligament	semim	semimembranosus (flexor tibialis internus)
iltb 2, 3	iliotibialis 2 and 3 origins		
iltr ant	iliotrochantericus anterior (iliofemoralis, pars)	sr	sacral ribs (cut)
		S 1, S 2	sacral vertebrae 1 and 2
iltr med	iliotrochantericus medius (iliofemoralis, pars)	tr	trochanter
iltr post	iliotrochantericus posterior iliofemoralis, pars)		

In the case of avian muscles the approximate reptilian equivalents are given in parentheses for convenience.

The origin of lizards

ROBERT L. CARROLL

Redpath Museum, McGill University, Montreal, Canada

The family Paliguanidae from the Upper Permian and Lower Triassic of South Africa provides the earliest evidence of the emergence of lizards from more primitive reptiles. This material indicates that the basic adaptive pattern of the Lacertilia was associated with small body size (a snout-vent length less than 150 mm) and an insectivorous diet. The presence of highly specialized epiphysial joints suggests that determinant growth was a primitive feature of lizards. A further adaptive complex involved the development of a streptostylic quadrate and a middle ear structure much more sensitive to airborne sounds than that of more primitive reptiles. Definitive features of the lacertoid shoulder girdle were established by the Permo-Triassic, but the pelvic girdle and tarsus evolved more slowly. Although features usually considered characteristic of lizards are expressed in both prolacertilians and millerosaurs, neither of these groups are closely related to the ancestry of squamates. Lizards and sphenodontids represent divergent stocks that probably shared a common ancestry among primitive eosuchians.

CONTENTS

INTRODUCTION

Among living reptiles, lizards have the widest geographical distribution, the largest number of species, and occupy the greatest diversity of environments. In these features, they are as successful, in an evolutionary sense, as any of the mammalian orders. In marked contrast with Tertiary mammals, however, lizards as a group show a remarkably conservative skeletal anatomy. Most living lizard groups retain a pattern already fully established by the late Jurassic.

Genera showing a skull configuration typical of lizards have been described from as early as the Upper Triassic (Robinson, 1962; Colbert, 1970). As a result of work by Broom (1925), it is generally agreed that lizards evolved from early diapsids such as the eosuchian *Youngina* by the loss of the lower temporal bar. There is, however, a considerable morphological gap between the Upper Triassic lizards and the late Permian or early Triassic eosuchians. The genus *Prolacerta* (Parrington, 1935; Camp, 1945) from the early Triassic *Lystrosaurus* zone of South Africa shows cranial features that suggest an intermediate position between eosuchians and primitive lizards. This genus has been emphasized in all recent discussions of the origin of lizards (e.g. Robinson, 1967). Recent description of

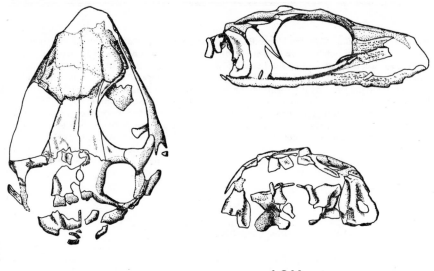

1 CM

Figure 1. Skull of *Paliguana whitei* in dorsal, lateral and occipital views. Albany Museum, Grahamstown, No. 3585; *Daptocephalus* or *Lystrosaurus* zone; ×2.

the postcranial skeleton by Gow (1975) demonstrates, however, that *Prolacerta* exhibits few of the specific characteristics expected in an ancestor of squamates and shows several features indicative of specialization toward a way of life quite distinct from that seen in primitive lizards.

Other lizard-like forms from the Permo-Triassic of South Africa have been described by Huxley (1868) and Broom (1903, 1926):

Paliguana whitei Broom (1903), Albany Museum, Grahamstown, no. 3585, either Upper Permian *Daptocephalus* or Lower Triassic *Lystrosaurus* zone (Fig. 1);

Palaeagama vielhaueri Broom (1926), McGregor Museum, Kimberley, no. 3707, *Lystrosaurus* zone (Fig. 2);

Saurosternon bainii Huxley (1868), British Museum (Natural History), no. 1234, Upper Permian, *Cistecephalus* or *Daptocephalus* zone (Fig. 3).

Because of the small size of these specimens and the difficulty of preparation, the original descriptions were very incomplete and they have not since been seriously considered in relationship to the origin of lizards. These specimens have

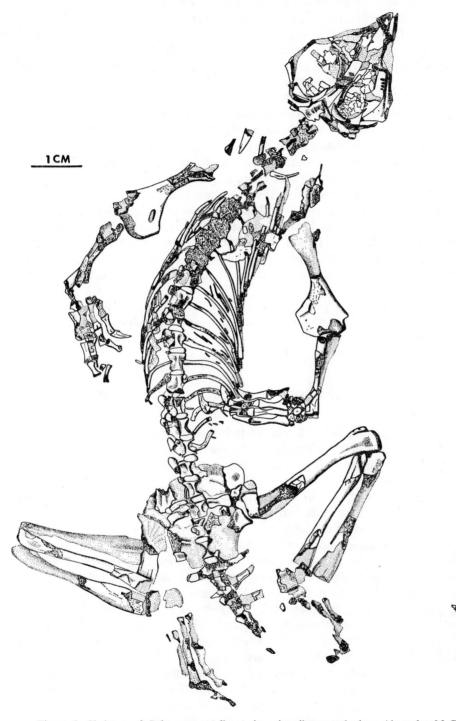

Figure 2. Skeleton of *Palaeagama vielhauri*, in primarily ventral view. Alexander McGregor Museum, Kimberley, No. 3707; *Lystrosaurus* zone; scale indicated on figure.

recently been further prepared and described (Carroll, 1975). In as far as the various specimens can be compared, all may be retained in a single family, the Paliguanidae. Of the presently known Permo-Triassic reptiles, these forms are unquestionably the most similar to lizards in all aspects of their skeleton.

Although more than a single genus is represented in this material, the size and general morphology are sufficiently consistent to permit a plausible composite reconstruction. Most features of the postcranial skeleton shown in Fig. 4 are taken from *Saurosternon*. The skull is that of *Paliguana*. Confirmation of the association of this type of skull and postcranial skeleton is provided by the type of *Palaeagama*, in which the entire skeleton is articulated, although the preservation of the individual elements is not sufficiently complete for restoration.

This material provides, for the first time, a basis for considering the nature of the earliest lizards and for discussing the evolution of this assemblage from more primitive reptiles. It enables us to evaluate the overall adaptive pattern of the ancestral lizards, and to see how this pattern has influenced the subsequent evolution of the group. Anatomical specializations which were developed by the end of the Permian were related to a particular way of life that separated lizards from other reptiles and established a pattern that has dominated lizard evolution to the present day.

In overall size and body proportions, the paliguanids resembled the smaller agamids and iguanids of the modern lizard fauna. It is probable that their general way of life was similar as well.

THE PALIGUANIDAE AND THE ORIGIN OF THE LACERTILIA
Skull

The skull of paliguanids is small relative to body size, and remarkably similar to those of the primitive gliding lizards *Kuehneosaurus* (Fig. 5) and *Icarosaurus* in both size and proportions. The posterior position of the pineal opening and the retention of postparietals and tabulars reflect its more primitive nature, but the general configuration of the skull and particularly the temporal region is basically similar to that of living iguanids and agamids. As in kuehneosaurs, the adductor musculature did not spread extensively over the lateral and dorsal surface of the parietal, but the posterior borders of both the postorbital and postfrontal are deflected downward. As in more primitive eosuchians, the supratemporal in *Paliguana* lay in a depression on the posteromedial surface of the parietal. In this position, it could have provided little, if any, support for the squamosal. In living lizards retaining a squamosal, this bone is supported medially by the expanded distal portion of the supratemporal. The supratemporal lies beneath the posterior ramus of the parietal, and is exposed both medially and laterally.

The Upper Jurassic lizard *Yabeinosaurus* (Endo & Shikama, 1942) appears to illustrate an intermediate condition. The supratemporal attaches proximally to the medial surface of the parietal, but extends distally a considerable distance to support the squamosal by its medial surface. There is clearly a progressive change in which the supratemporal becomes gradually more important in supporting the squamosal, enlarges posteriorly and ventrally, and, to a greater extent, laterally.

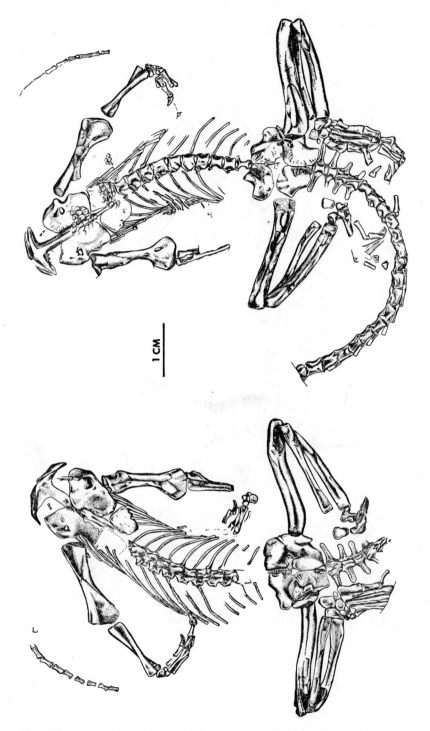

1 CM

Figure 3. *Saurosternon bainii*, British Museum (Natural History), No. 1234; counterparts; *Cistecephalus* or *Daptocephalus* zone; scale indicated on figure.

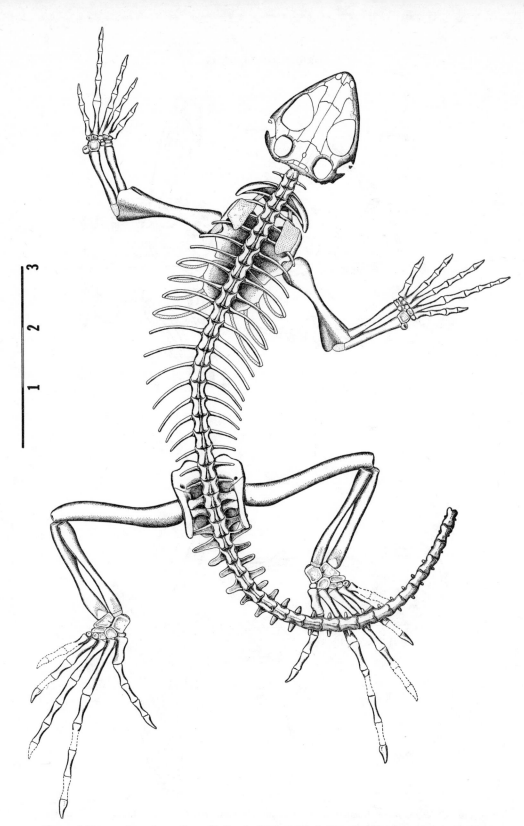

Figure 4. Restoration of a paliguanid lizard. Postcranial skeleton based on *Saurosternon bainii*, skull based on *Paliguana whitei*. Association confirmed by *Palaeagama vielhauri*.

Another significantly primitive feature of paliguanids, shared with the kuehneosaurs, is the nature of the fronto-parietal suture. It is slightly interdigitating and angles posterolaterally. Slight movement might have been possible between these bones, but there is no evidence of specialization to that end. Most modern lizards, and their Jurassic predecessors where known, show a nearly transverse suture between these bones, indicating a mesokinetic hinge. This is clearly a specialization that occurred within lizards, and was not a primitive feature of the group. This development is associated with the medial fusion of the frontals and/or parietals observed in most advanced lizards.

Unfortunately, the area of the external naris is not exposed in either of the skulls of paliguanids, precluding comparison with the pattern of either *Kuehneosaurus* or *Icarosaurus* where the narial opening is medial, or that of the majority of lizards, where it is paired. Unfortunately, the dentition of the Permo-Triassic

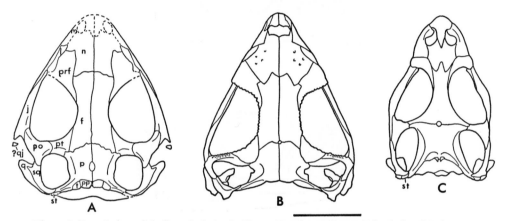

Figure 5. Dorsal view of skull roof of: A, the Upper Permian or Lower Triassic lizard *Paliguana;* B, the Upper Triassic genus *Kuehneosaurus* (from Robinson); C, a modern iguanid, *Crotaphytus*. Significant changes noted between A and B include: Loss of tabular and postparietal; anterior movement of parietal opening; elaboration of supratemporal on anterior or lateral face of posterior ramus of parietal. Between *Kuehneosaurus* and modern lizards, the median bones of the skull have become fused at the midline and a mesokinetic hinge forms between the frontal and parietals. The squamosal develops a ventral peg which inserts in the top of the quadrate, and the paroccipital process extends laterally to support the quadrate.

Scale is 1 cm in length. f, Frontal; j, jugal; l, lacrimal; m, maxilla; n, nasal; p, parietal; pf, postfrontal; po, postorbital; pp, postparietal; prf, prefrontal; q, quadrate; qj, quadratojugal; sq, squamosal; st, supratemporal; t, tabular.

forms is very poorly exposed. It is apparently subpleurodont. It certainly shows no evidence of being acrodont. Such an unspecialized pattern is what might be expected in the ancestors of the advanced lizard groups, but is also seen in the majority of primitive reptiles. None of the specimens shows any significant portion of the palate. The lower jaw of *Paliguana* has a retroarticular process, a feature common to lizards, but not well developed in primitive eosuchians.

In *Paliguana*, a trace of bone behind the short posterior ramus of the jugal may be a remnant of the quadratojugal, but the configuration of the quadrate and squamosal indicates that streptostyly was already fully established. As in *Kuehneosaurus*, the paroccipital process does not reach the quadrate, as it does in more advanced lizards.

The problem of hearing

The size and basic configuration of the quadrate of *Paliguana* resemble those of both the Upper Triassic kuehneosaurs and primitive living lizards of comparable size and habits (Fig. 6). It bears an anterior ridge that resembles that which in living lizards supports a tympanum. It has long been taken for granted that early lizards had a tympanum and were sensitive to airborne sounds since it has been widely accepted that these features were a common heritage of all primitive reptiles, both sauropsids and theropsids. There is now considerable reason to doubt that this faculty was well developed in primitive reptiles.

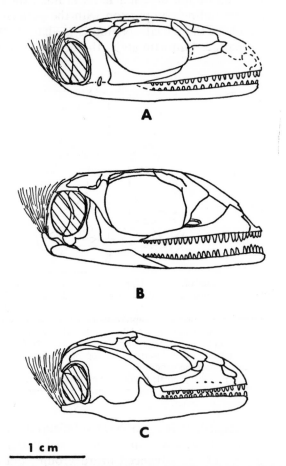

Figure 6. Skulls of: A, *Paliguana;* B, *Kuehneosaurus;* and C, *Crotaphytus;* in lateral view, showing similarity of the configuration of the quadrate. Tympanum and depressor mandibulae restored in the fossil genera on the basis of the pattern in *Crotaphytus;* ×2.

Romer (1967, 1970), Hotton (1959) and others have cited the presence of a well developed stapes in captorhinomorphs and other primitive reptiles as indicating that the middle ear in these forms was capable of transmitting airborne sounds, and restored a tympanum, supported by fleshy tissue, immediately behind the cheek region. The position and size of the tympanum differed according to the author, since there was no definite evidence from the configuration of either the cheek region or the stapes for its attachment.

A quite different opinion has been expressed by Watson. On several occasions (1954, 1957) he has suggested that the absence of an otic notch in captorhinomorphs and pelycosaurs indicated that the tympanum was absent and hence that these primitive reptiles were insensitive to airborne sound.

Recent work on the electrophysiology of the ear in modern lizards (Wever & Werner, 1970; Manley, 1973) provides a basis for reevaluating the question of hearing in early reptiles. Despite the absence of detailed information on the soft anatomy in early reptiles and the differences in their bony structure from living lizards, the basic principles of sound transmission and limitations of physiological response provide a frame of reference applicable to both fossil and living forms.

The apparatus for detecting airborne vibrations in modern lizards consists of a large, usually superficial, tympanum, supported anteriorly by the quadrate and posteriorly and ventrally by the tough skin and fascia of the neck region. Within the tissue of the tympanum is embedded the cartilaginous extracolumella. The extracolumella is attached to the end of a long thin stapes, or columella, which functions as a piston to transmit vibrations to the inner ear. The stapes is expanded medially as a round footplate, inserted in the oval window. Vibrations of the stapes are transmitted to the fluid of the inner ear. The middle ear operates as a mechanical transformer to magnify the force received at the tympanum in order to compensate for the difference in the relative impedance of the air and the fluid of the inner ear. In modern lizards this is achieved by having a large ratio between the area of the tympanum and that of the stapedial footplate (approximately 20 to 1 in *Crotaphytus*) and through the use of the extracolumella as a lever (much as the extra ossicles of the mammalian ear). This lever increases the force by a factor of approximately 2.

A notable feature of primitive reptiles, captorhinomorphs and pelycosaurs, is the massive appearance of the stapes and particularly the large size of the footplate. Assuming that the middle ear of primitive reptiles acted as a transformer to match the impedance of the air and inner ear fluid, the tympanum would have had to have a diameter several times as great as the height of the skull in order for the ear to have a sensitivity equal to that of modern lizards. The shaft of the stapes extends anteriorly at an angle of approximately 45° to the plane of the cheek in primitive captorhinomorph reptiles, further reducing the mechanical force that might be transmitted by a superficial tympanum. Rather than resting loosely in the oval window, the stapes in primitive reptiles is typically held tightly in place by an overlapping portion of the parasphenoid, greatly limiting the amplitude of vibration, if not precluding it entirely.

It might be suggested that the ear of primitive reptiles was simply much less effective in detecting airborne sounds than is that of living lizards. The overall structure, however, indicates that this was not its primary function. According to Manley (1973: 612), the type of ear structure seen in primitive reptiles, with a massive stapes having high inertia, would be ideal to detect tissue-conducted sounds from the substrate. The middle ear in primitive reptiles apparently evolved to detect groundborne, not airborne, sounds.

It is difficult to judge the amount or nature of airborne sounds in the Palaeozoic. In the absence of any evidence that either predators or prey of Palaeozoic reptiles were generating sounds of the intensity and in the frequency potentially detectable by these forms, it is probable that selection to produce greater acuity to airborne sound was not strong.

A factor generally ignored in evaluating hearing in early tetrapods is their body form and posture. This has been considered by Reisz (1975) who points out that most primitive tetrapods probably kept their heads close to the ground much of the time. This was necessary because of their short necks and the configuration of the limbs and girdles. The anatomy of the shoulder joint of captorhinomorphs (Holmes, 1977) is such that much muscular effort would be necessary to lift the anterior portion of the body above the substrate. It is unlikely that the head and shoulders were habitually held above the ground, as is the case with lizards. With the head and trunk on the ground, they would have been in an ideal position to hear the footfalls of approaching predators, or the scurry of insect prey. And, as Manley has indicated, this is the function for which the heavy stapes of early tetrapods is well suited.

An important feature of the early diapsids, as pointed out by Reisz, is that the lightly-built skull and long neck would have permitted them to maintain the head in a raised position. This is a clear advantage from the standpoint of better vision, but must have left them much less sensitive to groundborne sounds. The configuration of the temporal region and stapes of *Petrolacosaurus* led Reisz to suggest that, like captorhinomorphs, this early diapsid lacked a functional tympanum and had little sensitivity to airborne vibrations. One may suppose a fairly strong selective pressure toward developing greater sensitivity to airborne sounds if other factors tended toward freeing the head from the substrate, as evidently occurred in diapsid evolution.

Manley has suggested that a small tympanum may have been present in early reptiles, serving to relieve changes in pressure resulting from movement of the massive stapes. Although initial response would have been limited to loud, low-frequency sounds, acuity could be progressively increased by expanding the tympanum and reducing the weight and size of the stapes. Judging from *Youngina* (Fig. 13), primitive eosuchians had not progressed far in this direction, for the stapes is still massive and the quadrate is apparently not specialized to support a large tympanum. This pattern is in marked contrast to the condition seen in *Kuehneosaurus* and *Paliguana*. The stapes has not been described in either of these forms, but the quadrate has essentially the same configuration as in living lizards, with a distinct anterior rim for attachment of the tympanum. In the absence of other information, it may be assumed that detection of airborne sound was an important factor in the biology of these forms, although the acuity and range of frequency response may have been very limited by the standards of modern lizards.

A tympanum of sufficient size to detect a significant range of airborne sounds is only possible when the surface of attachment is far enough anterior so that the membrane is not constricted by the muscles of the neck region. The anterior position of the quadrate in *Paliguana* and *Kuehneosaurus* provides as much space for a tympanum as that seen in modern lizards. The development of the retroarticular process in these genera permits the depressor mandibulae to pass behind the tympanum without compressing it against the cheek.

The anterior position of the quadrate in early lizards should also be considered in relationship to changes in feeding patterns. A basic selective force operating on the pattern of the skull of early lizards was presumably their insectivorous habit. Robinson (1973) has pointed out two distinct strategies developed by lepidosaurs in their jaw mechanics. In primitive eosuchians and sphenodontids, the jaw musculature is massive and extends forward nearly half the length of the jaw. This provides a powerful, but relatively slow, crushing or slashing action. In early lizards, by contrast, the temporal region is shorter, and the area of jaw muscles correspondingly reduced. The force of the bite is less, but the speed is greater—a considerable advantage in dealing with agile insects. Shortening of the temporal region is achieved in early lizards by the more anterior position of the quadrate. Factors leading to more efficient hearing and a more rapid jaw movement apparently reinforced each other in modifying the position and configuration of the quadrate.

Both factors may also be associated with the loss of the lower temporal bar. What is known of the squamosal and quadratojugal in the earliest diapsid *Petrolacosaurus* (Reisz, 1975) indicates that they are insubstantial bones, with little if any development of areas for resisting longitudinal stress. It is clear from the work of Barghusen (1968) that the lower cheek in primitive reptiles was not important as an area for attachment of the adductor jaw musculature. Progressive thinning of this region could proceed without much reorganization of the remainder of the skull or changes in jaw musculature. One might expect, as in early pelycosaurs (Reisz, 1972), a reduction in ossification to accommodate increase in the mass of the lateral portion of the adductor musculature (originating higher on the cheek) leading first to thinning and later to elimination of the lower temporal bar. Additional lateral accommodation of the jaw musculature would be necessary in the antecedents of lizards in order to compensate for the short cheek region.

As pointed out by Robinson (1973), freeing of the quadrate from the jugal provided an acoustic barrier, separating the tympanum from the noise of the upper dentition. She points out further that the presence of a synovial joint between the quadrate and articular effectively shields the ear from noise of the jaw articulation and lower dentition.

On the basis of information currently available, selection for a larger area of tympanum attachment was the factor responsible for the dorsal elaboration of the quadrate, to a position where it was supported dorsally by the squamosal, parietal, and later by the paroccipital process. The dorsal extension of the quadrate is a necessary requisite for the development of the lacertoid type of streptostyly, but does not in itself necessitate movement of the quadrate relative to the skull roof. Streptostyly is typically discussed primarily in relationship to feeding mechanics (Frazzetta, 1962); the intimate association of the tympanum with the quadrate indicates the advisability of considering also the relationship of streptostyly to hearing. The anterior margin of the tympanum is firmly attached to the quadrate. Posteriorly the depressor mandibulae provides a barrier to anterior-posterior compression. Ventrally, however, the skin that supports the tympanum is also attached to the retroarticular process. If the quadrate were firmly attached to the skull roof, the quadrate-articular joint would act as a simple hinge; as the mouth is opened, dorsal movement of the retroarticular process would compress the ventral area of the tympanum (Fig. 7). The development of streptostyly allowed part of the action of opening the jaw to be transferred to the joint between the squamosal and the skull roof, thus distributing the compressional forces more evenly. The magnitude of this compression is difficult to assess, but it should be considered in evaluating the origin of streptostyly. In primitive lizards without a mesokinetic joint, action of the depressor mandibulae would tend to rotate the base of the quadrate posteriorly as the jaw is opened, and the contraction of the pterygoideus would pull it forward as the jaw is closed.

Frazzetta (1962) has observed that the articulating surface of the quadrate of varanid and other advanced lizards swings anteriorly, rather than posteriorly, as the jaw is opened, in contrast to the direction suggested here for primitive forms.

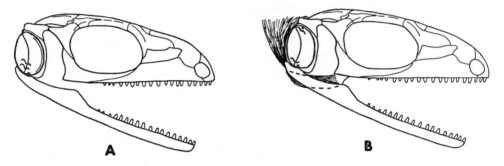

Figure 7. Restoration of *Paliguana* skull in lateral view, indicating possible distortion of the tympanum by the retroarticular process as the jaw is opened: A, with the quadrate rigid; B, with the quadrate movable on the squamosal (streptostyly). Depressor mandibulae and pterygoideus muscle are restored to show their opposing effects in rotation of the quadrate.

It should be noted that the development of mesokinesis in modern lizards produces a pattern of intracranial mobility far different from that seen in paliguanids and kuehneosaurs. As Frazzetta indicates, modern lizards have developed other mechanisms for protecting the tympanum from compression as the jaw is opened. The retroarticular process is angled ventrally from the plane of the jaw, and a special ligamentous band connects the anterior end of the basal unit with the lower jaw. There has thus been a continued interrelationship between the requirements of mastication and hearing in the anatomy of the quadrate and temporal region in the evolution of lizards.

Vertebrae and ribs

The vertebrae of *Palaeagama* and *Saurosternon* (Fig. 8A) are basically primitive in being amphicoelous, with large trunk intercentra. In most lizards, the centra are procoelous, with the intercentra absent for most of the trunk region. Among the Gekkonidae, however, most genera have amphicoelous vertebrae (Fig. 8B). Kluge (1967) has argued that procoely is primitive for lizards and that gekkonids represent a secondary achievement of the amphicoelous condition. Hoffstetter & Gasc (1969) emphasize the similarity of the vertebrae of amphicoelous gekkonids with those of more primitive lepidosaurs (Triassic lizards, the eosuchians and sphenodonts), as well as pointing out that the procoelous condition has been achieved separately in several lines of gekkonids, but by a different process than that seen in other lizards. The basic similarity of the vertebrae of paliguanids and amphicoelous gekkos supports the view that this pattern is primitive for squamates.

Unfortunately, none of the specimens of paliguanids shows the cervical vertebrae in ventral view, so that it is not possible to determine whether or not the intercentra were specialized as hypapophyses. In *Palaeagama*, the area of articulation between the atlas and axis resembles the more posterior zygapophyses and the two halves of the atlas arch are not as clearly separated as they are in sphenodontids and more primitive reptiles. These are features common to all advanced lizard groups. A feature of most modern lizards that is not expressed in paliguanids is caudal autotomy.

Figure 8. A. Restoration of posterior trunk vertebrae of *Saurosternon bainii* in lateral, anterior, dorsal and ventral views, ×3. B. Ventral view of trunk vertebrae of *Gekko gekko*, ×2. Note the similarity of the intercentra and the common presence of conspicuous foramina subcentrale.

The neural spines of paliguanids are low, as in most primitive lizards, but in strong contrast with those of *Prolacerta*. As in sphenodontids and several groups of primitive lizards (Hoffstetter & Gasc, 1969), there are small accessory articulating surfaces medial to the zygapophyses.

The known paliguanid specimens have 23 or 24 presacral vertebrae, two sacrals and approximately 70 caudals. According to the criteria of Hoffstetter and Gasc, nine vertebrae can be recognized as cervicals, but the position of the first rib attached to the sternum is uncertain. Most primitive lizards have eight cervicals and 24 presacrals seems to be the common count.

Unfused ribs are apparently present on all the vertebrae anterior to the sacrum except the atlas. Only *Dibamus* among modern lizards has ribs on the axis. In *Palaeagama*, extensions from several trunk ribs can be seen just behind the sternum. These probably consist of calcified cartilage, and represent the ventral portions of the ribs that were attached to the sternum. At least three, but possibly four, ribs are so attached. This pattern is seen in both lizards and sphenodontids. Ventral extensions of the ribs were apparently present in a number of eosuchians, but there is no evidence for their presence in *Prolacerta* or in any of the Middle Triassic forms frequently associated with the Lacertilia.

The two sacral ribs, and all those in the anterior portion of the tail, are firmly fused to the transverse processes, with no evidence of separation, such as can be seen in primitive or immature eosuchians.

Shoulder girdle and forelimb

The shoulder girdle is well exposed in ventral view in *Saurosternon*. The pattern is generally that of modern lizards. Of particular importance are the fenestration of the anterior margin of the scapulocoracoid and the presence of a substantial sternum. A sternum is present in several eosuchians, as well as in sphenodontids. The fenestration of the scapulocoracoid, however, shows a uniquely lacertoid configuration. The pattern of this bone throughout the Lacertilia has recently been discussed by Lécuru (1968a). Although there is much variation in the pattern of fenestration, a common feature is a strong, anteriorly-projecting bar, the procoracoid, separating the scapulocoracoid fenestra and the anterior coracoid fenestra. This is apparently the pattern seen in *Saurosternon*. As in lizards, there is a large supracoracoid foramen. In most lizards, a suture is present between the scapula and the coracoid. The bones are indistinguishably fused in *Saurosternon*.

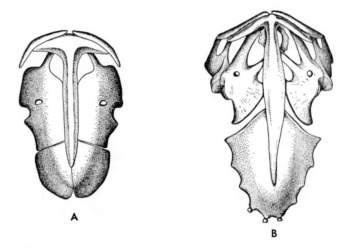

A   B

Figure 9. Ventral view of shoulder girdle of: A, *Saurosternon baini*, ×2; and B, *Iguana iguana*, ×1·5.

A very important feature of the scapulocoracoid is the short anteroposterior extent of the glenoid. In contrast with the condition in most primitive reptiles and the Triassic forms frequently allied with lizards—*Prolacerta*, *Tanystropheus* and *Macrocnemus*—the articulating surface is as short as that of modern lizards, and shows a similar posteroventral orientation.

As a result of the comprehensive work by Lécuru (1968b), comparison of the dermal shoulder girdle and sternum of *Saurosternon* can be made with all lizard groups. Unquestionably, the closest comparison lies with the iguanids (Fig. 9). The dermal shoulder girdle of that group does not, of course, differ greatly from the pattern of more primitive reptiles. The blades of both the clavicle and the interclavicle are narrow and oriented transversely to the long axis of the body. This pattern also appears in *Tupinambis*, *Xenosaurus* and, in an only slightly modified form, in *Varanus*.

The sternum of *Saurosternon* is paired, but this is probably a juvenile characteristic. It is short, relative to the scapulocoracoid, and the anterior margin is angled only slightly forward at the midline. As preserved, the areas for rib attachments are not evident. There is no evidence for the presence of the mesosternum.

The humerus appears primitive in having very wide extremities and a large entepicondylar foramen. In these characteristics, it resembles *Sphenodon* more closely than it does lizards. A very important feature is the nature of the articulation with the scapulocoracoid. The limited extent of the glenoid can have accommodated only a small condylar articulating surface. The proximal ends of the humeri are not preserved in *Palaeagama*, and the ends of the bone are not ossified in *Saurosternon*, but there was presumably a short epiphysial knob, as in modern lizards (Fig. 10). Such a joint would greatly facilitate movement at the shoulder, allowing the arm to be brought alongside, and even pulled under, the trunk, as in lizards and sphenodontids, but in strong contrast with primitive reptile groups. It is probable that this facility is associated with the presence of a sternum and sternal attachments of the ribs as in *Sphenodon* and lizards. The ability to raise the

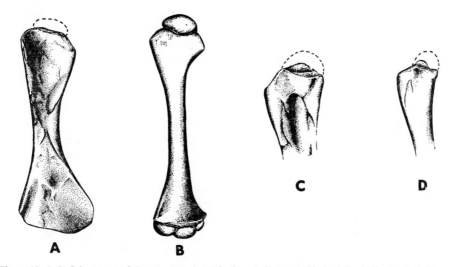

Figure 10. A. Left humerus of *Saurosternon bainii* in dorsal view, ×3, (dashed line indicates probable configuration of proximal articulating area), B. Humerus of *Iguana iguana*, ×2. C. Proximal end of the left femur of *Saurosternon bainii*, in ventral view, ×3 (dashed line indicates probable configuration of proximal articulating surface). D. Proximal end of left femur of *Iguana iguana* in ventral view (dashed line indicates extent of proximal articulating surface—this area was removed during cleaning of the bone).

forepart of the body off the substrate appears to be an important factor in the biology of lizards. The vast majority of modern lizards, and certainly the primitive members of the group, had short necks, so that a premium was placed on modifying the forelimbs so as to raise the head above the substrate to assist in extending their range of vision for prey capture and avoidance of predation.

The elbow joint in paliguanids may have retained the primitive pattern of widely placed extensor and flexor muscles, common to the captorhinomorphs (Holmes, 1977), so that the area of the entepicondyle remained fully ossified, surrounding the entepicondylar foramen. The ulna, like that of most living lizards, is notable for the ossification of a specialized distal portion that forms a ball-like surface for articulation with the ulnare. This is a feature also seen in sphenodontids, but not present in other groups of reptiles.

A restoration of the carpus of *Saurosternon* is shown in Fig. 11A. The vaguely spherical appearance of the elements is probably a result of the immaturity of the specimen. Less of the carpus is preserved in *Palaeagama*, but the individual elements fit together like a mosaic. This is a pattern seen in mature forms of most primitive reptilian groups. The carpus of *Saurosternon* is primitive in retaining all of the elements seen in captorhinomorphs, but is advanced in its small size relative to other limb proportions. Both the lateral and the proximomedial extent is reduced. The perforating foramen, strongly expressed in the concave margins of the intermedium and ulnare in primitive eosuchians, is lost.

The pattern of the squamate carpus has recently been described by Renous-Lécuru (1973) with illustrations of members of most lizard families. Except for the obviously specialized condition seen in chameleons and forms with reduced limbs, the pattern is remarkably constant. According to Renous-Lécuru and most other recent authors, there are five distal carpals, a single centrale, the ulnare, radiale and pisiform. The intermedium is generally small, and is absent in many genera belonging to diverse families. Sphenodontids and the amphisbaenid genus *Bipes* differ in having two centralia. In these forms the intermedium is large.

In both sphenodontids and lizards there are small epiphyseal elements associated with the epipodials, the carpals and the metacarpals.

The pattern suggested by Renous-Lécuru as being ancestral to that of lizards is essentially that seen in *Sphenodon* and also similar to that of the eosuchians. The pattern seen in *Saurosternon* is even closer to that of modern lizards in that the intermedium is reduced to a small element, although it still supports a small portion of the radius. If comparison is made with the carpus of representatives of most of the major lizard groups (Fig. 11 D–K) the remaining elements appear at first glance to be completely comparable. As the bones are identified by Renous-Lécuru and others, however, there is a discrepancy in the identification of the medial elements. Whereas the eosuchians and *Saurosternon* have in succession the radiale, medial centrale, and a distal carpal, no medial centrale is recognized in the lizards. The bone typically identified as the first distal carpal is not in line with the other distal carpals, but is more proximal in position, immediately medial to the lateral centrale. The area that appears serially homologous with the other distal carpals is identified as a large epiphysial element at the proximal end of the first metacarpal. Comparison with the more primitive forms suggests that this epiphysial element may have evolved from a former distal carpal. If this is the case, the element termed the first distal carpal in lizards would be homologous with the medial centrale of both eosuchians and sphenodontids. Functionally, the proximal end of the first metacarpal has intruded into the row of distal carpals. This is evident in most genera of lizards with normal limb proportions illustrated by Renous-Lécuru. It is a situation, perhaps analogous with that in the pes, where the proximal portion of the fifth metatarsal has intruded into the tarsal series. Haines (1969: 104) notes that the first metacarpal of a varanid is unique in showing growth from the proximal end.

If this interpretation of the carpals is correct, lizards as a group retain both centralia, but incorporate the first distal carpal into the head of the first metacarpal as an epiphysial element. In this respect, *Saurosternon* resembles the eosuchian or sphenodontid condition more closely than that of lizards.

As a result of the narrowing of the carpus, the heads of the metacarpals of *Saurosternon* overlap as in modern lizards. This tendency is less marked in sphenodontids. No important changes are noted in the configuration of the phalanges between eosuchians and lizards.

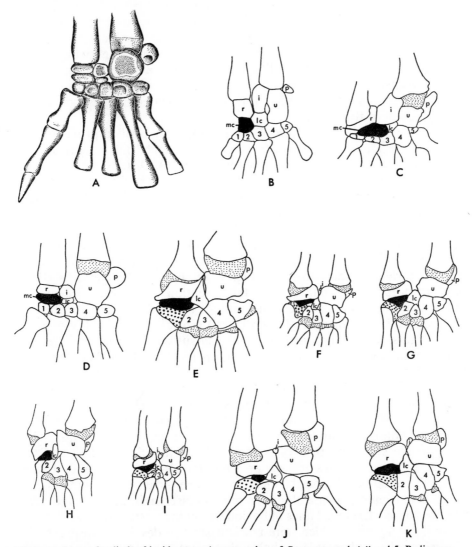

Figure 11. Lower forelimb of lepidosaurs: A, restoration of *Saurosternon bainii*, ×4·5; B, diagrammatic restoration of the Upper Pennsylvanian diapsid *Petrolacosaurus* (from Reisz, 1975); C, *Sphenodon;* D, *Saurosternon;* E, *Iguana;* F, *Uromastix*, an agamid; G, *Macroscincus*, a scincid; H, *Eublepharis*, a gekkonoid; I, *Zonosaurus*, a gerrhosaurid; J, *Lepidophyma*, a xantusiid; K, *Heloderma*; C and E–K are taken from Renous-Lécuru, 1973.

Numbers indicate distal carpals; i, intermedium; lc, lateral centrale; mc, medial centrale; p, pisiform; r, radiale; u, ulnare. Solid shading indicates medial centrale in primitive forms, and bone identified as distal carpal 1 in lizards. Epiphyses indicated by stippling. Open circles indicate area identified as the epiphysis of the first metacarpal in lizards. All are dorsal views of left limb.

Pelvic girdle and rear limb

The pelvic girdle of the Paliguanidae appears primitive in most respects. Like those of the captorhinomorphs and eosuchians, the puboischiadic plate is solid, with little evidence for the initiation of thyroid fenestration. The anterior medial portion of the pubis is apparently not twisted laterally, as is the case in modern lizards, although this is difficult to evaluate in flattened specimens. There are clearly a number of important changes in the bone and musculature of the pelvic girdle between the primitive captorhinomorph pattern and that exhibited by modern lizards. This subject is, however, a complex problem that cannot be considered in detail here. As in advanced eosuchians, the paliguanids have evolved toward the lacertoid condition in fusing the anterior caudal ribs to the vertebrae. Considering the major importance of the caudofemoralis musculature in lizard locomotion (Synder, 1954), this suggests greater emphasis on strong and rapid adduction of the femur.

Judging from the configuration of the incompletely ossified proximal end, the femur, like the humerus, apparently closely approached the pattern of modern lizards in the presence of a knob-like condyle (Fig. 10). Distally, the bone may also have been extended by a specialized epiphysial surface. The tibia shows no specifically lacertoid features, but the restoration of the articulation between the fibula and the tarsals suggests the presence of a crescentic articulating surface such as is present in the living iguanids.

The tarsus very well illustrates the mosaic of ancient and lizard-like features characteristic of the paliguanids. The bones can be clearly seen in both dorsal and ventral views in *Saurosternon* (Fig. 12 D, F). All the elements present in captorhinomorphs are retained, with the exception of the medial centrale. The astragalus and calcaneum are separate, and the former bone is marked by a notch of the perforating foramen. A large (lateral) centrale is retained, along with all five distal tarsals. The proportions of the tarsus as a whole, and the detailed structure of the constituent bones, however, are much modified from the pattern in primitive reptiles.

The extensive modification in the function of the tarsus and the role of the foot in locomotion between primitive tetrapods and lizards has recently been outlined by Robinson (1975). As was earlier pointed out by Schaeffer (1941), the role of the foot in locomotion is relatively passive in primitive tetrapods. The tarsus is large, relative to the crus, and the bones form a loose mosaic. Bending and twisting can occur between the crus and the proximal tarsals, and a hinge joint is present between the distal tarsals and the metatarsals. The tarsus was presumably capable of some dorso-ventral flexure along the proximodistal axis, as well as transversely. There are no well developed lines of flexion within the tarsus, nor does there appear to have been any areas which were particularly well developed to resist bending. As in living salamanders, it is probable that plantar flexion was produced by the flexor primordialis which ran down the back of the crus, and extended around the posterior ventral surface of the tarsus to insert on the metatarsals and phalanges. Because of the loose arrangement of the bones, the tarsus of primitive tetrapods, including captorhinomorph reptiles, was presumably not able to support very strong action of this muscle. In contrast with modern lizards, it is doubtful that plantar flexion played a significant role in forward propulsion.

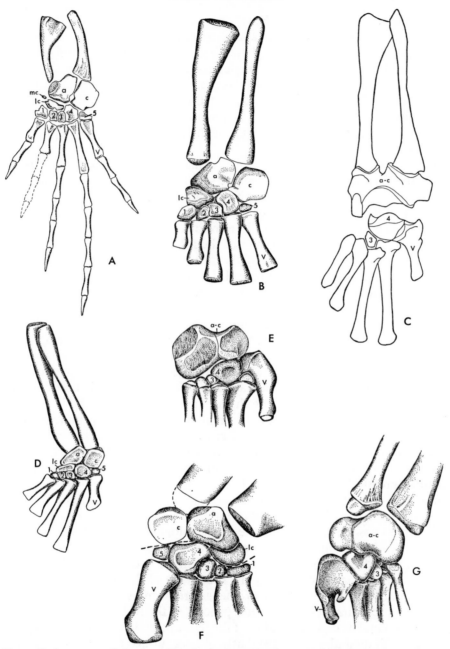

Figure 12. Lower rear limb of reptiles: A, the Pennsylvanian romeriid *Paleothyris*, ×2; B, the primitive eosuchian *Galesphyrus, Cistecephalus* zone, South Africa, ×1·5; C, diagrammatic outline of *Varanus* (from Robinson, 1975) to show articulating surface between the astragalocalcaneum and the fourth distal tarsal; D, *Saurosternon bainii*, ×2; E. *Iguana iguana*. Note progressive shortening of the tarsus relative to the crus and the presence of epiphyses in *Iguana* in positions occupied by distal tarsals in *Saurosternon*. A–E dorsal views. F. Ventral view of *Saurosternon bainii* (heavy dashed line indicates the position of the mesotarsal joint, lightly dashed line indicates outline of possible epiphyseal extension of the fibula). G. Ventral view of *Iguana iguana*. Numbers indicate distal tarsals; a, astragalus; c, calcaneum; a–c, astragalocalcaneum; lc, lateral centrale; mc, medial centrale; V, fifth metatarsal.

In modern lizards (Figs 12 C, E, G) the tarsus is much shorter relative to the crus and the pes. The calcaneum and astragalus are fused, and the centrale is presumably incorporated into the latter bone. The proximal tarsals are closely integrated with the crus, forming a single functional unit. The distal surface of the astragalocalcaneum forms a rounded hinge joint with the fourth distal tarsal. The third distal tarsal in lizards, and both the second and third distal tarsals in sphenodontids, are reduced in size. The remaining distal tarsals appear to be lost. The distal tarsals are closely bound with the metatarsals, and together operate as a functional unit.

One aspect of the lepidosaurian foot that has recently been considered in some detail is the function of the fifth metatarsal (Robinson, 1975). The structure of this element separates lizards and sphenodontids markedly from more primitive diapsids. Dr Robinson recognizes two, quite distinct, functions of the fifth metatarsal, one in relationship to the movement of the foot as a whole and the other limited to the unique properties of the fifth digit.

Like the calcaneum in mammals, the fifth metatarsal in modern lepidosaurs acts as a lever to lift the foot and propel the body forward. Two strong ligaments of the fibular subdivisions of the femoral head of the gastrocnemius muscle attach to tubercles on the plantar surface of the fifth metatarsal. Contraction of this muscle flexes the entire foot, lifting the posterior portion. This action is possible because of the position of the fifth metatarsal beneath the more medial elements of the foot, and the limited hinge action between the fourth distal tarsal (to which the fifth metatarsal is closely bound) and the fused astragalocalcaneum.

The rigidity of the tarsus and metatarsals of lizards which is necessary for the function of the lever action in lifting the foot and propelling the body forward was achieved at the cost of limiting the flexibility of the foot. In primitive tetrapods, arching of the tarsus and the proximal ends of the metatarsals could be used to assure a strong grip on the terrain. Development of the fifth metatarsal as an effective lever was possible only through strengthening the tarsus as a unit, which resulted in greatly limiting the possibility of flexure. Metatarsals 1–4 are tightly integrated, and preclude arching of the proximal portion of the foot.

To compensate for the inflexibility of the tarsus and metatarsals 1–4, the fifth digit in lizards has developed a unique role, analogous with the hallux or pollex of mammals. A hinge joint between the fifth metatarsal and the fourth distal tarsal allows the fifth digit to be brought into opposition with the first. The movement of the fifth digit is accomplished by the brevis slip of the peronaeus muscle and a tendon of the femorotibial head of the gastrocnemius. The brevis slip of the peronaeus muscle attaches to the outer process of the fifth metatarsal and adducts the metatarsal and its digit. A tendon of the femorotibial head of the gastrocnemius rotates the fifth digit about the distal articulating surface of the fifth metatarsal so that its ventral surface can face that of digit one.

The configuration of the tarsus and pes of *Saurosternon* indicates the initial development of the functions perfected by advanced lepidosaurs. The distal articulating surfaces of the tibia and fibula are nearly flat and in the same plane, in strong contrast to the much more ventral point of articulation of the tibia in captorhinomorphs. A limited hinge-like movement between the crus and the proximal tarsals might have been possible, but a tight association appears more probable. The ventral or posterior surface of the area of articulation of the astragalus and calcaneum with the fibula is recessed ventrally, as in modern lizards. The bony portion of the fibula ends bluntly, suggesting the presence in the living animal of a cartilaginous extension fitting into this recess, as in lizards.

All of the tarsals are smoothly rounded, in contrast with their more angular appearance in mature individuals of modern lepidosaurs. This rounded appearance may be partially a factor of immaturity (as was noted also for the carpals) and the bones may have fitted more closely in the adult. Most of the space of the tarsus is occupied, however, suggesting that the bones had achieved a relationship to one another that would not be likely to change with further ossification. The astragalus and calcaneum are certainly not fused to one another, and the centrale is maintained as a discrete element. It is closely integrated with the astragalus, however, indicating that the bones would have functioned as a unit in the living animal. Fusion of the centrale with the mediodistal surface of the astragalus of *Saurosternon* would produce the configuration of the astragalus seen in unspecialized lizards. The configuration of the adjacent surfaces indicates that the proximal tarsals articulated as a unit with the distal tarsals. Distal tarsal 4, as in modern lizards, is by far the largest, and intrudes somewhat into the row of proximal bones. Distal tarsal 3 is recessed into the medial surface of 4. Distal tarsals 1 and 2 remain as substantial elements, articulating with the centrale.

The proximal ends of the metatarsals are closely integrated with each other. The head of the fifth metatarsal extends medially beneath that of the fourth, which partially underlies the third, and so on.

The intrusion of the fifth metatarsal into the row of distal tarsals would have precluded the simple hinge movement that was possible between the tarsals and metatarsals in more primitive reptiles. This, together with the close relationship of the proximal tarsals with the crus, produces in *Saurosternon* a primitive expression of the mesotarsal joint common to modern lepidosaurs. This suggests that the function of the foot may have approached that seen in living lizards and sphenodontids.

The configuration of the fifth metatarsal of *Saurosternon* indicates that its role in producing opposition of the fifth digit had probably proceeded further in the direction of modern lizards than had its function as a lever to lift the foot. All primitive reptiles that retain a small body size have a digital pattern resembling that of primitive lizards. The phalangeal pattern count is 2, 3, 4, 5, 4 and the length of the digits increases from the first to the fourth, with the fourth especially long. The fifth is much shorter than the fourth. In primitive captorhinomorphs and small primitive eosuchians, the proximal ends of the metatarsals are not overlapping and the fifth metatarsal resembles metatarsals 1 to 4. Each metatarsal articulates with a corresponding distal tarsal. In *Saurosternon*, the fifth metatarsal is much shorter than the fourth, as it is in lizards, and has a well developed medially facing facet for articulation with the fourth distal tarsal. Articulation is still maintained with a distinct fifth distal tarsal, however. The proximal end is expanded laterally to form an outer process, as in lepidosaurs, suggesting the development of an area for insertion of the brevis slip of the peronaeus muscle to abduct the fifth digit. The distal articulating surface is much broader than is that of the other metatarsals, forming a wide surface for rotation of the proximal phalange of this digit so that it can be brought into opposition with the more medial digits.

The considerable overlapping of the proximal ends of the metatarsals and the marked asymmetry of the pes suggest that the grip of the digits was established as in primitive modern lepidosaurs.

The degree to which the fifth metatarsal of paliguanids such as *Saurosternon* was capable of acting as a lever to lift the back of the foot is more difficult to assess. Its efficiency depends to a great extent on how closely the distal elements of the tarsus were attached to the metatarsals. There is no way of estimating the amount of connective tissue that may have been present in this area. The presence of all the primitive distal tarsals, as well as the lack of fusion of the proximal tarsals suggests that the tarsus as a whole remained fairly flexible. No plantar tubercles are evident on the fifth metatarsal. They might have been ossified in more mature individuals, but their absence in this specimen suggests that there was not a strong attachment for the gastrocnemius as would be expected in a foot capable of strong plantar flexion. The potential for developing the fifth metatarsal as a lever is certainly evident, but the osteological evidence does not indicate that this role was fully developed.

Aside from the problem of the function of the fifth metatarsal in *Saurosternon*, there is also a question as to whether it should be considered 'hooked', a term coined by Goodrich (1916) but recently considered more systematically by Robinson (1975). Goodrich originally stated: 'the fifth metatarsal (of lepidosaurs) is quite peculiar, and differs from the others in that it is shortened and markedly hooked. The bent proximal end projects forwards (inwards) and also extends farther proximally than the remaining metatarsals, passing over the end of the fourth'.

The fifth metatarsal in *Saurosternon* certainly articulates with the fourth distal tarsal, and the facet for articulation faces medially, but the shaft of the bone does not actually appear to be bent. Compared with those of living lepidosaurs, the fifth metatarsal is only slightly specialized and could at most be termed 'incipiently hooked'.

The configuration and probable function of the tarsals and metatarsals of paliguanids clearly presages those of modern lepidosaurs. A logical extension of trends already evident would lead to the development of a truly hooked fifth metatarsal and the fusion of the proximal tarsals. It is interesting that these features of the tarsus, considered characteristic of all lizards, would have developed after the anatomy of the skull and the shoulder girdle had evolved to a nearly modern level.

Another feature of the foot which should be considered is the fate of the distal tarsals 1, 4 and 5. Although distal tarsals 1 and 2 are said to be missing in all lizards, examination of modern genera (for example *Iguana*, Fig. 12E) reveals the presence of bony elements proximal to metatarsals 1 and 2. These, of course, are epiphysial elements, but they are similar in size and position to the distal tarsals of *Saurosternon* and other early reptiles. Both an epiphysis and a distal tarsal are present proximal to the third metatarsal in *Iguana*. Except where histological study is possible, it would be very difficult to differentiate epiphyses and distal tarsals where these elements occupy essentially the same anatomical position. It might be argued, in fact, that, as in the case of the first metacarpal, the tissue primitively organized to form discrete tarsal elements has in modern lizards integrated with the metatarsals and functions as epiphyses.

The fate of the fifth distal tarsal should be viewed separately. Several alternatives may be considered. In some eosuchians (Harris & Carroll, 1977), the fifth distal tarsal unquestionably becomes incorporated with the fourth. In a series of specimens showing a long growth sequence, it can be seen that the fifth distal tarsal appears briefly as a separate element and then fuses to the lateral surface of the fourth. A similar fusion also occurs in a millerosaur (Gow, 1972: fig. 25). The function of the fifth metatarsal in lizards is such as to reduce the more lateral area of its proximal articulation. In this group, it seems unlikely that the fifth distal tarsal would fuse with the fourth; it seems more probable that it would be reduced and gradually disappear as a recognizable area of ossification. Rather than being completely lost, however, embryological evidence published by Sewertzoff (1908; quoted by Robinson, 1975) on the gekkonid *Tarentola* indicates that at least in this particular genus the fifth distal tarsal fuses with the fifth metatarsal. The intrusion of the fifth metatarsal into the area of the distal tarsals makes such an incorporation of tissue a reasonable interpretation.

Scales

In their scales, the Paliguanidae also straddle the line between primitive lepido-saurs and lizards. Dermal ventral scales, lost in modern lizards, are still retained, at least in small numbers. Typical squamate epidermal scales are also present, preserved as impressions in the area of the shoulder girdle in *Saurosternon* (Fig. 3), and appear as in modern genera.

EPIPHYSES, SIZE AND DIET

In modern lizards, the ends of the limb bones usually remain clearly distinct from the shaft, even in mature individuals. This characteristic is due to the presence of a separate area of ossification within the epiphyses, and the limitation of the area of growth to the epiphysial plate between the shaft and a large mass of tissue in the area of articulation (Haines, 1969). The presence of a secondary region of ossification is probably responsible for the elaboration of specialized condylar joints, particularly at the proximal ends of the humerus and femur, and the development of specialized areas of attachment or articulation at the ends of other limb bones.

In Palaeozoic reptiles such as captorhinomorphs and pelycosaurs, the ends of the bones are slow to ossify and in fossil material of immature individuals the shafts terminate in shallow depressions. With maturity, the articulating areas ossify so as to appear as indistinguishable continuations of the shaft. Growth occurs at the ends of the bones, with a thin covering of cartilage remaining at the surface of the articular area even in mature forms.

In the well-preserved and articulated skeleton of the late Permian paliguanid *Saurosternon*, the ends of most limb bones are incompletely ossified, but the shafts closely resemble those of modern lizards in which the epiphyses have not yet ossified, or have been removed by boiling. Articulation between the humerus and the glenoid, and between the femur and acetabulum requires the presence of knob-like condylar surfaces such as are present in modern lizards. Apparently these areas were cartilaginous at the time of death of this individual. Another specialized articular surface, at the distal end of the ulna, is ossified. The presence of large, specialized articular surfaces implies that the pattern of distinct epiphyses, common to lizards, had been established in *Saurosternon*.

As Haines (1969: 103) has noted, 'The epiphysial mechanism has always been understood to promote growth. It is possibly of equal biological importance as a precise mechanism for stopping growth—the fusion of epiphysial centres provides this.' Among reptiles, lizards and sphenodontids are unique in having determinate growth. Unlike populations of crocodilians and chelonians, individuals notably larger than the average species size are not encountered.

Except among the specialized varanoid lizards and their Mesozoic relatives, small body size is an important aspect of the biology of lizards. The need to limit body size in lizards appears to be related to dietary specialization. The vast majority of living lizards are insectivorous. The most important contribution to the subject is a paper by Pough (1973) on lizard energetics and diet. He studied members of the families Agamidae, Gerrhosauridae, Iguanidae and Scincidae, all of which contain both carnivores (primarily insectivorous) and herbivores. Pough found a very strong correlation between gross body size and diet that was applicable to all these groups. Species weighing less than 50-100 g are insectivorous, whereas those that weigh more than 300 g are almost all herbivorous. Juveniles of large herbivorous species tend to be insectivorous until they reach body weights of 50-300 g. Pough estimates that above 300 grams, the energy expended by a primitive lizard in capturing small insect prey is greater than that provided by the insect as food.

These primitive lizard families also appear to lack the anatomical and physiological characteristics necessary for the substained activity associated with the capture of larger, vertebrate prey. Plants would thus appear to be the only practical food source for larger members of the more primitive living lizard groups.

The more specialized families—Anguidae, Chamaeleontidae, Helodermatidae, Teiidae and Varanidae—do not include herbivorous species, although each family includes species weighing over 300 g. The chamaeleontids differ markedly from primitive lizards in their way of feeding, with the development of a specialized tongue, allowing them to capture insects with a minimum of movement of the entire body. Varanoids have a much more sophisticated anatomy and metabolism than more primitive forms, allowing them to capture large vertebrate prey with less relative expenditure of energy than is required by iguanids or agamids.

An herbivorous habit is very rare among living lizards. Only about 50 of the 2500 living species are primarily herbivorous. As in other primitive tetrapod groups, it is probable that the earliest lizards were initially unable to assimilate plant food. If their anatomy and metabolism made it difficult, if not impossible, to feed on either plants or larger vertebrates, selection would have forced them to retain a small body size. Faced with the necessity of remaining small by their dietary specialization, primitive lizards evolved the ability to terminate their growth at an appropriate size by completely ossifying the epiphyses.

The predominance of small insectivorous forms among primitive living lizard families supports the assumption that this was the pattern of the ancestral members of the group. This is further confirmed by the fossil record of Jurassic and Triassic

Table 1. Measurements of lizards from the Permian, Triassic and Jurassic

	Length of skull or lower jaw (mm)	Humerus (mm)	Femur (mm)	Skull plus trunk (mm)
UPPER JURASSIC				
Cordyloidea				
Paramacellodus oweni	26·5			
Saurillus obtusus	17·2			
Pseudosaurillus becklesi	26·2	14·6		
Lacertoidea				
Macellodus brodiei	23·6			
Meyasaurus fauri		19·4		
Scincoidea				
Becklesisaurus scincoides	44·4			
Anguoidea				
Dorsetisaurus purbeckensis	48	24		
Ilerdaesaurus crusafonti	18			
Gekkonoidea				
Yabeinosaurus tenuis	21·5	7·2	10	83
Ardeosaurus brevipes	14·5		9·1	59·1
Ardeosaurus digitatellus	21·3		10	76·6
Ardeosaurus cf. *digitatellus*	20·4			77·5
Eichstaettisaurus schroederi	21	10	14	85
Palaeolacerta bavarica	11·8	4·2		36·2
Bavarisaurus macrodactylus			23·8	83·1
?Agamidae				
Euposaurus cerinensis				
15681	11·2	5	7·5	35·7
15682	8·8	4·5	6	24·5
UPPER TRIASSIC				
Icarosaurus siefkeri	24·8	20·1	34·7	95
Kuehneosaurus latus	37·1			
UPPER PERMIAN AND LOWER TRIASSIC				
Palaeagama vielhauri	27	25	30	115
Saurosternon bainii		18·5	25	
Paliguana whitei	27			

Data from Hoffstetter (1964, 1965, 1967), Colbert (1970) and Cocude-Michel (1963).

lizards (Table 1). Almost all of these forms have a body size within the range of modern insectivorous genera. On the basis of this information, it is difficult to escape the conclusion that ancestral lizards were small (less than 150 mm in snout-vent length) and insectivorous.

PROBLEMS OF CLASSIFICATION
Position of the Paliguanidae among the lizard infraorders

On the basis of the skull, *Paliguana* and, by extension, the Paliguanidae, may be included in the Eolacertilia, as defined by Robinson (1967). She indicated the following features as being characteristic of the group: paired midline bones of the skull roof; large lacrimal; (numerous palatal teeth; ventral ramus of the opisthotic); loss of the ventral ramus of the squamosal; disappearance of the quadratojugal; atrophy of the posterior ramus of the jugal; slight contact of quadrate and pterygoid. (Those features in parentheses cannot be seen in the known specimens of paliguanids.) Another important feature that might be added to this definition is the absence of a mesokinetic joint between the frontals and parietals.

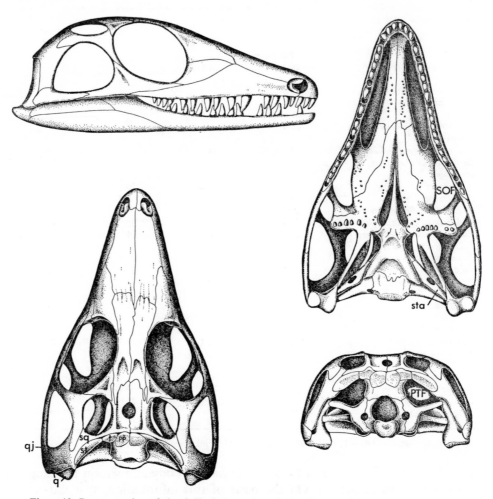

Figure 13. Reconstruction of the skull of *Youngina*, ×2, showing the primitive eosuchian pattern which is the closest known antecedent to the condition seen in paliguanid lizards. The genus is considerably more primitive than paliguanids in the configuration and orientation of the stapes, the limited exposure and posterior position of the quadrate, and the retention of a complete lower temporal bar.

pp, Postparietal; PTF, post-temporal fenestra; q, quadrate; qj, quadratojugal; SOF, suborbital fenestra; sq, squamosal; st, supratemporal; sta, stapes; t, tabular.

Unfortunately, the appendicular skeleton is not sufficiently well known in either *Kuehneosaurus* or *Icarosaurus* to permit extensive comparison with paliguanids. One important feature is the presence of amphicoelous vertebrae in all three genera. This is certainly a primitive feature for lizards in general, as are the configuration of the squamosal and the shape of the upper temporal opening.

No features of the paliguanids have been noted in which they are more specialized than known members of the advanced lizard infraorders. They could be ancestral to all of these groups. The pattern of the skull roof is most similar to that of the Iguania, but the resemblances are all in the retention of primitive features, and do not indicate especially close affinities. Some resemblances, also of a primitive nature, can be noted in the postcranial skeleton of iguanids as well, notably in the configuration of the dermal shoulder girdle.

Institut de Paléontologie, Paris No. 1908-5-2

One of the most striking features of the lizard fauna of the Upper Triassic is the presence of the highly specialized genera *Icarosaurus* and *Kuehneosaurus*, whose ribs extended laterally to support an extensive gliding surface. Even more surprising is the presence of a similar form from the Upper Permian or Lower Triassic of Madagascar. This species has yet to be named or described in detail, but an illustration is included here (Fig. 14) for comparison with the more typically lizard-like paliguanids, and to show the extent of radiation within the general lacertoid habitus in the Permo-Triassic. The bones of the skull are disarticulated, and the surface of the elements poorly preserved making restoration very difficult. The cheek region appears to be open, suggesting the development of a typically streptostylic condition. Most of the temporal region is missing however. Post-cranially, none of the features here cited as indicative of primitive lizards can be observed. The scapula is very tall and narrow with no evidence of anterior fenestration. A very large cleithrum is retained. This bone may simply be missing in other early lizard-like forms, but its presence would be surprising in an advanced relative of the lizards.

No sternum can be seen, although this bone might be obscured by other elements of the pectoral girdle.

The limbs show no evidence of the specialized epiphysial surfaces seen in paliguanids, although this feature might be less evident in a very mature individual.

It is interesting, and perhaps significant, that the sternum is not evident in *Icarosaurus*, and that the scapula is a tall narrow bone, without anterior fenestration. *Icarosaurus* does not appear to show specialized epiphysial articulating surfaces. In neither animal is the glenoid sufficiently well preserved to indicate the extent of the articulating surface. These features of the shoulder girdle may be related to the locomotor specialization of these gliding forms, and so do not necessarily indicate close relationship. Although the kuehneosaurs are among the earliest known lizards, some caution must be used in considering the anatomy of their appendicular skeleton as a model for the primitive lacertoid pattern.

In the absence of articulated cranial material, there may be some question as to whether or not the gliding form from Madagascar is even a lizard. It is at least possible that this adaptive pattern was achieved from an essentially eosuchian grade of development.

Prolacerta, Macrocnemus and Tanystropheus

Description of the paliguanids and recognition of their close relationship to subsequent lizards requires reevaluation of two other Palaeozoic groups that have

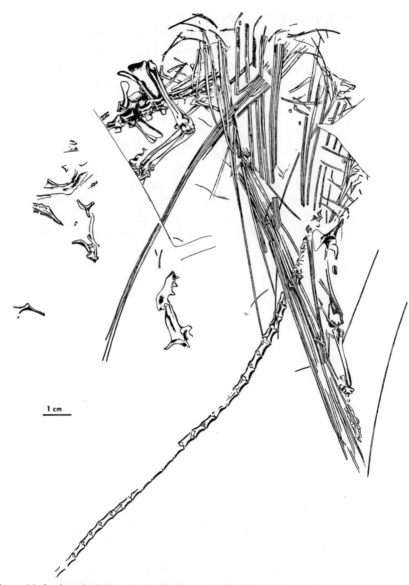

Figure 14. Institut de Paléontologie, Paris, no. 1908–5–2, Upper Permian or Lower Triassic of southwestern Madagascar; scale indicated on figure. Line truncating hand and anterior end of vertebral column marks the edge of one block. Skull elements to the left were drawn as a mirror image from the counterpart block. Skull bones near centre of figure belong to another specimen.

previously been considered close to the ancestry of squamates—the Prolacertiformes and the millerosaurs. The genus receiving the most attention in recent years as a possible lizard ancestor is *Prolacerta* (*Pricea*) described originally by Parrington (1935) and later by Camp (1945) on the basis of a particularly well preserved skull, and most recently by Gow (1975) who has added greatly to our knowledge by a description of the postcranial skeleton. The skull of *Prolacerta* is an almost ideal structural intermediate between that of primitive eosuchians such as *Youngina* and primitive lizards. There is a well developed upper temporal opening, but the lower temporal bar is incomplete, with the quadratojugal limited to a small splinter of bone. The quadrate is well developed dorsally. The squamosal retains an extensive ventral process, however, greatly limiting the motility of the quadrate. Robinson (1973) suggests that a tympanum, if present, would have been supported by the squamosal and quadratojugal, rather than by the quadrate. On the basis of these features of the cranial anatomy, and not considering other Permo-Triassic reptiles, *Prolacerta* seems almost ideally antecedent to lizards in general (see Robinson, 1967). There is not, however, a particularly close detailed resemblance of the skull with those of the Upper Triassic lizards *Kuehneosaurus* and *Icarosaurus*, as might be expected if *Prolacerta* were close to their ancestry. Of considerable significance is the greater size of the skull of *Prolacerta*, approximately twice the length of those of the paliguanids and kuehneosaurs. The facial region is much longer.

Despite the features of the skull which suggest affinities with lizards, the post-cranial skeleton, as recently described by Gow, shows none of the specialized features expected in the ancestors of lizards. There is no sternum. The endochondral shoulder girdle is not fenestrated anteriorly. The tarsus is specialized in a way suggestive of thecodonts, rather than lizards. The cervical vertebrae, clearly illustrated by Camp, are elongate and accompanied by extremely long thin ribs. The cervical vertebrae in all primitive lizards are short (Hoffstetter & Gasc, 1969) and it is only among the advanced anguimorphs that elongate cervicals have developed. On the basis of these features, it might be argued that *Prolacerta* was a somewhat specialized member of a basically primitive assemblage that gave rise to lizards. Other features, indicative of the general biology of *Prolacerta*, suggest rather that it was a member of a distinct radiation, not significant to the origin of lizards. Considering not only the paliguanids and kuehneosaurs, but all Jurassic lizards and the vast majority of Cretaceous and Cenozoic forms, it can be seen that small body size is a feature common to all primitive members of this group. As was discussed in the previous section, small size and determinant growth appear basic to the lizard habitus. The total body weight of *Prolacerta*, based on comparisons with modern forms (Pough, 1973: fig. 1), would have been approximately ten times that of the Triassic and Jurassic lizards. The limb joints are primitive, with no evidence for the presence of special ossified epiphyseal surfaces. Unlike the condition in the paliguanids, there is no evidence of growth limitation. In contrast with all primitive lizards, the head was elevated by the neck, rather than by raising the front part of the body by the forelimbs.

The reception of airborne sound and the incipient development of streptostyly certainly played important roles in the biology of *Prolacerta*, but these features were almost certainly developed independently of trends seen in the actual ancestors of lizards. The basic dichotomy of the stocks must lie among the primitive eosuchians, with the initial adaptation of the proto-lizard stock being a reduction in body size, with intensification of an insectivorous mode of life. The ancestors of prolacertids remained of moderate body size, and retained less specialized dietary habits. In terms of both its broad biological adaptation and phylogenetic relationships, *Prolacerta* is an eosuchian, with no significant affinities with lizards.

Two other genera, *Macrocnemus* (Kuhn-Schnyder, 1967, 1974) and

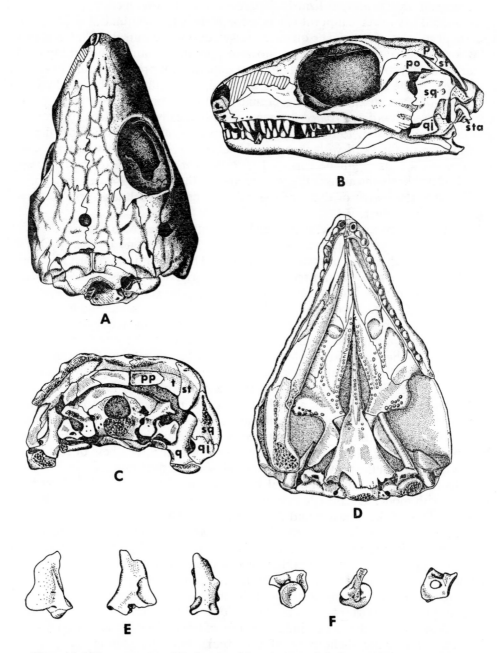

Figure 15. Millerosaurs: A and B, dorsal and lateral views of skull no. 14A from the Rubidge Collection; C and D, occipital and palatal views of skull no. 70 from the Rubidge Collection. Note absence of suborbital and post-temporal fenestrae that characterize *Youngina* and lizards, and great extent of postparietal, tabular and supratemporal. E. Left quadrate in lateral, medial, and posterior views, Bernard Price Institute no. 720. F. Stapes of same specimen in proximal, distal and oblique dorsal views. Skulls ×2, quadrate and stapes, ×4·5.

p, Parietal; po, postorbital; pp, postparietal; q, quadrate; qj, quadratojugal; sq, squamosal; st, supratemporal; sta, stapes; t, tabular.

Tanystropheus (Wild, 1973) have been classified as primitive lizards. The skull of these forms is fundamentally similar to that of *Prolacerta*, with a large quadrate and partial or complete loss of the lower temporal bar. Postcranially, they resemble *Prolacerta* in the elongation of the cervical vertebrae, carried to an extreme in *Tanystropheus*, and the configuration of the cervical ribs. The pelvis and the foot appear lacertoid, with the development of a thyroid fenestra and a hooked fifth metatarsal. Otherwise the postcranial skeleton lacks any of the distinctive lacertoid features seen in the paliguanids. Although the specific relationship of *Prolacerta*, *Tanystropheus* and *Macrocnemus* may be subject to dispute, all are part of one radiation, quite distinct from that associated with the origin of lizards.

Millerosaurs

The pattern seen in millerosaurs emphasizes the parallel development of features usually associated with lizards, among other groups of late Palaeozoic reptiles. The phylogenetic position of millerosaurs has long been debated. The presence of a single lateral temporal opening led the early describers to attribute them to the synapsids (Efremov, 1940; Romer & Price, 1940; Broom, 1948). Watson (1954) considered that they should be included among the sauropsids as possible structural antecedents to the diapsids. Romer suggested diapsid affinities in 1956, but in the third edition of *Vertebrate Paleontology* (1966) he indicated closer affinities with procolophonoids and pareiasaurs.

In the latest review of the group, Gow (1972) suggested that millerosaurs were directly ancestral to lizards. He indicated that the quadrate was quite freely movable in some genera, presaging the lacertoid condition, although remains of the lower temporal bar persist in all members of the group. Gow suggested that the reduction of the lower temporal bar and freeing of the quadrate preceded the development of the upper temporal opening in the ancestry of lizards. More recent information on the great antiquity of the diapsid condition (Reisz, 1975) and the cranial anatomy of other lizard-like forms provides a basis for further appraisal of the condition exhibited by the millerosaurs. As a result of the superb preparation of the South African specimens by Dr Gow, many features of their anatomy are exposed for detailed comparison. Of particular importance is the configuration of the quadrate and temporal region. A disarticulated quadrate of *Millerosaurus* (Fig. 15E) shows that it retains the general configuration seen in captorhinomorphs and is totally unlike those of lizards. The bone has no cephalic condyle for articulation with the squamosal. In contrast, the quadrate and squamosal are closely integrated and together articulate with the skull roof via the large supratemporal.

Both Watson and Gow have discussed the presence of an otic notch in millerosaurs. This structure appears quite different from that in lizards, however, for the tympanum was apparently supported by both the squamosal and quadratojugal, rather than by the quadrate. This bone certainly lacks the well developed conch common to both paliguanids and later lizards. Although it is possible that at an early stage in the origin of lizards the tympanum was supported by the squamosal, the definitive pattern of articulation between the quadrate and squamosal could not have developed until the ventral portion of the squamosal had atrophied to such an extent that it could have provided no space for tympanic attachment. The configuration of the quadrate seen in younginids is a more likely antecedent to that of lizards, since the bone is widely exposed both posteriorly and laterally, although there is not a clearly defined area for tympanic attachment.

The probable attachment of the tympanum to an extensive notch formed by the quadratojugal and squamosal in millerosaurs indicates evolution of the hearing apparatus above the level seen in primitive captorhinomorphs. Further evidence is provided by the small and light stapes whose internal structure shows extremely cancellous bone. The stapes is extremely short and thick, however, in contrast with the long thin columella of most lizards. The different configuration of the stapes may be partially explained by the original difference in the shape of the occiput in millerosaurs and ancestral lizards. In the latter group, as exemplified by *Kuehneosaurus* and *Paliguana*, the skull is very wide posteriorly, relative to its height, with the tympanum quite distant from the foramen ovalis. The millerosaur occiput is much narrower, so that a much shorter stapes can transmit vibrations from the surface of the skull to the inner ear. Such a short thick stapes is an unlikely antecedent to the condition seen in lizards, however, where emphasis is placed on a very large ratio between the area of the tympanum and that of the footplate.

The entire configuration of the occipital surface of the skull in millerosaurs is quite unlike that of either lizards or younginids. The great occipital extensions of postparietal, tabular, supratemporal and squamosal resemble the condition seen in primitive captorhinomorphs such as *Paleothyris*. The post-temporal fenestrae are small and ill-defined. Accepting that the condition seen in the known genera is primitive, it seems probable that millerosaurs are direct captorhinomorph derivatives and are not related to eosuchians. This interpretation is supported by the configuration of the temporal region.

The extent and nature of the lateral temporal opening is extremely variable from specimen to specimen among the South African genera, apparently independent of both taxonomic and growth factors. It is difficult to evaluate whether the long-term changes are acting to increase or restrict the extent of the opening among these forms. No millerosaur shows evidence of an upper temporal opening. Watson noted the peculiar sculpturing of the skull roof of millerosaurs and suggested that they might have lizard-like osteoderms, which, as in several lizard groups, might have covered areas of primitively extensive temporal openings. More complete preparation by Gow demonstrates that they are not distinct osteoderms, but that the pattern of the skull is produced by sculpturing of the dermal bones. The manner in which the squamosal articulates with the supratemporal would seem to preclude development of an orthodox upper temporal opening in the group. On the basis of the analysis of the origin of diapsidy in *Petrolacosaurus* (Reisz, 1975) it seems unlikely that any group which first developed a well defined lateral opening would subsequently develop an upper opening of the type seen in eosuchians and lizards. It seems probable that the evolution of an upper temporal opening was preceded by a stage in which the area of origin of the several components of the adductor jaw muscles showed a tendency to separate, with the development of gaps between the origins of the different units. With the prior development of a well-defined lateral opening, the lateral muscle may have become elaborated, but would show less tendency to separate from those more medial in position.

In addition to the presence of two pairs of temporal openings, primitive diapsids exhibit other openings in the skull that differentiate them from their captorhinomorph ancestors. One, already mentioned, is the post-temporal fenestra. A small opening is present in romeriids, but its margins are poorly defined as a result of the low degree of ossification of the paroccipital processes. In eosuchians, including *Petrolacosaurus* from the Upper Pennsylvanian and *Youngina*, these openings are large and well defined, but in millerosaurs they remain as in romeriids. Eosuchians are also distinguished from their antecedents by the presence of a large opening between the palatine and ectopterygoid, the suborbital fenestra. In millerosaurs there may be a small foramen, as in romeriids, but there is certainly not extensive fenestration.

There are some features of the postcranial skeleton, such as the fusion of the anterior caudal ribs to the transverse processes, in which millerosaurs parallel the eosuchians, but none are sufficient to indicate close relationship. Millerosaurs show some reduction in the extent of the glenoid and the size of the humeral head from the captorhinomorph pattern. Perhaps this joint allowed a somewhat lizard-like posture, but the nature of the preservation does not allow detailed comparison. There is no evidence of a specialized epiphysial condyle on the humerus. The sternum is apparently not ossified in this group. The pelvic girdle retains the primitive plate-like appearance of captorhinomorphs, although the relative length of the ischium is reduced. The fifth distal tarsal is incompletely fused with the fourth, but the tarsus otherwise retains a primitive configuration. The fifth metatarsal is not hooked, nor is it much shorter than the fourth.

It seems certain that millerosaurs are not at all close to the ancestry of lizards, but retain an essentially captorhinomorph grade of evolution. It is probable that the lower temporal opening developed independently, as did the elaboration of an otic notch. The mobility of the quadrate was certainly achieved separately from that seen in lizards, for the relationship of the bone with the skull roof is totally different.

The development of a movable quadrate and a specialized surface for the support of the tympanum in Prolacertiformes and millerosaurs, as well as in the ancestors of lizards, is somewhat analogous to the situation encountered in the origin of paired fins in vertebrates (Westoll, 1958). Similar functional requirements have led to the elaboration of similar structures in several groups, none of which originally showed any evidence of their initiation.

Sphenodontidae

In the course of this study, it has been natural to refer to *Sphenodon* for comparison. The lizard-like features of *Sphenodon* have long been recognized, but under currently accepted notions of its taxonomic position, these must be assumed to have developed as a result of convergent evolution, rather than indicating close affinities with ancestral squamates. New information on the anatomy and biology of early lizards makes it necessary to reconsider the phylogenetic position of this genus.

The skeletal anatomy of terrestrial sphenodontids shows a very conservative pattern from the Upper Jurassic to the present. What little is known of the Upper Triassic genera suggests that little change in the postcranial skeleton has occurred since the Middle Mesozoic. The peculiar configuration of the temporal region and the dentition in the Triassic genus *Glevosaurus* (Robinson, 1973) are the only described features in which early sphenodontids differ significantly from the living genus.

Sphenodontids resemble lizards from the late Mesozoic and Cenozoic in the following specialized characteristics:

1. determinant growth;
2. specialized articulating surfaces of the long bones;
3. specialized joint between the ulna and ulnare;
4. presence of a sternum;
5. attachment between sternum and ribs;
6. fenestration of the pelvic girdle;
7. fusion of astragalus and calcaneum;
8. loss of distal tarsal 1 and 5;
9. hooking of fifth metatarsal;
10. caudal atotomy.

On the basis of the paliguanids, it must be assumed that characteristics 6, 7, 8, 9, and 10 have been achieved separately in lizards and sphenodontids, although the functional basis for the specialization in foot structure is established in ancestral lizards.

Sphenodontids are more primitive than most modern lizards, but resemble paliguanids in other features:

1. amphicoelous vertebrae and intercentra throughout the trunk region (also seen in most gekkos);
2. presence of entepicondylar foramen in humerus;
3. presence of ventral dermal scales;
4. retention of all primitive elements in carpus (distal carpal 1 not modified as an epiphysis).

Sphenodontids are more primitive than paliguanids, retaining a pattern close to younginids, in the nature of the atlas and its articulation with the axis, and in the absence of anterior fenestration of the endochondral shoulder girdle.

The most significant differences between sphenodontids and lizards are in the anatomy of the skull. Except for *Glevosaurus*, the temporal region shows a typically diapsid configuration. Even in that genus, the proportions of the skull follow the pattern of primitive diapsids (such as *Youngina*) in strong contrast to the relative shortening of the temporal region seen in paliguanids and kuehneosaurids. A very interesting feature is that *Glevosaurus* has apparently achieved a large area of tympanic support without the conspicuous anterior movement of the quadrate seen in ancestral lizards. The pattern in *Glevosaurus* suggests that sensitivity to airborne sounds was improved in at least some early sphenodontids, but possibly separately from the trend in lizards. The reduction of the lower temporal bar presumably occurred independently in *Glevosaurus*, but this area is only weakly ossified in primitive diapsids such as *Petrolacosaurus* and *Tangasaurus*.

Description of a Lower Triassic rhynchosaur (Carroll, 1977) emphasizes the absence of any specialized features indicative of close relationship with sphenodontids. The often cited dental similarity between rhynchosaurs and sphenodontids is based on a total misconception of the nature of the teeth and their implantation in rhynchosaurs. As has been demonstrated by Malan (1963) and Chatterjee (1974) the teeth in this group are set in deep sockets and arranged in multiple tooth rows. This is totally distinct from the acrodont pattern seen in all sphenodontids. The skulls of primitive rhynchosaurs are diapsid, like the majority of sphenodontids, and have similar proportions to those of younginid eosuchians, but these are certainly primitive characteristics. The postcranial skeleton of rhynchosaurs is highly specialized in a way totally distinct from that of sphenodontids and lizards. There is no evidence of specialized epiphysial joints, and the posture remains primitive. Although a hooked fifth metatarsal is evident in the Lower Triassic rhynchosaurs, the tarsus is peculiar in the manner of movement of the centrale into the row of proximal tarsals.

There is no evidence that supports close affinities between sphenodontids and rhynchosaurs. The specimens currently available suggest rather that sphenodontids are more closely related to ancestral lizards than they are to any other group of Permo-Triassic reptiles. The presence of so many similar features of the postcranial skeleton suggests a common origin from among the eosuchians. Both lizards and sphenodontids show a similar manner of adaptation to the habitus of small quadrupedal insectivores, emphasized by the practice of determinate growth and similar anatomy and posture of the limbs. As pointed out by Robinson (1973), a basic difference can be seen in the proportions of the skull and mechanics of the jaw musculature in the two groups. This suggests divergence of the two groups prior to the perfection of the middle ear, but after the attainment of specialized joint surfaces.

Sphenodontids might either be classified as a distinct order of lepidosaurs, or be associated with the Lacertilia as a separate suborder within the Squamata. It is certainly advisable to avoid the term Rhynchocephalia in referring to sphenodontids, since this name implies association with the quite unrelated rhynchosaurs.

SUMMARY

(1) Consideration of the anatomy of the Paliguanidae emphasizes several characteristics as being basic to the adaptation of lizards:
small size;
determinate growth;
bony epiphyses elaborated into specialized articulating surfaces;
insectivorous diet;
head held above ground;
streptostylic quadrate;
ear sensitive to airborne sound;
fenestrate scapulocoracoid;
large sternum to which anterior trunk ribs are attached.

(2) Other features typical of modern lizards are not fully developed in the early members of the group, although their initiation is evident:
specialization of the tarsus with fusion of astragalus and calcaneum;

hooking of fifth metatarsal;

fenestration of the pelvic girdle.

(3) Amphicoely is primitive for squamates. Procoely has evolved separately in gekkonoids and other groups of modern lizards.

(4) The bone in the wrist of lizards typically identified as the first distal carpal is probably homologous with the medial centrale in other reptile groups. The first distal carpal has become incorporated into the proximal end of the first metacarpal as an epiphysis.

(5) Distal tarsals one and two may have become incorporated into their corresponding metatarsals as epiphyses. The fifth distal tarsal may either have been lost entirely, or been incorporated into the proximal end of the fifth metatarsal.

(6) The Permo-Triassic Paliguanidae may be included within the Eolacertilia as the most primitive known lizards and the possible antecedents of all more advanced members of the suborder.

(7) Neither the Prolacertiformes nor the millerosaurs are closely related to the ancestry of lizards.

(8) Sphenodontids and lizards probably share a common origin among eosuchians initially specialized by their small size and determinate growth for an insectivorous way of life.

ACKNOWLEDGEMENTS

I wish to thank Dr Cruickshank for providing facilities at the Bernard Price Institute, Johannesburg, where this work was initiated. Dr Kitching was most helpful in providing stratigraphic information for the South African diapsid reptiles. I very much appreciate the cooperation of Dr Gow, who allowed me to study the millerosaur specimens which he had so painstakingly prepared, and for showing me work in progress on the postcranial skeleton of *Prolacerta*. I wish to thank him particularly for generously permitting me to quote his findings on the anatomy of that genus. Dr Pamela Robinson was very helpful in allowing me to see her unpublished work on Upper Triassic lizards, and in discussing various anatomical and taxonomic problems. The careful preparation of the type of *Saurosternon bainii*, on which are based many of the conclusions regarding the postcranial skeleton of the paliguanids, was carried out by Mr Croucher at the British Museum (Natural History). Many of the drawings were completed by Mrs Pamela Gaskill. Ms Winer's patient typing of numerous drafts of the paper is much appreciated. Support for this work was provided by grants from the Faculty of Graduate Studies and Research, McGill University, the National Research Council of Canada and the Merrill Trust.

REFERENCES

BARGHUSEN, H. R., 1968. The lower jaw of cynodonts (Reptilia, Therapsida) and the evolutionary origin of mammal-like adductor jaw musculature. *Postilla*, No. 116: 1–49.

BROOM, R., 1903. On the skull of a true lizard (*Paliguana whitei*) from the Triassic beds of South Africa *Rec. Albany Mus.*, *1:* 1–3.

BROOM, R., 1925. On the origin of lizards. *Proc. zool. Soc. Lond.*: 1–16.

BROOM, R., 1926. On a nearly complete skeleton of a new eosuchian reptile (*Palaeagama vielhaueri*, gen. et sp. nov.). *Proc. Zool. Soc. Lond.*: 487–91.

BROOM, R., 1948. A contribution to our knowledge of the vertebrates of the Karroo beds of South Africa. *Trans. R. Soc. Edinb.* (*B*), *61:* 577–629.

CAMP, C. L., 1945. *Prolacerta* and the protorosaurian reptiles. *Am. J. Sci., 243:* 17–32, 84–101.

CARROLL, R. L., 1975. Permo-Triassic 'lizards' from the Karroo. *Palaeont. afr. 18:* 71–87.

CARROLL, R. L., 1976. *Noteosuchus*—the oldest known rhynchosaur. *Ann. S. Afr. Mus. 72:* 37–57.

CHATTERJEE, S., 1974. A rhynchosaur from the Upper Triassic Maleri Formation of India. *Phil. Trans. R. Soc., 267*(13): 209–61.

COCUDE-MICHEL, M., 1963. Les Rhynchocéphales et les sauriens des calcaires lithographique (Jurassique supérieur) d'Europe occidentale. *Nouv. Archs Mus. Hist. nat. Lyon, 7:* 1–187.

COLBERT, E. H., 1970. The Triassic gliding reptile *Icarosaurus. Bull. Am. Mus. nat. Hist., 143:* 89–142.

EFREMOV, I. A., 1940. Die Mesen-Fauna der permischen Reptilien. *Neues Jb. Miner. Geol. Paläont. (Abt. B), 84:* 379–466.

ENDO, R. & SHIKAMA, T., 1942. Mesozoic reptilian fauna in the Jehol mountainland, Mandchoukuo. *Bull. cent. natn. Mus. Mandchoukuo, 3:* 1–19.

FRAZZETTA, T. H., 1962. A functional consideration of cranial kinesis in lizards. *J. Morph., 3*(3): 287–320.

GOODRICH, E. S., 1916. On the classification of the Reptilia. *Proc. R. Soc. Lond.* (B), *89:* 261–76.

GOW, C. E., 1972. The osteology and relationships of the Millerettidae (Reptilia: Cotylosauria). *J. Zool., Lond., 167:* 219–64.

GOW, C. E., 1975. The morphology and relationships of *Youngina capensis* Broom and *Prolacerta broomi* Parrington. *Palaeont. afr., 18:* 89–131.

HAINES, R. W., 1969. Epiphyses and sesamoids. In C. Gans *et al.* (Eds), *Biology of the Reptilia, 1:* 81–115. London: Academic Press.

HARRIS, J. & CARROLL, R. L., 1977. *Kenyasaurus*, a new eosuchian reptile from the early Triassic of Kenya. *J. Paleont.* (in press).

HOFFSTETTER, R., 1964. Les Sauria du Jurassique supérieur et specialement les Gekkota de Bavière et de Mandcourie. *Senckenberg. biol., 45:* 281–324.

HOFFSTETTER, R., 1965. Les Sauria (=Lacertilia) du Jurassique supérieur du Montsech (Espagne). *Bull. Soc. geol. Fr., 7*(7): 549–57.

HOFFSTETTER, R., 1967. Coup d'oeil sur les sauriens (=lacertiliens) des couches de Purbeck (Jurassique supérieur d'Angleterre). (Résumé d'un mémoire), *Colloques int. Cent. natn. Rech. scient., 163:* 349–71.

HOFFSTETTER, R. & GASC, J. P., 1969. Vertebrae and ribs of modern reptiles. In C. Gans *et al.* (Eds), *Biology of the Reptilia, 1*(A): 201–310. London: Academic Press.

HOLMES, R., 1977. The osteology and musculature of the pectoral limb of small captorhinids. *J. Morph.* (in press).

HOTTON, N., 1959. The pelycosaur tympanum and early evolution of the middle ear. *Evolution, 13:* 99–121.

HUXLEY, T. H., 1868. On *Saurosternon bainii* and *Pristerodon mckayi*, two new fossil lacertilian reptiles from South Africa. *Geol. Mag., 5:* 201–5.

KLUGE, A. G., 1967. Higher taxonomic categories of gekkonid lizards and their evolution. *Bull. Am. Mus. nat. Hist., 135:* 3–59.

KUHN-SCHNYDER, E., 1967. Das Problem der Euryapsida. *Colloques Int. Cent. natn. Rech. scient., 163:* 335–48.

KUHN-SCHNYDER, E., 1974. Die Triasfauna der Tessiner Kalkalpen. *Neuj Bl. naturf. Ges. Zürich:* 1–119.

LÉCURU, S., 1968a. Remarques sur le scapulo-coracoïde des lacertiliens. *Annls Sci. nat. (Zool.), 10*(12): 475–510.

LÉCURU, S., 1968b. Étude des variations morphologiques du sternum, des clavicules et de l'interclavicule des lacertiliens. *Annls. Sci. nat. (Zool.), 10*(12): 511–44.

MALAN, M., 1963. The dentitions of the South African Rhynchocephalia and their bearing on the origin of the rhynchosaurs. *S. Afr. J. Sci., 59:* 214–20.

MANLEY, G. A., 1973. A review of some current concepts of the functional evolution of the ear in terrestrial vertebrates. *Evolution, 26:* 608–21.

PARRINGTON, F. R., 1935. On *Prolacerta broomi*, gen. et sp. nov., and the origin of the lizards. *Ann. Mag. nat. Hist., 16*(10): 197–205.

POUGH, F. H., 1973. Lizard energetics and diet. *Ecology, 54:* 837–44.

REISZ, R., 1972. Pelycosaurian reptiles from the Middle Pennsylvanian of North America. *Bull. Mus. comp. Zool., Harv., 144*(2): 27–61.

REISZ, R., 1975. *Petrolacosaurus kansensis* Lane, the oldest known diapsid reptile: 185 pp. Unpublished PhD. thesis, Dept. of Biology, McGill University, Montreal.

RENOUS-LÉCURU, S., 1973. Morphologie comparée du carpe chez les Lepidosauriens actuels (Rhynchocéphales, Lacertiliens, Amphisbéniens). *Gegenbaurs morph. Jb., 119:* 727–66.

ROBINSON, P. L., 1962. Gliding lizards from the Upper Keuper of Great Britain. *Proc. geol. Soc.*, No. 1601: 137–46.

ROBINSON, P. L., 1967. The evolution of the Lacertilia. *Colloques int. Cent. natn. Rech. scient.*, No. 163: 395–407.

ROBINSON, P. L., 1973. A problematic reptile from the British Upper Trias. *J. geol. Soc., 129:* 457–79.

ROBINSON, P. L., 1975. The functions of the hooked fifth metatarsal in lepidosaurian reptiles. *Colloques int. Cent. natn. Rech. scient.*, 461–483.

ROMER, A. S., 1956. *Osteology of the reptiles:* 772 pp. Chicago: University of Chicago Press.

ROMER, A. S., 1966. *Vertebrate Paleontology*, 3rd ed.: 468 pp. Chicago: University of Chicago Press.

ROMER, A. S., 1967. Early reptilian evolution re-viewed. *Evolution, 21*(4): 821–33.

ROMER, A. S., 1971. Unorthodoxies in reptilian phylogeny. *Evolution, 25*(1): 103–12.

ROMER, A. S. & PRICE, L. I., 1940. Review of the Pelycosauria. *Geol. Soc. Am.*, special paper no. 28: 1–538.

SCHAEFFER, B., 1941. The morphological and functional evolution of the tarsus in amphibians and reptiles. *Bull. Am. Mus. nat. Hist., 78*, art. 6: 395–472.

SNYDER, R. C., 1954. The anatomy and function of the pelvic girdle and hind limb in lizard locomotion. *Am. J. Anat., 95:* 1–46.

WATSON, D. M. S., 1954. On *Bolosaurus* and the origin and classification of reptiles. *Bull. Mus. comp. Zool., Harv., 3*(9): 299–449.

WATSON, D. M. S., 1957. On *Millerosaurus* and the early history of the sauropsid reptiles. *Phil. Trans. R. Soc. (B)*, 240(673): 325–400.

WESTOLL, T. S., 1958. The lateral fin-fold theory and the pectoral fins of ostracoderms and early fishes. In T. S. Westoll (Ed.), *Studies on fossil vertebrates presented to David Meredith Seares Watson:* 180–211. London: Athlone Press.

WEVER, E. G. & WERNER, Y. L., 1970. The function of the middle ear in lizards: *Crotaphytus collaris*, Iguanidae. *J. exp. Zool., 175:* 327–42.

WILD, R., 1973. Die Triasfauna des Tessiner Kalkalpen. *Tanystropheus longobardicus* (Bassani) (Neue Ergebnisse). *Schweizerische palaeont. Abh., 95:* 1–162.

Intercentra: a possible functional interpretation

F. R. PARRINGTON

University Museum of Zoology, Cambridge

The retention of intercentra in the cervical region of cynodonts is discussed. It is suggested that if they could be lowered the neck could be bent through a greater amount without either reducing the strength of the cervical column or stretching the nerve cord as much as necks with primitive centra but no intercentra. The eventual loss of intercentra in the cervical region of mammals and other amniotes is explained as a result of the development of other and better ways of achieving flexibility.

The first fossil collected by the writer in South Africa in 1930 was a skull of the cynodont _Galesaurus planiceps_ Owen which was picked up in the brickfields donga at Harrismith, O. F. S. It was, in fact, the third known specimen since Owen described this, the type cynodont, in 1859, the second recorded specimen having been called _Glochinodon detinens_ by van Hoepen in 1916. The skull was especially interesting in having retained a hyoid bone and one stapes, and in having an incisor erupting. But also of great interest was the fact that the proatlas, atlas, axis and the two succeeding vertebrae were well preserved (Parrington, 1934). They showed the structure of the axis, the presence of well developed intercentra, and also the presence of anapophyses which had not been recorded in any cynodont before but were recognized by Broili & Schröder (1936) in other material.

We now know that cervical intercentra were retained in at least three cynodont genera (_Galesaurus_, _Thrinaxodon_ and _Cynognathus_) but dorsal intercentra have never been recorded (Jenkins, 1971). Intercentra, so far as they are understood, appear to have been structures which in early tetrapods either allowed the vertebral column to be especially flexible by forming additional centra in each body segment, or were components of rhachitomous vertebrae which were especially adapted to allow considerable twisting as well as bending of the column (Parrington, 1967). The question arises, therefore, why did the cynodonts retain these structures in the cervical region after they had lost them in the thoracic and lumbar regions? To understand this problem it would seem necessary to understand the function of cervical vertebrae.

The theriodonts as a whole have large heads and it would seem clear that they carried their heads above the line of the back. In this, as in having seven cervical vertebrae, the cynodonts resemble most modern mammals. The upward sloping neck (Jenkins, 1971) carrying the heavy head was subject to two stresses, tension dorsally and compression ventrally; these stresses had to be resisted by the vertebrae and the possibility is, therefore, that the ventrally situated intercentra were concerned with the compression. But if they were so concerned the way in which they acted is obscure. Straightforward amphicoelous or amphiplatyan centra without intercentra would appeared to have served equally well.

If a series of amphicoelous or amphiplatyan vertebrae, characteristic of primitive reptiles, are flexed ventrally neighbouring centra hinge about their ventral edges and the zygapophyses and anapophyses will be parted to some degree and the nerve cord will be stretched accordingly (Fig. 1A). If such vertebrae are separated

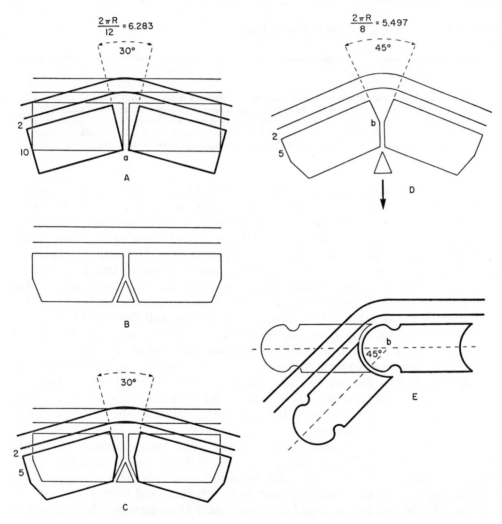

Figure 1. Diagrammatic representation of the actions of centra and intercentra and the resulting stretching of the nerve cord. A. Represents two adjacent centra which might be amphicoelous or amphiplatyan moved through 30° at a point (a) with the resulting stretch of the nerve cord which in life would be distributed along the cord. The introduction of intercentra (B) would appear to be of no advantage (C) unless the intercentra could be squeezed or pulled down (D) when the centra would hinge against each other at a higher point (b) with the result that the stretch of the nerve cord and the separation of the zygapophyes would be less. An alternative way of achieving the same result is to form a ball and socket joint (E) and this would have the additional advantages of allowing lateral flexure. The amounts of stretch for the given units refer to the bottom of the nerve cord.

by ventral intercentra the result would appear to be much the same (Fig. 1B, C), the nerve cord being stretched by about the same amount if the centra tilt through the same angle. If, however, the intercentra were squeezed or pulled down a sufficient amount the centra will hinge about their lowest line of contact, that is a line which passes through the original position of the apex of the intercentrum (Fig. 1D). The result of this dorsal transposition of the hinge is that, though the centra may have been moved through the same or even a greater amount, the neural arches and spines will be separated by a smaller amount and the nerve cord will be stretched less according to the height of the hinge. The presence of the intercentra, if they could be moved ventrally, served to keep the interlocking devices functioning through a greater angle and the nerve cord would be stretched to a lesser degree accordingly. In short the neck would be stronger for a given amount of bending with less danger of the neck being 'drawn'. This may well be the reason for the retention of cervical intercentra after those of the thoracic and lumbar regions had been lost, there being a greater need for flexing the neck.

It is surely no great supposition that wedge shaped intercentra, which were certainly lined with cartilage, could be squeezed ventrally when the epaxial muscles were relaxed and the heavy head lowered. When the head was raised the intercentra, it can reasonably be supposed, were raised again into their normal positions either by tendons or, less likely, by special muscles.

It will be noticed that the diagrams show that while withdrawn intercentra allow the neck to be bent through a greater angle without stretching the nerve cord to the same extent as otherwise would be the case, the result is to bend the cord more tightly. But if the coils of such a snake as the Puff adder (*Bitis arietans*) when it is about to strike are anything to judge by, the nerve cord can be greatly coiled without coming to harm.

If the presence of cervical intercentra were advantageous as has been suggested, the question next arises as to why they were ever abandoned. It is quite certain that they were not present in the Upper Triassic mammal *Megazostrodon* and almost certain that they were not present in *Eozostrodon*. The answer is that there were other, probably better, ways of achieving flexibility.

The cervical vertebrae of *Megazostrodon* have been shown to have centra which are not round in end view but are distinctly compressed dorsoventrally. This achieves the same effect as raising the hinge line closer to the nerve cord (Jenkins & Parrington, 1976). Similar shaped centra have been found among the remains of *Eozostrodon*; they are distinguished from those of the contemporaneous lizards by the fact that they are platycoelous and not amphicoelous. The delicacy of the enlarged neural arches of *Megazostrodon*, allowing for an enlarged spinal cord and the necessary blood supply, will account for the presence of only one well preserved cervical vertebra of *Eozostrodon* in the Cambridge collection. The centrum measures 2·0 mm in width but only 1·0 mm in depth and is oval shaped.

Yet another way of achieving the same results is by developing ball and socket joints, that is forming opisthocoelous or procoelous centra, modifications common in later reptiles and mammals. Here (Fig. 1E) the point of rotation will be the centre of the ball. This requires either some constriction of the neck of the ball or a shortening of the edges of the socket. Both developments are found. It should be noticed that this form of joint gives the animal great lateral as well as dorso-ventral flexure and is thus a better joint.

Plainly it would be of considerable interest if the movements of vertebral columns could be examined by X-ray cinematography. But this is not easy and it is likely that a form with comparatively large intercentra would be necessary.

Intercentra occur in various reptiles to-day. To quote from Hoffstetter & Gasc (1969) 'The intercentrum is a highly variable vertebral component' and 'The intercentrum tends to disappear in the trunk except in amphicoelous vertebrae'. This remark is not opposed to the views advanced here. In the cervical region of reptiles, they claim, it generally forms a midventral spine which may migrate anteriorly or posteriorly and may even fuse with the centrum. But they say that in the Lepidosauria the intercentra are well developed and separate elements. Plainly there are various functions which intercentra perform.

If the foregoing suggestions are acceptable (and especially if they find support by practical experiments on living animals) a modest step forward has been gained in understanding the significance of the great variety of form found in vertebrae. But problems still remain. Thus for many years Professor Stanley Westoll diligently sought a specimen of *Osteolepis macrolepidotus*, a crossopterygian fish, among the often scrappy remains of the form occurring in a quarry near Scrabster in Caithness, hoping that one of the disrupted specimens might show two or three conjoined vertebrae. Eventually, when he was appointed to the chair in Newcastle University, he came across a beautiful specimen which had been among the collections of the Geological Department of that University for many years. It showed very beautifully that certainly almost all of the vertebral column was essentially rhachitomous. Work carried out in Professor Westoll's laboratory by him and Dr S. M. Andrews (Andrews & Westoll, 1970) showed that the same condition existed in the crossopterygian fish *Eusthenopteron*, although for some unknown reason the vertebrae were widely spaced. If, as has been claimed, rhachitomous vertebrae allow considerable twisting as well as flexure of the vertebral column, why should they occur in fishes? Any fish such as *Osteolepis* or *Eusthenopteron* which twisted its body while swimming would immediately tend to revolve about its longitudinal axis as it moved forward and although the advantages of being able to do this are not obvious, such a movement might facilitate either the taking of prey from below or the avoidance of predators.

Again, the gastrocentrous vertebrae of the amphibian *Seymouria* and its allies consist of well developed notochordal centra and large crescentic intercentra. The neural arches, which are low and swollen, and have only very small spines, rest entirely on the centra and the considerable spaces above the intercentra must have, it is presumed, been filled by cartilage. The neural arches are interesting in having zygapophyses which are not only strengthened by their thickening but also in having their articulating surfaces horizontal, thus allowing lateral flexure to an unusual degree but resisting twist. It may be suggested that the large intercentra, like those of embolomerous amphibia, allow lateral flexure to an unusual degree thus extending the length of the stride. But the earliest reptiles are small animals and thus had their own weight problems. Also the occurrence of intercentra is often uncertain.

REFERENCES

ANDREWS, S. M. & WESTOLL, T. S., 1970. The postcranial skeleton of *Eusthenopteron foordi* Whiteaves. *Trans. R. Soc. Edinb.*, *68:* 207–329.

BROILI, F. & SCHRÖDER, J., 1936. Beobachtungen an Wirbeltieren der Karroo formation. XVII, Uber Cynodontia-Wirbel. *Sber. bayer. Acad. Wiss.:* 61–76.

HOEPEN, E. C. N. VAN, 1916. Preliminary notice of new reptiles of the Karroo Formation. *Ann. TransV. Mus.*, *5*, no. 3, suppl. 2: 2 p.

HOFFSTETTER, R. & GASC, J-P., 1969. Vertebrae and ribs of modern reptiles. In C. Gans (Ed.), *Biology of the Reptilia*, *1:* 201–310. London: Academic Press.

JENKINS, F. A., 1971. The postcranial skeleton of African cynodonts. *Bull. Peabody Mus. nat. Hist.*, *36:* 1–216.

JENKINS, F. A. & PARRINGTON, F. R., 1976. The postcranial skeletons of the Triassic mammals *Eozostrodon, Megazostrodon* and *Erythrotherium*. *Phil. Trans. R. Soc.* (*B*).: *273:* 387–431.

OWEN, R., 1859. On some reptilian fossils from South Africa. *Q. Jl geol. Soc. Lond.*, *16:* 49–63.

PARRINGTON, F. R., 1934. On the cynodont genus *Galesaurus* with a note on the functional significance of the changes in the evolution of the theriodont skull. *Ann. Mag. nat. Hist.*, *13*(10): 38–67.

PARRINGTON, F. R., 1967. The vertebrae of early tetrapods. *Colloques Int. Cent. nat. Rech. scient.*, No. 163: 269–79.

Index of scientific names

Subject index